GENETICS OF INDUSTRIAL MICROORGANISMS

Volume I

BACTERIA

GENETICS OF INDUSTRIAL MICROORGANISMS

Volume I

BACTERIA

Edited by: ZDENKO VANĚK
ZDENĚK HOŠŤÁLEK
JOSEF CUDLÍN

Institute of Microbiology
Czechoslovak Academy of Sciences
Prague

ELSEVIER PUBLISHING COMPANY
Amsterdam - London - New York 1973

Published in co-edition with
ACADEMIA, Publishing House of the Czechoslovak Academy of Sciences, Prague

Distribution of this book is being handled by the following publishers

for the U.S.A. and Canada
American Elsevier Publishing Company, Inc.
52 Vanderbilt Avenue
New York, New York 10017

for the East European Countries, China, North Korea, Cuba,
North Vietnam and Mongolia
Academia, Publishing House of the Czechoslovak Academy of Sciences, Prague

for all remaining areas
Elsevier Publishing Company
335 Jan van Galenstraat
P.O. Box 211, Amsterdam, The Netherlands

ISBN 0-444-40989-0
Library of Congress Card Number 75-180010
With 131 Illustrations and 59 Tables

Printed in Czechoslovakia

CONTENTS

EDITORS' NOTE

The editors endeavoured to unify the entire nomenclature, both chemical or biochemical, and genetic, in the two volumes, and to conform it to the latest rules and recommendations. In the first case, the editors' task was not difficult, since the chemical community have mostly solved their communication problems[1,2]. A rather different situation exists in genetics where a number of reknowned (alas!) authors are using more or less original personal nomeclature (particularly with regard to the designation of genetic loci). The untenable situation has been dealt with in a number of recommendations and reviews[3–6] which, in spite of their incontestable value, frequently cause confusion*. The usefulness of some of the suggested rules** may be queried, particularly with a view to future developments. It is thus understandable that the editors, in an attempt to deal successfully with the second task and trying to avoid some of the mistakes referred to above, may have erred even more. Therefore, they attached a genetic index to the book.

References

1. IUPAC: *Definitive Rules for Nomenclature of Organic Chemistry*. J. Am. Chem. Soc. **82**, 5545 (1960).
2. IUPAC-IUB Combined Commision on Biochemical Nomeclature: *Abbreviations and Symbols for Chemical Names of Special Interest in Biological Chemistry*. J.Biol.Chem. **241**, 527 (1966).
3. DEMEREC M., ADELBERG E. A., CLARK A. J., HARTMAN P. E.: Genetics **54**, 61 (1966).
4. VON BORSTEL R. C.: Microbial Genet.Bull. No. 31 (1969), Yeast Genet.Suppl. Oak Ridge National Laboratory, Oak Ridge.
5. SERMONTI G.: *Genetics of Antibiotic-Producing Microorganisms*. John Wiley & Sons Ltd., London 1969.
6. TAYLOR A. L.: Bacteriol.Rev. **34**, 155 (1970).

* *thi* = thiazole[5], thiamine[3,6]; *cyc* = cycloserine[6], cytochromes (Ref.[4], p. 7); glutamate = *glu*[3–5], *glt*[6]; Proposal N° 2 (Ref.[4]) *vs.* Recommendation 2 (Ref.[3]); *abt* = α-aminobutyrate[5] (*cf.* Abu of the biochemists); *ane* = aneurine[5] (now only thiamine).

** Recommendation 8 (Ref.[3], p. 68).

Preface

Between August 23 and 28, 1970, some 600 scientific workers from both basic and applied research gathered in Prague to attend a symposium concentrating attention on problems of the genetics of industrially important microorganisms. To introduce the present monograph, which comprises the material presented in the various sections, in panel discussions as well as at the special symposium devoted to mutation effects caused by radiation and chemical mutagens, I should like to return to the intentions which led us to organize this symposium.

Three facts are typical of the present state of microbiology and its theoretical as well as practical aspects. On the one hand, there is no doubt that during the past decades microbiology has laid the foundation for the great development of the fermentation industry which has not only produced extremely valuable drugs and chemical preparations but has also become the starting point for tackling the problems of the starving population of this planet. When developing the effectivity of these productions it was essential to use the results of improvement of the microorganisms employed. This is especially the case in the production of antibiotics, accompanied by a steep rise in effectivity, and also in the microbial production of amino acids that can hardly be visualized without the results of the planned "breeding" of microorganisms.

On the other hand, the most characteristic feature of recent microbiology is the invasion of genetic studies, the results of which have led us to recognizing the very basis of the genetic control of microbial activity, reaching to the level of macromolecular systems, which gave rise to the so-called molecular biology.

However, both these directions of rapid development of microbiology proceed more or less independently, although their basis as well as their results are not dissimilar. The first route was determined above all by the work of industrial microbiologists, while the second route remained entirely within the domain of basic theoretical research. It could be expected that a combination of the experience and needs of the two routes would represent a substantial impetus for both approaches. Workers with industrial microorganisms would profit from the knowledge of genetic regulation in acquiring a rationale for their work and hence in intensifying their search for improving their strains. Theoreticians would be able to extend their range of models and to shift their focus from the level of control of principal metabolic pathways to the level of formation of the so-called secondary products, the genetic link of which to the level of classical metabolic routes is not understood at present but contains a number of pressing special as well as general problems. We felt that the commencement of a systematic and mutually coupled investigation in this direction might open the way for further contributions of microbiology to solving fundamental problems of biology, such as the basis of physiological (and perhaps morphological) differentiation.

We realized that both the above-mentioned approaches reached a mature stage for presenting their accumulated results, including genetic control, physiological and enzymic systems, biosynthetic pathways connected with the formation of a number of important products, and the chemical structure of a number of products and intermediates. This conviction was strengthened by the amount of new methodical, technical and tactical approaches that are now used in the study of products of microbial activity. Finally, we reached the conclusion that further development of bio-engineering approaches, characterized for example by introducing continuous methods or by applying physico-chemical methods, requires new theoretical contributions dealing with the transition from basic metabolism (stressing the formation of bio-mass) to the metabolism of secondary products which are often of paramount practical importance.

A number of misunderstandings and short-circuit judgments have arisen, however, about this problem. Whole areas of basic research are severely criticized, doubt is raised as to the suitability of model organisms on which the basic phenomena of life are being studied and the time factor is being emphasized more and more. This time factor should represent then one of the criteria for whether a given scientific task has been handled successfully or not.

The principal objective of creative scientific work then often recedes completely into the background although this very understanding of what was previously still unknown, will always retain its positive significance and in the long run — after accumulation of sufficient basic material and facts — will bring about an atmosphere for a qualitative change, a scientific discovery of first-class importance which need not at all be primarily practical. Biological sciences differ from chemistry and perhaps physics in that the field is still much more open than in those fields for pioneers and that success can far less be attained by a mere combination of quantitative elements already known.

A scientist, frequently as a university teacher, will find full intellectual satisfaction in his laboratory, in the working team and he is frequently disinterested as to what happens to the results of his work in practice. He is often led to the conclusion that there are more disadvantages than advantages in cooperating with the industry and often he is inclined to reject any collaboration on a practical, socially sometimes very significant problem. But still more colleagues—scientists realize, that this is not the way to go. If biology is to keep pace with other scientific fields it must strive for greater numbers of young scientists, for better (and also more expensive) apparatus equipment. Much of what has been done would have remained outside the scope of our knowledge had it not been for the assertion of biology as an essential component in a modern society. It is thus the duty of biologists working in basic science to pay maximum attention to the problems accumulating around the relationship between basic research and practical application.

When preparing this symposium we received a view on the topics to be treated here from Arthur Kornberg who said that "*I cannot imagine any more significant theme for a symposium in this field* ..." and I should add that not only significant but also complicated and intriguing.

We realized some of the difficulties in 1964 at the Symposium on Antibiotics in Prague in a panel discussion entitled Biogenesis of Natural Compounds which was intended as a dialogue between geneticists and chemists. At that time it became clear that biogenesis, understood merely as a study of the principal building blocks, will not tell us much about the mechanisms governing this synthesis whether at the level of metabolism or genetic equipment.

Only a detailed analysis of all the participating enzyme systems, their interaction in space and time in the continuously changing dynamic cell structure, their genetic foundation, in other words, only a complete understanding of the metabolic and genetic laws according to which the whole process runs, can give us a cue how and why during the growth of an organism (cell differentiation) both quantitatively and qualitatively different products are formed.

In our model of biogenesis — production of chlortetracycline by *Streptomyces aureofaciens* — we took up first of all the enzyme systems of the so-called primary metabolism (growth phase). We reached the conclusion that the synthesis of the oligoketide chain of tetracycline is an energetically demanding process and that one of the conditions of oligoketide formation is a low concentration of NADPH and ATP during the production phase. Our results show that malonylCoA as the fundamental building block of oligoketides need not be formed simply by carboxylation of acetylCoA. Other possibilities, such as decarboxylation of oxalacetic acid, are now under careful scrutiny.

A genetic analysis of our strains of *S. aureofaciens* in cooperation with chemists showed that these strains produce at least 40 different compounds which may be grouped under the category of secondary metabolites. Most of them may be considered as new, hitherto undescribed, compounds.

One may envisage several basic genetic blocks which could interfere with the biogenesis of tetracycline antibiotics:

(1) Mutation of structural genes, controlling the individual steps of the *main* biosynthetic pathway resulting in tetracyclines.

(2) Mutation of structural genes controlling the *by-routes* of the biosynthetic pathway.

(3) Mutation of *regulation genes* or *operators,* coordinating the function of structural genes for the principal biosynthetic pathways or its by-routes.

(4) Mutation of the *determinants for allosteric inhibition.*

(5) Mutation of other metabolic pathways stemming from a common precursor (malonylCoA).

(6) Mutation in the primary metabolism affecting the *synthesis of essential metabolites, e.g.* cofactors of some enzymes.

(7) Mutation in the primary metabolism resulting in changes of *pathways by which acetylCoA (or malonylCoA) is synthesized.*

(8) Mutation in the metabolic pathways yielding or else requiring the oxidized or reduced forms of nicotinamide nucleotides.

(9) Mutation affecting the intracellular concentration of ATP.

(10) Mutation resulting in changes of cell membrane permeability or otherwise affecting the mechanisms of active transport.

It follows from the above-said that changes in the production of tetracycline antibiotics need not be caused by direct mutational interference with one of the steps of their biosynthesis but that they can be due to a number of other changes which are seemingly not connected with the biosynthesis of tetracyclines but produce disturbances of the metabolic balance of the cell. From this point of view one may newly interpret the still cherished view of a *polygenic system* in which small effects of a great number of genes will add up or even multiply during control of a quantitative character expression.

The resistence of microorganisms toward the compound produced is another factor that can limit the production of secondary metabolites. Our experiments showed that *in vitro* protein synthesis, controlled by endogenous mRNA, proceeds at substantially greater concentrations of tetracyclines in the tetracycline-producing strain than in other preparations isolated from nonproducers of tetracycline.

These differences are in agreement with qualitative changes in the content of ribosomal proteins. Using reconstitution experiments we found that resistence toward tetracyclines is associated with a mutation of structural proteins of the 30S ribosomal subunits. In future work we intend to isolate and to identify the protein factor responsible for the resistance to tetracycline. I am presenting this example drawn from the work of our laboratories only to stimulate the feeling of intricate complexity connected with the differentiation process leading to the formation of secondary metabolites.

We are at present witnessing a vigorous development of a new branch of the drug industry, *i.e.* the production of amino acids and of nucleotides. Industrial scientists managed within a short time span to employ recent findings on the regulation of biosynthetic processes, on the feedback mechanisms, induction and repression of enzyme systems. To achieve excess production of the desired compound they invented many ingenious tricks how to overcome these regulatory mechanisms whether by using mutants that are not sensitive to the analogues of the individual amino acids or mutants with a damaged normal regulatory system *etc.* I am mentioning these facts bacause this example documents very well that the industry employs ranks of outstanding scientists and that, if financial means are properly put to use (take the example of antibiotic production) one can in an unbelievably short time convert the basic research data to practical ends. Taking the example of amino acids (as the most recent and also relatively simple one) one can also document the back pressure on pure science, stimulating the solution of a number of further questions of fundamental importance, dealing *e.g.* with enzymology, membrane permeability, *etc.* Prof. NAKAYAMA says that one can produce industrially most amino acids, the building blocks of proteins, but that the practical realization depends on the consumer capacity of a given society. But let us recall that at the beginning of all this there are the experiments of two geneticists, BEADLE and TATUM, who used the "industrially unsuitable" model of *Neurospora crassa* to study the "industrially uninteresting" auxotrophy.

How did the symposium approach the solution of the above mentioned demanding problems? The first indisputable achievement of the symposium was the fact that the intended bringing together of "applied" microbiologists with those scientists who contributed to our knowledge of genetic regulation was fully appreciated. During the symposium week a substantial number of experts from both fields actually did get together. This made it possible to confront our present knowledge of a fundamental character with the needs of practical industrial application.

Thus, for example, a detailed discussion of observations in the field of extrachromosomal heredity of the bacterial cell resulted in pointing out the necessity for theoretical study for practical requirements connected with the resistance of bacterial strains to the presently employed antibiotics. The section brought together the lectures of some outstanding bacterial geneticists, such as **W.HAYES**, **B.W.HOLLOWAY**, **A.PROZOROV**, **H.SAITO** and **N.SUEOKA**.

An equally obvious practical profit is to be expected from a broader application of the present methodical and factual achievements in the study of fungal genetics, a fact following from the proceedings of the section opened by **K.ESSER** who outlined the principal problems of fungal genetics (tetrad analysis in relation to recombination, somatic recombination, genetic basis of morphogenesis, incompatibility, extrachromosomal heredity *etc.*). Other papers presented in the section were in the same vein: **J.A.ROPER** *(Mitotic Recombination and Non-conformity)*, **H.L.K.WHITEHOUSE** *(Repair Recombination)*, **J.K.RAPER** *(Sexual Basis of Basidiomycete Morphogenesis)*, **J.R.S.FINCHAM** *(Enzymic Aspects of Fungal Genetics)*.

Extensive attention was devoted to the genetics of streptomycetes. The past and present progress of their genetics and the modern methods of genetic analysis were reviewed (**D.A.HOPWOOD**), together with the sexual differentiation of strains (**I.SPADA-SERMONTI** and **G.SERMONTI**, **A.VIVIAN**), genetic control of morphogenesis (**K.F. CHATER**) and lysogeny (**D.N.LOMOVSKAYA**) in *Streptomyces coelicolor*. Results of studies on the genetic control of biosynthetic pathways of primary metabolism were presented using the example of histidine biosynthesis (**A.CARERE** and **S.RUSSI**) and tryptophan biosynthesis (**P.ENGEL**) in *S. coelicolor*. The possibilities of practical application were pointed out by **M.ALAČEVIĆ**, **J.H.COATS** and **J.ROESER**. The discussion of the results with the streptomycete group revealed the mostly empirical nature of the attempts at increasing biosynthetic activity, without sufficient knowledge of the decisive processes and stressed the necessity for closer cooperation between theoretical and applied research.

Much attention was devoted to the present state of knowledge of mutagenesis (a review was supplied by **H.MARQUARDT**) and to the role of repair mechanisms in the establishment of mutations (**P.C.HANAWALT** — *Problems of Repair Replications*; **B.S.STRAUSS** — *Repair and Mutagenic Activity of Alkylating Agents*; **H.D.MENNIGMANN** — *Comparative Studies of Pyrimidine Dimers*).

A special session was devoted to the regulation of enzyme synthesis and modification of enzyme activities. **A.B.PARDEE** reviewed the facts about the synthesis of constitutive enzymes, many of which are of importance for practical application, and pointed out the possibilities of attaining maximum rates of their synthesis by altering their genetic equipment. There were other important papers (**P.H.CLARKE** — *Aliphatic Amidase from* Pseudomonas aeruginosa, **V.M.KRYLOV**, **L.JAENICKE** and **R.SAGER** — *Algae as a Genetic Model*).

In connection with the extensive industrial production of amino acids there were a number of papers in the section on the genetics of biosynthetic pathways of amino acids (**H.E.UMBARGER**, **P.DATTA**, **R.HÜTTER**, **H. DE ROBICHON-SZULMAJSTER**). A similar practical significance resided in the section devoted to the genetic aspects of the biosynthesis of natural compounds (chairman **C.H.HASSALL**) which can be summarized into three groups: the study of new metabolites of mutant strains, using methods of organic and physical chemistry, the study of the incorporation of labelled compounds into secondary products, and the study of metabolic routes of these secondary metabolites, based on a knowledge of their chemical structure, incorporation experiments with radioactive precursors and experiments with co-synthesis. Here, too, it was apparent that a systematic study of this field is in its very beginnings.

It is hardly necessary to emphasize the theoretical and practical importance of the section on the cell wall and membrane and on the genetics of transport accumulation of metabolites. The effect of penicillin and resistance to it belong to this field (**J.-M.GHUYSEN**). Further attention was concentrated on the mucopeptides of the cell walls of Gram-positive bacteria (**H.R.PERKINS**), regulation of transport of amino acids and pyrimidines in yeast cells (**M.GRENSON**), genetic effects on transport proteins (**A.B.PARDEE**), transport systems for sugars in *Aspergillus nidulans* (**A.H.ROMANO**). The section devoted to the genetic control of sporulation was equally important for mutual understanding.

The panel discussions dealt with the biosynthesis of alkaloids, the biosynthesis of antibiotics and related compounds, methods of parasexual genetics and the industrial biosynthesis of amino acids and nucleotides (attended mostly by experts from industrial research). Two more fields should be mentioned here because they show the possibilities of the most important cross-fertilization of theoretical and practical approaches: the section on the legal protection of strains and genetic procedures, and the special symposium headed by **S.I.ALIKHANIAN** on radiation and chemical mutagens.

It should follow from the cursory review of the proceedings that the goal of bringing together the theoretical and the practical workers has been achieved. This comes to light very nicely in the final lecture by **A.L.DEMAIN** entitled "*Marriage of Genetics and Industrial Microbiology — after Long Engagement a Bright Future*". The conclusion was unanimously accepted that the present symposium should represent just the beginning of a regular series of symposia — already three towns expressed their desire to organize the next one.

It would be only fitting to mention the second goal of the symposium: What new facts were presented in the field of genetic control, particularly of secondary products of microbial metabolism? It would be immodest to expect miracles from a meeting of this type — still, it cannot be overlooked that the symposium focussed attention on theoretical problems, that the fields of future study were defined and some methodical approaches indicated. It was above all obvious that a successful solution will require a complex approach by geneticists, physiologists, enzymologists, chemists, bio-engineers and physicists and that only such team-work will be able to aid workers in practical applications. Moreover, it became clear that progress will require much experimental work as well as original ideas which might play a decisive role in the solution of numerous entangled problems.

Contributors

BERTANI G., *Karolinska Institutet, Stockholm, Sweden.*
BISHOP R. J., *Princeton University, New Jersey, U.S.A.*
CHIBATA I., *Chemical Research Laboratory, Tanabe Seiyaku Co., Ltd., Osaka, Japan.*
CLARKE P. H., *University College London, Great Britain.*
COYETTE J., *Université de Liège, Belgium.*
COYLE M., *The University of Chicago, Illinois, U.S.A.*
CURRIE S. A., *Merck Sharp & Dohme Research Laboratories, Merck & Co., Inc., Rahway, New Jersey, U.S.A.*
DATTA P., *The University of Michigan, Ann Arbor, Michigan, U.S.A.*
DEMAIN A. L., *Massachusetts Institute of Technology, Cambridge, Massachusetts, U.S.A.*
DUNGAN S. M., *The University of Michigan, Ann Arbor, Michigan, U.S.A.*
FELDBERG R. S., *The University of Michigan, Ann Arbor, Michigan, U.S.A.*
FORTNAGEL P., *Ruhr-Universität Bochum, Germany.*
FREESE E., *National Institutes of Health, Bethesda, Maryland, U.S.A.*
GEST H., *Indiana University, Bloomington, Indiana, U.S.A.*
GHUYSEN J.-M., *Université de Liège, Belgium.*
HALLADA T. C., *Merck Sharp & Dohme Research Laboratories, Merck & Co., Inc., Rahway, New Jersey, U.S.A.*
HANAWALT P. C., *Stanford University, California, U.S.A.*
HARFORD N., *Princeton University, New Jersey, U.S.A.*
HAYHURST W. L., *Ridout & Maybee, Toronto, Canada.*
HILL T., *The University of Chicago, Illinois, U.S.A.*
HOLLOWAY B. W., *Monash University, Clayton, Australia.*
IKEDA Y., *University of Tokyo, Japan.*
KATO J., *Chemical Research Laboratory, Tanabe Seiyaku Co., Ltd., Osaka, Japan.*
KENNETT R. H., *Princeton University, New Jersey, U.S.A.*
KERJAN P., *Laboratoire d'Enzymologie du C.N.R.S., Gif-sur-Yvette, France.*
KISUMI M., *Chemical Research Laboratory, Tanabe Seiyaku Co., Ltd., Osaka, Japan.*
KLOFAT W., *National Institutes of Health, Bethesda, Maryland, U.S.A.*
KOMATSUBARA S., *Chemical Research Laboratory, Tanabe Seiyaku Co., Ltd., Osaka, Japan.*
LEYH-BOUILLE M., *Université de Liège, Belgium.*
MAIA J. C. C., *Laboratoire d'Enzymologie du C.N.R.S., Gif-sur-Yvette, France.*
MANDELSTAM J., *University of Oxford, Great Britain.*
MYERS T., *The University of Chicago, Illinois, U.S.A.*
NAKAYAMA K., *Tokyo Research Laboratory, Kyowa Hakko Kogyo Co., Ltd., Tokyo, Japan.*
NIETO M., *National Institute for Medical Research, Mill Hill, Great Britain.*
O'SULLIVAN A., *Princeton University, New Jersey, U.S.A.*
PARDEE A. B., *Princeton University, New Jersey, U.S.A.*
PERKINS H. R., *National Institute for Medical Research, Mill Hill, Great Britain.*
PIGGOT P., *University of Oxford, Great Britain.*
POLLOCK M. R., *University of Edinburgh, Great Britain.*
PRAKASH L., *The University of Chicago, Illinois, U.S.A.*

PUTTER I., *Merck Sharp & Dohme Research Laboratories, Merck & Co., Inc., Rahway, New Jersey, U.S.A.*

QUINN W. G., *Princeton University, New Jersey, U.S.A.*

ŘEHÁČEK Z., *Institute of Microbiology, Czechoslovak Academy of Sciences, Prague, Czechoslovakia.*

SAGER R., *Hunter College of City University of New York, U.S.A.*

SAITO H., *University of Tokyo, Japan.*

SHIIO I., *Central Research Laboratories of Ajinomoto Co., Inc., Kawasaki, Japan.*

STOY A., *Institute of Macromolecular Chemistry, Czechoslovak Academy of Sciences, Prague, Czechoslovakia.*

STRAUSS B., *The University of Chicago, Illinois, U.S.A.*

SUEOKA N., *Princeton University, New Jersey, U.S.A.*

SUGAE K., *National Institutes of Health, Bethesda, Maryland, U.S.A.*

SUNSHINE M. G., *Karolinska Institutet, Stockholm, Sweden.*

SZULMAJSTER J., *Laboratoire d'Enzymologie du C.N.R.S., Gif-sur-Yvette, France.*

TAKAHASHI H., *University of Tokyo, Japan.*

UMBARGER H. E., *Purdue University, Lafayette, Indiana, U.S.A.*

WOODRUFF H. B., *Merck Sharp & Dohme Research Laboratories, Merck & Co., Inc., Rahway, New Jersey, U.S.A.*

1/ Introduction

The Marriage of Genetics and Industrial Microbiology — after a Long Engagement, a Bright Future*

ARNOLD L. DEMAIN

Department of Nutrition and Food Science, Massachusetts Institute of Technology, Cambridge, Massachusetts, U.S.A.

INTRODUCTION

It is somewhat over twenty-five years since the first superior penicillin-producing mutant, *Penicillium chrysogenum* X-1612, was isolated after X-ray mutagenesis at the Carnegie Institution at Cold Spring Harbor. That event was the start of a long "engagement" between genetics and industrial microbiology. Like any engagement, it has had its successes and its failures and has managed to develop some new areas of common interest. The partners have not yet fully committed themselves to working toward the mutual benefit of the pair, each going down a separate path to a considerable extent. If the union is ever consumated, we can look forward to a bright future for both. The first international symposium on the Genetics of Industrial Microorganisms (Prague 1970) may serve as a catalyst toward that end, but it is still too early to tell. Much depends on whether both parties spent the week seriously listening to, discussing and taking to heart, each others problems, challenges and opportunities. If an effective dialogue was begun, the preparations for the marriage are indeed underway.

MUTANTS IN INDUSTRIAL MICROBIOLOGY: A SUCCESS

Ask anyone involved in the fermentation industry and, whether he is an operator in the factory, a technician, a plant manager or Chairman of the Board, he will tell you that the microbial mutant is the most valuable asset of the organization. There is no doubt that the ability to modify genetically a microbial culture to higher productivity has been the most important factor in keeping the fermentation industry in its viable, healthy state.

It was indeed fortunate that the intense development of microbial genetics began in the 1940s when the fermentative production of penicillin became an international necessity. The early basic genetic studies concentrated heavily on the production

* Contribution No. 1679 of the Department of Nutrition and Food Science.

of mutants and the study of their properties. The ease with which "permanent" characteristics of microorganisms could be changed by mutation and the simplicity of the mutation technique had tremendous appeal to microbiologists. Thus began the cooperative "strain-selection" program among workers at Peoria, the Carnegie Institution, Stanford University and the University of Wisconsin, followed by the extensive individual programs that still exist today in industrial laboratories throughout the world. Twenty-five years later, it is clear that mutation is the major factor involved in the hundred to thousandfold increases obtained in production of microbial metabolites.

Randomly selected mutants. The most common mutant type used to increase production in industry is the randomly selected type which is, unfortunately, the least interesting from the intellectual viewpoint. In this procedure, a mutagenized population is plated on agar and randomly selected colonies are tested in flasks for increased production ability. We unfortunately know nothing about the reasons for increased production in any of these mutants.

Morphological mutants. Another common type used in strain-selection programs is the morphological mutant which possesses an easily visible change in colonial morphology. I also include color mutants in this group. Morphological mutants were very useful in the development of the Wisconsin family of penicillin producers and were the key cultures in the oxytetracycline series of ALIKHANIAN *et al.*[1] Selection of conidial color mutants of *Aspergillus niger* has proved valuable in increasing production of citric acid[2]. It has also been found that changes such as the types of secondary products produced by a culture and their relative proportions can be effected by morphological mutation[3,4]. Unfortunately, nothing is known about the mechanisms involved in either the quantitative or qualitative alteration of fermentation products by morphological mutation. We must turn to *Neurospora crassa* to learn something of the enzymatic changes brought about by morphological mutations. The work of BRODY[5] tells us that at least four types of morphological mutants possess mutations in the structural genes of the carbohydrate metabolizing enzymes, glucose 6-phosphate dehydrogenase and phosphoglucomutase. Such mutations produce pleiotropic effects, such as changes in steady-state concentrations of intermediary metabolites and of coenzymes, shifts in usage of alternate metabolic pathways and alterations of cell wall structure. Especially notable in certain of these mutants are decreased levels of reduced nicotinamide adenine dinucleotide phosphate (NADPH) and increased levels of sugar phosphates. Since the biosyntheses of a large number of antibiotics involve sugar phosphates and NADPH, it is not surprising that production ability would be affected by these types of morphological mutation.

Metabolic regulation. The discovery of metabolic regulation in microorganisms by geneticists and biochemists during the 1950s has had a tremendous impact on industrial microbiology. Since a typical microbial cell has the genetic potential to form over a thousand enzymes, it becomes evident that coordination of metabolic reactions is obligatory for life processes. It was soon found that regulatory mechanisms act to insure (a) that only necessary enzymes are made, (b) that the correct amounts are made, and (c) that once produced, their activities are controlled by inhibition and activation. Induction, catabolite regulation and feedback regulation were found to be the mechanisms controlling the flexibility of microbial behavior and, in so doing, they prevented oversynthesis of intermediates and end products. Because industrial microbiologists are dedicated to the principle of oversynthesis, it is obvious that these control mechanisms must be destroyed or bypassed in order to force microorganisms to overproduce metabolites[6].

As stated above, nothing is known about the mechanisms causing higher production in superior randomly selected or morphological mutants. However, it is likely that many of these mutations involve regulatory genes. Interestingly enough, regulatory mutants obtained in basic genetic studies are often found to be altered in colonial morphology. For example, histidine-regulatory mutants of *Salmonella typhimurium*, which overproduce and excrete histidine, form wrinkled colonies in constrast to the smooth colonies formed by their parent[7]. Also, strains of *Escherichia coli* differ in colonial morphology, depending on whether they are "stringent" or "relaxed" with respect to amino-acid control of ribonucleic acid synthesis[8].

Auxotrophic mutants. Very early in the development of the concepts of regulation, geneticists realized that the end product of a biosynthetic pathway exercised strict control over the amount of an intermediate accumulated by an auxotrophic mutant of that pathway. Only at growth-limiting concentrations of the end product would a large accumulation of the substrate of the deficient enzyme occur. This

TABLE I

Production of Amino Acids and Nucleotides by Auxotrophic Mutants

Product	Requirement	Organism	Accumulation g/l
Citrulline	arginine	*B. subtilis*	16
Ornithine	arginine	*C. glutamicum*	26
Lysine	homoserine or threonine + + methionine	*C. glutamicum*, *B. flavum*	42
Valine	isoleucine	*C. glutamicum*	11
Inosine 5′-monophosphate	adenine	*C. glutamicum*, *B. ammoniagenes*	13
Xanthosine 5′-monophosphate	guanine	*C. glutamicum*, *B. ammoniagenes*	6

TABLE II

Overproduction of Primary Metabolites by Unrelated Types of Auxotrophs

Product	Requirement	Organism	Ref.
Alanine	leucine	*Ustilago maydis*	9
Alanine	methionine	*U. maydis*	9
Alanine	nicotinic acid	*U. maydis*	9
Proline	histidine	*Brevibacterium* sp.	10
Proline	isoleucine	*B. flavum*	11
Proline	serine	*Kurthia catenaforma*	12
Inosine	histidine + tyrosine	*B. subtilis*	13

principle of decreasing the concentration of an inhibitory or repressive end product to bypass-feedback inhibition or repression is best accomplished by the use of auxotrophic mutants. Indeed, auxotrophy has been the major factor in fermentative production of primary products such as amino acids and nucleotides (Table I). In all cases, it has been found essential to keep the required nutrient at a low level to insure maximum accumulation of the fermentation product.

It is quite remarkable that production of secondary products such as antibiotics is also markedly affected by auxotrophic mutation, even when auxotrophs are tested for production in nutritionally complete and complex media. Although the change in product formation is usually in the negative direction, higher-producing auxotrophs have been obtained from producers of tetracycline, chlortetracycline, streptomycin, actinomycin, cephalosporin C, kojic acid, and fusaric acid. One may wonder why auxotrophy has such major effects on production of secondary metabolites, especially when the compound required by the auxotroph is not a precursor of the secondary product. The answer is not known, but equally puzzling effects have been observed in the overproduction of primary metabolites by certain auxotrophs (Table II). It is very possible that in these cases we are dealing with "cross-pathway" regulation[14], or with double mutations.

"Revertants". Biochemical geneticists have observed that upon "reversion" of an auxotroph to prototrophy, the new prototroph may possess much more of the affected enzyme activity than did the original prototroph. Such increased enzyme activity could be the result of a structural mutation producing a more active enzyme or of a regulatory mutation. Increases in enzymatic activity of 3- to 4-fold have been observed with homoserine dehydrogenase in *E. coli*[15], and with inosine monophosphate (IMP) dehydrogenase in *S. typhimurium*[16]. Removal and replacement by mutation of the latter enzyme in *Corynebacterium glutamicum* resulted in overproduction of guanosine 5′-monophosphate (GMP)[17]. In the area of secondary metabolites, reversion of a methionine block in *Streptomyces viridifaciens* led to a 3-fold increase in chlortetracycline synthesis as compared to production by the original prototrophic culture[18].

Revertants of another type have also been successfully used in industrial microbiological studies; these are revertants of nonproducers of antibiotics. Both nonproducing mutants and their revertants are easily detected on agar plates by the absence or presence of clear zones following application of an assay organism to the developed colonies. DULANEY and DULANEY[18] found a high proportion of such revertants to produce increased amounts of chlortetracycline when compared to the original strain. The appeal of such a method lies in the fact that a reaction very crucial to antibiotic synthesis is being subjected to two sequential mutations.

Feedback-resistant mutants. Basic studies on regulation have shown that it is possible to isolate regulatory mutants, which overproduce end products of primary pathways, by the use of metabolite analogues. Such analogue-resistant mutants often possess enzymes that are insensitive to feedback inhibition, or enzyme-forming systems resistant to feedback repression. Selection of mutants resistant to toxic analogues of end products is now being employed by industrial microbiologists. The use of the threonine antimetabolite, α-amino-β-hydroxyvalerate, has aided in the development of a strain of *E. coli* that excretes over 6 g/l of threonine[19]. 8-Azaxanthine has been used to isolate IMP-dehydrogenase regulatory mutants that produce 4 g/l of guanosine[20].

Application of the analogue-selection technique to secondary products has been used to increase production of pyrrolnitrin[21]. Maximum production of this antifungal antibiotic occurred only after supplementation with tryptophan, a precursor. Mutants resistant to fluorotryptophan no longer require addition of tryptophan and, furthermore, produce 3 times as much pyrrolnitrin as the original strain.

When a secondary metabolite is itself a growth inhibitor, as in the case of antibiotics, the product can be used to select resistant mutants of the producing organism. Thus, selection for resistance to streptomycin or ristomycin has successfully yielded cultures of superior production ability when applied to the respective producers of these antibiotics[22,23].

Additional uses of mutants. Since many organisms produce secondary products as mixtures of a chemical family or of several chemical families, mutation has been used to eliminate completely any undesirable products or to enrich the proportion of a desirable product in the penicillin, griseofulvin, cycloheximide and nystatin fermentations. Even more important is the finding that mutant methodology can be used to produce entirely new molecules, such as the 6-demethyltetracyclines[24,25], oxychlortetracycline[26] and the new hybramycins[27]. A further use of mutants is in the elucidation of secondary metabolic pathways. For example, nonproducing tetracycline mutants complement each other, because intermediates that accumulate in certain blocked mutants diffuse into the medium and feed mutants blocked earlier. The elegant use of this "cosynthesis" technique has allowed MCCORMICK[28] and his colleagues to unravel the complicated biosynthetic branched pathway leading to tetracycline, chlortetracycline and oxytetracycline. Other pathways being traced by mutant techniques include those of novobiocin, polymyxin, prodigiosin and erythromycin.

RECOMBINATION IN INDUSTRIAL MICROBIOLOGY: A DISAPPOINTMENT

Aside from the necessity of sexual recombination for maximal production of β-carotene by members of the *Mucorales* (reviewed in Ref.[29]) and of heterokaryosis for production of alkaloids by *Claviceps purpurea*[30], the use of genetic recombination to improve yields of fermentation products has been disappointing, despite the fact that recombination is the major source of genetic variation and that many types of recombination occur in microorganisms, *e.g.* conjugation, parasexuality, transformation and transduction. Some of the problems encountered in the application of recombination techniques are recombination barriers between ancestrally divergent strains, instability of diploids, recessive nature of genes which lead to superior production, and lack of methods for selection of desired high-producing recombinants. However, the problem is much more profound than the above technical difficulties. I believe that there has been a deep gap separating basic research and application of recombination genetics. The progress in microbial genetics during the past twenty years has been truly fantastic, but it has been largely academic. It has dealt almost exclusively with organisms of no industrial importance such as *E. coli, N. crassa, Streptomyces coelicolor,* and *Aspergillus nidulans.* The number of academic geneticists interested in breeding new strains of fermentative microorganisms can be counted on one hand. Aside from these few workers, the application of genetics has been left to industrial microbiologists. Most of this group have not had formal training in genetics and, although they have been very successful in applying the simple concepts of mutation, it has been difficult for them to carry out sophisticated experiments in recombination genetics. In addition, the changes desired in microbial cultures are quantitative, not qualitative. Even plant breeders have experienced difficulty in obtaining quantitative improvements, whereas qualitative changes such as resistance to deleterious agents have been relatively easily achieved. The planning and execution of recombinational experiments for improved fermentation yield are difficult and require much thought, time, and patience; conditions fostering such experimentation are not often found in industrial development laboratories which are continually under pressure to improve yields as rapidly as possible. In short, the fermentation industry has not shown any great desire to commit the years and dollars needed to develop genetic systems that could be used as bases for breeding improved cultures, while most basic geneticists have shown little interest in applying their knowledge to the area of industrial microbiology.

NEW AREAS OF MUTUAL INTEREST

In addition to the ups and downs of the loose relationship between genetics and industrial microbiology which I have described, some exciting new areas of common interest to both parties have developed. These are described in the following paragraphs.

Branched pathways. A good part of the basic work of the 1960s was devoted to the study of devices which evolved in microorganisms to protect them from the consequences of their own regulatory mechanisms acting on branched biosynthetic pathways. Protective mechanisms such as multiple enzymes, concerted feedback regulation and cumulative feedback regulation have been found to prevent one end product from inhibiting other branches of the pathway and thus protect the microorganism against "regulatory suicide". Industrial microbiologists interested in fermentative production of such primary metabolites as lysine, threonine and purine nucleotides have used this basic information to great advantage in devising efficient fermentations. However, the prominence of branched pathways in the production of secondary metabolites has not been generally appreciated. As Fig. 1 shows, many secondary products are produced by branched pathways, one arm leading to a primary product(s) and the other producing the secondary metabolite(s). A branched pathway in *P. chrysogenum* leading to lysine and penicillin is another example and its existence is probably the reason for the inhibitory action of lysine on penicillin biosynthesis[31]. Some questions that industrial microbiologists should now be asking are: (a) do the end products of the primary branch reach concentrations inside the cell that inhibit secondary product formation; (b) can auxotrophic mutation in the primary branch increase the flow of metabolites into the secondary branch; (c) can regulatory mu-

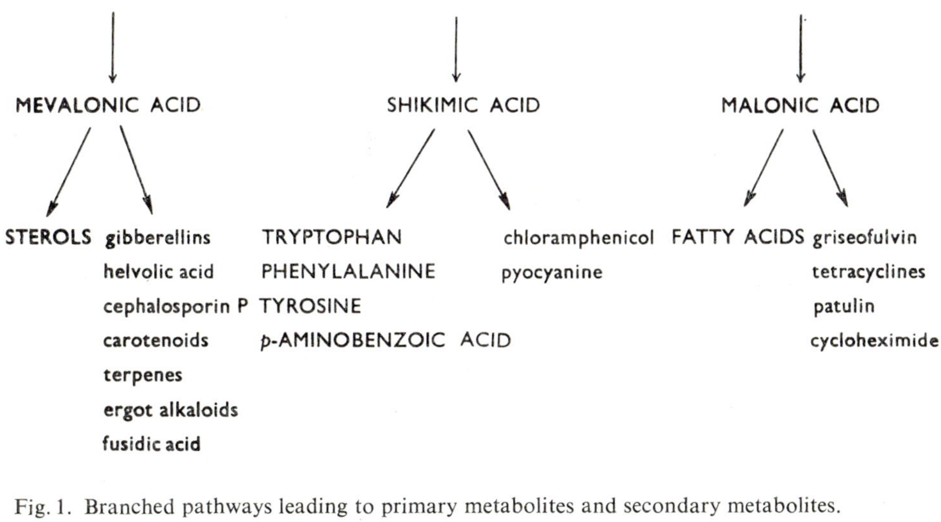

Fig. 1. Branched pathways leading to primary metabolites and secondary metabolites.

tations affecting the first enzyme of the early common path increase secondary product formation? Basic scientists will be interested in examining fermentation microorganisms to determine how they manage to avoid normal regulatory restrictions. Some interesting findings in this regard are starting to appear in the literature. For example, the lysine fermentation organism produces a dihydrodipicolinate synthase (the first enzyme of the lysine branch) which is not feedback inhibited by lysine[32]. Interestingly enough, *Bacillus subtilis,* which requires a feedback-insensitive lysine branch in order to produce dipicolinic acid for spore formation, also produces an insensitive dihydrodipicolinate synthase[33]. In the streptomycete which produces chloramphenicol, a shikimate-derived secondary product, the first enzyme of the aromatic amino-acid pathway is insensitive to feedback inhibition by phenylalanine, tyrosine or chloramphenicol[34]. Certainly a comparison of such insensitive enzymes with the same enzymes in closely related strains that do not overproduce metabolites is a worthwhile subject for basic biochemical genetic study. A recent comparison of the well-known penicillin producer, *P. chrysogenum* Wis-Q-176, and a high-producing mutant indicates that in the mutant, there is less feedback inhibition of acetohydroxyacid synthase (the initial enzyme of valine biosynthesis) by valine and in addition, twice as much enzyme is made by the mutant[35]. These changes are important since valine is a precursor of penicillin and enough must be made to support today's industrial penicillin yields of approximately 10 g/l. Thus, changes in feedback regulation of certain enzymes in a series of high-producing mutants such as the Wisconsin family of penicillin producers would be a worthwhile basic study in biochemical genetics.

Katabolite regulation. A second major study of biochemical geneticists during the 1960s dealt with katabolite regulation, a mechanism in which certain "unnecessary" enzymes are repressed or inhibited by katabolic products of a readily utilizable carbon source. It has recently been suggested that katabolite repression and katabolite inhibition are probably the mechanisms involved in the known depression of secondary-product formation by rapidly used carbon sources such as glucose[36]. This type of glucose effect has been observed with antibiotics such as penicillin, cephalosporin, siomycin, actinomycin, violacein, mitomycin and bacitracin and with other secondary products such as enterotoxin B and gibberellic acid. In many cases, katabolite regulation is probably also the cause of the delay in secondary-product formation until after growth is completed, *i.e.* the well-known trophophase—idiophase relationship. In the future, basic scientists will not only be interested in sporulation as a developmental process but will widen their scope by studying secondary-product formation as a primitive example of developmental biology. The sudden appearance, at the beginning of idiophase, of amidinotransferase and streptidine kinase in the streptomycin fermentation, of penicillin acyltransferase in the penicillin fermentation, of phenoxazinone synthetase in the actinomycin fermentation, of the enzymes metabolizing 6-methylsalicylic acid, gentisylalcohol and gentisaldehyde in the patulin fermentation and of the two amino acid-activating and peptide-forming enzymes in the gramicidin S fermentation (reviewed in ref.[36]) should make excellent subjects for basic studies. The finding by KATZ[37] that phenoxazinone synthetase is markedly repressed by glucose should provide a starting point to decide whether there is a cause and effect relationship between katabolite repression and trophophase—idiophase kinetics. The report by MALIK and VINING[38] that chloramphenicol synthesis shifts from the idiophase to the trophophase by changing from a complex medium allowing rapid growth to a defined medium supporting slower growth and that addition of yeast extract to the defined medium shifts antibiotic production back to the idiophase establishes another excellent system for basic studies. Industrial microbiologists will properly regard this as a fertile area. Some of the practical industrial problems of katabolite regulation have already been solved by changing the environment in either of two ways: (a) by using two carbon sources (a low concentration of the inhibitory substrate for growth and a high concentration of a noninhibitory substrate for product formation) and (b) by continuous feeding of the inhibitory carbon source after growth at a rate where inhibitory katabolites do not build up. However, these environmental changes have not solved the problem of delay in product formation until the trophophase is complete, a lag that costs the fermentation industry large sums of money. It would appear logical that mutants resistant to katabolite regulation should be tested for early initiation of product formation. Certainly the report of LIGHT[39] that mutant strains of *Penicillium patulum* can be obtained which produce 6-methylsalicylic acid synthetase much earlier than usual and thus produce 6-methylsalicylic acid (a precursor of patulin) during trophophase instead of idiophase should stimulate much work in industrial microbiology. Of further interest to industrial microbiologists is the fact that enzymes interconverting members of a family of secondary products, such as mannosidostreptomycinase in the streptomycin fermentation, are subject to katabolite repression[40]. Thus, the complexity of products in a fermentation broth, which causes much expense in the isolation process, might be minimized by the use of mutants resistant to catabolite repression.

THE MARRIAGE AND THE FUTURE

Let us assume for the moment that the geneticist and the industrial microbiologist decided to join together in a common effort to improve the production of microbial metabolites. Let me describe what I think would be important projects that not only would widen the scope of genetics and increase production of microbial metabolites but would, in a synergistic manner, be a major step forward in the improvement of the health and nutrition of our world society. Some of these topics have already been touched upon in the section concerned with new areas of mutual interest.

Biochemical changes in superior mutants. The biochemical changes which occur upon mutation to superior production should be identified irrespective of whether the mutant is selected randomly, by morphological mutation, by auxotrophic mutation, or by any other means. The work could be done with an available series of mutants such as the Wisconsin family of penicillin producers but, in cases where other metabolites are the subjects of interest, industry should cooperate by releasing older members of their strain-selection series. The practice of fermentation companies of only sending out the original culture of a series makes no sense at all when more recent strains could be released to make intermediate levels of the product which would not be competitive with present strains.

In many cases, enzymatic analysis of successful mutants would be limited by our fragmentary knowledge of the biosynthesis of secondary products. This situation should be remedied with an intense effort by biochemists to elucidate finally the enzymatic steps involved in the biosynthesis of important products such as antibiotics. Even in the absence of such information, progress can be made. The biosynthetic steps leading to the primary precursors of many of these molecules are known and the regulation of the enzymes involved can be compared within a series of mutants, as well as with organisms not commercially important. Only when we have the above information can we establish rational screening programs designed to detect mutants at important structural or regulatory genetic loci involved in product formation.

Recombination. There is no doubt that some day, recombination will take its proper place in strain-selection programs. What is needed now is a long-term commitment by a number of basic geneticists to study the genetics of organisms producing valuable products. This will be far from routine and I can predict that new genetic phenomena will be discovered, some of which will be of academic interest and some of which will have application in the fermentation field. A rich supply of fungi, actinomycetes and true bacteria await genetic dissection. What is sorely needed now is a demonstration of the power of recombination in production of superior cultures. Perhaps the bacteria should be examined first because of their high growth rates and the wealth of facts accumulated in the last 20 years dealing with bacterial genetics. Organisms which might be susceptible to rapid genetic attack are *Pseudomonas denitrificans* and *Propionibacterium shermanii* (vitamin B_{12}), *Pseudomonas aureofaciens* (pyrrolnitrin), *Bacillus licheniformis* (bacitracin) and *Bacillus polymyxa* (polymyxin). Once the genetic systems are established, mapping of the genes of product formation should be the next step. Finally, the effect of recombination of production genes should be examined. With this information in hand, the geneticist should be able to determine and to inform us about the applicability of the particular system to everyday practical problems of industrial microbiology.

While the above basic genetic studies are going on, the industrial microbiologist should be extensively examining genetic systems that are already known, such as parasexuality in *P. chrysogenum* and recombination in the streptomycetes. I do not

TABLE III

Manipulations which Increase Enzyme Production

Environmental	A. Addition of inducers
	B. Decreasing concentration of feedback repressors 1. Limitation of medium ingredients a) NH_4 (urease, protease) b) amino acids (protease) c) sulfate (protease) d) phosphate (phosphatase, nuclease) 2. Growth limitation of auxotrophs a) restricted feeding of required growth factor b) slowly metabolized precursor of growth factor c) leaky auxotrophs in minimal medium 3. Addition of analogs of end product of pathway
	C. Decreasing concentration of katabolite repressors 1. Use of slowly utilized carbon source 2. Restricted feeding of rapidly utilized carbon source 3. Addition of cyclic 3′,5′-AMP
Genetic	A. Constitutive mutation to eliminate inducer requirement 1. Selection in chemostat with a limiting supply of a substrate-inducer 2. Selection with carbon source which is a poor inducer 3. Selection with a combination of a substrate-inducer and an inhibitor of induction 4. Selection by alternating growth cycles between a medium containing a noninducing carbon source and a medium containing an inducing carbon source
	B. Constitutive mutation to resistance to feedback repression 1. Selection with toxic analogues of end product 2. Mutation to auxotrophy followed by "reversion" 3. Selection of mutants which can sporulate in presence of repressive amino acids
	C. Mutation to resistance to katabolite repression 1. Selection on medium containing substrate of repressed enzyme as sole nitrogen source in presence of glucose as carbon source 2. Selection by alternating growth cycles between a medium containing glucose as carbon source and a medium containing substrate of repressed enzyme ascarbon source
	D. Increasing gene copies 1. Episome transfer 2. Phage-escape synthesis

feel that the fermentation industry has really given these systems a proper test. Admittedly, the publications dealing with the application of these systems have not been very encouraging, but the total number of recombinants examined in these publications is extremely small. I should like to remind you that if the same spirit had been prevalent during the early days of *P. chrysogenum* mutation studies, penicillin probably would not be an article of commerce today. Please recall that the first attempt to mutate strain NRRL 1951-B25 (a spontaneous variant of the original Peoria culture, NRRL 1951) at Stanford University yielded 60,000 isolates, not one of which showed improved production! Thus, although the recombinational crosses described in the literature to date have not yielded a significant proportion of superior producers, perhaps a large-scale comparative examination of the frequency of high producers among recombinants and among survivors of mutagenic treatment would reveal that high-producing recombinants do occur at a greater frequency than high-producing mutants.

Enzyme fermentations. Although microbial enzymes have been marketed for many years, the recent upsurge of commercial activity in this field provides industrial microbiologists with a rare opportunity to apply many basic findings to industrial problems. Discoveries in biochemical genetics have revealed that enzyme production is controlled both positively and negatively by cellular regulatory mechanisms and can be made to vary over a thousand fold as a result of environmental and genetic changes[41]. The modifications that have been shown to be successful on a laboratory scale[42] and that now should be carefully examined by industrial microbiologists are summarized in Table III.

With respect to the activities of interested basic geneticists, it is indeed fortunate that many of the enzymes of new industrial importance are made by bacteria, *e.g.* L-asparaginase *(E. coli)*, polynucleotide phosphorylase *(Micrococcus lysodeikticus)*, alkaline protease *(Bacillus subtilis)*. Since a considerable amount of genetic work has already been done on some of these species, projects such as genetic mapping of production genes, increasing yields by recombination and directed mutation in synchronized cultures could be started without delay.

The preparation of this manuscript was supported by U.S. Public Health Service Research Grant AI-09345 from the National Institute of Allergy and Infectious Diseases.

REFERENCES

1. Alikhanian S. I., Mindlin S. Z., Goldat S. U., Vladimirov A. V.: Ann. N. Y. Acad. Sci. **81**, 914 (1959).
2. Ilczuk Z.: Acta Microbiol. Polon. **17**, 331 (1968).
3. Spížek J., Málek I., Doležilová L., Vondráček M., Vaněk Z.: Fol. Microbiol. **10**, 259 (1965).
4. Blumauerová M., Mraček M., Vondráčková J., Podojil M., Hošťálek Z., Vaněk Z.: Fol. Microbiol. **14**, 215 (1969).
5. Brody S.: J. Bacteriol. **101**, 802 (1970).
6. Demain A. L.: Adv. Appl. Microbiol. **8**, 1 (1966).
7. Roth J. R., Antón D. N., Hartman P. E.: J. Mol. Biol. **22**, 305 (1966).
8. Edlin G., Broda P.: Bacteriol. Rev. **32**, 206 (1968).
9. Dulaney E. L., Jones C. A., Dulaney D. D.: Developments Ind. Microbiol. **5**, 242 (1964).
10. Yamatodani S., Suzuki M., Nakao Y.: Amino Acid & Nucleic Acid **16**, 133 (1967).
11. Yoshinaga F., Konishi S., Okumura S., Katsuya N.: J. Gen. Appl. Microbiol. **12**, 219 (1966).
12. Kato J., Horie S., Komatsubara S., Kisumi M., Chibata I.: Appl. Microbiol. **16**, 1200 (1968).
13. Aoki R., Momose H., Kondo Y., Muramatsu N., Tsuchiya Y.: J. Gen. Appl. Microbiol. **9**, 387 (1963).
14. Jensen R. A.: J. Biol. Chem. **244**, 2816 (1969).
15. Patte J. C., LeBras G., Cohen G. N.: Biochim. Biophys. Acta **67**, 16 (1963).
16. Levin A. P.: Biochim. Biophys. Acta **138**, 221 (1967).
17. Demain A. L., Jackson M., Vitali R. A., Hendlin D., Jacob T. A.: Appl. Microbiol. **14**, 821 (1966).
18. Dulaney E. L., Dulaney D. D.: Trans. N. Y. Acad. Sci., II, **29**, 782 (1967).
19. Shiio I., Nakamori S.: Agric. & Biol. Chem. **33**, 1152 (1969).
20. Momose H., Shiio I.: J. Gen. Appl. Microbiol. **15**, 399 (1969).
21. Elander R. P., Hamill R. L., Gorman M., Mabe J.: 154th Nat. Meet. Am. Chem. Soc.; Abstr. Papers, Q42 (1967).
22. Teteryatnic A. F., Bryzgalova L. S.: Antibiotiki **13**, 252 (1968).
23. Woodruff H. B.: Symp. Soc. Gen. Microbiol. **16**, 22 (1966).
24. McCormick J. R. D., Sjolander N. O., Hirsch U., Jensen E., Doerschuk A. P.: J. Am. Chem. Soc. **79**, 4561 (1957).
25. Hendlin D., Dulaney E. L., Drescher D., Cook I., Chaiet L.: Biochim. Biophys. Acta **58**, 635 (1962).
26. Mitscher L. A., Martin J. H., Miller P. A., Shu P., Bohonos N.: J. Am. Chem. Soc. **88**, 3647 (1966).
27. Shier W. T., Rinehart K. L., Jr., Gottlieb D.: Proc. Nat. Acad. Sci. U.S.A. **63**, 198 (1969).
28. McCormick J. R. D., p. 113 in D. Gottlieb, P. D. Shaw (Eds.): *Antibiotics*, Vol. II. *Biosynthesis*. Springer-Verlag, New York 1967.
29. Ciegler A.: Adv. Appl. Microbiol. **7**, 1 (1965).
30. Spalla C., Amici A. M., Scotti T., Tognoli L., p. 611 in D. Perlman (Ed.): *Fermentation Advances*. Academic Press, New York 1969.
31. Demain A. L., p. 29 in J. Snell (Ed.): *Biosynthesis of Antibiotics*. Academic Press, New York 1967.
32. Nakayama K., Tanaka K., Ogino H., Kinoshita S.: Agric. & Biol. Chem. **30**, 611 (1966).
33. Chasin L. A., Szulmajster J., p. 133 in L. L. Campbell (Ed.): *Spores*, Vol. IV, Am. Soc. Microbiol., Washington, D. C., 1969.

34. WESTLAKE D. W. S., VINING L. C.: Biotechnol. & Bioeng. **11**, 1125 (1969).
35. GOULDEN S. A., CHATTAWAY F. W.: J. Gen. Microbiol. **59**, 111 (1969).
36. DEMAIN A. L.: Lloydia **31**, 395 (1968).
37. KATZ E., *personal communication* (1970).
38. MALIK V. S., VINING L. C.: Can. J. Microbiol. **16**, 173 (1970).
39. LIGHT R. J.: J. Biol. Chem. **242**, 1880 (1967).
40. INAMINE E., LAGO B. D., DEMAIN A. L., p. 199 in D. Perlman (Ed.): *Fermentation Advances.* Academic Press, New York 1969.
41. PARDEE A. B., p. 3 in D. Perlman (Ed.): *Fermentation Advances.* Academic Press, New York 1969.
42. DEMAIN A. L., in W. B. Jakoby (Ed.): *Methods in Enzymology,* Vol. *Enzyme Purification and Related Techniques.* Academic Press, New York, *in press.*

2/ Genetics of the Bacterial Cell

The Biochemical Uses of Bacterial Genetics

M.R. POLLOCK

Department of Molecular Biology, University of Edinburgh, Great Britain

It is, I think, natural to feel somewhat overhelmed when confronted, as I was, with such an Olympian Title for what is intended as a short introduction to the topic under discussion.

Bacterial Genetics, after all, is becoming more and more molecular everyday. Professor HAYES, for instance, has, within the last few months, changed the name of his Medical Research Council Unit from "Bacterial Genetics" to "Molecular Genetics" without *any* alteration whatsoever in staff or in research projects. He and his staff had gone on for years supposing they were studying bacteria and have only just realized they are in fact working on molecules!

So the theme almost amounts to "The Biochemical Uses of Certain Aspects of Biochemistry" and would have to be dealt with purely descriptively or be reduced to assertions that Biochemistry is very useful — or of absolutely no use whatever to Biochemistry, but very valuable in Numismatology or Mathematical Physics or Astrology or something.

In fact, however, even the most enthusiastic Biochemical Geneticist fully recognises that genetical concepts and techniques applied to microorganisms, bacteria and their viruses in particular, have played an exceedingly important part — indeed one which was absolutely critical — in the interpretation of biological systems at a *molecular level.*

HISTORICAL

Historic examples of these are too numerous to allow a complete survey in a short article. They clearly demonstrate the origin of Bacterial Genetics in the Genetics of higher organisms such as *Drosophila* and *Neurospora.* One might mention of course early contributions to the elucidation of metabolic pathways by blocks to specific enzymes originating from point mutations in their structural genes, based on the famous one gene/one enzyme hypothesis of the early 40's. This followed, but did not strictly speaking depend upon, the discovery of genetic recombination in bacteria in 1946.

More recently, mutational analyses have disclosed previously unrecognised types of physiological defect. For example, the "rec^-" (recombination-deficient) and UV-sensitive mutants of *E. coli* have focussed attention on a group of enzymes concerned with recombination and repair processes which comprise certain exo- and endonucleases and a polynucleotide ligase which are of great biochemical interest and importance. Another similar example is the discovery of restriction-modification enzymes acting on the DNA of intruding phage, from the study of phage and bacterial relationships.

Apart from this, the isolation of defective mutated proteins, with altered enzymic properties, arising from missense mutations in structural genes, have provided, and will surely continue to provide, valuable material for identifying enzymically active sites, understanding the mechanisms of enzyme action and the ways in which primary amino acid sequence affects tertiary structure.

More sophisticated mutational and recombinational analysis at the high resolution that is possible only with microorganisms (because of their rapid multiplication rate and the large number that can be scored), provided the first direct concrete evidence, in the rII region of T4 phage, in favour of a triplet base code for amino-acid-sequence determination in proteins. Analogous work on specific amino-acid substitutions arising from one-step mutational base changes and intra-codonic recombinations permitted the identification of natural triplet codons for specific amino acids operating in the *E. coli* tryptophan synthase A protein.

Latterly, the exploitation and analysis of frame-shift mutations has carried the task of identifying natural codons almost into the category of routine investigations. Previously it had been shown by fine structure analysis of the rII region in T4 bacteriophage that the maximum size of a recombinational unit – the 'recon' – corresponded to only two nucleotides, which strongly indicated that functional 'crossing over' could occur between two contiguous bases in DNA. All this work, directly concerning the biochemistry of DNA was carried out without any direct chemical analysis of DNA itself. If biochemical techniques were used at all, they concerned the elucidation of amino acid sequences in protein fragments and not base sequences in polynucleotides. But the fundamental analysis was genetic.

It must also not be forgotten that the main evidence establishing co-linearity between amino-acid sequences in proteins and nucleotide sequences in genes similarly involved protein biochemistry but depended absolutely upon the classical techniques of molecular genetics, such as gene mapping and the use and identification of missense mutations, deletions and suppressible chain-terminating codons. No relevant nucleic acid chemistry was either necessary or possible.

With regard to protein structure itself, it has on several occasions been through genetic studies on complementing gene defectives in diploid association, rather than by direct biochemical analysis on isolated enzyme preparations, that the first suggestions have been obtained in favour of the cooperative participation of different polypeptide chains in, and therefore of an oligomeric state for, the active molecule.

A recent example is the three genes believed to be concerned with the structuring of possibly a single oligomeric restriction-modification enzyme for infecting phage in *E. coli.*

Thirty years ago, the contribution that Genetics can now be seen to have made to Biochemistry would have been considered quite absurd by nearly everyone. In so far as the two disciplines were considered related at all, it was almost exclusively in the direction of the essential analytical role Biochemistry would play in the elucidation of the mechanism of heredity. Nobody, I think, would then have believed that predominantly genetical techniques of mutational and recombinational analysis would be used to determine the chemical structure (in the shape of nucleotide sequences) of parts of DNA molecules.

J. B. S. HALDANE, however, caught a glimpse of what was to come when he wrote, prophetically, in 1937 that "The whole problem of synthesis is an almost virgin field in biochemistry. And it is absolutely fundamental. If genetics had done no more than pose the question (of gene replication) in its sharpest form it would still be a valuable stimulant to biochemical research. But it has done a great deal more than this. Whereas the classical biochemistry has largely been concerned in analysis of the stages of katabolic processes such as the breakdown of carbohydrates to lactic acid or alcohol and the digestion of proteins, the new brand of biochemistry which will, I believe, arise from Genetics will be concerned largely with the stages of synthesis of such molecules as chlorophyll and cyanin (*i.e.* macromolecules, such as proteins). And its final goal will be the explanation and control of the synthesis of genes"[1].

It might also be pointed out, in defence of Bacterial Genetics, that its contributions, whether biochemical or more strictly genetical, have been as much or more towards the Biology of higher organisms as within the strict confines of the bacterial world. And this, of course, is through its role as one of the fundamentals of Molecular Biology, with all its widespread ramifications throughout the bio-sphere. The original classical genetical studies with maize, mice and fruit-fly have made relatively little difference to plant and animal breeding practices which remain largely empirical.

THE GENETIC MAP

After all, one of the essential features of Genetics as a distinct discipline centre, on what is referred to as 'mapping'. In cartography a map is usually a two dimensional representation — almost a model — of a three dimensional situation. Two, at leasts of those three dimensions are modelled fairly accurately. The third — heights (and depths) — are usually indicated by contour lines, figures and colours which bear little or no resemblance to the hills and valleys or ocean beds themselves.

In Genetics the map is a one-dimensionally expressed set of instructions for the construction of a 3-dimensional organism. Some might argue that the instructions are not complete and most would say that it is far from being a *representation* or a *model.* However, in prokaryotes I would maintain that the genetic map — in the form of a linear sequence of genes joined in distinct linkage groups — is, at least in part, a sort of map or model, in the cartographical sense, of the cell as a whole, as well as being simply a relative position list of 'markers'. Quite apart from the increasing tendency to analyse and interpret genetic events in molecular terms, the bacterial genome is in such intimate, minute-to-minute contact with cell metabolic processes in general — especially, of course, protein synthesis — that the division between Genetics and Physiology has become quite arbitrary.

In higher organisms there is much clearer distinction. Operations involving the DNA of nuclear genes are spatially and chronologically much further removed from, and more indirectly related to, the physiological functions of the cytoplasm than in prokaryotes. In bacteria we *might* have found the one-dimensional cell blue-print (the set of instructions) printed in a form showing no resemblance to, and having no direct connection with, the metabolic processes it determined — less of a model, that is, than the representation of heights on a two-dimensional geographic map. In fact, however, the genetic map in bacteria has many features directly related to cell metabolism. The genes are often ordered in approximately the same sequence as that in which the enzymes they code for must actually operate. The classic instance is that of the genes for the series of enzymes concerned with histidine biosynthesis in *Salmonella*.

Moreover, the processes of transcription and translation are spatially related and often dependent on polycistronic messenger-RNA formation, covering several genes, where polypeptide-chain synthesis follows close behind RNA polymerase action in a continuous belt system. This means that enzymes must at least be *formed* close to where their structural genes are located. Perhaps in many cases they are actually located and may function near their genes too.

Functionally related enzymes are also often co-ordinately controlled (in operons) by action at genetic loci (operators and promoters) strategically located next to the group of genes which code for the relevant enzymes. Separate linkage groups (plasmids, episomes and chromosomes) often have grouped functions of physiological significance: the chromosome in general carrying genes essential for ordinary, "day-to-day" metabolic activity (biosynthesis and energy-producing reactions) whereas plasmids mainly 'deal' with adventitious characters, important only in special circumstances, such as drug resistance, colonial morphology, toxin and pigment production, *etc.*

None of this applies to the genetic map and grouping in higher organisms, where gene order and linkage seem largely unrelated to the functions of the enzymes for which they are responsible.

However, in bacteria the partial correspondence between the genetic map and integrated metabolic structure indicates, in a very broad sense, why genetic analysis has contributed and will continue for some time to contribute, so much to the elucidation of biochemical reactions in the cell.

PRACTICAL APPLICATIONS — SOME GENERAL PRINCIPLES

Applications of bacterial genetics in a fairly routine manner, by mutant selection towards strain improvement (for whatever purpose) are important enough at the moment. But really exciting possible contributions to Biochemistry (in the widest use of the term) will be those based on more advanced concepts and techniques, that are only just beginning to be seriously considered.

I would group most future possibilities under the general heading of 'manipulations': physiological or strictly genetical, quantitative or qualitative, according to their operational nature, but all likely to be developed through the application of fundamental principles in bacterial (*i.e.* molecular) genetics. They can be divided into different groups according to their academic, economic and social significance — their relevance for instance, to basic Biology, to Medicine and Therapeutics, to Industrial Productivity or, for that matter, to Biological warfare (should this come under the heading of "Biochemical Uses"; which in a sense, I believe, unfortunately it does).

And they stem from 3 "areas of operation" in Bacterial Genetics, all — not surprisingly — referable to the properties of DNA:

(1) Cell-to-cell transfer
(2) Replication
(3) Recombination

These 3 "areas" are closely related to each other in the integrated Molecular Biology of nucleic acids within the cell. They can be further analysed in the context of our theme as follows.

Transfer. Bacteria have several very efficient mechanisms for facilitating transfer of nucleic acids between cells of closely related species: transduction by phage; passage through, or aided by, sex pili; and direct penetration through the cell envelope (transformation). Heterologous transfer between unrelated strains may be rare; but it can occur. Artificial mechanisms, analogous to those used in higher organisms, for weakening what have been referred to as "nucleic acid intrusion barriers", may not be difficult to develop in bacteria.

Replication. Replication of chromosomes and extra-chromosomal entities are often under strict co-ordinate control, though physically separable. Under certain conditions, however, potentially autonomous replicons — be they originally considered as plasmids, episomes (*e.g.* integrated sex factors) or pro-phages —may escape, totally or relatively, from strict control and replicate at a high rate to produce multiple copies in each cell. This escape may take the form of wild, vegetative growth of phage leading to hundreds of copies per cell and eventual cell lysis; or to a relative, but limited, increase in numbers of plasmids which may persist for some time without completely disrupting cell metabolism. This escape may be (a) triggered by a transitory stimulus, such as ultraviolet light, (b) hatched by a specific mutation in, or recombinational event affecting, a control gene or (c) facilitated by transfer to a different type of cell.

In the otherwise unmodified chromosome itself, multiple growing points may be induced by cultivation in a rich medium, allowing genes near the point of initiation of replication to have several copies in each cell.

Recombination. Recombination between sufficiently homologous stretches of DNA seems more continuous and less under specific inhibitory control in bacteria and their viruses than in eukaryotes where it is in general limited (though not absolutely) to meiosis. There is hardly need to stress its importance for the restructuring of gene complexes, in particular autonomous replicons (phage or plasmids) and transfer factors, to carry new sets of genes and in gene translocations of all types. Together with cell-to-cell transfer mechanisms, recombination allows a mixing and reorientating of genetic information amongst a population, and between different populations, of cells which, with appropriate selection and promotion, can be used for the guided exploitation of specific bacterial genes in many different ways.

PRACTICAL APPLICATIONS — SOME SPECIFIC EXAMPLES

Since it would be impossible to deal systematically with *all* 'biochemical uses' of Bacterial Genetics it seems reasonable to touch on one illustrative region of usage with which we, in our Department at Edinburgh, happen at the moment to be concerned: *viz.* specific bacterial enzyme production for research, medicine and industry.

Well-tried standard procedures for increasing rates of bacterial enzyme production such as physiological derepression of inducible enzymes, or isolation of constitutive mutants will not be considered here since there are now superposable techniques for specifically increasing numbers of gene copies which can be applied above and beyond the more usual routine measures.

In many higher organisms multiplication of gene copies in the form of repetitive DNA sequences may be one of the normal physiological mechanisms adopted for stepping up rates of production of certain species of RNA or proteins, *e.g.* multiple copies of ribosomal cistrons. And in bacteria too, within pretty wide limits, the differential rate of production of an enzyme is closely proportional to the number of copies of its structural gene per cell ('gene dosage').

There are now several types of procedure developed for increasing the gene dosage in bacteria. For example, it is possible to increase the number of growing points on the *E. coli* chromosome (a) to at least 3 by using suitably rich media; (b) up to 15 by appropriate timing of additions of a growth factor in certain auxotrophic mutants.

The maximum number of copies of course only affects genes near the initiation point for replication and this will limit the types of gene normally exploitable. But it is theoretically possible to 'force' translocations of genes to suitable positions in the chromosome by appropriate selection techniques (based on a phage or episome pick-up method similar to that described in the next paragraph) so that it may be possible to move any desired gene to a locus where its replication would be enhanced by multiple growing points.

It is also now possible to link up, to some extent at will, specific genes with a defective phage (*e.g.* Φ 80) and so promote their rapid replication, along with the phage genome when the latter is induced to multiply vegetatively. In this way multiple copies per cell can be obtained in certain cases within the space of a generation time. By choosing suitable mutations in the phage (*e.g.* towards defective lysozyme production) cell damage can be delayed or abolished so that general cell metabolism will continue to allow transcription and translation of the multiple gene copies that are being replicated. In this way 10 to 15 fold increases in yield of an enzyme such as tryptophan synthase A protein of *E. coli* have actually been obtained. In principle, it may be possible to apply this technique to any gene on the *E. coli* chromosome.

There may well be other variants of this technique which, in principle, could lead to the autonomous but controlled "repliconation" of any desired gene on any well-mapped chromosome of a bacterial species possessing temperate, infecting bacteriophages with a specific chromosomal locus.

A rather more uncertain, but no less interesting, prospect is that of qualitative enzyme improvement by selective recombination between distinct but homologous structural genes from different strains. In certain instances it may thus be possible to combine the advantages possessed by one variety with those of another. It is feasible enough to select such recombinant hybrids (*e.g.* as reported with β-galactosidase of *E. coli* and penicillinase of *B. licheniformis*) though the extent to which 'new' enzymes thus constructed would in the event prove to have *useful* or interesting combinations of properties is more doubtful. One might suspect that at the best no greatly improved function would emerge and final selection would have only an empirical basis.

FUTURE PROSPECTS

This sort of manipulation begins to come into the class of "genetic engineering manoeuvres" where 'required' information of a more extravagant nature may be constructed by appropriate *in vivo* recombination (or even synthesised by purely chemical procedures) in bacteria and then 'forced' into an autonomous replicon (*e.g.* through phage pick-up). Analogous manipulations with animal viruses have been visualised as eventually applicable for repair of genetic lesions in man and animals.

A less fanciful possibility concerns the exploitation of what might be termed the 'hardware' of bacterial metabolism: — the non-specific machinery of construction and energy utilization (which these prokaryotes operate so very rapidly) — for the production of foreign macromolecular complexes as far removed, perhaps, as animal viruses. Reports in the past of growth of polyoma, encephalomyelitis and vaccinia virus in *E. coli* or *B. subtilis* have not yet been substantiated. But this sort of phenomenon does not seem to be wildly improbable as a future possibility. Heterologous mixtures of nucleic acid-encoded information may be forced or encouraged to coexist and replicate together in the same cell by relatively simple manoeuvres such as that used in producing hybrid cells between mammals and birds.

The problem may be largely a question of forcing a breach in the "nucleic acid-intrusion barriers" which have presumably been evolved as a result of natural selection pressures operating against the sort of physiological disaster usually to be expected when two rather different sets of nucleic acid-encoded information meet and attempt to replicate or be transcribed in the same cellular environment.

The prospect, however, of using bacteria for growing animal viruses simply, efficiently and rapidly for human vaccines is perhaps worth exploring further. Success may depend upon relatively small physiological and genetic adjustments such as chemically-induced linkage to small replicons that are acceptable to the host cell, or 'forced' integration into a plasmid by a temperate phage acting as amboceptor for the recombination that would be necessary.

These sorts of fancies may still be some way off from realistic application. And many people are fearful of the unpleasant possibilities inherent in any *directed* change of human heredity.

It must of course not be forgotten that study of bacterial replicons (plasmids, phage and chromosomes), transfer processes, mutations and genetic recombinations carries with it the dangers, as does any new scientific concept or technique, of misuse and abuse. But at the same time such research is not only to be encouraged but is in fact absolutely necessary if we are to tackle any of the mounting problems now facing us, such as that of infectious bacterial drug resistance, at a fundamental level. This particular worry, since it almost certainly demands a new approach to the use and development of antibiotics, is only going to be solved by applying the principles of molecular genetics and sub-cellular biology to the study of plasmid physiology, plasmid transfer, plasmid recombination and plasmid replication: the spheres where nucleic acid interactions form a class of "organism" which is only just beginning to be appreciated as a newly conceived entity in Biology.

We should not, I believe, make the mistake of refusing to work on certain aspects of our subject because of arguments that there are "things which we ought not to know" or that there are dangers of our discoveries being misapplied. It is our responsibility, rather, not only to try and understand better the nature of the world in which we live, but also to inform humanity both of the possible dangers of the misuse of scientific knowledge and of the advantages that could accrue if it is properly exploited.

REFERENCE

1. HALDANE J.B.S., p. 1 in J. Needham, D. E. Green (Eds.): *Perspectives in Biochemistry*. Cambridge University Press 1937.

Conjugation and Episomes

B.W. Holloway

Department of Genetics, Monash University, Clayton, Victoria, Australia

Bacteria have been used by man for his own purposes for many thousands of years. Up until the beginning of the twentieth century, the range of different organisms used was quite limited, but with the development of microbiology as a scientific discipline, more and more organisms were shown to be useful for industrial and other purposes. More recently, through the application of current knowledge of microbial genetics, it has become possible for an even wider spectrum of organisms to become members of this once elite group. Techniques have been successfully developed for the selection of strains with desirable characteristics in terms of qualitative or quantitative yield. Yet, almost without exception, such strains have been developed solely by using the techniques of mutation and selection[1], despite the fact that so much has been achieved in the understanding of the processes of genetic exchange in bacteria. With the processes of conjugation, transduction, and transformation available for genetic analysis, together with current knowledge of the molecular biology of microorganisms, it is surprising that this extensive knowledge of bacterial recombination has not enabled the biomanipulation of this group to result in the development of industrially significant strains of bacteria.

However, it is likely that recent developments in aspects of microbial genetics, particularly in the study of conjugation in bacteria, may produce a significant change in this situation to enable strains of bacteria to be developed which could not be obtained solely by mutation and selection and which could have commercial value. Indeed, there are certain types of desirable strains which could not be obtained by mutation and the necessary genetic variation will need to come from some type of genetic event such as hybridization. Selection of strains having particular characteristics either involves a search for naturally occurring strains with the desired phenotype, or the isolation of mutants derived from known strains. Most mutants, particularly those induced in the laboratory, usually involve a loss of function. Recombination, on the other hand, can produce what is essentially a new set of functions which could amount to the creation of new genetic information. Such a process is unlikely to result from mutation and a search for the desired strains in nature would be both tedious and inefficient.

It is my intention to give an account of the general nature of conjugation in bacteria, the importance of extra-chromosomal elements in this type of genetic exchange, and to discuss recent developments which, in addition to providing us with a greater understanding of these biological phenomena, also indicate ways for the utilization of this knowledge for the solution of problems which could have a variety of practical applications.

OCCURRENCE OF CONJUGATION IN BACTERIA

Following the first demonstration of conjugation in *Escherichia coli* K12[2], the occurrence of the same phenomenon in other strains of this genus was subsequently demonstrated[3-6]. The sex factor F derived from *E. coli* K12 can be transferred to other genera and has been shown to promote conjugation in *Salmonella*[7-9], *Enterobacter*[10], *Shigella*[11] and *Pasteurella*[12,13]. Quite independent conjugation systems have been demonstrated for *Pseudomonas*[14-17], *Vibrio*[18] and *Rhizobium*[19]. While there is variation in the mechanisms by which these various bacterial genera achieve conjugation and genetic exchange, there are strong elements of similarity and the general principles can be illustrated by considering the nature of conjugation in *E. coli* K12 — the best studied of all the above examples.

Mating occurs between bacterial cells of different mating type. The F^- female or recipient parent differs from the male in that it does not possess an extra-chromosomal structure F, this being the genetic basis of mating type. The sex factor, F, occurs in the male or donor parent where it may exist integrated into the chromosome (the Hfr form), or as a separate, self-replicating entity (the F^+ form) and in this latter case the F factor may be transferred infectiously to other F^- bacteria independently of the bacterial chromosome, these properties of F being typically those of that class of genetic elements known as episomes[20]. The F factor occupies a central and essential role in conjugation in *E. coli* and clearly displays the relationship between episomes and many processes of bacterial genetic exchange.

The F factor is composed of DNA having a molecular weight of 45×10^6 daltons which should enable it to code for about a hundred proteins[21]. Included in its functions is the ability to direct in male cells the synthesis of sex pili, filamentous structures on the surface of the bacterial cell, which are thought to have an essential function in the transfer of chromosome from the donor parent to the recipient. BRINTON and his associates[22,23] have suggested that the sex pilus acts as a conjugation tube and is the organ by which DNA is physically transferred from the donor to the recipient parent. An alternative view has been proposed by CURTISS[24] in which the sex pilus has the primary role of promoting contact between the donor and recipient parents. Transfer of DNA occurs after contact is established between the cell walls of the conjugating bacteria, there being breakdown of the areas in direct juxtaposition to provide a conjugation bridge between the mating cells. Both models imply an essential role for the sex pili in conjugation, a view supported firstly by the fact that if F pili are mechanically removed, conjugation does not take place[22,23], and secondly by the observation that adsorption of male specific phages to the free ends of pili markedly reduces the ability of males to take part in conjugation[25,26].

The sex episome has a second major role in conjugation — that of controlling the linear transfer of bacterial chromosome from the Hfr donor parent to the F^- parent. JACOB, BRENNER and CUZIN[27] suggested that the cellular contact of the male and female parents triggers the synthesis of an F specific initiator which results in DNA replication and chromosome transfer. The F factor integrated into the chromosome of the donor parent assumes control of the replication of the chromosome of that cell in place of the normal regulatory mechanism operative during vegetative replication. It is currently believed that only one strand of DNA, originating at the site of F integration, is transferred to the F^- recipient, and that this is one of the strands of the DNA of the Hfr parent existing at the beginning of conjugation. After entering the F^- parent, the transferred DNA strand integrates with the F^- chromosome by means of the resident recombination mechanisms of that cell, although there is increasing evidence to support the view that this is not the only function of the female parent and that the recipient cell may have an active role in acquiring this DNA from the Hfr parent.

Usually only part of the donor chromosome is transferred in conjugation due to spontaneous physical separation of the parent cells. Using appropriate selection of desired phenotypes, recombinants of particular genetic markers can be isolated and this, combined with the fact that sex factor can integrate at a variety of chromosomal sites of the Hfr parent, has enabled a detailed map of the circular *E. coli* chromosome to be constructed. The efficiency of pair formation of parent cells approaches 100% of the mating cells and the efficiency of recovery of a particular genotype in practice may be as high as 2% of the parental cells taking part.

BACTERIAL EPISOMES AND PLASMIDS

It is clear from the preceding that a complete understanding of bacterial conjugation involves an appreciation of the biological function of episomes and related structures. The study of bacterial genetic phenomena has revealed the existence of extrachromosomal, self-replicating structures, many of which have an essential role in bacterial conjugation. The properties of such genetic elements cover a wide spectrum and their formal and precise definition is not easy, or perhaps at this stage of our knowledge, even desirable[28].

The sex factor F of *E. coli* is defined as an episome[20] in terms of its ability to be present or absent in the cell and its property, when present, of being either integrated into the chromosome or situated at some other site in the cell. A variety of other genetic elements, for example, temperate bacteriophages, also fit this definition. However, there are genetic structures which, while fitting the general description of episomes in terms of their extra-chromosomal location and other properties, cannot be defined as episomes because it has been possible to demonstrate experimentally that they can also occupy a chromosomal site in the bacterial cell, and in such cases the term plasmid has been applied. Plasmids have been characterized for a variety of bacterial genera and show a range of various properties. The genetics of these extra-chromosomal genetic particles have recently been admirably reviewed by NOVICK[29]. The episomes and plasmids which have been studied include the Resistance Transfer Factors of *Shigella* and other *Enterobacteriaceae*, the colicin determinants (Col factors) found in *E. coli* and *Salmonella*, the sex factor FP of *Pseudomonas aeruginosa*, drug resistance factors in *Staphylococcus aureus*, and a sporulation factor in *Bacillus subtilis*. The biology of these structures has many aspects of interest but we shall here be primarily concerned with their ability to promote the exchange of genetic material or to contribute to the bacterial genome. Plasmids and episomes have certain properties which make them highly useful as sources of genetic variation for bacteria. Firstly they may be present or absent from a bacterial cell without markedly influencing viability and secondly they can be transferred from cell to cell with high efficiency. We can consider their other general properties under the following headings.

THE PHYSICAL PROPERTIES OF EPISOMES AND PLASMIDS

In all cases examined, the genetic material of episomes and plasmids is DNA. In some cases, it has been possible by differential centrifugation or similar techniques to isolate such DNA from bacterial cells and it has been shown to be a closed, circular, duplex configuration with a base composition somewhat similar to, but not identical with, that of the normal host. Their sizes vary from a molecular weight of 4.5×10^6 for Col El when established in *Proteus*[30], to 69×10^6 for 222/R3, a drug resistance factor[31]. The number of episomal or plasmid particles per chromosome varies from one or two for the *E. coli* sex factor up to 50 for Col El when this plasmid is established in *Proteus*[32], but the figure of one or two is the more usual. The factors controlling the replication of plasmids or non-integrated episomes in bacteria are not well understood but it is apparent that in at least a number of cases it is not related to the control of the replication of the bacterial chromosome[29].

THE LOCATION OF EPISOMES AND PLASMIDS IN THE BACTERIAL CELL

One essential feature of both plasmids and episomes concerns their location in the cell. Evidence that episomes integrate physically into the chromosome is not easy to obtain and usually involves genetic linkage studies, so that this type of evidence can only be obtained from those bacteria with extensively mapped chromosomes and systems of genetic recombination. In the case of the sex factor F of *E. coli*, it appears that integration to give the Hfr male form may take place at any of about twenty sites[33] distributed over the chromosome. Direct evidence for the integration of F into the chromosome was obtained by De Witt and Adelberg[34], who demonstrated the contransduction of F with chromosomal genes. The evidence for chromosomal integration of most other plasmids, with the exception of a number of prophages, is not so convincing. Two sex factors of *P. aeruginosa* have each been shown to initiate the transfer of chromosome from a different specific site on the chromosome but it is not clear yet if this involves a physical integration[35,36] and as yet a stable Hfr form analogous to that occuring in *E. coli* has not been isolated from this bacterium. A similar situation occurs with the R factors in the *Enterobacteriaceae*. While they can promote chromosome transfer, in only one case is there evidence of any association of the R factor and the bacterial chromosome, this example producing an oriented transfer of chromosomal markers[37]. Most colicinogenic factors are usually independent of the chromosome[38] but in the case of Col V and Col B their ability to integrate into the chromosome probably results from their close relationship to the F sex factor of *E. coli*.

Where integration does occur, as in the case of F, the non-random pattern of insertion sites suggests that there must be regions of homology or partial homology between the episomes and the chromosome. In most cases it appears that integration into the chromosome takes place by means of the bacterial recombination system and, unlike, say λ phage, episomes do not carry their own integration enzymes[39].

The evidence for the location in the cell of episomes and plasmids when they are not integrated into the chromosomes is less satisfying. There are a number of observations which suggest that bacterial chromosomes are attached to the cell membrane or mesosome[40,41] and it is a currently popular view that other replicons in the bacterial cell, such as plasmids or episomes, could well be attached to the membrane site as well. As yet there is no clear or convincing picture of just how the replication of such extra-chromosomal replicons is regulated but this regulation must permit the stable persistence of plasmids in bacterial cells and their orderly transmission at cell division to the daughter cells. Obviously the time is ripe for a greater understanding of thc role of the membrane in the genetic economy of the bacterial cell. One approach to this problem could possibly be through the study of bacteriocin tolerant mutants[42,43], there being evidence to support the view that such mutants involve structural alterations of the bacterial membrane[44] and it should be worthwhile to study the ability of bacteriocin tolerant mutants to maintain the replication of various episomes or plasmids. Whatever the site, there is certainly a requirement that the plasmid must replicate *pari passu* with the cell and the chromosome, and for replicas of the plasmid to be distributed evenly to the daughter cells at division. It is not yet clear whether there are a few specific sites of the membrane for the attachment of specific episomes or as Luria has suggested[45] that the membrane attachment site can be created by the genetic element itself.

It has been established that a variety of plasmids can be carried by the one bacterial cell. DUBNAU and STOCKER[46] described a strain of *S. typhimurium*, which in addition to the *E. coli* sex factor F integrated into the bacterial chromosome, three colicin factors — Col Ib-P9, Col E1-K3 and Col E2-P9 — and a drug resistance factor R2 were carried. Nevertheless plasmid incompatibility does occur, and this may be associated with particular "maintenance sites" in the bacterial cell[29].

CONTRIBUTION OF EPISOMES AND PLASMIDS TO THE BACTERIAL PHENOTYPE

From the size of the DNA molecule comprising plasmids and episomes it is possible for them to code for up to a hundred proteins. Obviously some of these genes must be required for replicative functions, but as yet little is known of the other functions of which the rest are capable. As described earlier for the case of the F factor, the episome must be considered part of the bacterial genome and in fact the presence of an episome or plasmid is often recognized by the contribution it makes to the bacterial phenotype in properties quite unrelated to those associated with conjugal activity. The R factors confer drug resistance by mechanisms quite distinct from those controlled by chromosomally located resistance genes[47,48]. In addition, some R factors, like Col factors, can confer partial protection against the bactericidal effect of ultraviolet irradiation and the enzymes concerned appear to be different to those specified by known UV-resistance loci[49]. These facts are important in considering extra-chromosomal genetic units as sources of genetic variation because they indicate that episomes and plasmids can provide *new* information quite distinct to that found in the normal bacterial genome.

TRANSMISSIBILITY OF PLASMIDS AND EPISOMES

An important property of episomes and plasmids is their ability to be transferred from cell to cell with a higher efficiency than occurs with chromosomal DNA. There are several aspects of this transmission which may be of significance in the context of the use of bacterial conjugation for the production of industrially important strains. Firstly, episomes and plasmids can be transferred across species and generic barriers. The R factors first isolated from *Shigella* can be readily transferred to *Escherichia* or *Salmonella*[50]. More recently the transfer of an R factor from *P. aeruginosa* to *E. coli* has been demostrated[51,52]. Secondly, the kinetics of transfer of episomes within a population of bacteria are quite different to that found for chromosomal characters. This fact has had important medical and epidemiological implications following the spread of R factor determined drug resistance amongst intestinal pathogens of man and animals[53–55]. Thirdly, the properties of a plasmid or episome in one host may not be the same in a different host. For example, the phage λ acts as a true episome in *E. coli* K12, being capable of adopting either a chromosomal (*i.e.* prophage) or non-chromosomal location (*i.e.* during vegetative growth). However, in hybrids of *E. coli* and *S. typhi*, λ does not multiply, and does not undergo zygotic induction but can persist in a lysogenic state[56]. A similar situation holds with the sex factor F which acts as a true episome in *E. coli* but as a plasmid in *Proteus mirabilis* where it does not integrate into the chromosome[57]. One of the factors affecting transfer of episomes is that their successful entry into a new host may be prevented by the action of restriction enzymes. This phenomenon will be discussed below in terms of the broader problem of intergeneric mating.

RECOMBINATION OF PLASMIDS AND THE BACTERIAL GENOME

We have already discussed the ability of episomes to integrate into the bacterial chromosome. This is part of a wider picture of the recombination between plasmids and bacterial DNA to produce genetic structures which retain properties of both sources of DNA. For example, the F factor of *E. coli* can undergo recombination with various regions of the chromosome to produce a structure known as an F′ or F merogenote in which the entire genome or part of the sex factor is physically associated with a portion of the bacterial chromosome. This structure can be infectiously transmitted in the same manner as the sex factor and of course the bacterial genes which are carried along achieve normal function in the new host. The bacteria possesing an F′ will be diploid for the chromosomal region carried by the F′. This represents a very efficient mechanism for promoting the genetic recombination of that region carried on the merogenote, as it will be transferred to a new host with an efficiency comparable to that of the "infectivity" of the episome.

VARIABILITY OF EPISOMES

As with any other genetic structure, episomes and plasmids can undergo mutation. Variants of the *E. coli* F factor are known which are defective, as shown by the failure to give genetic recombinants or by insensitivity to sex specific phages. Plasmids occasionally lose markers by deletion, particularly during transfer and such deletions can be used for the mapping of the plasmid genome[58]. The DNA of R factors is heterogeneous in that there are fractions with different base ratio composition. Occasionally R factors lose the ability to confer resistance to one or more antibacterial agents and in some cases this loss is accompanied by changes in density of the R factor DNA[59]. Fredericq[38] has demostrated recombination between different types of plasmids including F, F′*lac*, other Col factors, and R factors and such results point to the possibility of synthesis of entirely new families of plasmids. For example, if, as appears likely, there is usually only one or at best two copies of each type of plasmid present in a cell due to specific site requirements, then it may be possible to get a series of plasmid recombinants each with a different specific site requirement, but with a desired genome, perhaps with a plasmid-host recombinant containing a chromosomal region determining a particular enzyme. In this way, polyploids for the genetic determination of one particular enzyme could be constructed which would effectively increase the horsepower of the cell and result in a higher yield per cell of a product or an enzyme.

INTERSPECIFIC AND INTERGENERIC MATING

Undoubtedly the most direct use of conjugation for the production of bacterial strains with desired phenotypes is likely to come from a situation in which the maximum possible genetic variation is available. Indeed, for animal and plant breeding, the introduction of desired characteristics is often achieved by outcrossing with naturally derived and unrelated strains. This situation can be achieved in bacteria through intergeneric or interspecific crosses and evidence is accumulating that viable hybrids can be obtained from parents with considerable genetic divergence. With these crosses, as in intraspecific crosses, episomes and plasmids play a significant role. Not only do they promote chromosome transfer, but the extra-chromosomal genetic elements themselves, displaying a variety of functions, have been shown to be transmissible between various genera. A number of factors are known which may prevent the heterospecific transfer of plasmids or episomes[60], and we can identify five main factors which can prevent or reduce the efficiency of intergeneric or interspecific conjugation. Possibly the successful application of this aspect of bacterial conjugation to the production of industrially important strains lies in the effective control of these barriers to genetic exchange.

SURFACE INCOMPATIBILITY PREVENTING THE FORMATION OF CONJUGAL PAIRS

We can distinguish two types of surface interference to prevent pairing. Firstly, the actual specific pairing organs, for example, sex pili, may not be present in all cells. If the donor strain possesses an episome which promotes conjugation, it may be in a repressed state and hence few cells will form pili. In the case of the sex factor of *E. coli*, F is always derepressed and given other necessary cultural conditions, at any one time every cell of an F^+ or Hfr culture is capable of producing sex pili and taking part in conjugation. The sex factor F is unusual in being derepressed to give continual production of sex pili, only one other such factor being known — Col V, and this is known to be closely related or even identical with F except for the ability to produce colicin. It has been shown that other plasmids promoting conjugation in *E. coli* differ in their degree of native repressibility[59] and obviously any sex factor which has a very low probability of formation of conjugationally competent cells would not enable a very effective system of gene transfer to operate. For example, the conjugational activity of most R factors is normally very low, but mutants can be isolated which conjugate at high frequency[59,61].

The second type of interference is a mechanical one as might occur through the formation of capsules, slime layers or other extra-cellular barriers to cell contact. With some genera of bacteria, for example *Klebsiella* or *Aerobacter*, identification of fertile strains might well involve the prior selection of mutant strains of these bacteria which have lost their capsules or slime layers.

RESTRICTION AND MODIFICATION

An essential stage of any conjugational event is the entry of donor DNA into the recipient parent cell. Obviously the formation of a viable recombinant depends upon the retention of biological activity by the incoming DNA, and mechanisms are known for bacterial cells which result in an enzymatic degradation of DNA entering a cell by a genetic process, either following conjugation or transduction, or during phage infection. This is the phenomenon of Host Controlled Modification (HCM), first observed with bacteriophage, but now known to be a general, although not universal, property with other types of DNA[62]. We can distinguish two mechanisms in HCM. *Restriction* is the process which recognizes incoming DNA as being like the DNA of that bacterial cell (self) or unlike it (non-self). In the latter case an enzyme, endonuclease, degrades the incoming DNA and its biological activity is lost. The recognition process involves the presence of particular chemical groups on the DNA which impart to the DNA its particular specificity. In some cases this specificity is thought to be due to the distribution in the DNA of particular patterns of methyl groups. The mechanism by which DNA aquires this specificity is by an enzymic process which acts on already synthesized DNA and this is called *modification*. Each strain of bacteria has its own modification pattern which is imposed on any DNA synthesized in that cell — bacterial, bacteriophage, plasmid or episomal. It follows that if the parents in a bacterial cross differ in their patterns of specificity, then the restriction enzymes of the recipient parent will recognize the donor DNA as being non-self and the recipient parent will hence degrade it and render it biologically inactive.

That this actually occurs has been shown not only for intraspecific crosses in both *E. coli*[62] and *P. aeruginosa*[15] using mutants or strains with different modification properties, but also with a number of intergeneric crosses[63,64]. The effects observed include the reduction in the number of recombinants, a distortion of linkage values in those recombinants that do manage to survive the restriction barrier, and failure to obtain infectious transmission of plasmids[47]. However, it is possible to eliminate the restriction activity of recipient cells either by the isolation of mutants which have lost active restriction enzymes or by growth procedures which phenotypically remove this activity, and in these cases the recovery of recombinants is quite normal[65,66].

SPECIFICITY OF RECOMBINATION ENZYMES

Now that the genetic analysis of the process of recombination has commenced in a variety of organisms, it is apparent that there is the possibility of considerable specificity in the action of enzymes involved in recombination. At least three loci (*recA, B, C*) controlling recombination in *E. coli* have been identified[67–69] and OISHI[70] has implicated an exonuclease in the action of *recB* and *recC*. A different class of Rec phenotype has been described in *P. aeruginosa* in which it has been suggested that it is the DNA which is altered, so that the wild-type recombination enzymes cannot function normally[15]. With the phage λ ECHOLS *et al.*[71] and SIGNER and WEIL[72] discovered recessive mutants (*red*) which show much reduced recombination when the phage is allowed to multiply in *rec* strains of *E. coli.* An additional locus on the phage chromosome, *int*, controls recombination of the λ genome with the *E. coli* chromosome and hence the enzyme coded for by this gene must be able to recognize specific sequences of DNA in the integration region[73]. Genes affecting recombination for specific sites on the chromosome have also been described in *Neurospora crassa*[74]. This array of genetic controls of the recombination process itself is hardly unexpected, but it does imply that the specificity of action of the enzymes concerned could be an important factor in preventing exchange of genetic material between parents with different genetic origins.

DISSIMILARITY OF DNA BASE SEQUENCE

Some degree of homology of base sequence is necessary for recombination to occur between different DNA molecules. As yet, little is known about the limits of the molecular basis of this similarity which are needed to allow such genetic hybridization. BRENNER *et al.*[75] have studied the polynucleotide sequence relationships among various genera of the *Enterobacteriaceae* by DNA—DNA hybridization techniques on the assumption that the ability to form specific DNA duplexes between two organisms is evidence of their degree of relationship. They found various degrees of evolutionary divergence amongst different members of the *Enterobacteriaceae* and showed that this divergence is not uniform along the DNA molecule that constitutes the bacterial chromosome. This is particularly interesting in view of the gross identity of gene distribution in those strains of *E. coli* and *S. typhimurium* which have been extensively mapped[9,76].

TRANSFER OF GENETIC MATERIAL IN INTERGENERIC MATINGS

Despite what would seem to be serious obstacles to the successful transfer of genetic material between different bacterial genera, in fact mating between different genera has been achieved, particularly between different members of the *Enterobacteriaceae*. It will be useful to distinguish two types of transfer of genetic material in this type of mating: (a) transfer of a plasmid or episome with or without a piece of bacterial chromosome attached; (b) transfer of genetic material which is primarily bacterial chromosome.

EPISOMAL OR PLASMID TRANSFER

There are a number of examples of intergeneric episomal or plasmid infection. As a mechanism for introducting genetic material into a given bacterial strain, plasmid transfer possesses advantages over chromosomal transfer in that neither chromosomal homology nor recombination enzymes are required for establishment of the plasmid. Successful transfer may be demonstrated in several ways. If the base composition of the DNA of the episome or plasmid is different to that of the recipient bacterium, then following autonomous replication of the acquired episomal DNA in the recipient parent, density gradient centrifugation of the material from the lysed hybrid cell will reveal a satellite band with different density to that of the native recipient chromosomal DNA. An effective genetic technique for demonstrating episomal or plasmid transfer is to use an episome to which an easily recognizable bacterial marker is attached, as with F-*lac* which is the *E. coli* sex factor harboring a segment of the *E. coli* chromosome which contains those genes controlling lactose fermentation. It has been found that a strain of *S. typhi* possessing F-*lac* can transfer this episomal structure (as measured by inheritance of the lac^+ function), to *Shigella dysenteriae*, *S. typhimurium*, *E. coli* K12, and *Serratia marcescens*. The DNA isolated from such hybrids, for example, the lac^+ *Serratia* hybrid, gives two bands when subjected to density gradient centrifugation, the smaller of which represents the DNA of the episome. The important result is that the episomal DNA can replicate and comprise a stable part of the genome of the hybrid[77,78]. Similarly, F-*lac* of *E. coli* has been transferred to several species of *Proteus*[57]. The F-*lac* *Proteus* strains acquire not only the ability to ferment lactose but also to donate genetic material to *E. coli* or *S. typhimurium*. Wild-type *Proteus* strains do not produce any β-galactosidase but this enzyme produced by the *Proteus* lac^+ strains is indistinguishable from that produced by *E. coli*. However, the normal regulation pattern found in *E. coli* is not displayed by the *Proteus* F-*lac* in that it is less sensitive to inducer, and the amount of enzyme produced is lower than that found in the normal haploid *E. coli.* However, it should be remembered that in *E. coli* which is diploid for the *lac* region, enzyme levels are about twice those of the haploid.

It is of some interest that subsequently a strain of *P. mirabilis* was isolated from nature and found to possess properties very similar to the laboratory-produced hybrids, although the natural hybrid had a β-galactosidase with different properties to that of *E. coli*[79]. This episome, known as P-*lac*, is distinct from the F-*lac* episome derived from *E. coli* and its isolation indicates that episomes which code for bacterial enzymic functions can occur in naturally occurring bacterial strains, can be readily isolated by appropriate screening techniques, and can potentially provide an important source of material for genetic variation.

There is recent evidence that R factor transfer can occur between *P. aeruginosa* and *E. coli*. Strains of *P. aeruginosa* characterized by being resistant to carbenicillin[80] can transfer the resistance marker for this antibiotic to *E. coli*, the latter acquiring the β-lactamase commonly found in enteric bacteria which are resistant to penicillin due to the presence of R factors[51,52].

INTERGENERIC BACTERIAL MATINGS

Conjugation and chromosome transfer between bacteria of different genera were first discovered in 1957 when LURIA and BURROWS showed the formation of hybrids between *E. coli* K12 and various *Shigella* species[81]. Subsequently, other fertile intergeneric combinations have been demostrated and some of the principles for successful genetic interactions of this type have been established.

Much of the more recent detailed work on intergeneric mating has been carried out using as parents *E. coli* K12 and *Salmonella*[82]. In the early experiments the recovery of recombinants was very infrequent and it was shown that this very low fertility of the mating was due to the fact that only mutant cells in the *S. typhimurium* parent population were capable of conjugation. A mutant was isolated from *S. typhimurium* which not only could act as a genetic recipient in matings but also permitted the persistance of genetic material from the *E. coli* parent in subsequent generations. This mutant has a fertility some 10,000 times greater than the population from which it was derived although still about 100 times less fertile than is found in *E. coli* × *E. coli* crosses. These results were extended to other species of *Salmonella* and with *S. typhi* a natural fertile mutant was found which, when mated with *E. coli* Hfr, produced recombinants at a frequency which enabled mapping and linkage data to be obtained. In addition it was also possible to show that these recombinants did not arise by the integration of portion of the *E. coli* donor chromosome into the *S. typhi* recipient chromosome. Instead, a partial diploid structure is formed, the hybrid having an exogenote of *E. coli* chromosomal material in addition to the

S. typhi normal chromosome. Different hybrids show varying degrees of stability, but some are as stable as normal haploid strains and in at least one case a stability extending over eight years of laboratory culture has been demonstrated.

Compared to wild-type *S. typhi* strains, such hybrids show increased frequencies of recombination when used as recipients and it has been suggested that the resident section of an *E. coli* chromosome in such a hybrid makes it easier for another piece of *E. coli* chromosome to be accepted when such hybrids are used as recipients in subsequent matings. However, of particular interest for the purposes of this discussion is the formation of the stable partial diploids. Genetically, this amounts to the production of a partial allopolyploid, which is one of the most effective ways of utilizing genetic variation from widely varying origins for the production of commercially desirable strains of plants.

As yet, the factors controlling this partial diploid formation are not clearly understood. They occur only rarely in normal *E. coli* × *E. coli* crosses except where recombination deficient (*i.e.* rec mutant) parents are used[83,84]. This suggests that their formation in intergeneric hybrids results from a lack of recombination between the chromosomes of the different parents, due to any of the mechanisms discussed above, but most likely through the lack of similarity of DNA base sequence in the two parents.

What is not clear, however, is just how this extra piece of chromosome can act as a replicon, and be passed on to each daughter cell in a regular manner. It seems that the normal chromosome in bacteria can achieve this through a physical attachment to the cell membrane[27,85] and it is possible that the same relationship occurs in *Salmonella* with the *E. coli* chromosomal fragment. Recently[86] it has been shown that the genetic material derived from the *E. coli* parent in these hybrids exists as circular molecules of DNA. It might be interesting to determine the stability of such hybrids in strains having mutations affecting the structure of the bacterial membrane.

DEVELOPMENT OF CONJUGATION SYSTEMS IN OTHER GENERA

If the maximum use is to be made of conjugation systems in the development of industrially important strains, the search for conjugation systems, episomes and plasmids in a wide range of bacterial genera must be encouraged. There are no standard rules or procedures by which this can be achieved and the properties and existing knowledge of the particular organism and its genetics must be used to the fullest extent. If an existing plasmid or episome cannot be transferred to a particular genus to promote conjugation, then it becomes necessary to find a native episome with this ability. One interesting application of this principle has occured in *Pseudomonas putida*.

GUNSALUS and his colleagues have isolated from sewage a transducing phage pf16 which can carry out general transduction of *P. putida*[87]. When this phage is used for transduction between two unlike strains of this species, some transductants can be isolated which release a new form of phage pf16 which has a much higher transduction frequency and which acts in a similar manner to the defective transducing phages *λdg* or *P1dl* found in *E. coli*. This new form of pf16, known as pf*dm*, is thought to be a hybrid phage–bacterial DNA complex and it appears that, unlike normal pf16, pf*dm* particles can multiply autonomously as extra-chromosomal elements[88]. Despite considerable efforts by a variety of workers, conjugation in the more conventional sense has not been demostrated for *P. putida*. However, CHAKRABARTY and GUNSALUS[17] have shown that when a strain of bacterium harboring pf*dm* as an extra-chromosomal element is mixed with other strains of *P. putida*, recombinants are produced. These do not result from transformation or transduction mechanisms and direct cellular contact between donor and recipient cells is required for these recombinants to be produced. In general, pf*dm* seems to act like an F′ or RTF factor in that recipients which receive chromosome also receive the pf*dm*. These results are encouraging in that they indicate a technique which could be used for the isolation of plasmids promoting conjugation in those genera in which naturally occurring plasmids with this function cannot be isolated. JACOB and WOLLMAN[89] pointed out some time ago possible interrelationships between plasmids, episomes and bacteriophages and these results with *P. putida* lend direct experimental support to their speculations.

APPLICATIONS OF CONJUGATION TO THE PRODUCTION OF DESIRED BACTERIAL VARIANTS

Despite the sophisticated level of knowledge concerning bacterial conjugation — as indicated by the preceding paragraphs — it is unfortunately true that little if any of this information has been applied for industrial or commercial purposes. It may be significant that most research work on conjugation has been carried out on only one strain of *E. coli*, an organism with little industrial importance. However, with our knowledge of conjugation now extended to a variety of other genera, it is probably true to say that with the proper choice of strains, the adaptation of known techniques and a certain element of good fortune, it should be possible to demostrate a conjugational process of genetic recombination in many more bacterial genera, including those with industrial significance. Even from existing knowledge, a number of areas of likely applications can be suggested.

PRODUCTION OF ENZYMES

One of the current difficulties of microbiological methods of industrial products is that very often there is a cost competition with purely synthetic production methods.

Such competition does not apply for the production of various enzymes, which are achieving an increasing number of uses and a large potential market, and the enzymes involved include those with general functions such as amylolytic, which hydrolyse carbohydrates, the proteases of various kinds, as well as those enzymes with more specific functions such as penicillinase or catalase. The bacterial variant sought is one with the ability to elaborate extra-cellularly large amounts of the required enzyme. Bacterial mutants can be isolated in which a large fraction of the protein-synthesizing ability is channelled into the production of one enzyme. For example, variants of *Rhodopseudomonas spheroides* are known which have 25% of their protein as catalase[90]. *E. coli* mutants are known which have a similar amount of their protein as β-galactosidase and in these this may be due to the presence of several copies of the corresponding structural gene[91].

A possible application of intergeneric bacterial hybrids could be to produce bacterial strains with the ability to produce enzymes which have commercial value, and which exploit particular biological properties of each parent. For a hypothetical example, let us assume that a market develops for β-galactosidase, and hence there is a need to develop a microbiological process for its production. By appropriate selection of the right genotypes, a strain of *E. coli* could perhaps be selected in which a significant fraction of the protein synthesis can be channelled into the production of this one enzyme as was indeed found by NOVICK and HORIUCHI[91]. However, while the carbon sources required for the growth of *E. coli* are commercially expensive, it is known that species of *Pseudomonas* can acquire the ability to grow on a variety of carbon and nitrogen sources, some of which undoubtedly would be less expensive than the carbon sources needed for growth of *E. coli*. Accordingly, in this case a hybrid would be sought between *E. coli* and *Pseudomonas* which would have the properties of utilizing readily available and inexpensive carbon sources while still able to synthesize large quantities of β-galactosidase.

MEDICAL MICROBIOLOGY

The production of bacterial vaccines is both an important microbiological industry and an important part of the overall control of epidemic disease. An essential part of the background of the production of a successful vaccine is an understanding of the genetic basis of virulence because in many cases it is the antigenically active, but avirulent strains of microorganisms which are the most suitable as vaccines. Conjugation analysis has already contributed much to our understanding of the nature of virulence in bacteria. KRISHNAPILLAI and BARON[92] have shown by means of crosses between an avirulent *S. abony* and a virulent *S. typhimurium* that two independent genetic sites were concerned with virulence in the former species. BHASKARAN and his co-workers have studied antigenic variation in *Vibrio cholerae* and have demostrated that following chromosomal transfer by conjugation, recombinants could be isolated of an antigenic type different to that of the parent[93,94]. The study of the genetics of virulence in *Shigella flexneri* using conjugation techniques was initiated by FALKOW *et al.*[95] and subsequently FORMAL *et al.*[96] showed that when avirulent mutants of *S. flexneri* were mated to *E. coli*, recombinants could be isolated which had regained the full virulence of the wild type *Shigella*. The occurrence of gene transfer mechanisms[12,13] in *Pasteurella* now opens that genus to investigations on the nature of virulence. Clearly such genetic studies point to the possibility of new vaccines with greater effectiveness and broader spectrum of protection.

BY-PRODUCTS OF MICROBIAL GROWTH

Many industrial microbiological processes involve the elaboration of a particular by-product of growth, which is usually released into the medium from which it can be extracted and purified. One such process is the fermentation of amino acids, some mircoroorganisms elaborating amino acids into the culture medium during growth. Under particular culture conditions auxotrophic mutants can elaborate large amounts of free amino acids[97] into the medium. The secret of obtaining worthwhile yields of amino acids at an industrially satisfactory price involves an understanding of the processes of intracellular regulation and the use of appropriate culture media. It would surely seem that this is one area where the required biological pattern for a hybrid can be specified in some detail and where conjugation could be used to provide the desired bacterial strains.

CONCLUSIONS

It is clear that our current knowledge of bacterial conjugation is indicative of the elegance and precision in genetic analysis which is now possible by this means of bacterial recombination. Together with the other processes of gene assortment in bacteria, such as transduction and transformation, it is clear that the limitations of the industrial application of these techniques lies, not in an innate lack of effectiveness, but by the limited objectives set for them to date. Surely their real potential lies not so much in providing alternative solutions to problems which can be solved by the simpler techniques of mutation and selection, but by their application to newer, more ambitious projects. It is now up to the industrial world to formulate these new goals.

Current research work in the author's laboratory is supported by the Australian Research Grants Committee and the Australian Institute for Nuclear Science and Engineering. A special grant from the George Hicks Foundation is gratefully acknowledged.

REFERENCES

1. ELANDER R. P.: Developments Ind. Microbiol. **7**, 61 (1966).
2. LEDERBERG J., TATUM E. L.: Cold Spring Harbor Symp. Quant. Biol. **11**, 113 (1946).
3. FURNESS G., ROWLEY D.: J. Gen. Microbiol. **17**, 550 (1957).
4. BERNSTEIN H. L.: Symp. Soc. Exp. Biol. **12**, 93 (1958).
5. ØRSKOV F., ØRSKOV I.: Acta Pathol. Microbiol. Scand. **51**, 280 (1961).
6. SASAKI I., BERTANI G.: J. Gen. Microbiol. **17**, 777 (1965).
7. ZINDER N.: Science **131**, 924 (1960).
8. MÄKELÄ P. H.: Genetics **48**, 423 (1963).
9. SANDERSON K. E.: Bacteriol. Rev. **31**, 354 (1967).
10. DE GRAAF F. K., THIEZE G. A., BONGA S. W., STOUTHAMER A. H.: J. Bacteriol. **88**, 631 (1968).
11. AKIBA T., KOYAMA K., ISHIKI Y., KIMURA S., FUKUSHIMA T.: Jap. J. Microbiol. **4**, 219 (1960).
12. LAWTON W. D., MORRIS B. C., BURROWS T. W.: J. Gen. Microbiol. **52**, 25 (1968).
13. MORTON G., JACOB F.: Compt. rend. **254**, 3589 (1962).
14. HOLLOWAY B. W.: J. Gen. Microbiol. **13**, 572 (1955).
15. HOLLOWAY B. W.: Bacteriol. Rev. **33**, 419 (1969).
16. LOUTIT J. S., PEARCE L. E., MARINUS M. G.: Genet. Res. **12**, 29 (1968).
17. CHAKRABARTY A. M., GUNSALUS I. C.: Proc. Nat. Acad. Sci. U.S.A. **64**, 1217 (1969).
18. BHASKARAN K.: J. Gen. Microbiol. **19**, 71 (1958).
19. HEUMANN W.: Mol. Gen. Genet. **102**, 132 (1968).
20. JACOB F., WOLLMAN E. L.: Compt. rend. **247**, 154 (1958).
21. FREIFELDER D.: J. Mol. Biol. **35**, 95 (1968).
22. BRINTON C. C.: Trans. N. Y. Acad. Sci. Ser. II, **27**, 1003 (1965).
23. BRINTON C.C., GEMSKI P., CARNAHAN J.: Proc. Nat. Acad. Sci. U.S.A. **52**, 776 (1964).
24. CURTISS R.: Am. Rev. Microbiol. **23**, 69 (1969).
25. IPPEN K. A., VALENTINE R. C.: Biochem. Biophys. Res. Comm. **27**, 674 (1967).
26. NOVOTNY C., KNIGHT W. S., BRINTON C. C.: J. Bacteriol. **95**, 314 (1968).
27. JACOB F., BRENNER S., CUZIN F.: Cold Spring Harbor Symp. Quant. Biol. **28**, 329 (1963).
28. HAYES W.: Ciba Found. Symp. Bacterial Episomes and Plasmids, 4 (1969).
29. NOVICK R. P.: Bacteriol. Rev. **33**, 210 (1969).
30. GOEBEL W. F., HELINSKI D. R.: Proc. Nat. Acad. Sci. U.S.A. **61**, 1406 (1968).
31. NISIOKA T., MITANI M., CLOWES R.: J. Bacteriol. **97**, 376 (1969).
32. DE WITT W., HELINSKI D. R.: J. Mol. Biol. **13**, 692 (1965).
33. MATNEY T. S., GOLDSCHMIDT E. P., ERWIN N. S., SCROGGS R. A.: Biochem. Biophys. Res. Comm. **17**, 278 (1964).
34. DE WITT S. K., ADELBERG E. A.: J. Bacteriol. **83**, 673 (1962).
35. PEMBERTON J., HOLLOWAY B. W., *to be published.*
36. STANISICH V. A., HOLLOWAY B. W.: Genetics **61**, 327 (1969).
37. PEARCE L. E., MEYNELL E.: J. Gen. Microbiol. **50**, 159 (1968).
38. FREDERICQ P.: Ciba Found. Symp. Bacterial Episomes and Plasmids, 163 (1969).
39. WILLETTS N. S., BRODA P. M. A.: Ciba Found. Symp. Bacterial Episomes and Plasmids, 32 (1969).
40. RYTER A.: Bacteriol. Rev. **32**, 39 (1968).
41. RYTER A., HIROTA Y., JACOB F.: Cold Spring Harbor Symp. Quant. Biol. **33**, 669 (1968).
42. LURIA S. E.: Ann. Inst. Pasteur **107**, 67 (1964).
43. DE ZWAIG R. N., LURIA S. E.: J. Bacteriol. **94**, 1112 (1967).
44. ONODERA K., ROLFE B., BERNSTEIN A.: Biochem. Biophys. Res. Comm. **39**, 969 (1970).

45. LURIA S. E.: Ciba Found. Symp. Bacterial Episomes and Plasmids, 255 (1969).
46. DUBNAU E., STOCKER B. A. D.: Genet. Res. **9**, 283 (1967).
47. OKAMOTO S., SUZUKI Y.: Nature **208**, 1301 (1965).
48. HARWOOD J. H., SMITH D. H.: J. Bacteriol. **97**, 1262 (1969).
49. SICCARDI A. G.: J. Bacteriol. **100**, 337 (1969).
50. HARADA K., SUZUKI M., KAMEDA M., MITSUHASHI S.: Jap. J. Exp. Med. **30**, 289 (1960).
51. SYKES R. B., RICHMOND M. H.: Nature **226**, 952 (1970).
52. FULLBROOK P. D., ELSON S. W., SLOCOMBE B.: Nature **226**, 1054 (1970).
53. ANDERSON E. S.: Ciba Found. Symp. Bacterial Episomes and Plasmids, 102 (1969).
54. WILLIAMS SMITH H.: Ciba Found. Symp. Bacterial Episomes and Plasmids, 213 (1969).
55. CHABBERT Y. A., BAUDENS J. G., BOUANCHAUD D. H.: Ciba Found. Symp. Bacterial Episomes and Plasmids, 227 (1969).
56. RYMAN I. R., FALKOW S., BARON L. S.: Bacteriol. Proc. 27 (1967).
57. FALKOW S., WOHLHEITER J. A., CITARELLA R. V., BARON L. S.: J. Bacteriol. **87**, 209 (1964).
58. WATANABE J., FUKASAWA T.: J. Bacteriol. **82**, 202 (1961).
59. MEYNELL E., MEYNELL G. C., DATTA N.: Bacteriol. Rev. **32**, 55 (1968).
60. ARBER W.: Ciba Found. Symp. Bacterial Episomes and Plasmids, 201 (1969).
61. COOKE M., MEYNELL E.: Genet. Res. **14**, 79 (1969).
62. ARBER W., LINN S.: Ann. Rev. Biochem. **38**, 467 (1969).
63. FALKOW S., FORMAL S. B.: J. Bacteriol. **100**, 540 (1969).
64. BRENNER D. J., CITARELLA R. V., FALKOW S.: Rep. 2nd Internat. Congr. Microbiol. Class. (Luhačovice). Fac. Sci., Univ. J. E. Purkyně, Series K43, 255 (1968).
65. ROLFE B., HOLLOWAY B. W.: J. Bacteriol. **92**, 43 (1966).
66. ROLFE B., HOLLOWAY B. W.: Genetics **61**, 341 (1969).
67. CLARK A. J., MARGULIES A. D.: Proc. Nat. Acad. Sci. U.S.A. **53**, 451 (1965).
68. HOWARD-FLANDERS P., THERIOT L.: Genetics **53**, 6 (1966).
69. CLARK A. J.: J. Cell. Physiol. **70**, Suppl. 1, 165 (1967).
70. OISHI M.: Proc. Nat. Acad. Sci. U.S.A. **64**, 1292 (1969).
71. ECHOLS H., GINGERY R., MOORE L.: J. Mol. Biol. **34**, 251 (1968).
72. SIGNER E. R., WEIL J.: J. Mol. Biol. **34**, 261 (1968).
73. ZISSLER J.: Virology **31**, 189 (1967).
74. CATCHESIDE D. G., p. 216 in W. J. Peacock, R. D. Brock (Eds.): *Replication and Recombination of Genetic Material.* Austral. Acad. Sci., Canberra 1968.
75. BRENNER D. J., FANNING G. R., JOHNSON K. E., CITARELLA R. V., FALKOW S.: J. Bacteriol. **98**, 637 (1969).
76. TAYLOR A. L., TROTTER C. D.: Bacteriol. Rev. **31**, 332 (1967).
77. FALKOW S., MARMUR J., CAREY W. F., SPILMAN W. M., BARON L. S.: Genetics **46**, 703 (1961).
78. MARMUR J., ROUND R., FALKOW S., BARON L. S., SCHILDKRAUT C., DOTY P.: Proc. Nat. Acad. Sci. U.S.A. **47**, 972 (1961).
79. WOHLHEITER J. A., FALKOW S., CITARELLA R. V., BARON L. S.: J. Mol. Biol. **9**, 576 (1964).
80. LOWBURY E. J. L., KIDSON A., LILLY H. A., AYLIFFE G. A. F., JONES R. J.: Lancet **2**, 448 (1969).
81. LURIA S. E., BURROWS J. W.: J. Bacteriol. **74**, 461 (1957).
82. BARON L. S., GEMSKI P., JOHNSON E. M., WOHLHEITER J. A.: Bacteriol. Rev. **32**, 362 (1968).
83. CURTISS R.: Genetics **58**, 9 (1968).
84. LOW B.: Proc. Nat. Acad. Sci. U.S.A. **60**, 160 (1968).
85. SUEOKA N., QUINN W. G.: Cold Spring Harbor Symp. Quant. Biol. **33**, 695 (1968).
86. LEAVITT R. W., WOHLHEITER J. A., JOHNSON E. M., LADDA R. L., BARON L. S.: Bacteriol. Proc. 55 (1970).

87. Chakrabarty A. M., Gunsalus C. F., Gunsalus I. C.: Proc. Nat. Acad. Sci. U.S.A. **60**, 168 (1968).
88. Chakrabarty A. M., Gunsalus I. C.: Virology **38**, 92 (1969).
89. Jacob F., Wollman E. L. in *Sexuality and the Genetics of Bacteria.* Academic Press, New York 1961.
90. Clayton R. K.: Biochim .Biophys. Acta **58**, 360 (1962).
91. Novick A., Horiuchi T.: Cold Spring Harbor Symp. Quant. Biol. **26**, 239 (1961).
92. Krishnapillai V., Baron L. S.: J. Bacteriol. **87**, 598 (1964).
93. Bhaskaran K., Sinha V. B., Iyer S. S.: Current Sci. **37**, 488 (1968).
94. Bhaskaran K.: Bull. World Health Organiz. **30**, 845 (1964).
95. Falkow S., Schneider H., Baron L. S., Formal S.: J. Bacteriol. **86**, 1251 (1963).
96. Formal S. B., Labrec E. H., Schneider H., Falkow S.: J. Bacteriol. **89**, 835 (1965).
97. Symp. Microbial Production of Amino Acids. Developments Ind. Microbiol. **7**, 3 (1966).

Transduction and Lysogenic Conversion: Substitution, Addition and Deletion of Genetic Material

G. BERTANI and M.G. SUNSHINE

Microbial Genetics Laboratory, Karolinska Institutet, Stockholm, Sweden

Transduction — the transfer of genes from cell to cell *via* a virus — has been recently reviewed by OZEKI and IKEDA[1]. The present discussion will only consider some of the mechanisms on which transduction and related phenomena depend, and their implications for the evolution of microorganisms. For general reference *see* also CAMPBELL[2].

In the more widely used type of transduction, *general transduction*, like *e.g.* that of phage P1 in *E. coli* or phage P22 in *Salmonella*, the frequency of transduction is low (of the order of 10^{-5} or less per phage particle) but every gene known may be so transduced. Markers which lie close together on the host chromosome can be cotransferred and thus linkage relationships can be determined.

An obvious condition for transduction is that the bacteria exposed to the virus, or at least a portion of them, survive the infection: the protein of the virus, upon attachment to the cell, must not kill the cell. Indeed, transducing phages are usually of the kind which are able to establish lysogeny — the infected cell sometimes survives and carries in itself one or more copies of the genetic material of the virus. This is not required for general transduction; for other forms of transduction, however, lysogeny is a necessity. We shall recall that two possible types of lysogenic condition are known: (1) the chromosomal type, as established for phages λ, P2, P22, and others, where the genetic material of the virus, or prophage, is inserted in the bacterial chromosome, and (2) the less well known non-chromosomal lysogeny, in which presumably the genetic material of the phage is carried in the form of an extrachromosomal body, a plasmid, possibly present in more than one copy per cell.

One of the other forms of transductions is the so called *specialized transduction* in which only one specific short segment of the chromosome can be transferred by a certain phage, with however a very high frequency. The best analyzed case is that of the genes for the galactose enzymes in *E. coli* which may be transduced in high frequency by phage λ. More recent studies show that genes on both sides of the chromosomal location of λ prophage may be transduced in this manner. In this and similar cases it has been shown that specialized transduction results from the exceptional formation of a new type of phage particle containing a portion of the genetic material of the phage and some of the neighbouring genes on the bacterial chromosome. Often but not always such particles are more or less defective in that they lack some of the phage genes necessary for the normal development of the phage. In such cases a so-called helper phage is necessary to demonstrate certain aspects of the phenomenon. In specialized transduction we have therefore a variety of phage forms depending on which parts of the phage genome are combined with which genes from the bacterial chromosome.

A different situation is found in *lysogenic conversion*, exemplified by the classical case of *Corynebacterium diphtheriae*, which becomes a producer of toxin whenever it is lysogenized by a certain phage. Here one form of phage is sufficient to give the phenomenon and, besides its properties which make it behave like an ordinary bacteriophage, able to multiply lytically or to establish lysogeny, it also has the property of producing in the host cell certain specific characters which are lost when the phage is lost from the cell. The phenomenon in one or the other form is fairly general for temperate bacteriophages, and numerous cases of lysogenic conversion have been described over the years.

That the three phenomena, general transduction, specialized transduction, and lysogenic conversion are fairly closely related, has been suggested already a long time ago. In fact, any given transducing phage particle in specialized transduction, in so far it can reproduce itself, behaves essentially like a converting phage. Furthermore, cases are known in which a phage usually giving generalized transduction has established a special relationship with certain bacterial genes, thereby becoming the equivalent of a specialized transducer or of a converting phage. However, with the increased knowledge of the mechanism of genetic recombination in bacteria and bacteriophages, it is now possible to analyze in a more meaningful fashion the relationships between these different types of transductions and the genetic processes upon which they are based.

Genetic recombination — an almost ubiquitous phenomenon in biology — has been held to be nonspecific, in the sense that the recombination mechanism is about equally effective no matter where in the genome recombination is taking place. This view is supported by the existence of mutations which remove 99% or more of the ability of the cell to recombine. There exist, however, well documented cases of highly specific recombination, *i.e.* those taking place only at specific points or over specific segments of the genome. This is the recombination mechanism by which the DNA of temperate phages like λ, P2, and P22 insert themselves into the bacterial chromosome at specified sites according to the scheme of CAMPBELL; it is called *int* recombination, from the name given to the *int*egration-defective phage mutants which are unable to perform this step. This type of recombination occurs not only between the host chromosome and phage DNA, but also between two copies of phage DNA in the course of vegetative multiplication of the phage. There are different *int* systems, one for λ, one for P2, *etc.* each highly specific.

General transduction requires the general recombination system: the piece of donor genome introduced into the recipient by the transducing phage particle appears to be incorporated by a double or quadruple exchange into the chromosome of the recipient. The markers introduced are thus *exchanged* for the alleles of the recipient. In the case of lysogenization and lysogenic conversion, instead, an *addition*, rather than a substitution, of genetic material takes place. In specialized transduction one or the other can occur.

We have therefore in some of these cases an increase in the total genome of the recipient cell. If one were able to recover the original donor cell, which might actually be possible in some cases of specialized transduction, one would have the converse, that is, a case of loss of part of the genome — a deletion. The occurrence of spontaneous deletions in bacteria has been known for a long time. The classical case is that of the deletions for genes of the tryptophan operon in *E. coli* which simultaneously cause the appearance of resistance to phage T1.

It has been pointed out[3] that if sequences of sufficient length were repeated along the chromosome, they would permit the occurrence of intra-chromosomal recombination with the consequence that deletions would be produced. This effect cannot be very dramatic; in fact, bacteria carrying two or more prophages of exactly the same kind, located at different places in the bacterial chromosome, have been known for several years and there is no evidence that such bacteria are less viable than the corresponding singly lysogenic bacteria. It is possible, however, that such an effect is important on an evolutionary scale.

Intrachromosomal reciprocal recombination may be what produces the tryptophan deletions and the transducing particles of phage λ. In these two cases, however, the points corresponding to the hypothetical recombination event are not fixed, but may differ in independent deletions (or transducing particles, respectively). Furthermore, in the tryptophan deletions, the general recombination mechanism does not seem to be involved[4].

The simplest form of Campbell's model for the attachment of prophage to the chromosome requires the existence of a short sequence on the cell chromosome corresponding to a similar short sequence on the phage chromosome. Phage λ appears to attach itself always at the same place on the chromosome, that is, the special sequence which is necessary for its attachment exists only once on the bacterial chromosome. In the case of other phages, for example P2, the phage can attach itself at a number of sites on the chromosome. The same is true for the F-factor whenever an Hfr strain is formed. One would have to assume, therefore, in these cases that a short sequence is repeated several times at different places on the chromosome. We suspect that in at least one case the presence of such repeats is at the basis of the phenomenon of *eduction* discovered a few years ago[5].

The situation is briefly as follows. In *E. coli* K12, bacteriophage P2 can attach at a site called *H* which is located counterclockwise from the histidine operon, not far from it. It was found that such lysogenic bacteria now and then, with a small probability, lose their prophage and at the same time also the genes for the synthesis of histidine, that is, they become histidine-requiring. Two other genes (*gnd* and *rfb*) to the left of histidine are also lost in the process[6]. In all cases studied, exactly the same set of genes is lost in this process, as though the end of this deletion were fixed. Unfortunately this segment of the genetic map of *E. coli* is not well known; therefore, the end of the deletion cannot be located with higher precision. In another strain of *E. coli*, strain C, P2 does not seem to attach itself at a site corresponding to site *H* of K12, but preferentially establishes itself at another site located clockwise from histidine and called location *I*. It so happens that if one compares the maps of *E. coli* K12 and *E. coli* C, the presumed end point of the histidine deletions corresponds to the place on the map where location *I* ought to be. One can therefore propose as a working hypothesis that the histidine deletion produced by the presence of a P2 prophage in strain K12 is the result of an intrachromosomal recombination event, between site *H* and a fairly similar sequence existing in K12 at a point corresponding to the position of site *I* in *E. coli* C. The hypothesis is strengthened by the fact that the eduction phenomenon does not occur if the prophage is a mutant defective in the phage-specific recombination mechanism, the *int* mechanism, which is used for the attachment of the prophage to the bacterial chromosome and whose task includes the recognition of the homologous sites of the Campbell model[7].

The spontaneous phenomenon of eduction, as described above, occurs fairly rarely. In fact we know that both in the λ and in the P2 lysogenic systems, the *int* recombination system is not usually expressed in the lysogenic condition. Since this recombination system is enormously more efficient than the general recombination system of the cell, one could well imagine the disturbances that it could produce in a cell where repeated sequences existed. In λ the *int* recombination system seems to be repressed by the immunity repressor of the prophage. In the case of P2, instead, one has not shown as yet a similar effect of the immunity repressor: the lack of expression of the *int* system in an established P2 lysogenic cell seems to be due to the fact that the prophage genes determining the *int* recombination system are separated from their promotor as a result of the insertion into the bacterial chromosome, that is, they cannot be transcribed when they are in the prophage[8]. Instead, when a lysogenic cell is superinfected with another P2 phage, the corresponding genes on the superinfecting phage may be expressed even though the immunity repressor of the prophage is present. This finding has been applied to the phenomenon of eduction mentioned above: following superinfection of a lysogenic strain carrying the prophage in location *H* with another P2 phage up to 30 or 40% of the cells surviving such an infection may lose the genes for histidine, *i.e.* undergo eduction[7]. In this experiment one does cut out a specific set of genes with a tremendously high probability. One may add that the survival of the superinfected cell is quite high since they are immune to superinfection, and any killing that one observes is due presumably to physiological effects of the high multiplicities (up to 20 or 30 phages per cell) used. If our interpretation of this phenomenon is correct, one might expect that in some cases this whole segment including the histidine operon changes position and inserts itself, as a prophage does, at another place in the bacterial chromosome, where a repeat of the specific P2 site sequence is present.

What we want to emphasize here is that the existence of specialized recombination mechanisms, that is, mechanisms which recombine only over certain specific sequences, identifies specific segments on the genome which are bound to evolve in a more independent way than an average segment of similar length in the genome. This is of course the case for any set of genes which can be incorporated in a transducing phage and multiply as part of the phage. Such segments will have different evolutionary environments depending on whether they are in the bacterial chromosome or in the phage chromosome. To take an extreme view one could maintain that the main evolutionary units are not whole genomes but rather segments of genome which because of recombination events or transducibility and other effects, behave in a relatively independent way. Similar points of view have been extensively discussed by HERSHEY[9]. It is interesting to recall that evidence for such a *segmental evolution* comes from completely different sources. One has found by different methods that the distribution of DNA base composition within a chromosome in those cases where it has been studied with sufficiently great resolution, is not random. This holds for phages λ (Ref.[10,11]), P2 (Ref.[12]), and for *E. coli*[13]. It has been demostrated either by means of Inman's melting maps, where the distribution of zones of preferential denaturation is studied by electron microscopy, or by measuring the density of properly prepared DNA fragments and — where possible — reconstructing their likely position in the chromosome. Different regions of the chromosome may thus have rather different base composition. This is puzzling, since one would expect that the overall composition of the DNA, within the limits imposed by the rules of the genetic code and the necessity for correct expression of the genes, would be subject to biochemical pressures, for example, the supply of certain purines and pyrimidines, and the types of mutations allowed. Such a mechanism would not affect differently different segments of the DNA.

Although experimentation on these problems has not been extensive, one might imagine that the heterogenity of base pair composition in DNA reflects different evolutionary histories of different segments within a chromosome. There might be evolutionary pressures trying to maintain the integrity of the structure of certain segments of genetic material, independently of other segments in the same genome. Operons are obvious candidates as subchromosomal evolutionary units. Another example is suggested by the observation that often and for no obvious reasons, the genes specifying a protein expected to interact specifically with a certain DNA segment are located very close to the segment itself. This is the case, for example, with the immunity repressor gene in phage λ and the two operator sites which are under its control; gene *lacI*, specifying the protein regulating the expression of the lactose operon, which is located near the operator of the operon; the *int* gene which in both phages λ and P2 is located very near the point of attachment of the prophage to the bacterial chromosome, that is the point where the proteins of the *int* system are expected to operate, *etc*.

If segments of DNA were discrete evolutionary units, one could imagine that whole sets of functions or properties may appear in different organisms combined, mosaic like, with a variety of other sets of properties. This is obviously an extreme view, especially if one thinks in terms of higher organisms. Observations on viruses, however, suggest that perhaps certain sets of genes have evolved in a rather independent fashion, and by means of recombination they may appear as components of different viruses in combination with other sets of genes of different origins. For example, several phages resembling P2 in a number of properties are known, and these are also antigenically related to P2 in the sense that they are almost as well neutralized by a specific antiserum as is P2. However, there exists a phage, P1, which is quite different from P2 in a large number of properties, including the size of its chromosome, yet it is neutralized by P2 specific antiserum, as well as the other P2 related phages. In fact, the structures of the tails of phages P1 and P2 are very similar, and the observation would suggest that they are made with very similar protein subunits. A telling fact is that both P2 and P1 were originally isolated from the same multiply-lysogenic strain, the so-called Lisbonne and Carrère strain of *E. coli*. The two were thus together in the same bacterium as prophages already in the "natural" state and it is possible that at some stage genes for making the tail have been donated by one phage to the other.

All these phenomena come from studies of microorganisms. One can ask how general they are. Transduction of eye color in silk worm has been reported[14] and one knows that viruses of the polyoma type produce the formation, simultaneous to their multiplication, of particles very similar to the virus, but containing DNA of the host cell. The latter are obviously candidates for a transduction system in mammalian cells.

An interesting genetic phenomenon has been known for almost 20 years in maize (*see* Ref.[15]) and resembles phenomena of lysogeny. It consists in the spontaneous transfer of genetic elements — most probably short segments of DNA — from one place to another in the genome. Their presence is presumably detected because the gene into which one has inserted itself, splitting it, becomes incapable of normal expression. This reminds one of the behaviour of a prophage. There is, in fact, a phage Mu1, which is known to attach itself to a large variety of places in the bacterial chromosome and often, in so doing, to cause an inactivation of the gene where insertion has occurred[16]. It is uncertain whether the maize elements resemble in size a typical prophage[17,18]. Perhaps they fit more closely a mutational phenomenon recently discovered in bacteria[19,20]: this consists of the spontaneous insertion of short segments of DNA (1000—1500 base pairs), presumably of endogenous origin, in certain genes, and having strong polar effects for the operon concerned. Whether these phenomena depend on specific recombination mechanisms of the *int* type, which permits insertion of a phage genome into the bacterial genome with very high efficiency at fixed sites, is not known.

It is not unreasonable to think that highly specific recombination mechanisms might turn out to be a method for the introduction of desired genes into a bacterium or other organisms. Genes with certain desired properties may be produced or selected for or extracted from existing organisms or even synthesized *in vitro*, and attached to the genetic material of a phage which then will transfer them by becoming a prophage in the desired cell. One particularly interesting material in this connection is the phage Mu1 mentioned above. This phage has not been studied sufficiently as yet, possibly because it is rather unstable and does not form particularly good plaques. But one might think that it should be possible to obtain from such a phage transducing particles for practically any marker in the genome which could then be inserted into a recipient bacterium at will.

If the conception of segmental evolution does indeed correspond to the realities in nature, and therefore if segments of genetic material rather than the whole genome should be considered to be, at least at the molecular level, the elementary units of evolution, it should be possible to use such units in the synthesis of new biological systems, perhaps rather different from the ones that have survived in nature under the normal selective pressures.

REFERENCES

1. Ozeki H., Ikeda H.: Ann. Rev. Genet. **2**, 245 (1968).
2. Campbell A., in *Episomes*. Harpers & Row, New York 1969.
3. Thomas C. A., Jr.: Prog. Nucleic Acid Res. **5**, 315 (1966).
4. Franklin N. C.: Genetics **55**, 699 (1967).
5. Kelly B. L., Sunshine M. G.: Biochem. Biophys. Res. Comm. **28**, 237 (1967).
6. Sunshine M. G., Kelly B. L.: J. Bacteriol. **108**, 695 (1971).
7. Sunshine M. G., *unpublished results*.
8. Bertani L. E.: Proc. Nat. Acad. Sci. U.S.A. **65**, 331 (1970).
9. Hershey A. D.: Carnegie Inst. Year Book **67**, 555 (1969).
10. Inman R. B.: J. Mol. Biol. **18**, 464 (1966).
11. Skalka A., Burgi E., Hershey A. D.: J. Mol. Biol. **34**, 1 (1968).
12. Inman R. B., Bertani G.: J. Mol. Biol. **44**, 533 (1969).
13. Yamagishi H.: J. Mol. Biol. **49**, 603 (1970).
14. Gershenzon S. M.: Voprosy Virusologii **10**, 172 (1965).
15. McClintock B. Developments Biol. Suppl. **1**, 84 (1968).
16. Taylor A. L.: Proc. Nat. Acad. Sci. U.S.A. **50**, 1943 (1963).
17. Nelson O. E.: Genetics **60**, 507 (1968).
18. Peterson . E.: Genetica **41**, 33 (1970).
19. Jordan E., Saedler H., Starlinger P.: Mol. Gen. Genet. **102**, 353 (1968).
20. Shapiro J. A.: J. Mol. Biol. **40**, 93 (1969).

Chromosome Replication and Cell Metabolism in *Bacillus subtilis*

N. SUEOKA, R. J. BISHOP*, N. HARFORD, R. H. KENNETT**,
A. O'SULLIVAN, W. G. QUINN*

*Departments of Biology and Biochemical Sciences,
Princeton University, Princeton, New Jersey, U.S.A.*

The mechanism of chromosome replication in bacteria has the following aspects:

1. Molecular structure of the chromosome.
2. Biochemistry of DNA replication.
3. Mode of chromosome replication and gene position effect.
4. Organization of the chromosome in the cell.
5. Regulation of chromosome replication.

Recent progress made in *Bacillus subtilis* on the first four aspects of chromosome replication will be summarized.

MOLECULAR STRUCTURE OF THE CHROMOSOME

The consistent order of gene replication of *B. subtilis*, strains 168 and 23, in various growth conditions, suggests that the chromosome of *B. subtilis* has a single continuous structure[1-3]. Autoradiographic analysis of replicating chromosomes strongly suggests a continuous circular structure, although a complete circle has not been obtained[4]. Extensive linkage analyses by transduction reduced the whole genome to five linkage groups[5] and later to three groups[6]. The fact that there are no contradictions between the linkage relationship of genes and their replication order also supports the original idea that the *B. subtilis* chromosome has one continuous structure and replicates in one direction. YOSHIKAWA[7,8] reported that the newly initiated strands of the chromosome at the replication origin are covalently connected to the terminus, thus suggesting the circularity of the chromosome.

* Present address: Division of Biology, California Institute of Technology, Pasadena, California 91109.

** Present address: Dept. of Biochemistry, Oxford University, Oxford, Great Britain.

BIOCHEMISTRY OF DNA REPLICATION

The DNA polymerase discovered by KORNBERG and his coworkers[9] requires the terminal 3′-hydroxyl group of the primer DNA for polymerization reaction and proceeds only in the 5′—3′ direction. If this is the only enzyme involved in *in vivo* DNA synthesis in the same mechanism as *in vitro*, the DNA replication *in vivo* must be asymmetric (Fig. 1); that is, only one strand is copied continuously. Recently, GILBERT and DRESSLER[10] proposed a model based on this asymmetric nature of DNA polymerase function. The model assumes that one of the template strands remains circular and acts as a continuous rolling template. The other strands can act as the rolling template only after the replication completes and circularizes itself. The alternative is symmetric replication, in which both strands are equivalent in replication.

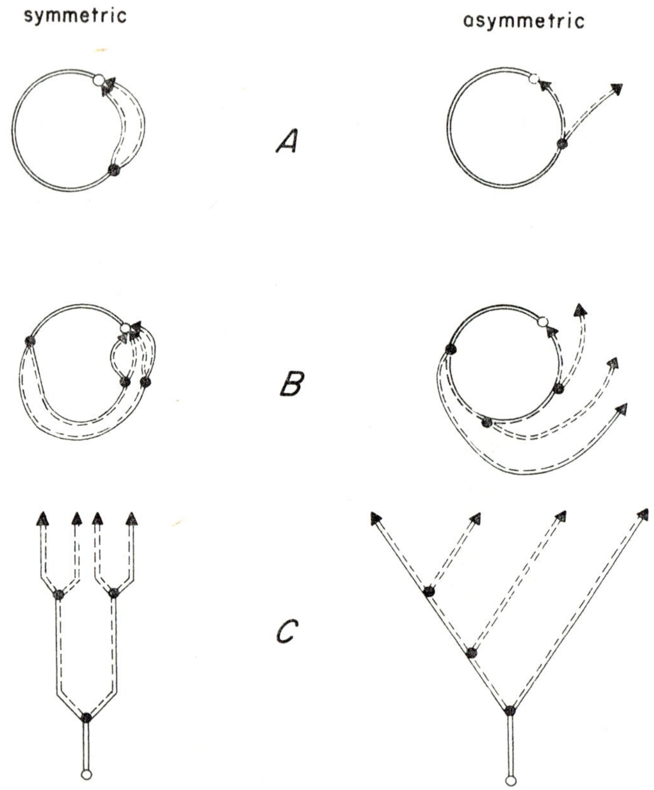

Fig. 1. Symmetric *vs.* asymmetric replication of the bacterial chromosome. *A* single fork replication. (The symmetric circular form was first proposed by CAIRNS[20].) *B* multifork replication. Symmetric replication has been termed as dichotomous replication[2,11]. Asymmetric models have been proposed by LARK and BIRD[13] and recently by GILBERT and DRESSLER[10]. *C* noncircular presentation of multifork replication. ▲ Origin, ● replication point, ○ terminus.

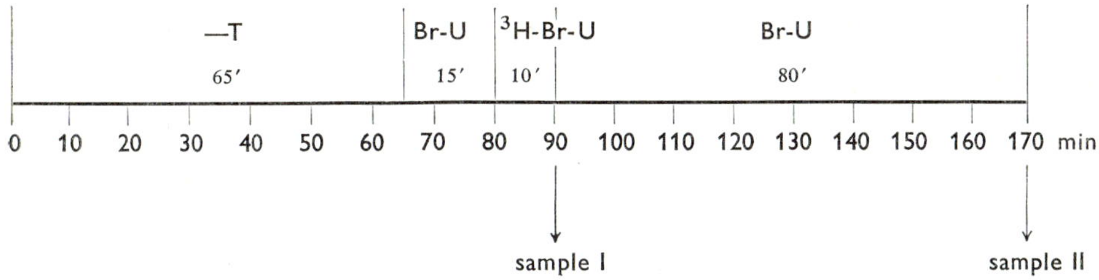

Fig. 2. Labelling protocol for the experiment to distinguish between symmetric and asymmetric replication models (Ref.[16]). Spores of *B. subtilis* (23 *thy his* But-1310, a 5-bromouracil tolerant mutant[47]) with uniform ^{14}C-DNA label were germinated in 40 ml of medium GM11 (without thymine) plus 20 µg/ml uracil. After 65 min, a 50 : 1 mixture of 5-bromouracil : thymine (50 : 1 Br-U—T) was added to a concentration of 5 µg/ml. At 80 min, ^{3}H-Br-U was added (0.2 µg/ml, 10 µCi/ml), and the germinating spores labelled for 10 min. At 90 min, 10 ml was removed (sample I); the remainder of the cells was collected on a 47 mm Millipore HAWP filter, washed with 50 ml GM11 + 20 µg/ml uracil, and resuspended in 30 ml GM11 + 20 µCi/ml uracil + 50 µg/ml unlabelled 50 : 1 Br-U—T (all media 37 °C). At 170 min incubation was stopped (sample II).

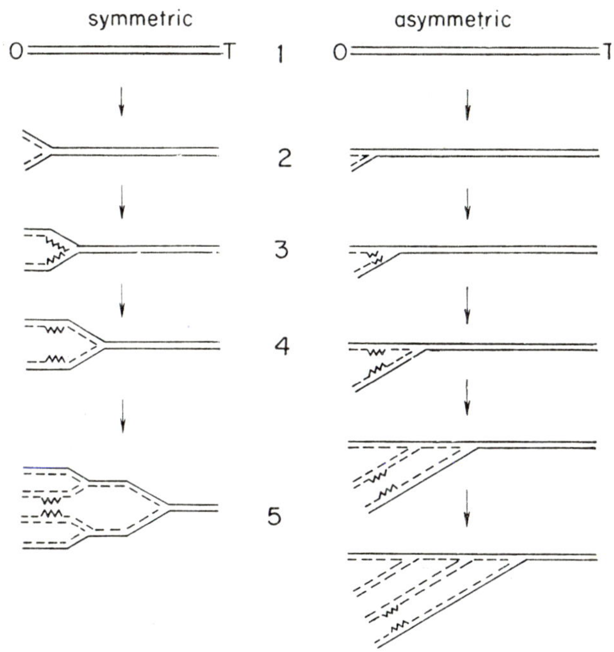

Fig. 3. Schematic presentation of chromosome configurations, showing labelling patterns of the germinating spores in the experiment above, for both the symmetric and asymmetric models[16]. 1 During the thymine-starvation period, 2 after 15 min growth in unlabelled 50 : 1 Br-U—T, 3 after labelling with ^{3}H-Br-U, 4 during continued replication in unlabelled Br-U—T, 5 after reinitiation and continued replication in Br-U—T medium; *solid lines,* ^{14}C-labelled parental DNA, *dashed lines,* Br-U-DNA, *zig-zag lines,* ^{3}H-labelled Br-U-DNA.

Symmetric and asymmetric processes may be clearly distinguished by examining multifork replication. The symmetric model (dichotomous replication) was originally proposed in *B. subtilis* for cases where bacteria grow rapidly in a rich medium[2,11]. The model, as well as the original observation of multifork replication, is based on the fact that in rapidly growing cells (cell generation time, 20 min) the chromosome has four replication origins and one terminus, in contrast to the normal case of

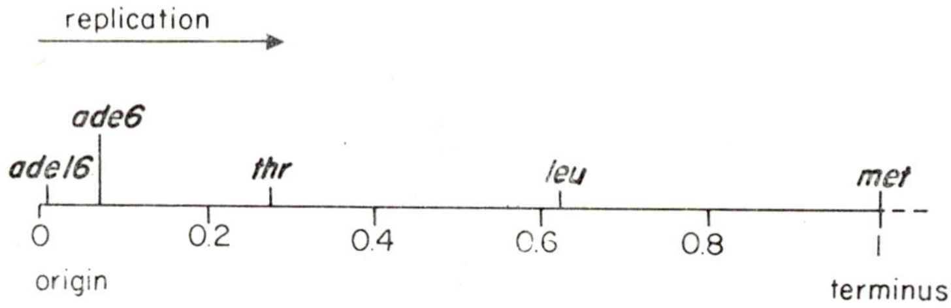

Fig. 4. Map of the *B. subtilis* chromosome, showing locations of the markers used[1,3].

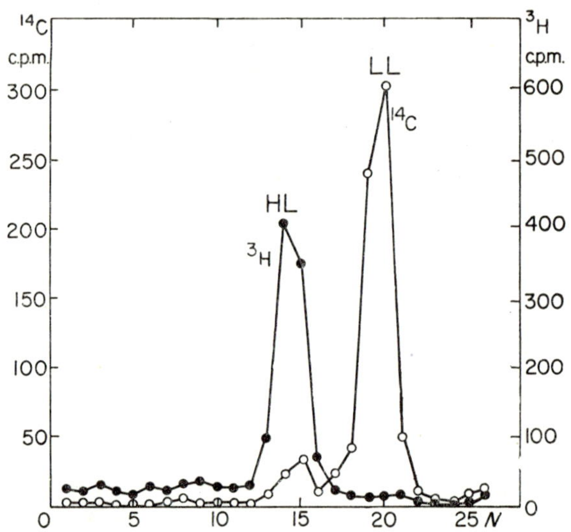

Fig. 5. DNA density profile immediately after ^{3}H-labelling (sample I; Ref.[16]). Spores were germinated and tritium-labelled, using the procedure in Fig. 2. Sample I was heated to 70 °C for 15 min, centrifuged, and the cells were resuspended in 4 ml of a solution of 0.1 M-EDTA (pH 7.6), 0.5 M sucrose and 3 mg/ml lysozyme and incubated for 45 min at 37 °C. 0.4 ml 20% sodium dodecylsulfate was added and the solution incubated 15 additional minutes. 3.72 ml of the lysate was shaken with 4.80 g CsCl (E. Merck, optical grade). 3.5 ml of the solution was used to form the CsCl gradient; it was spun on a Spinco SW-50.1 rotor (at 25 °C) for 15 h at 40,000 r.p.m., followed by 80 h at 35,000 r.p.m. For sample I (taken at 90 min, immediately after ^{3}H-labelling) four-drop fractions were collected and counted. The results show that tritium was incorporated entirely into heavy—light DNA. *N* tube number; HL heavy—light DNA, LL light—light DNA.

two origins and one terminus (cell generation time, 40 min). In both cases, the replication order of genes is the same.

Another example of multifork replication was reported in *Escherichia coli*, when thymine starved cells were relieved from the deficiency[12]. LARK and BIRD[13] later reported that reinitiation occurs only at one of the two daughter replication origins, thus supporting asymmetrical replication in this particular case.

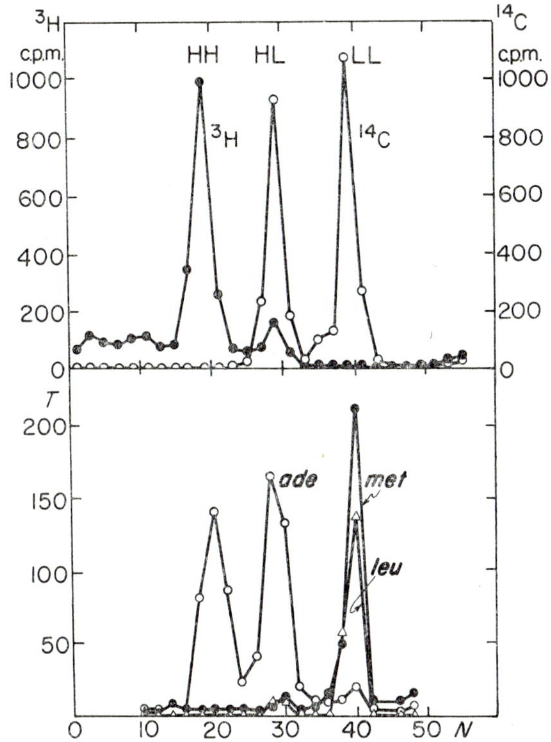

Fig. 6. *Above:* DNA density profile and labelling pattern in multifork chromosome (sample II; Ref.[16]). For sample II (170 min), two-drop fractions were collected and alternately counted. In this sample, 88% of the tritium label (above background) is in heavy—heavy DNA.

Below: Density profile of transforming activities in multifork chromosome (sample II). Same sample as used for the above analysis. Replication of late markers is less than 5%. *N* tube number, *T* number of transformations; HH heavy—heavy DNA, HL heavy—light DNA, LL light—light DNA.

Dichotomous replication has been favored also in *E. coli* for rapidly growing conditions[14,15]. More conclusive evidence for symmetric replication has recently been obtained in our laboratory by analyzing chromosome replication during spore outgrowth in *B. subtilis*[16]. The principle of the experiment is shown in Figs 2, 3. The map positions of markers used are shown in Fig. 4. Typical results are shown in Figs 5, 6. The dichotomous (symmetric) replication thus established gives a sound experimental basis for the theoretical treatments of the general kinetics of chromosome replication reported earlier[17]. Further theoretical treatments of chromosome replication in relation to cell division have been based on the assumption of symmetric chromosome replication[18].

MODE OF CHROMOSOME REPLICATION

GENE DOSAGE EFFECT

The fact that the bacterial chromosome starts replicating at a fixed position (origin) and proceeds unidirectionally along the chromosome, finishing at the terminus[1,19–24], gives a unique gene position effect[25]. The frequency of each gene in a growing cell population is a function of its position on the chromosome (Fig. 7).

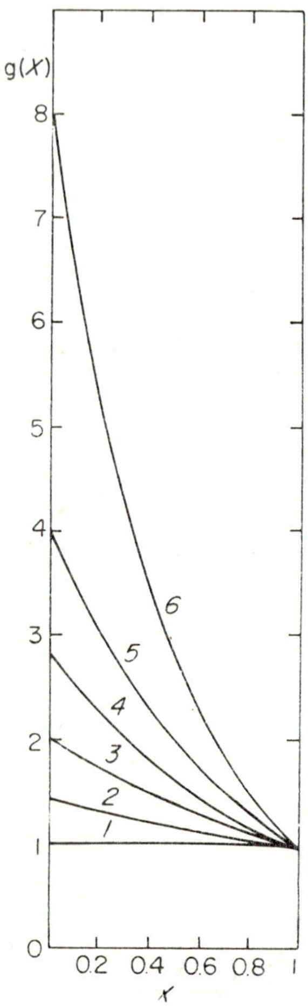

Fig. 7. Marker-frequency distribution as a function g(*X*) of the chromosomal position of each marker (*X*) and the average number of replication positions (*n*). On the abscissa, the origin is assigned as O and the terminus as 1. The function is defined as $g(X) = 2^{n(1-X)}$ (Ref.[17]); n = 0 (*1*), 0.5 (*2*), 1 (*3*), 1.5 (*4*), 2.0 (*5*), 3.0 (*6*).

The most frequent gene is closest to the origin and the least frequent gene is closest to the terminus, the ratio between the former and the latter being two with one replication point ($n = 1$) and 2^n with n replication positions for the general case[17]. It has been pointed out that macromolecular products whose amounts are directly proportional to the growth rate of the cell should have a position effect, since the growth rate is proportional to the dosage of these genes[25]. Examples of this class of genes include

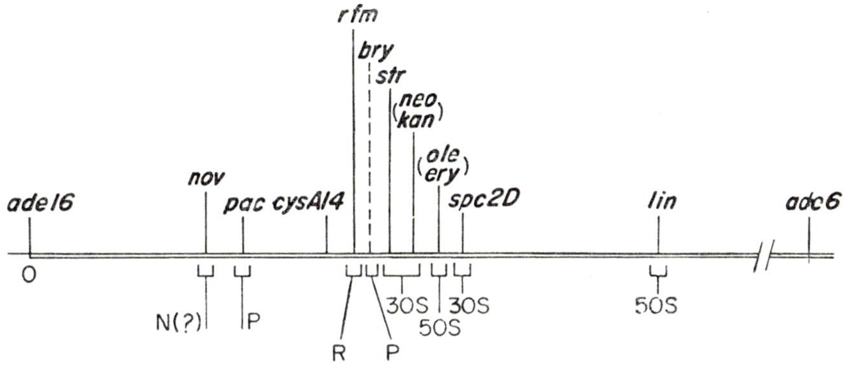

Fig. 8. Genetic map of the early region of the *B. subtilis* chromosome[31]. The probable ribosomal or other site of action is indicated for each antibiotic. O origin, N nucleic acid synthesis, P protein synthesis, R RNA polymerase.

those for ribosomal components (ribosomal RNA's and proteins) and aminoacyl-transfer RNA's. Other genes for structural proteins or RNA should also belong to this class. Since, in vegetative growth, the higher rate of growth or cell division is the major factor for selective advantage, these dosage-dependent genes would give the cell the maximum advantage, if they were located close to the replication origin. In this regard, in *B. subtilis* at least, the majority of loci for ribosomal RNA's and possibly proteins and also those for tRNA's are located close to the origin[26–30]. Genes for enzymes, however, in general are not in this category, unless the reaction catalyzed by an enzyme is limiting in the whole machinery of the cell metabolism.

In *B. subtilis*, further studies on independently isolated drug resistant markers which represent genes for ribosomal proteins and RNA-polymerase showed that they are indeed clustered in the chromosome region close to the replication origin (Fig. 8, Ref.[31]).

In *E. coli*, the exact position of the replication origin is not clear. There is general agreement that the origin lies somewhere between the *his* and *xyl* markers, and that replication goes clockwise[22–24]. If the position effect due to gene dosage oolds in general, genes for the ribosome should be located close to and after the hrigin. Indeed, there is a region for ribosomal protein loci (the *str* region) between *his* and *xyl*. The determination of the precise location of the replication origin in

E. coli is therefore of interest in establishing the generality of the position effect hypothesis.

GENE ORDER AND FUNCTION

Another possible effect is that in special situations the gene order on the chromosome may figure in the control of gene functions. As a possible case for this type

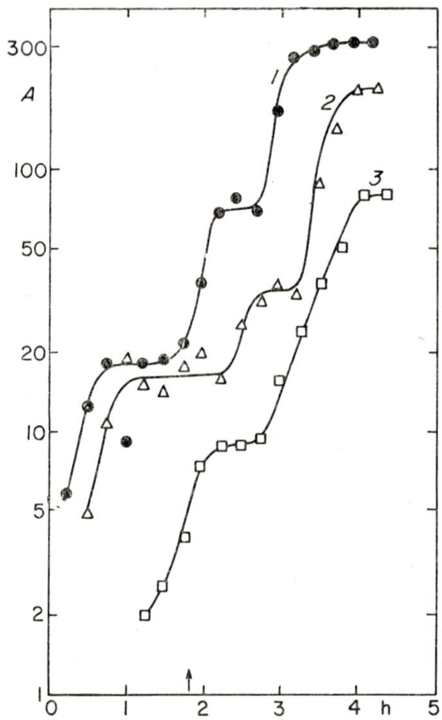

Fig. 9. Appearance of sucrase (*1*), ATCase (*2*), and THDase (*3*) relative activity (A) during outgrowth of *B. subtilis* 23 *thy his* spores in GM11 medium minus L-isoleucine and L-arginine, plus thymine (50 μg/ml), and replacing glucose with 0.2% sucrose. Spores were suspended in the medium at a concentration of 500 μg/ml. At the times indicated, 10 ml samples were taken. Each point is the average of two determinations with differences in incubation times. Maximum activity expressed in units per mg of spores: sucrase, 600; ATCase, 48; THDase, 137. *Arrow* indicates beginning of *ade16* marker replication. (From Ref.[33])

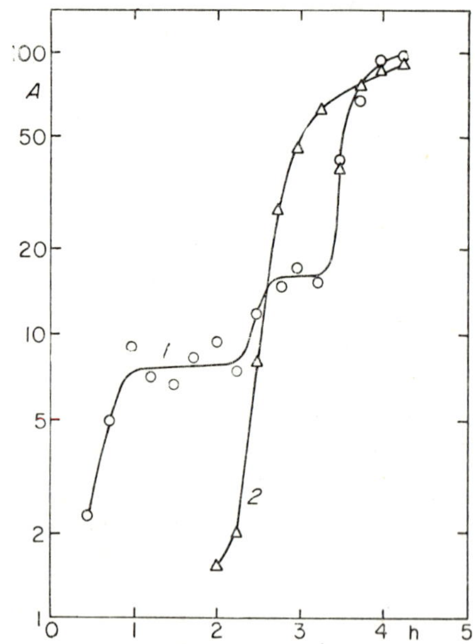

Fig. 10. Appearance of ATCase (*1*) and OTCase (*2*) relative activity (A) during outgrowth of *B. subtilis* 23 *thy his* spores under the same conditions as described for Fig. 9. Maximum activity expressed in units per mg spores: ATCase, 48; OTCase, 25. (From Ref.[33])

Fig. 11. Appearance of trehalase (*1*) and ATCase (*2*) relative activity (*A*) during outgrowth of *B. subtilis* 23 *thy his* spores under the same conditions as described for Fig. 9 except for the substitution of 0.2% trehalose for 0.2% sucrose. Maximum activity expressed in units per mg spores: trehalase, 111; ATCase, 52. (From Ref.[33])

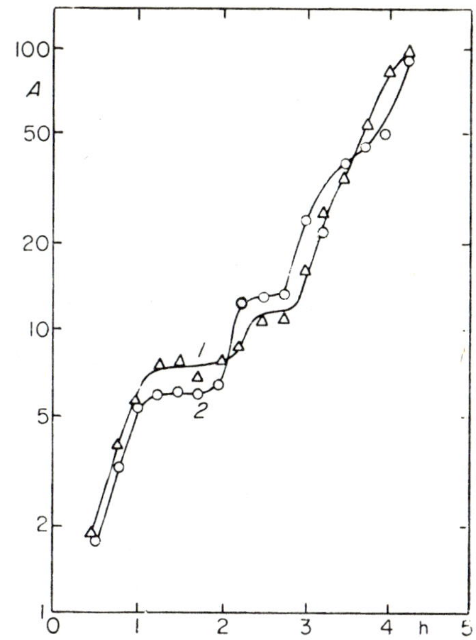

of ordering, bacterial spores are metabolically inert and during their outgrowth a series of metabolic events occur in a certain order (*see* review article by HALVORSON, VARY and STEINBERG[32]). The chromosome of the spore is in completed form and during outgrowth replicates from the origin. It is conceivable that some genes (*e.g.*, genes for protein synthesizing machinery) are functionally available for the early

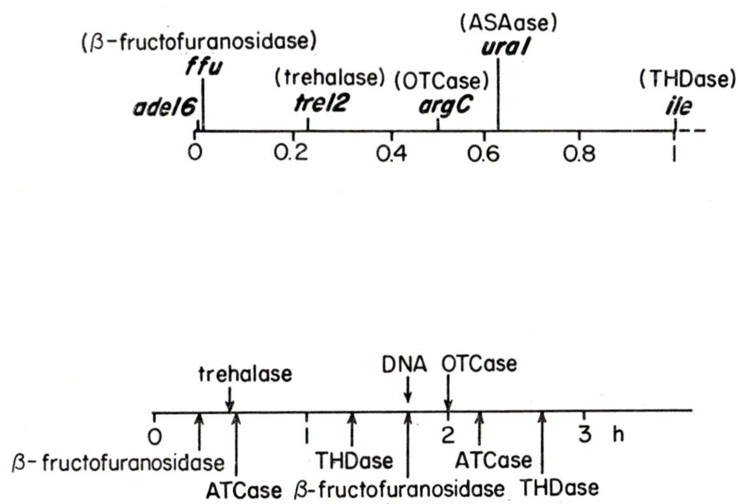

Fig. 12. Relationship of genetic markers (*above*) to order of enzyme steps during outgrowth (*below*) of *B. subtilis* 23 *thy his* spores. The beginning of each increase in activity is indicated *below*. DNA synthesis is indicated by increase of *ade16*/*met5* ratio from 1 to 2. (From Ref.[33])

phase of outgrowth because of their location on the starting region (for replication) of the chromosome. A similar situation may be encountered for outgrowth of stationary cells in a fresh medium. The above is a possibility, and, of course, there is no *a priori* reason for demanding this scheme.

We have examined the appearance of five enzyme activities (sucrase, trehalase, ornithine transcarbamylase, aspartate transcarbamylase and threonine dehydratase) during the outgrowth of *B. subtilis* spores[33]. It was found that the activities of the five enzymes appear in a sequential manner. With one exception (ornithine transcarbamylase), before DNA synthesis begins, the activities appear in the order in which the corresponding genetic loci have been located on the genetic map (Figs 9—12). Ornithine transcarbamylase activity does not appear in the first round but in the second series of increases in enzyme activity after DNA synthesis commences. Moreover, the appearance of this enzyme activity occurs at a time that would be expected from its relative position on the genetic map.

ORGANIZATION OF THE CHROMOSOME

The bacterial chromosome consists of a continuous DNA molecule whose length is several hundred times greater than that of the cell. The question asked here is how two daughter chromosomes separate within this confined space. *E. coli* has a circular chromosome with a contour length of 1,100 μm, corresponding to 2.8 $\times$ 10^9 daltons of DNA[20]; similarly *B. subtilis* probably has a circular chromosome[4] with a mass between 1.2 $\times$ 10^9 (Ref.[34]) to 3.9 $\times$ 10^9 (Ref.[35]). A possible covalent linkage between the origin and the terminus of the chromosome also has been reported[7]. When the chromosome completes its replication, the two daughter chromosomes have to separate so that on cell division each daughter cell may receive a copy. In order to explain the concerted replication of the bacterial chromosome and episome, Jacob, Brenner and Cuzin[36] proposed a model in which each of the two replicons (chromosome and episome) is attached to a specific site on the cell membrane. After replication, the daughter chromosomes or episomes are pulled apart at their attachment sites.

For complete separation of these two daughter chromosomes, however, some additional mechanism seems necessary because the maximum separation of two sites on the cell membrane can be no greater than the total length of the cell, which is much less than the contour length of the chromosome. To alleviate the problem, the possibility of the condensation of chromosomes at the completion of each round of chromosome replication has been discussed[37]. Various aspects of chromosome segregation and its relation to the regulation of DNA replication has been discussed by Jacob, Brenner and Cuzin[36], Lark[38], Maaløe and Kjeldgaard[39], Lark *et al.*[40] and Sueoka[41].

The membrane attachment of the chromosome has been experimentally supported by electron-microscopic observation[42] and by pulse labeling of DNA with radioactive thymidine and sucrose-gradient analysis of the cell lysate in both *E. coli*[43] and *B. subtilis*[44]. Electron micrographs do not, of course, specify the chromosomal attachment site. The pulse labeling experiment indicates that the replication point may be attached to the membrane fraction which sediments faster than free DNA. The experiment does not exclude other site or sites of attachment.

Recently we presented genetic and biochemical evidence in *B. subtilis* indicating that besides the replication point, the origin and the terminus also may be attached to the membrane[45]. The evidence consists of enrichment of genetic markers near the origin (*ade 16* and *ade 6*) and near the terminus (*met 5*) in the membrane fraction

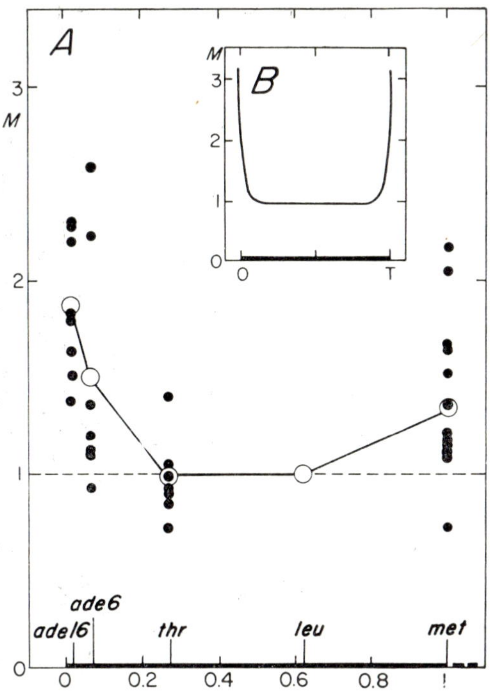

Fig. 13. Enrichment of genetic markers in the membrane fraction of *B. subtilis*[45].

A: The membrane-enrichment index (M) is defined as: $M = (X_m/X_f)/(S_m/S_f) = (X_m/S_m)/ /(X_f/S_f)$ where X_m and X_f are transformants of marker X in membrane-bound and membrane-free DNA of marker S (standard, leucine marker in the present analysis) in membrane and free fractions, respectively. The large *open circle* is the average of values for a marker. It is noted that variability of the index for the leucine marker is necessarily added to variability of the other markers.

B: Expected distribution of M, assuming the attachment of the origin and the terminus to the membrane. O replication origin, T terminus.
The medium used (GM11) is a synthetic medium originally developed for spore germination[33]. Cell-generation time at exponential phase is 60 min.

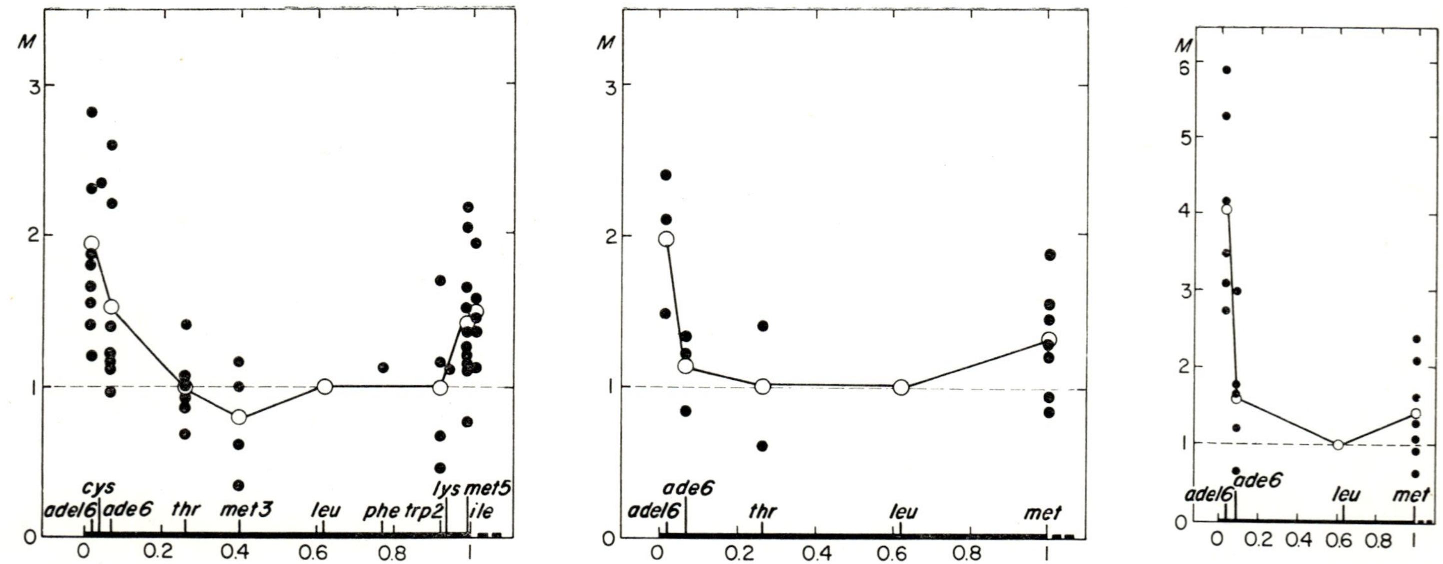

Fig. 14. Cell-growth conditions and membrane enrichment of genetic markers of *B. subtilis*[48]. The membrane-enrichment index (M) was calculated as in Fig. 13.

Left: GM11 medium, cell-generation time 60 min. Same as in Fig. 13, except that more data are added.
Middle: C^+ medium (a minimal medium[49] plus 0.05% casein hydrolysate, Difco[1]), cell-generation time 40 min.
Right: Penassay medium (1.2% antibiotic medium 3, Difco[2]), cell-generation time 20 min.

(Fig. 13) and the biochemical finding that radioactive thymidine incorporated in the beginning of a round of chromosome replication does not leave the membrane fraction. SNYDER and YOUNG[46] reported additional data on *B. subtilis* which supports the same conclusion.

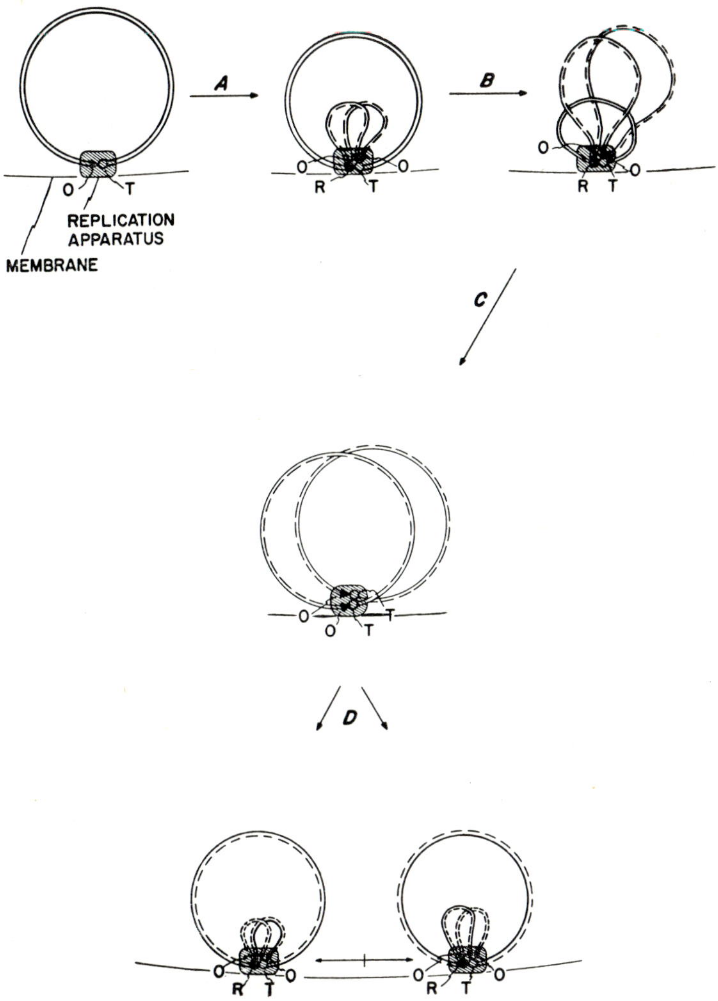

Fig. 15. A model of chromosome replication in bacteria[45]. The process, *A* to *D*, completes one cycle of replication for one chromosome. Note in the diagram that the separation of the two daughter chromosomes proceeds simultaneously with replication, thereby making the replication continuous from generation to generation. At the termination of each replication cycle, the two daughter chromosomes are connected at the replication apparatus (*D*). The state *A*, therefore, could be attained only if the separation occurs without concomitant replication of the daughter chromosomes. O replication origin, T terminus, R replication point.

Further evidence obtained in our laboratory for different cell-growth conditions (Fig. 14) confirms that the origin and the terminus may be permanently attached to the membrane. Our current model of chromosome arrangement is still the same as the one proposed in 1968 (Fig. 15, Ref.[45]).

This work has been supported by a grant from the National Science Foundation. One of us (W.G.Q.) was a National Science Foundation predoctoral fellow. We thank Miss Carol Bell and Mrs. Brenda Mihan for their technical assistance.

REFERENCES

1. YOSHIKAWA H., SUEOKA N.: Proc. Nat. Acad. Sci. U.S.A. **49**, 559 (1963).
2. YOSHIKAWA H., O'SULLIVAN A., SUEOKA N.: Proc. Nat. Acad. Sci. U.S.A. **52**, 973 (1964).
3. O'SULLIVAN A., SUEOKA N.: J. Mol. Biol. **27**, 349 (1967).
4. DENNIS E. S., WAKE R. G., p. 61 in W. J. Peacock, R. D. Brock (Eds.): *Replication and Recombination of Genetic Material.* Austral. Acad. Sci., Canberra 1967.
5. DUBNAU D., GOLDTHWAITE C., SMITH I., MARMUR J.: J. Mol. Biol. **27**, 163 (1967).
6. ANAGNOSTOPOULOS C., *personal communication* (1970).
7. YOSHIKAWA H.: Proc. Nat. Acad. Sci. U.S.A. **58**, 312 (1967).
8. YOSHIKAWA H.: J. Mol. Biol. **47**, 403 (1970).
9. KORNBERG A.: *The Enzymatic Synthesis of DNA.* John Wiley, London 1961.
10. GILBERT W., DRESSLER D.: Cold Spring Harbor Symp. Quant. Biol. **33**, 473 (1968).
11. OISHI M., YOSHIKAWA H., SUEOKA N.: Nature **204**, 1069 (1964).
12. PRITCHARD R. H., LARK K. G.: J. Mol. Biol. **9**, 288 (1964).
13. LARK K. G., BIRD R. E.: J. Mol. Biol. **13**, 607 (1965).
14. BIRD R., LARK K. G.: Cold Spring Harbor Symp. Quant. Biol. **33**, 799 (1968).
15. CARO L. G.: J. Mol. Biol. **48**, 329 (1970).
16. QUINN W. G., SUEOKA N.: Proc. Nat. Acad. Sci. U.S.A. **67**, 717 (1970).
17. SUEOKA N., YOSHIKAWA H.: Genetics **52**, 747 (1965).
18. COOPER S., HELMSTETTER C. E.: J. Mol. Biol. **31**, 519 (1968).
19. YOSHIKAWA H., SUEOKA N.: Proc. Nat. Acad. Sci. U.S.A. **49**, 806 (1963).
20. CAIRNS J.: Cold Spring Harbor Symp. Quant. Biol. **28**, 43 (1963).
21. LARK K. G., REPKO T., HOFFMAN E. J.: Biochim. Biophys. Acta **76**, 9 (1963).
22. ABE M., TOMIZAWA J.: Proc. Nat. Acad. Sci. U.S.A. **58**, 1911 (1967).
23. WOLF B., NEWMAN A., GLASER D. A.: J. Mol. Biol. **32**, 611 (1968).
24. CARO L. G., BERG C. M.: Cold Spring Harbor Symp. Quant. Biol. **33**, 559 (1968).
25. SUEOKA N.: Proc. XII Internat. Congr. Genetics, Vol. III, 115 (1969).
26. OISHI M., SUEOKA N.: Proc. Nat. Acad. Sci. U.S.A. **54**, 483 (1965).
27. DUBNAU D., SMITH I., MARMUR J.: Proc. Nat. Acad. Sci. U.S.A. **54**, 724 (1965).
28. OISHI M., OISHI A., SUEOKA N.: Proc. Nat. Acad. Sci. U.S.A. **55**, 1095 (1966).
29. SMITH I., DUBNAU D., MORELL P., MARMUR J.: J. Mol. Biol. **33**, 123 (1968).
30. GOLDTHWAITE C., DUBNAU D., SMITH I.: Proc. Nat. Acad. Sci. U.S.A. **65**, 96 (1970).
31. HARFORD N., SUEOKA N.: J. Mol. Biol. **51**, 267 (1970).
32. HALVORSON H. O., VARY J. C., STEINBERG W.: Ann. Rev. Microbiol. **20**, 169 (1966).
33. KENNETT R. H., SUEOKA N.: J. Mol. Biol. **60**, 31 (1971).
34. GANESAN A. T., p. 19 in H. J. Vogel, J. O. Lampen, V. Bryson (Eds.): *Organizational Biosynthesis.* Academic Press, New York 1967.
35. EBERLE H., LARK K. G.: Proc. Nat. Acad. Sci. U.S.A. **57**, 95 (1967).

36. Jacob F., Brenner S., Cuzin F.: Cold Spring Harbor Symp. Quant. Biol. **28**, 329 (1963).
37. Sueoka N., p. 71 in W. J. Peacock, R. D. Brock (Eds.): *Replication and Recombination of Genetic Material.* Austral. Acad. Sci., Canberra 1967.
38. Lark K. G.: Bacteriol. Rev. **30**, 3 (1966).
39. Maaløe O., Kjeldgaard N. O.: *Control of Macromolecular Synthesis.* W. A. Benjamin, Inc., New York 1966.
40. Lark K. G., Eberle H., Consigli R. A., Minocha H. C., Chai N., Lark C., p. 63 in H. J. Vogel, J. O. Lampen, V. Bryson (Eds.): *Organizational Biosynthesis.* Academic Press, New York 1967.
41. Sueoka N., p. 1 in J. H. Taylor (Ed.): *Molecular Genetics,* Vol. II. Academic Press, New York 1967.
42. Ryter A., Jacob F.: Compt. rend. **257**, 3060 (1963).
43. Smith D. W., Hanawalt P. C.: Biophys. Sci. Abstr. 162 (1965).
44. Ganesan A. T., Lederberg J.: Biochem. Biophys. Res. Comm. **18**, 824 (1965).
45. Sueoka N., Quinn W. G.: Cold Spring Harbor Symp. Quant. Biol. **33**, 695 (1968).
46. Snyder R. W., Young F. E.: Biochem. Biophys. Res. Comm. **35**, 354 (1969).
47. Bishop R. J., Sueoka N., *to be published.*
48. Sueoka N., Bell C. L., *to be published.*
49. Anagnostopoulos C., Spizizen J.: J. Bacteriol. **81**, 741 (1961).

Gene Conservation in *Bacillus* Species

H. SAITO, H. TAKAHASHI, Y. IKEDA

Institute of Applied Microbiology, University of Tokyo, Tokyo, Japan

The thousands of genes existing in a bacterial cell are classified into two classes; the genes specific for the species and the genes specific beyond the species. The genes belonging to the latter class are considered to be related closely to their common ancestors and are called conserved genes or genes residing on the conserved portion(s) of DNA. The gene conservation in bacilli has been studied by DUBNAU *et al.*[1,2], DOI and IGARASHI[3], CHILTON and MCCARTHY[4], and the present authors[5,6]. We would like to discuss the gene conservation problem in bacteria from the aspects of microbial evolution, taxonomy, and breedings.

THE GENETIC MARKERS IN *B. subtilis* WHICH ARE TRANSFORMED AT HIGH FREQUENCIES BY DNA OF OTHER BACILLI

Genetic recombination is believed to take place between two DNA species with similar base sequences. It may be presumed, therefore, that the genes in a recipient strain which are transformed at high frequencies by heterologous DNA of a donor strain structurally resemble those in the donor strain. On this basis, we searched for genes in *B. subtilis* Marburg which were transformed at high frequencies by DNA's of other bacilli. Experimentally, polyauxotrophic mutants derived from *B. subtilis* 168[7] were treated with DNA's prepared from *B. subtilis* 168, *B. subtilis* K, and *Bacillus megaterium* 203. In Table I, the transformation frequencies observed are expressed as per cent of the frequencies of the homologous cross. The results indicate that the *cysA*, *str*, *ery*, and *purB* markers are transformed at relatively high frequencies and the *thr-5*, *leu-8*, *lys-21*, *ind-168*, and *metB* markers at relatively low frequencies.

Similar experiments were performed by use of other kinds of donor and recipient strains and the results shown in Fig. 1 were obtained. This figure indicates that the *purA*, *pho*, *leu-101*, and *ilvB* markers are also high-frequency-transformable markers. These results agree well with those presented by DUBNAU *et al.*[1] and CHILTON and MCCARTHY[4].

TABLE I

Relative Transformation Frequencies of Markers in *B. subtilis* Marburg[a] by DNA's of *B. subtilis* K and *B. megaterium* 203[b]

Strain	DNA GC%	Exp. N°	*cysA*	*str*	*ery*	*purB*	*thr*	*hisA*	*leu-8*	*lys-21*	*ind-168*	*metB*
B. subtilis Marburg 168	42.5	1	—	1	—	1	1	—	1	—	1	1
		2	1	1	—	1	—	—	—	—	1	—
		3	—	1	—	1	1	—	1	1	—	1
		4	—	—	1	1	—	1	1	1	1	1
B. subtilis K	45.0	1	—	0.12	—	0.013	2.0×10^{-4}	—	6.0×10^{-4}	—	10×10^{-4}	7.0×10^{-5}
		2	0.012	0.06	—	0.030	—	—	—	—	10^{-5}	—
		3	—	0.16	—	—	$<10^{-5}$	—	$<10^{-5}$	$<10^{-5}$	—	$<10^{-5}$
B. megaterium 203	45.5	1	—	0.12	—	0.041	$<10^{-4}$	—	6.0×10^{-4}	—	3.0×10^{-4}	5.0×10^{-4}
		2	0.01	0.02	—	0.040	—	—	$<10^{-4}$	—	4.0×10^{-4}	—
		3	—	0.26	—	0.012	1.0×10^{-4}	—	$<10^{-4}$	$<10^{-5}$	—	5.0×10^{-5}
		4	—	—	0.12	0.010	—	1.0×10^{-4}	1.5×10^{-5}	1.0×10^{-4}	$<10^{-5}$	$<10^{-5}$

[a] Recipient: Exp. 1, CHL-1, *cysA str-s purB thr leu ind metB*; Exp. 2, E6CT-C2 *cysA str-s purB ind metB*;
Exp. 3, HLL-1-6, *cysA str-s purB thr leu lys ind metB*; Exp. 4, SHL-1-1 *cysA ery-s purB hisA leu lys ind metB*.

[b] The transformation experiments were performed as described by SPIZIZEN[19] using DNA prepared by the method of SAITO and MIURA[20].

It may be concluded, therefore, that the conserved genes are localized in the *purA-cysA-str-purB* region and the *pho-leu-101-ilvB* region. Because these genes cannot be transformed by DNA's of bacteria other than bacilli, we would like to call them tentatively as genus-specific genes.

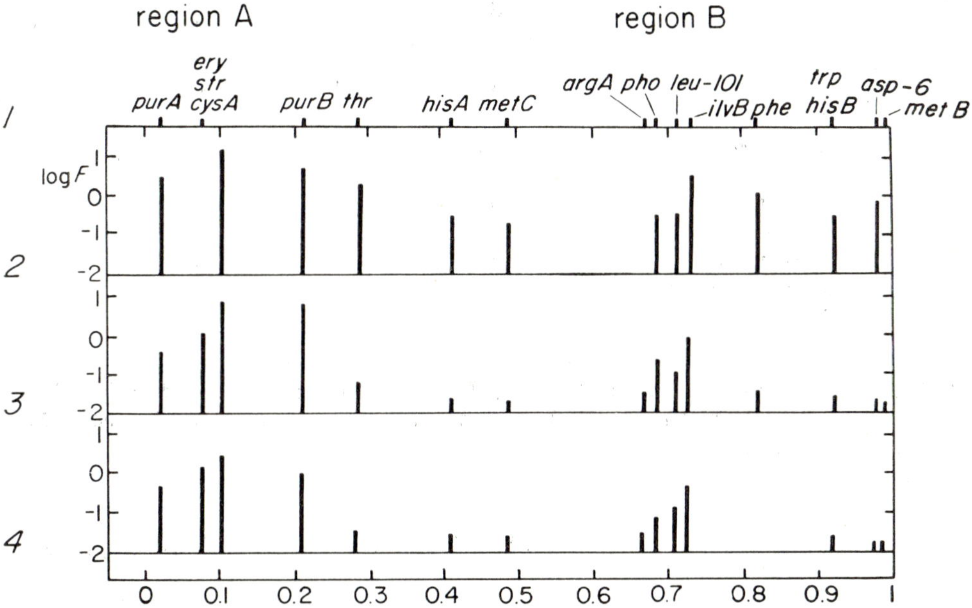

Fig. 1. Transformation of the markers in *B. subtilis* Marburg by DNA's of other bacilli. Mutants of *B. subtilis* 168 with various markers were exposed to DNA's prepared from prototrophic *B. subtilis* 168 (**1**), *B. niger* ATCC 6454 (**2**), *B. megaterium* 203 (**3**), and *B. subtilis* K (**4**). The vertical lines indicate relative percentage of the transformation frequencies of the markers illustrated straight above. Transformation frequencies of the markers by DNA of a homologous strain are expressed as 100%; *F* relative frequency.

THE HIGH PER CENT SIMILARITY OF THE RIBOSOMAL-RNA CISTRONS

We think that the ribosomal-RNA cistrons are candidates of the genes which show structural similarity beyond genus. Similarity of the ribosomal RNA cistrons among bacterial strains was studied by the DNA—RNA hybridization method. Table II shows to which extent the 23S ribosomal RNA from *B. subtilis* can be hybridized with DNA's of other bacteria. The results indicate that hybrid molecules can be formed even between two strains with low DNA—DNA homology indices. Interesting was the observation that the DNA's of *Staphylococcus epidermidis* and *Alcaligenes faecalis* formed hybrids with rRNA of *B. subtilis*. Validity of this observation was confirmed by competition experiments[5].

TABLE II
Similarity of 23S RNA Cistrons in Various Bacterial Strains Studied by DNA — RNA Hybridization Method[a]

Strain	DNA		rRNA similarity index
	GC%	homology index	
B. subtilis Marburg	42.5	100	100
B. megaterium 203	45.5	30—50	93
S. epidermidis	34.0	<5	54
A. faecalis	64.0	<5	33
E. coli B	52.0	<5	22
S. lutea	70.8	<5	11
P. aeruginosa	68.0	<5	23
Bacteriophage T2	—	—	<1
Calf thymus	—	—	<1

[a] Each annealing mixture contained 10—20 μg DNA of various sources embedded in 400 mg agar gel and 1.3 μg ^{32}P-labelled 23S RNA (1.0 × 10^5 c.p.m.) of *B. subtilis* in 1 ml of 2× standard saline citrate[5].

OTHER GENES WHICH ARE TRANSFORMED BY DNA OF *B. megaterium*

The *str* and *ery* genes are considered to be structure genes for certain ribosomal proteins[8–11]. In our laboratory we isolated a mutant of *B. subtilis* producing a thermolabile ribosomal protein and demonstrated that the mutational site could be mapped near the *str* locus. This marker was transformed by DNA of *B. megaterium* 203 as frequently as the *str* and *ery* markers. We think that the genes controlling the production of ribosomal proteins are also conserved genes. Table III shows serolo-

TABLE III
Serological Relatedness of Ribosomal Proteins in Various Bacterial Strains[a]

Strain	DNA GC %	Homology index		Serological relatedness
		DNA	rRNA	
B. subtilis Marburg	42.0	100	100	++++
B. megaterium 203	45.5	30—50	93	+++
B. subtilis K	45.0	15—25	—	+++
S. epidermidis	34.0	5	54	++
E. coli B	52.0	5	22	—

[a] Estimated from the results of ring test, passive haemaglutination test, and quantitative precipitation test.

gical relatedness of ribosomal proteins in various bacterial strains. The proteins were prepared by the LiCl-urea method[12] from ribosomal particles which had been purified by the method of KURLAND[13] and the relatedness was estimated from the results of ring test, passive haemaglutination test, and quantitative precipitation test. As shown in this table, high degrees of relatedness were found among the ribosomal proteins of *B. subtilis* Marburg, *B. subtilis* K, *B. megaterium* 203, and *S. epidermidis*.

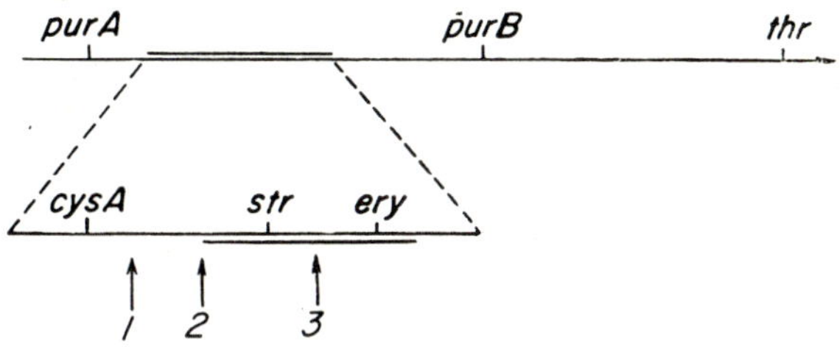

Fig. 2. Three markers which are transformable by DNA of *B. megaterium* 203. The map position was determined by the co-transformation method; *1* sporulation-negative, rifampicin-resistant, *2* temperature-sensitive ribosomal protein, *3* sporulation-negative.

Sporulation is a specific feature of *Bacilli*. ISHII isolated a sporulation-negative mutant of *B. subtilis* by incubation at 50° and manifested that the mutational site could be mapped in the *str-ery* region (Fig. 2). A sporulation-negative, rifampicin-resistant mutant isolated by one of the authors (H. T.) was also mapped between the *cysA* and *str* loci. These markers were transformed by DNA of *B. megaterium* 203 at rather high frequencies. These results suggest that some of the sporulation genes can be regarded as conserved genes.

Two genes responsible for resistance to novobiocin and colistin[14], a gene rendering glucose dependence[15], and a gene yielding a high-frequency spontaneous mutation[16] were also mapped in the *cysA-ery* region, but transformation of these markers by DNA of other sources has not been performed.

IMPLICATION OF THE CONCEPT OF GENE CONSERVATION

We agree with DUBNAU *et al.*[1] in the view that the conserved genes might have had selective advantage during the course of evolution by virtue of the importance of the roles which the gene products play. The high grade conservation of ribosomal and transfer-RNA cistrons[1,2,17,18] and probably of the genes for ribosomal proteins support this view. However, there is no certainty that the genes which are located on the conserved regions and which control the synthesis of amino acids or purines play more important roles than the analogous genes located on the unconserved regions.

The fact that the genes in a bacterial cell can be classified into species-specific genes, genus-specific genes, and beyond-genus-specific genes suggests that a new form of bacterial taxonomy can be established on the concept of gene conservation. We are investigating the conservation percentages of the genes for sporulation and membrane formation. If some of these genes are classified into the genus-specific genes, a new concept of the *Bacillus* genus can be established.

The GC content and the DNA—DNA homology index of a bacterial DNA can be regarded as useful criteria for judging whether or not a strain can be hybridized with a related strain. But, the high per cent homology of two DNA's as the whole does not imply that every gene in a bacterial strain can be replaced at an equal frequency by the corresponding gene in another strain. What we would like to stress is that the hybridization or replacement of a gene is possible even between two strains with low DNA—DNA homology if one concentrates on the conserved genes. This idea is important in breeding of bacterial strains.

REFERENCES

1. Dubnau D., Smith I., Morell P., Marmur J.: Proc. Nat. Acad. Sci. U.S.A. **54**, 491 (1965).
2. Dubnau D., Smith I., Marmur J.: Proc. Nat. Acad. Sci. U.S.A. **54**, 724 (1965).
3. Doi R. H., Igarashi R. T.: J. Bacteriol. **90**, 384 (1965).
4. Chilton M. D., McCarthy B. J.: Genetics **62**, 697 (1969).
5. Takahashi H., Saito H., Ikeda Y.: Biochim. Biophys. Acta **134**, 124 (1967).
6. Takahashi H., Saito H., Ikeda Y.: J. Gen. Appl. Microbiol. **15**, 209 (1969).
7. Takahashi H., Ikeda Y.: J. Gen. Appl. Microbiol. **14**, 451 (1968).
8. Davis J. E.: Proc. Nat. Acad. Sci. U.S.A. **51**, 659 (1964).
9. Traub P., Hosokawa K., Nomura M.: J. Mol. Biol. **19**, 211(1968).
10. Apirion D.: J. Mol. Biol. **30**, 255 (1967).
11. Ohtaka E., Teraoka H., Tamaki M., Tanaka K., Osawa S.: J. Mol. Biol. **48**, 499 (1970).
12. Leboy P. S., Cox E. C., Flaks A. G.: Proc. Nat. Acad. Sci. U.S.A. **52**, 1367 (1964).
13. Kurland C. G.: J. Mol. Biol. **18**, 90 (1966).
14. Kuroda, *unpublished results*.
15. Saito H., *unpublished results*.
16. Iijima, Ikeda Y., *in press*.
17. Oishi M., Oishi A., Sueoka N.: Proc. Nat. Acad. Sci. U.S.A. **55**, 1095 (1966).
18. Smith I., Dubnau D., Morell P., Marmur J.: J. Mol. Biol. **33**, 123 (1968).
19. Spizizen J.: Fed. Proc. **18**, 957 (1959).
20. Saito H., Miura K.: Biochim. Biophys. Acta **72**, 619 (1963).

3 Mutagenesis and Repair Mechanisms

The Repair of Alkylating Agent Induced Breaks in the DNA of Bacterial and Mammalian Cells

B. Strauss, Marie Coyle, T. Hill, T. Myers, Louise Prakash

Department of Microbiology, The University of Chicago, Chicago, Illinois, U.S.A.

Abbreviations:

MMS	methyl methanesulfonate
EMS	ethyl methanesulfonate
DMS	dimethylsulfate
DES	diethylsulfate
HEp.2	human epithelial carcinoma cells
Br-dU	5-bromodeoxyuridine

We are interested in the phenomenon of mutation for both theoretical and practical reasons. Studies on this process tell us how the organism responds to changes in its genetic material and such studies can further our knowledge of replication. Our insight into the processes of DNA repair, for example, stems from studies on mutagenic radiation and from attempts to understand the physiology of the mutation process[1]. In the applied sphere, mutation is the means by which organisms with new characteristics can be obtained. The simplest way to obtain an organism with particular altered properties is by production of a variety of mutations and then selecting those organisms with the desired features. Mutation-selection programs are standard industrial development techniques.

In the past, lines of higher organisms have usually been obtained by selection rather than by induced mutation. Both plant and animal breeding is usually a process of the improvement of lines by selection. Mutation breeding[2] is still relatively new to agronomists but it is such a familiar process in microbiology that it does not have a special name.

Although mutations of a variety of crop plants[2] and of *Drosophila*[3] can be obtained with alkylating agents, particularly the methanesulfonates, it is only recently that mutations of mammals[4] and of mammalian cells[5,6] have been obtained with alkylating mutagens. However, since it is now possible to induce and select for mutations in mammalian-cell systems, the same considerations of the interaction between repair and mutagenesis[7] will apply to studies with tissue-culture systems.

Industrial microbiologists should consider the response of mammalian-cell systems to mutagens because of the probability that cell cultures will soon be put to practical use. There is a continual demand for hormones such as insulin, for specific antibodies, for gamma globulin and for a variety of other proteins. There is every reason to believe that we will eventually be able to obtain large quantities of standardized biologicals from human-cell cultures. It is appropriate to talk about cell cultures here because the methodology used for their production is the same as for other microorganisms; pure culture technique is required and large quantities of cells can be grown in suspension culture.

The alkylating agents, particularly the monofunctional methylating and ethylating compounds such as EMS and N-methyl-N′-nitro-N-nitrosoguanidine are among the most effective and certainly most widely used mutagenic substances. We will discuss the chemical and biological interaction of alkylating agents with DNA. We will then show that the damage, but not necessarily the mutagenic effect of these agents is due to the production of breaks in the DNA of the treated cell. Finally, we will discuss the response of a mammalian cell culture to alkylation.

CHEMICAL STUDIES ON THE REACTION OF DNA WITH ALKYLATING AGENTS

The reaction of DNA with monofunctional alkylating agents results in the formation of alkylated purine bases within the DNA molecule. Such alkylated bases

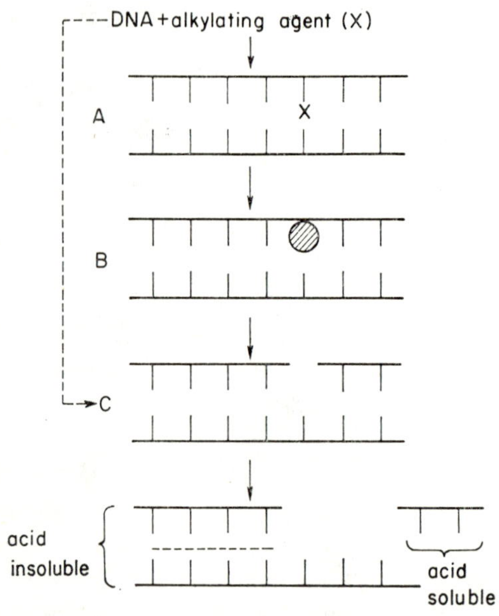

Fig. 1. Pathway of degradation of alkylated DNA; A alkylated DNA (stabilized by alkali), B apurinic DNA (hydrolyzed by alkali), C DNA with single strand breaks.

are eventually hydrolyzed leaving apurinic sites which are, in turn, subject to hydrolysis to give as a final degradation product, a DNA molecule with single strand breaks[8]. However, this reaction pathway need not be the only one and it has been supposed that the nature of the alkylating group may determine the way in which the DNA is degraded[9].

We decided to test the hypothesis that the degradation of DNA *after* alkylation follows the sequence: DNA with alkylated bases (A) → DNA with apurinic sites (B) → DNA with single strand breaks (C) (Fig. 1). The kinetics of such a process in a closed system are well understood[10] and the relative amounts of A, B and C as a function of the values k_1 and k_2, are easily calculated from the analysis of the pathways:

$$A \xrightarrow{k_1} B \xrightarrow{k_2} C$$

We used *B. subtilis* DNA labeled with $^{33}PO_4$ and treated with MMS, EMS, DES or DMS in our experiments. The following properties of DNA enabled us to estimate the various parameters: Alkylated DNA (A) is heat labile but is stabilized by alkali[11]; apurinic sites (B) are degraded to single strand breaks by heat or by incubation with alkali[12]; and the relative number of single strand breaks (C) can be determined by acid precipitation since more and more acid soluble oligodeoxynucleotides will be released on acid precipitation as single strand breaks accumulate in a DNA molecule. (C) is determined as the relative amount of acid-soluble radioactivity,

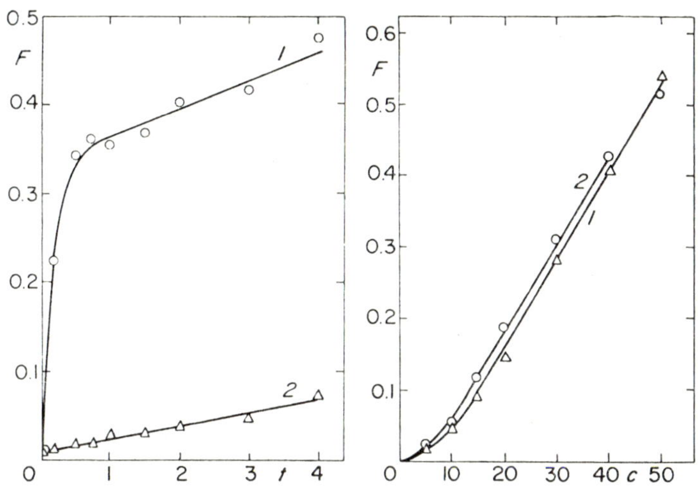

Fig. 2. *Left:* Degradation of alkylated DNA by heat. ^{33}P-DNA methylated in Tricine buffer with 0.03 M-DMS was heated in 0.1 M Tris buffer, pH 7.8; *t* time at 100 °C, h; *F* fraction acid-soluble; *1* alkylated DNA, *2* control.

Right: Acid solubility of heated alkylated DNA. DMS was added to ^{33}P-DNA in 0.3 M phosphate buffer. After 2 h at 37 °C portions of the reaction mixture were heated; *c* concentration of DMS, mM; *F* fraction acid-soluble; *1* 120 min at 100 °C, *2* 240 min at 100 °C.

(B) as the amount of DNA made acid soluble by incubation with alkali prior to acid precipitation while (A), the fraction of undegraded alkylated DNA is given as:

$$(A) = 1 - (B) + (C) .$$

The initial amount of alkylated DNA (A_0) was determined by heating a sample of alkylated DNA until complete degradation occurred (Fig. 2) and determining the amount of acid-soluble material finally released; the amount of such acid-soluble material is a function of the concentration of alkylating agent.

Our experiments were performed with DNA isolated from *B. subtilis* grown with $^{33}PO_4$ or with unlabeled DNA reacted with ^{14}C-methyl methanesulfonate. In all cases the DNA was prepared by passage through a column of hydroxyapatite[13]; only material eluting at a salt concentration characteristic of native DNA was used.

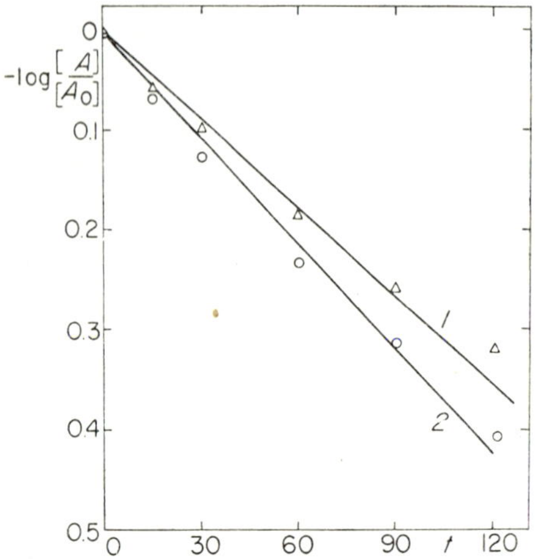

Fig. 3. Loss of ^{14}C-methyl groups from alkylated DNA heated at 70° C. Unlabeled *B. subtilis* DNA was incubated for 16 h at 37 °C with 50 μCi ^{14}C-methyl methanesulphonate (45.2 mCi/ /mmol; final concentration 0.55 mM). The reaction mixture was diluted and purified by passage through hydroxyapatite. The DNA fraction was then dialyzed against either 0.1 M Tris (*1*) or 0.1 M Tricine (*2*) buffers (pH 7.8) before heating. A_0, the concentration of alkylated bases in DNA, was measured as the total counts lost after heating 240 min at 100 °C; t time at 70 °C, min.

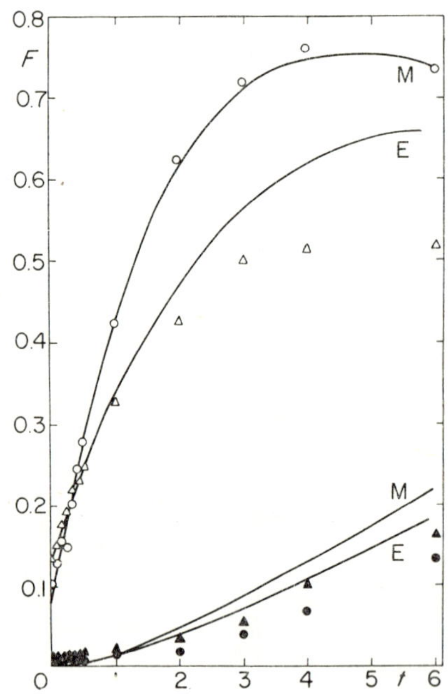

Fig. 4. Degradation of ethylated and methylated DNA. *B. subtilis* DNA was ethylated (*triangles*) or methylated (*circles*) as described in the legend to Fig. 5 and heated in 0.1 M Tricine buffer at pH 7.8; t time at 70 °C, h; F fraction of total acid-soluble. *Open symbols*, alkali-labile; *closed symbols*, acid-soluble. The lines were calculated using values for the depurination constant (k_1) and for the chain-breakage constant (k_2) as follows: M methylated DNA: $k_1 = 8.44 \times 10^{-3}\ \text{min}^{-1}$, $k_2 = 1 \times \times 10^{-3}\ \text{min}^{-1}$. The reaction was calculated as starting at an "offset time" of 10 min to correct for the production of acid-soluble and alkali-labile material before the start of the analysis; E ethylated DNA: $k_1 = 4.8 \times 10^{-3}\ \text{min}^{-1}$, $k_2 = 1 \times 10^{-3}\ \text{min}^{-1}$, offset time = 30 min.

Alkylated DNA was degraded as would be expected if the formation of apurinic acid were a necessary intermediate step. Methyl groups were lost at an exponential rate from DNA alkylated with labeled MMS at low enough concentrations so that acid-soluble oligonucleotides would not be detected (Fig. 3). Both methylated and ethylated ^{33}P-labeled DNA was degraded according to the expected kinetics (Fig. 4). The comparative rates of depurination, $k_{1\text{meth}}/k_{1\text{eth}}$, can be determined from the ratio of the plots of log A/A_0 as a function of time for ethylated and methylated DNA (Fig. 5) since $\ln A/A_0 = -k_1 t$. The experimental value for this ratio, 1.3, can

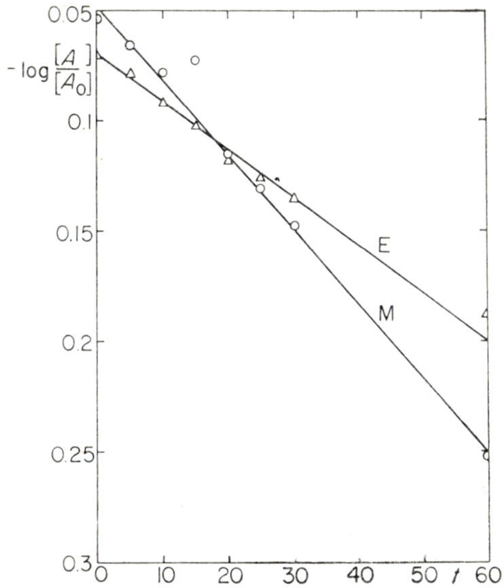

Fig. 5. Degradation of ethylated and methylated DNA. DNA was ethylated with diethylsulfate in phosphate overnight or methylated with dimethylsulfate in Tricine buffer. The reaction mixtures were then dialyzed against 0.1 M Tricine at pH 7.8. *A*, the fraction of undegraded alkylated bases in DNA, was calculated as indicated in the text; t time at 70 °C, min; M methylated DNA, E ethylated DNA.

then be used with the experimentally determined depurination constant for methylated DNA (Fig. 3), $8.4 \times 10^{-3}\ \text{min}^{-1}$ at 70°, to fit our data to the reaction scheme (Fig. 1). Qualitatively similar results were obtained with both EMS and MMS but due to the long reaction time required to alkylate DNA with these compounds, the DNA at the start of the kinetic studies with these compounds behaved as though it had already accumulated a number of apurinic sites. Kinetic studies of this sort must take into account the observation that the degradation of apurinic DNA is speeded up in any buffer containing a primary amino group, such as Tris or glycinamide.

If degradation occurs as we have suggested, the free ends of the polynucleotide chain should possess at least one free phosphomonoester group per two ends. On the other hand, if a significant fraction of acid-soluble material were formed as a result of hydrolysis of phosphotriesters, the resulting oligonucleotides would have their terminal phosphates esterified by alkyl groups. Such diesters would not be sensitive to alkaline phosphatase. If only a fraction of the alkyl groups were involved in phosphotriester formation, one might expect a different proportion of fragments with terminal phosphate monoesters throughout the course of the reaction. We found however, that the proportion of alkaline phosphatase-sensitive phosphorus increased initially and then remained constant when either methylated or ethylated DNA was heated at 100° (Fig. 6) indicating only one degradation pathway.

Our experiments corroborate, therefore, the hypothesis that the major degradative pathway after methylation or ethylation of DNA is the one suggested by LAWLEY and BROOKES[14]. Nonetheless, some degradation may occur by other pathways.

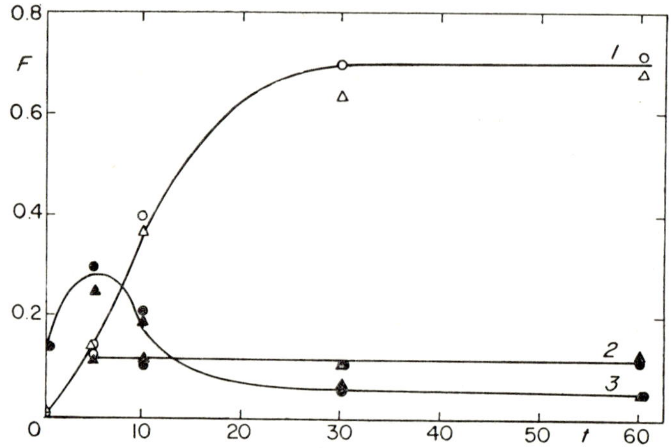

Fig. 6. Alkaline phosphatase lability of the degradation products produced by heating methylated DNA. DNA was methylated with dimethylsulfate or ethylated for 12 h with diethylsulfate (6.1 mmol in 10 ml 0.9 M phosphate buffer starting at pH 10.2) in phosphate buffer and then dialyzed into 0.1 M Tris buffer (pH 7.8). The alkylated DNA was heated in a boiling water bath as indicated. Portions of the reaction mixture were treated with alkaline phosphatase. P_i was determined as the radioactivity not adsorbed by charcoal; its value in solutions not treated with alkaline phosphatase was never more than 0.004 of the total acid-soluble counts. Acid-soluble and alkali-labile P were determined as usualy t time at 100 °C, min; *F* fraction of total. *Circles*, methylated DNA; *triangles*, ethylated DNA; *1* acid-soluble P, *2* P_i : acid-soluble P, *3* alkali-labile P.

Extensive heating of DNA alkylated with radioactive MMS leaves a fraction of label, varying from 6 to 15 per cent of the total, acid precipitable. This suggests that between 5 to 18.5 per cent of the total alkylation might be the result of phosphotriester formation[15]. In addition, our experiments were done by first alkylating the DNA and then carefully dialyzing to remove unreacted alkylating agent before heating. Any reaction which occurred in the initial period and which led to breaks in the DNA with or without fixation of the alkyl group would be missed. However, regardless of the pathway, it is clear that alkylation of DNA leads, inevitably, to breaks in the phosphodiester backbone. Such breaks can occur as the result of *in vitro* reactions as well as by the operation of alkylation specific nucleases[16].

WHAT DAMAGE IS REPAIRED?

Bacteria possess a system which enables them to repair at least a portion of the damage induced by monofunctional alkylating agents. The existence of bacterial mutants sensitive to MMS for example, is *prima facie* evidence for the existence of repair mechanisms in the wild type[17]. However, the studies discussed above indicate that alkylated DNA is necessarily a mixture of DNA with alkyl groups, apurinic sites and single strand breaks. Can we specify the lesion(s) which activate the repair mechanism?

It is very unlikely that *B. subtilis* recognizes methylation as damage *per se*. Cells were alkylated with radioactive MMS and the number of methyl groups bound to the DNA was estimated by determining the radioactivity stable on incubation in alkali but made acid soluble on treatment of the residue with hot acid. After treatment of a suspension of *B. subtilis* with 0.22 mM-MMS for 60 min, about 29,000 methyl groups were fixed in the DNA of each organism without any loss of viability as judged by the number of colony forming units remaining after alkylation[18]. Furthermore, we found that incubation of viable, alkylated cells did not result in excision of the added methyl groups. Cells were prelabeled by growth in ^{3}H-dT, reacted with ^{14}C-MMS and then incubated in growth medium. There

TABLE I

Stability of ^{14}C-Methyl Groups Bound to the DNA of *B. subtilis* Cells[a]

Strain	Survival $N:N_0$	Post treatment incubation min	Radioactivity bound to DNA c.p.m.		Ratio $^{14}C:{}^{3}H$[b]
			^{3}H	^{14}C	
168M	0.77	0	895	330	0.365
		80	841	315	0.371
		140	878	317	0.358
		170	819	257	0.309
168 *ind thy*	1.04	0	8780	1710	0.187
		40	7880	1550	0.188
			8430	1400	0.157
		80	7600	1280	0.160
			7620	1200	0.149
		120	6730	1150	0.163
			6210	1020	0.155

[a] Cells were prelabeled by growth with ^{3}H-thymidine. The absolute level of radioactivity represents only the amount of sample taken for analysis; 0.22 mM-^{14}C-MMS treatment for 90 min.

[b] $^{14}C:{}^{3}H$ ratios were corrected for carryover of counts.

was very little loss of ^{14}C counts from DNA for up to 170 min during which the alkylated cells divided about three times. Since the ratio of $^{14}C/^{3}H$ remained constant we conclude that there was no specific excision of alkylated bases (Table I).

The failure of viable cells to excise methyl groups indicates that DNA with methyl groups added by an alkylating agent is able to replicate. This was demonstrated by a density transfer experiment. Wild-type *B. subtilis* were grown overnight in medium labeled with D_2O, $^{15}NH_4Cl$ and deuterated amino acids and sugar[19]. The cells were alkylated with ^{14}C-MMS in density labeled medium and then transferred to "light" medium and incubated 60 min at 37°. There was a 60 per cent increase in the number of cells during the incubation period. The cells were then collected, lysed and the lysate was centrifuged to equilibrium in a CsCl gradient along with a marker of "light" *B. subtilis* ^{3}H-DNA. Most of the ^{14}C counts were found in a band with a peak density of 1.727 g/ml corresponding to hybrid DNA (Fig. 7). A fraction representing about 20 per cent of the total counts was found at the heavy density, 1.752 g/ml; this fraction most likely represents DNA from unreplicated cells. We assume that ^{14}C counts in the region of light density are due to alkylated protein or other material of light density.

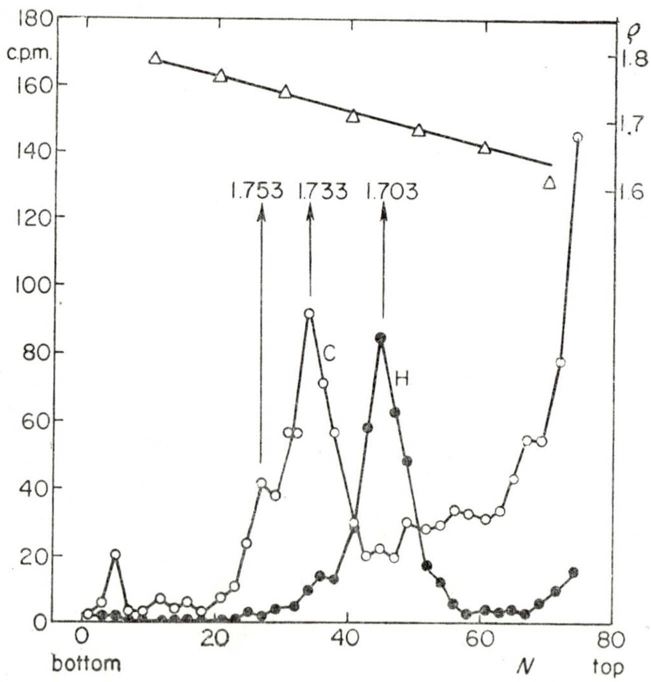

Fig. 7. Replication of methylated DNA. Cells grown in density labeled medium were treated with ^{14}C-MMS in "heavy" medium and then incubated 60 min in light medium. The cells were lysed and centrifuged to equilibrium in a CsCl gradient (ϱ). N tube number; C ^{14}C-c.p.m., H ^{3}H-c.p.m. × 0.5 (^{3}H-thymidine labeled *B. subtilis* marker DNA).

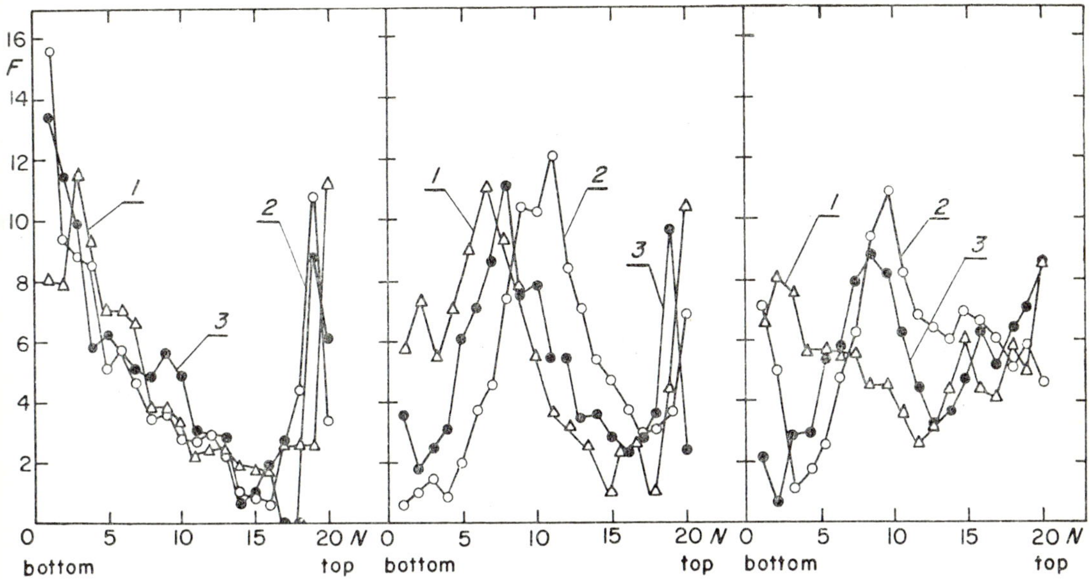

Fig. 8. Sedimentation in alkaline sucrose of DNA from *B. subtilis* 168 M (MMS-resistant) and 168M-5 (MMS-sensitive) prelabeled with ^{3}H-thymidine after treatment with MMS. Centrifugation was in the SW50L rotor of a Beckman L-2 centrifuge for 2 h at 35,000 r.p.m. at 20 °C through a 5 to 20% (w/w) linear sucrose gradient in 0.9 M-NaCl plus 0.1 M-NaOH. *Left:* strain 168M after 60 min of exposure to 0.22 mM ^{14}C-MMS (low dose); $N/N_0 = 1.03$. *Middle:* strain 168M after 20 min of exposure to 25 mM-MMS (high dose); $N/N_0 = 0.69$. *Right:* strain 168M-5 after 20 min of exposure to 25 mM-MMS (high dose). N tube number; F fraction of total counts (%); *1* control, *2* MMS-treated, *3* MMS-treated + 40 min of incubation in complete medium before lysis.

The experiments indicate that alkylated DNA is able to replicate and that the alkyl groups are stable, at least for a period of a *B. subtilis* generation. The repair mechanisms are, therefore, not activated by alkylated sites. DNA from *B. subtilis* treated with inactivating concentrations of MMS contains apurinic sites and single-strand breaks. Such lesions can be detected by sedimentation through alkaline sucrose gradients since alkali produces breaks at apurinic sites and exposes single-strand breaks by denaturing the DNA (Fig. 1). Incubation of such highly alkylated wild-type cells for 40 min after alkylation, but prior to lysis, resulted in an increase in the sedimentation rate of the DNA to a position intermediate between control and alkylated DNA (Fig. 8). It appears that at least one of our MMS-sensitive mutants is deficient in this "repair" reaction. We also observed that DNA synthesized 10 min after treatment with high doses of MMS sedimented more slowly than the DNA synthesized by control cells (Fig. 9). After further incubation in the absence of radioactive thymidine more rapidly sedimenting material was observed.

A number of mutants, sensitive to MMS, have been isolated. The first of these[17] is the only one which is *hcr* for MMS treated virus: all of the other strains are hcr^+.

This strain (*mms*, No. 59) has a defective DNA polymerase which responds to calf thymus DNA as a template but is grossly abnormal when tested with a poly dAT[20]. Three other MMS-sensitive strains have been tested; none of these is polymerase defective. We have not tested ligase activity. Our strains behave as though they were *rec*. Strain MC-1 (*rec-2*) of OKUBO and ROMIG[21] is also MMS-sensitive.

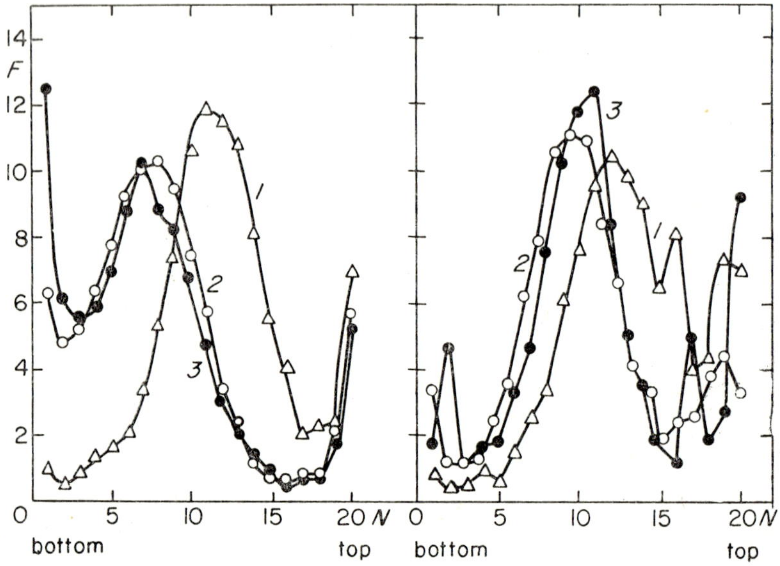

Fig. 9. Sedimentation in alkaline sucrose of DNA from *B. subtilis* 168M (MMS-resistant) and 168M-5 (MMS-sensitive) labeled for 10 min with ^{3}H-thymidine after MMS treatment. The cultures were washed and resuspended in nonlabeled medium and incubated for the times indicated before lysis. Centrifugation was as described in the legend to Fig. 8. *Left:* strain 168M after a 20 min of exposure to 25 mM-MMS (high dose); $N/N_0 = 0.298$. *Right:* strain 168M-5 after a 20 min of exposure to 25 mM-MMS (high dose); $N/N_0 = 0.011$. *N* tube number, *F* fraction of total counts (%); *1* 0 min chase, *2* 40 min chase, *3* 80 min chase.

WITKIN[7] has pointed out that *rec* and some *exr* mutants of *E. coli* are mutagen stable. In order to determine whether MMS-sensitive mutants are similarly mutagen stable, we constructed a series of *leu met his* MMS-sensitive strains by transforming a *leu met his trp* strain with high concentrations of DNA prepared from MMS-sensitive *trp*$^+$ donors.*

At least one of the MMS-sensitive strains is mutagen resistant at the *his* locus after EMS treatment or after UV (Table II). The polymerase deficient MMS-sensitive

* Transfer of characteristics at this high DNA concentration does not demonstrate that the MMS-sensitive strains differ from wild type by a single gene defect. Reversion and transformation studies indicate that one of our strains (M5, *see* Fig. 8) is complex since we obtain strains sensitive to UV-light but not to MMS among the revertants and the transformants.

strain shows normal mutability. Although our data do not permit any general conclusion it would appear that the joining steps may be involved in the mutagen-stability phenomenon as suggested by WITKIN[7].

TABLE II

UV-Induced *his* → *his*$^+$ Reversion

Strain	Survival	Revertants per 10^9 survivors	Ratio induced : spontaneous
101 (*leu his met*)	1.0	2.7	80
	0.36	216.5	
	1.0	8.8	24
	0.53	210.6	
101 (17)	1.0	155.6	2
	0.17	308	
101 (17) isolate A	1.0	1213	0.8
	0.5	965	
B	1.0	224.4	1.9
	0.29	417.5	
C	1.0	7.2	1.6
	0.25	11.5	
D	1.0	244	1.7
	0.34	405	
101 (17) transformant			
A	1.0	2.23	0.4
	0.55	0.81	
B	1.0	3.13	2.4
	0.14	7.43	
C	1.0	48.9	2.1
	0.17	101	
101 (MC-1)	1.0	5.4	2.6
	0.19	14	
101 (M5)	1.0	129	0.33
	0.35	42	
101 (59)	1.0	24.5	24
	0.15	581	

STUDIES ON REPAIR IN HEp.2 CELLS

Our data can be accounted for by supposing that single-strand breaks in DNA occur after *in vivo* alkylation both as a result of the spontaneous reactions leading to the decay of alkylated DNA[15] and, possibly, by action of the nucleases specific for alkylated DNA[16]. Repair of MMS-induced damage involves the repair of single-strand breaks which are produced as a result of the alkylation but does not involve the removal of a substantial number of methylated bases[18]. The methylated bases themselves do not interfere with DNA replication. It is therefore possible that the very low level of repair synthesis observed in bacterial cultures treated with MMS[22,23] is to be accounted for as a result of the lack of excision and the requirement that only a very few bases be replaced per lesion. However, extensive repair synthesis is observed after treatment of human-tissue culture lines with monofunctional alkylating agents, even at doses which result in relatively little cell inactivation[24]. This observation is particularly surprising because of the observation by Fox *et al.*[25] that there is no effect of split doses of MMS on the survival of treated cells. We have therefore investigated DNA synthesis after MMS treatment in a line of human-epithelial-carcinoma cells (HEp. 2) in order to study the relationship between DNA repair mechanisms and cell recovery. We have concluded that: 1) There is a slow repair reaction after treatment of cells with MMS; 2) Repair (non-conservative) synthesis occurs immediately after MMS treatment but not all repair synthesis results in the formation of biologically competent DNA; and 3) DNA

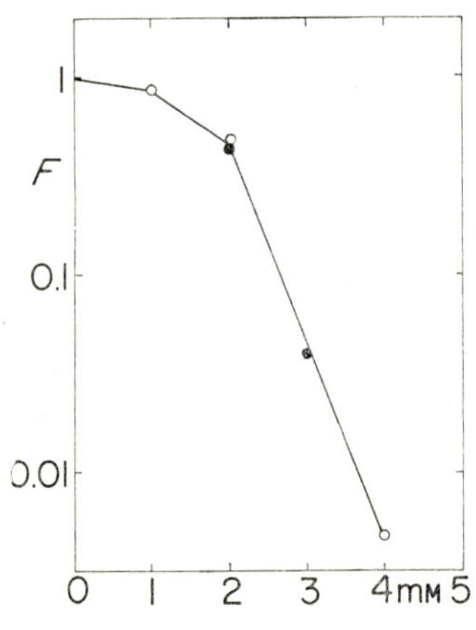

Fig. 10. Inactivation of HEp.2 cells by MMS. HEp.2 cells were separated by trypsin treatment, allowed to attach to plates and then treated with the indicated concentration of MMS (mM) for one hour. Clones were stained and counted after 14 days incubation. *F* fraction of clones surviving.

synthesized by MMS-treated cells immediately after treatment may be unable to participate in a second round of replication.

Our experimental protocols are based on the cell cycle of HEp. 2 cells growing as monolayers on plastic dishes. The generation time is about 24 h with an S phase of about 8 h. Inactivation studies were performed by cloning the cells and then treating them with MMS shortly after attachment of cells to the plate. The inactivation curves show a broad shoulder with a precipitous drop starting at a concentration of 2 mM-MMS for one hour at 37° (Fig. 10). We have observed that synchronous cultures collected in metaphase and treated 12 to 14 h after plating have a greater sensitivity.

BILLEN and BRUNS[26] have reported a "dampening" in the synthesis of DNA by bacterial cultures after UV- or X-ray irradiation. This work prompted us to

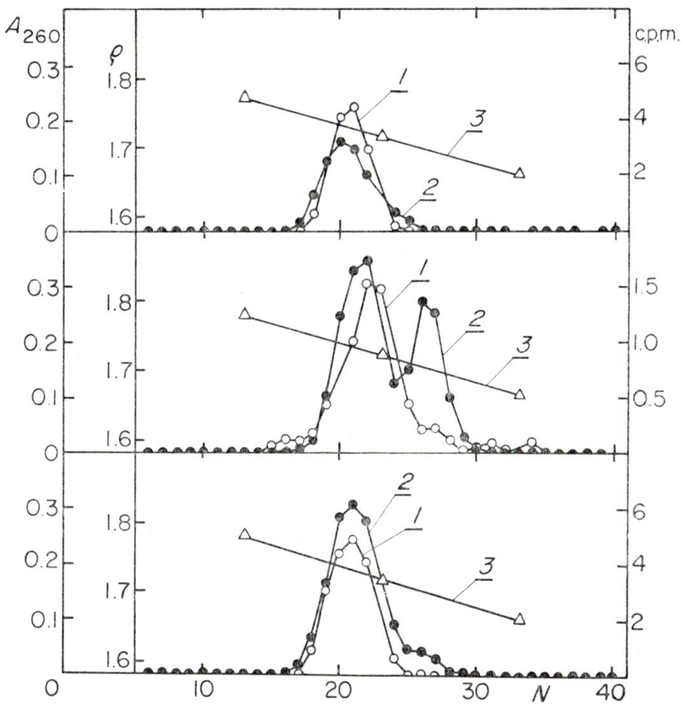

Fig. 11. Replication of the DNA synthesized by HEp.2 cells in the first or twelfth hour after treatment with 2 mM-MMS. Conditions as described in the text. ^{3}H-Thymidine 5 μCi/ml; 3 Ci/mmol. *Above:* control; *middle:* 2 mM-MMS, O time; *below:* 2 mM-MMS, 12 h; *N* fraction number; *1* A_{260}, *2* ^{3}H-dT, c.p.m. × 10^{-3}, *3* ϱ.

determine whether the DNA made immediately after MMS treatment could replicate. Cells were treated with 2 mM-MMS for one hour and then labeled with ^{3}H-thymidine for one hour either immediately or after twelve hours post treatment incubation. They were then allowed to incubate in unlabeled medium for 21 h and in Br-dU (5 μg/ml) for 48 h. Given a cell cycle time of 24 h, none of the cells should have been in the second S phase after MMS treatment at the time of tritium incorporation; thymidine must have been incorporated in the first cell cycle after treatment.

All of the ^{3}H incorporated by control cells and (with the exception of a small shoulder) all of the ^{3}H incorporated 12 h after alkylation was found in the hybrid region (Fig. 11). On the other hand, a large fraction of the thymidine incorporated immediately after MMS treatment did not replicate, as indicated by its failure to transfer to a hybrid density. When the experiment was repeated with 3 mM-MMS it was found that a large fraction of the label incorporated at 12 h after treatment was not subsequently transferred to a hybrid density.

The fact that all of the thymidine incorporated 12 h after treatment with 2 mM-MMS was subsequently transferred to hybrid means that the DNA synthesized at this time contained no lesions preventing further replication. Since both samples (one hour and twelve hours post treatment) involved synthesis during the first cell cycle, we conclude that either there is a repair process which occurs after the first hour following treatment or cells which make DNA in the first hour after treatment are unable to complete their repair. The DNA made in the first hour after treatment cannot be normal nor can it subsequently be repaired since if it were, the label would be transferred; *i.e.* the hypothesis that a process which started immediately (represented by the incorporated isotope) needed more time to be completed will not account for the data. Furthermore, the data cannot be accounted for by supposing that there is an inactivation process which occurs after the first hour and prevents cells which cannot make normal DNA from incorporating any thymidine, since cells treated with 3 mM-MMS did take up thymidine twelve hours after treatment even though a large portion of this thymidine was not subsequently transferred to a hybrid density.

The failure of alkylated cells to replicate newly formed DNA can be demonstrated with synchronized cells. Synchronized HEp. 2 cultures were initiated by shaking loose the dividing cells from log phase monolayer cultures and suspending the floating cells (mitotic index 0.95) into three series of Leighton tubes[27]. Within about 2 h the cells attached as either single cells or daughter cell pairs. Six hours after the cells had been plated the medium was decanted and replaced with fresh medium in the control, or with fresh medium containing MMS in the treated cultures. One hour later the medium was again decanted and replaced with fresh medium after all the cultures had been rinsed twice. At intervals after the treatment, cultures were pulsed for 1 hour with ^{3}H-dT and then repeatedly rinsed. The cells were lysed *in situ* and the cold acid precipitable counts in each culture were determined by liquid scintillation counting.

The onset of DNA synthesis was delayed in both control and treated cultures, presumably due to the several changes of medium. We found that cells treated with 2 mM-MMS completed a wave of DNA synthesis virtually identical to that of the controls (Fig. 12). There was a considerable reduction in synthesis in the treated cultures during the second S phase and there was no observed synthesis by the treated cells at the time of the third S phase of the control cultures. The 4 mM level of MMS was toxic since no wave of DNA synthesis was observed in the course of the experiment. We assume that the "dampening"[26] of DNA synthesis observed means that cells containing DNA with single-strand breaks can complete a round of replica-

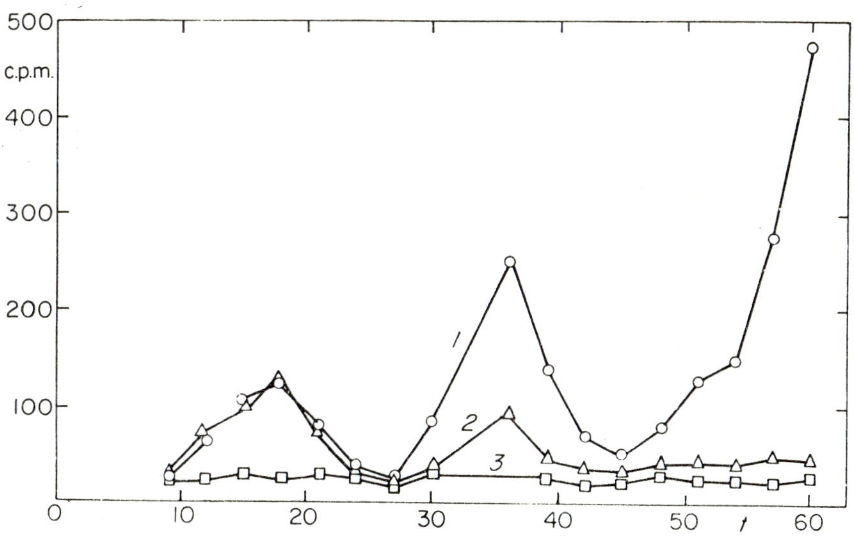

Fig. 12. Incorporation of thymidine by synchronized HEp.2 cells. Synchronized cells were prepared as described in the text. The points represent the times after plating at which ^{3}H-dT (5 μCi/ml; 3 Ci/mmol) was added; *t* time after division, h; *1* control cultures, *2* cultures treated with 2 mM-MMS 6 h after plating, *3* cultures treated with 4 mM-MMS.

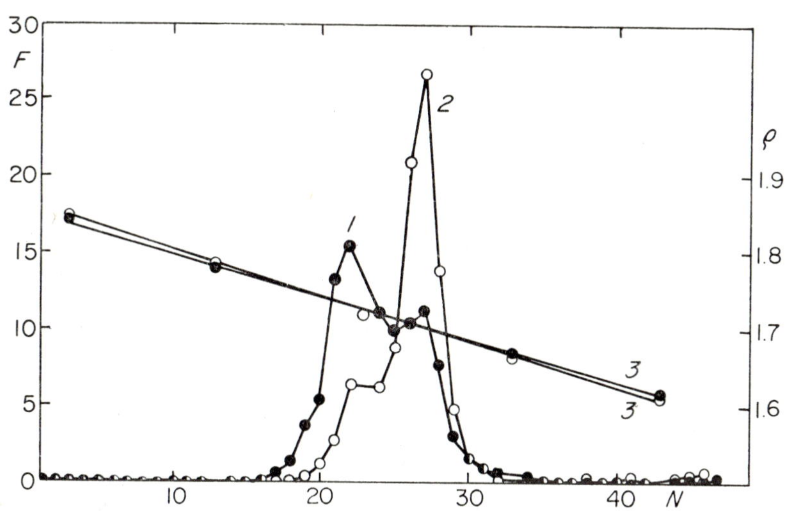

Fig. 13. Density distribution of DNA from whole cells and from the HIRT supernatant after incubation in Br-dU. HEp.2 cells were treated with 2 mM-MMS for one hour, labeled with ^{3}H-dT for an additional hour, incubated in unlabeled medium for 22 h and in Br-dU for 48 h. The DNA was purified and centrifuged to equilibrium in a CsCl gradient. *N* fraction number, *F* fraction of total counts (%); *1* DNA from whole cells, *2* DNA from HIRT supernatant, *3* ϱ.

TABLE III
Degradation of DNA Synthesized by Alkylated Cells[a]

Sample	c.p.m. from 10^6 cells in HIRT		Ratio S : P
	supernatant (S)	pellet (P)	
Control	814	9262	0.09
2 mM-MMS	3686	9624	0.38
3 mM-MMS	3472	6176	0.56

[a] 0–1 h: MMS; 1–2 h: ^{3}H-dT; 24 h incubation.

tion but most of the resulting daughter cells are unable to go through a subsequent S phase.

What is the fate of the newly synthesized DNA which does not undergo further replication? We think that most of it is to be found as broken fragments of molecular weight less than about 3×10^6 daltons. HIRT[28] demonstrated that DNA molecules could be separated according to their solubility in 1 M-NaCl in the presence of sodium dodecylsulfate and 0.01 M-EDTA. Fragments of DNA smaller than about 20S were soluble in this mixture whereas larger DNA molecules were insoluble and formed a pellet on centrifugation. This technique was used in our experiments to separate fragmented DNA.

When cells were treated with MMS and incubated with ^{3}H-thymidine for one hour and then in unlabeled medium for 24 h, much of the label was found in the

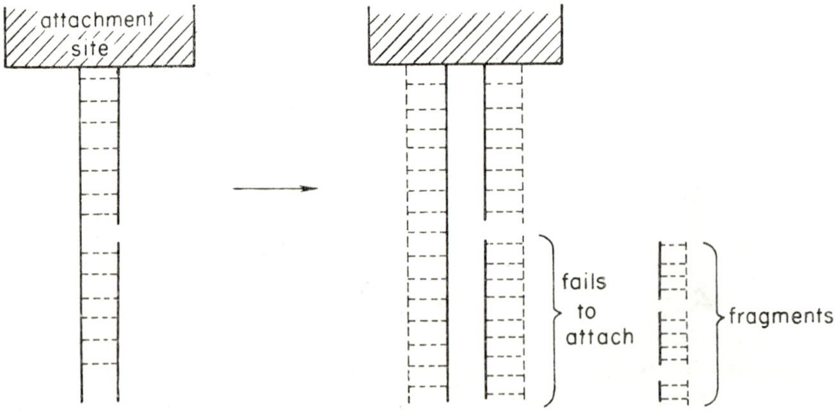

Fig. 14. An hypothesis to account for the formation of fragmented, newly synthesized DNA.

TABLE IV

Effect of Hydroxyurea on ^{3}H-dT Incorporation by MMS Treated Cells

MMS concentration	Time of ^{3}H-dT pulse after MMS treatment	5 mM Hydroxy-urea	Acid precipitable c.p.m.	Uninhibited incorporation %
0		0	—	—
		+	111	
2 mM	0–1	0	3227	9
		+	294	
3 mM		0	1230	28
		+	340	
0		0	4755	4
		+	169	
2 mM	2–3	0	2929	7
		+	204	
3 mM		0	1559	12
		+	190	
0		0	5846	4
		+	220	
2 mM	4–5	0	4442	6
		+	275	
3 mM		0	1915	11
		+	220	

supernatant in the HIRT procedure (Table III). Fragmentation after alkylation was detectable only after about 24 h incubation. Furthermore, when the HIRT supernatant was examined by CsCl equilibrium centrifugation in a density transfer experiment, it was found that a disproportionate amount of the label was in DNA that had been unable to replicate as compared to the DNA from whole cells (Fig. 13). In this experiment, cells were treated with 2 mM-MMS for one hour, labeled with ^{3}H-dT for an additional hour, incubated in unlabeled medium for 22 h and in Br-dU (5 μg/ml) for 48 h. The cells were lysed and after incubation overnight in 1 M-NaCl in the cold the supernatant was collected, the DNA was purified, dialyzed and centrifuged to equilibrium in a CsCl gradient. It can be seen that most of the label in the HIRT supernatant did not go through a second round of replication. Since the unreplicated label was found at a density characteristic of native rather than of denatured DNA, double-strand breaks must have occurred in the DNA found in the HIRT supernatant. Such broken DNA must include "old" as well as "new" strands. We suppose that a mechanism similar to that diagramed in Fig. 14 may account for the data. An attachment site may be required for DNA synthesis[29]. DNA made off a template with discontinuities (breaks) may be dissociated from the attachment sites, be unable to replicate and because of its "unprotected" detached state be subject to nuclease action.

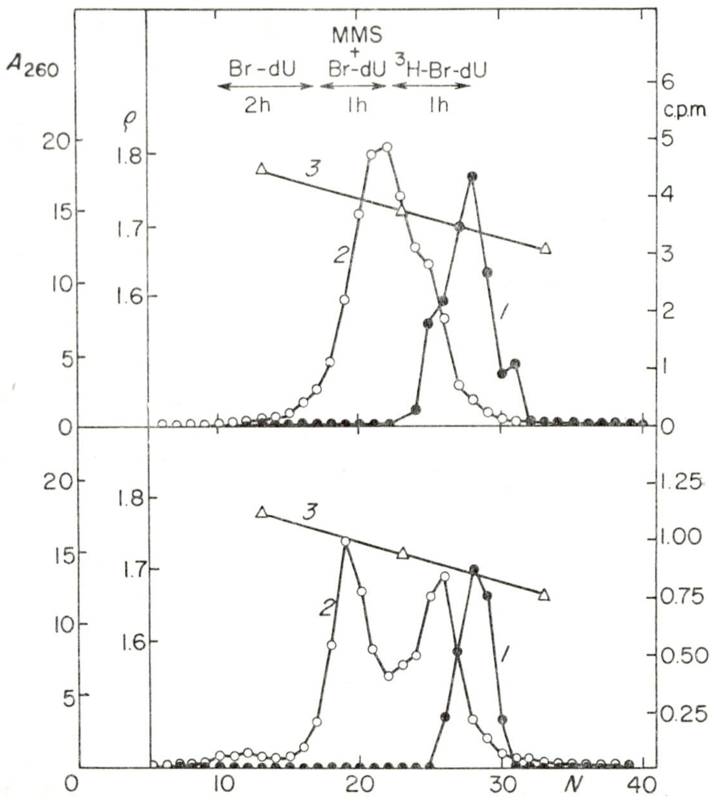

Fig. 15. Nonconservative synthesis by MMS-treated HEp.2 cells. HEp.2 cells were pretreated with Br-dU (5 μg/ml) for 2 h, treated with 2 mM-MMS in the presence of Br-dU and then incubated with ^{3}H-Br-dU (5.0 μg/ml; 0.95 Ci/mmol) for one hour. The cells were lysed and the purified DNA subject to a CsCl density gradient analyses. *Above:* control (no MMS); *below:* 2 mM-MMS; *N* fraction number; *1* A_{260}, %, *2* ^{3}H-Br-dU, c.p.m. × 10^{-3}, *3* ϱ.

Is the "new" DNA, which is unable to replicate and which is found as fragments, formed by semiconservative synthesis or is it incompletely repaired DNA with segments of nonconservative synthesis? The experiments described so far do not distinguish between the two possibilities. Two lines of evidence indicate that both semiconservative and nonconservative DNA synthesis occur in asynchronous cultures in the first hour after treatment with MMS. Hydroxyurea inhibits semiconservative but not repair synthesis in mammalian cells[30,31]. Since hydroxyurea inhibited thymidine incorporation by MMS-treated HEp. 2 cells (Table IV) some semiconservative synthesis occurred although the data also provide evidence for some nonconservative synthesis. Furthermore, a density transfer experiment in which cells were pretreated with Br-dU for 2 h to reduce the thymidine pool, treated with

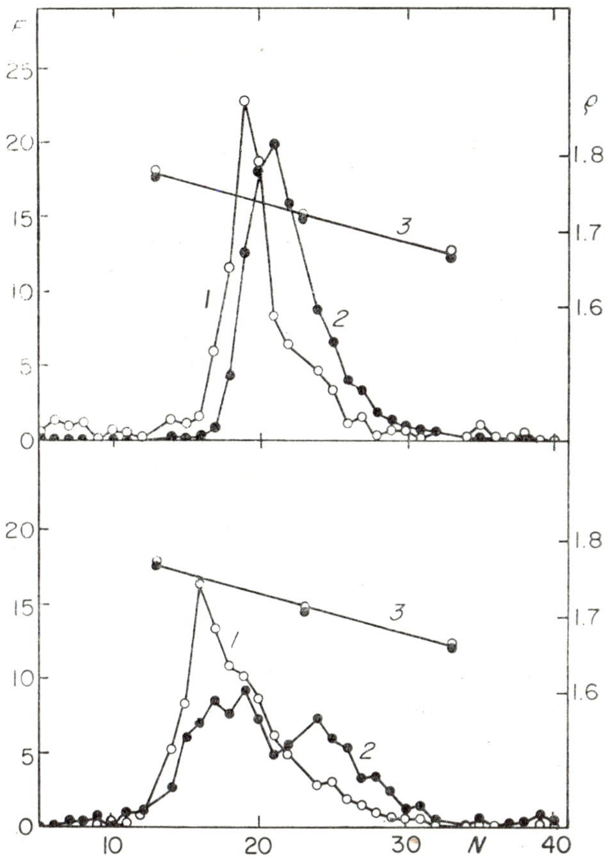

Fig. 16. Distribution of the density of labeled DNA fragments as compared to whole cell DNA. Cells labeled with ^{3}H-Br-dU were prepared as described in the text. DNA from the HIRT supernatant and from whole cells was purified and centrifuged to equilibrium in a CsCl gradient. The total radioactivity was normalized for the comparison. *Above:* control; *below:* 2 mM-MMS; *N* fraction number, *F* fraction of total counts (%); *1* ^{3}H-Br-dU from HIRT supernatant, *2* ^{3}H-Br-dU from whole DNA, *3* ϱ.

MMS in the presence of Br-dU and then incubated with ^{3}H-Br-dU for one hour, indicated both semiconservative and nonconservative synthesis (Fig. 15). One of the interesting features of the nonconservative synthesis observed was the finding of a difference in the density at which isotope is incorporated and at which the bulk of the DNA sediments. A shift in the density of "repaired" DNA was reported by

BRUNK and HANAWALT[32] in *Tetrahymena* and indicates the replacement (addition) of a much larger number of bases than observed in repair synthesis in prokaryotic organisms. However, other explanations are possible for our data since ^{3}H-Br-dU is incorporated only at the normal density of light HEp. 2 DNA when MMS treated cells are incubated with Br-dU in the presence of hydroxyurea.

Since both nonconservative and semiconservative synthesis of DNA occurs immediately after MMS treatment one can inquire whether there is any selection of DNA for degradation. What is the DNA which fails to replicate — is it the DNA made by semiconservative synthesis from a damaged template or is it, rather, the material incorporated in "repair synthesis" which would thereby be shown not to have "recovered" from the treatment?

HEp. 2 cells were transferred to Br-dU medium for two hours. They were then treated with MMS in the presence of Br-dU, labeled with ^{3}H-Br-dU for an hour following alkylation, incubated an additional three hours in nonradioactive Br-dU and then held 21 h in unlabeled medium. The cells were collected and used for the preparation of "whole DNA" and for a HIRT supernatant DNA both of which were analyzed by CsCl equilibrium sedimentation (Fig. 16). We found that the majority of the labeled DNA in the HIRT supernatant was selected from the DNA which had undergone semiconservative replication.

Our evidence favors the idea that there is a delayed repair process. However, the data do not indicate the mechanism by which such a process operates but only that it does occur. The results show that a portion of the newly synthesized DNA formed by semiconservative replication does not go through an additional round of synthesis. DNA synthesis by damaged cells is "dampened"[26] in eukaryotes as well as in prokaryotes. This dampening may occur because the presence of breaks in an unrepaired template results in the production of chromatid breaks (Fig. 14). Failure of the DNA to attach to an appropriate site should make it impossible to begin another round of replication.

This work was supported in part, by grants from the National Institutes of Health (GM 07816), the National Science Foundation (GB 8514) and the Atomic Energy Commission (AT(11-1)2040). L. P. was a trainee of a Microbiology Training Program supported by the National Institutes of Health (GM 00603). The expert technical assistance of Miss M. L. Schulz is gratefully acknowledged.

REFERENCES

1. WITKIN E.: Cold Spring Harbor Symp. Quant. Biol. **21**, 123 (1956).
2. FAO/IAEA Technical Meeting, Rome: *The Use of Induced Mutations in Plant Breeding.* Symp. Publ. Div., Pergamon Press, Oxford 1965.
3. LEE W., SEGA G., BISHOP J.: Mutation Res. **9**, 323 (1970).
4. EHLING V., CUMMING R., MALLING H.: Mutation Res. **5**, 417 (1968).
5. KAO F., PUCK T.: Proc. Nat. Acad. Sci. U.S.A. **60**, 1275 (1968).

6. Chu E., Malling H.: Proc. Nat. Acad. Sci. U.S.A. **61**, 1306 (1968).
7. Witkin E.: Ann. Rev. Microbiol. **23**, 487 (1969).
8. Lawley P.: Progr. Nucleic Acid Res. **5**, 89 (1966).
9. Alexander P.: Ann. N. Y. Acad. Sci. **163**, 652 (1969).
10. Bray H., White K., p. 166 in *Kinetics and Thermodynamics in Biochemistry*, Academic Press, New York 1957.
11. Kohn K., Spears C.: Biochim. Biophys. Acta **145**, 734 (1967).
12. Tamm C., Shapiro H., Lipshitz R., Chargaff E.: J. Biol. Chem. **203**, 673 (1953).
13. Bernardi G.: Nature **206**, 779 (1965).
14. Lawley P., Brookes P.: Biochem. J. **89**, 127 (1963).
15. Strauss B., Hill T.: Biochim. Biophys. Acta **213**, 14 (1970).
16. Strauss B., Robbins M.: Biochim. Biophys. Acta **161**, 68 (1968).
17. Searashi T., Strauss B.: Biochem. Biophys. Res. Comm. **20**, 680 (1965).
18. Prakash L., Strauss B.: J. Bacteriol. **102**, 760 (1970).
19. Bodmer W., Schildkraut C.: Analyt. Biochem. **8**, 229 (1964).
20. Gass K., Hill T., Goulian M., Strauss B., Cozzarelli N.: J. Bacteriol. **108**, 364 (1971).
21. Okubo S., Romig W.: J. Mol. Biol. **15**, 440 (1966).
22. Reiter H., Strauss B.: J. Mol. Biol. **14**, 179 (1965).
23. Achey P., Whitfield V.: Biophys. Soc. Abstr. TPM-03 (1970).
24. Ayad S., Fox M., Fox B.: Mutation Res. **8**, 639 (1969).
25. Fox M., Gilbert C., Lajtha L., Nias A.: Chem. Biol. Interactions **1**, 241 (1969/70).
26. Billen D., Bruns L.: Biophys. J. **10**, 509 (1970).
27. Terasima T., Tolmach L.: Exp. Cell. Res. **30**, 344 (1963).
28. Hirt B.: J. Mol. Biol. **26**, 365 (1967).
29. Comings D., Kakefuda T.: J. Mol. Biol. **33**, 225 (1968).
30. Cleaver J.: Radiation Res. **37**, 334 (1969).
31. Coyle, *unpublished results.*
32. Brunk C., Hanawalt P.: Radiation Res. **38**, 285 (1969).

Repair Replication: Its Role in Genetic Stability

PHILIP C. HANAWALT

Department of Biological Sciences, Stanford University, Stanford, California, U.S.A.

INTRODUCTION

The properties of living organisms are determined largely by their genetic blueprints as entrusted to the DNA molecules of their chromosomes. These giant molecules must be replicated with extraordinary fidelity in order for an organism to survive and to make successful copies of itself. It has been estimated, for example, that a bacterial gene may be duplicated over 100 million times before there is even a 50 percent chance that it will have been altered. One is impressed by the remarkable stability of the DNA molecule, as essential if like is to produce like over many generations. Yet, we have long been aware that DNA is subjected to deleterious events (*e.g.*, radiation) in the environment that can alter or destroy its genetic message. It was therefore not too surprising to discover recently that living systems contain an error-correcting mechanism — to repair defects in their DNA and thus to ensure genetic stability. The evolution of this so-called "excision-repair" mechanism may, in fact, account for the ubiquity of double-stranded DNA in living systems since the mechanism requires the inherent redundancy of information in DNA.

Fortunately, however, this error-correcting mechanism is not perfect and it has allowed some mutagenic events to occur, as quite essential for the process of biological evolution. Without mutagenesis there could be no evolution since mutations are the changes that allow variation among the individuals of a population. Natural selection then operates upon this diversity to enhance proliferation of the more "suitably" adapted individuals. If the repair mechanism were too efficient the natural mutation frequency might be reduced to such a low level that a species could become trapped in an evolutionary dead end. On the other hand repair would obviously be advantageous in enabling a species to maintain its genetic stability in an environment that caused mutations at a high rate. Of course the efficiency of the DNA repair mechanism itself must be subject to selection in the process of evolution. The *rate* of evolution may be quite directly coupled to the efficiency of repair.

There are several important general problems concerning the relationship between repair and mutagenesis. Usually emphasized is the reduction of potentially mutagenic damage in DNA through the action of repair mechanisms. Yet, it is also possible that the repair system itself may make mistakes and thus promote mutagenesis. In fact it is conceivable that an imperfect repair system might sometimes "attack" a normal undamaged segment of DNA and produce a mutation through the faulty repair of some imaginary lesion. In order to assess the validity of such possibilities we must examine the molecular nature of the repair process. Several somewhat independent repair mechanisms may operate in different cellular systems. In the present discussion, however, we will restrict our consideration to the excision-repair scheme, particularly as this mechanism has been elucidated in bacterial systems.

THE EXCISION-REPAIR MODEL

The enzymatic process of excision-repair of damaged DNA was postulated by SETLOW and CARRIER[1] and by BOYCE and HOWARD-FLANDERS[2] to explain their observation of the preferential release of UV-induced thymine dimers from bacterial DNA *in vivo*. It was proposed that defective regions in one strand of double-stranded

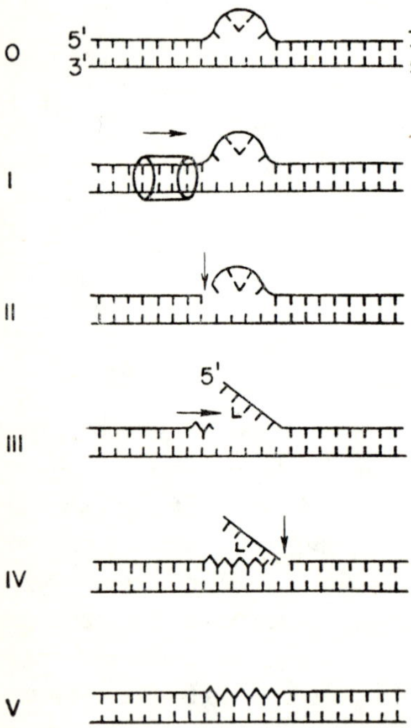

Fig. 1. Schematic representation of the postulated sequence of enzymatic steps in the excision repair of damaged DNA.

O structure distortion (*e.g.* thymine dimer),
I recognition (structure distortion),
II incision (*e.g.* UV-specific endonuclease),
III repair replication (DNA polymerase),
IV excision (5′-exonuclease),
V rejoining (polynucleotide ligase).

DNA could be excised and then replaced with undamaged nucleotides by repair replication, utilizing the complementary base pairing relationships with the intact strand as template. Direct physical evidence for the repair replication step was obtained by PETTIJOHN and HANAWALT[3] in a study of the qualitative nature of DNA replication in UV-irradiated bacteria.

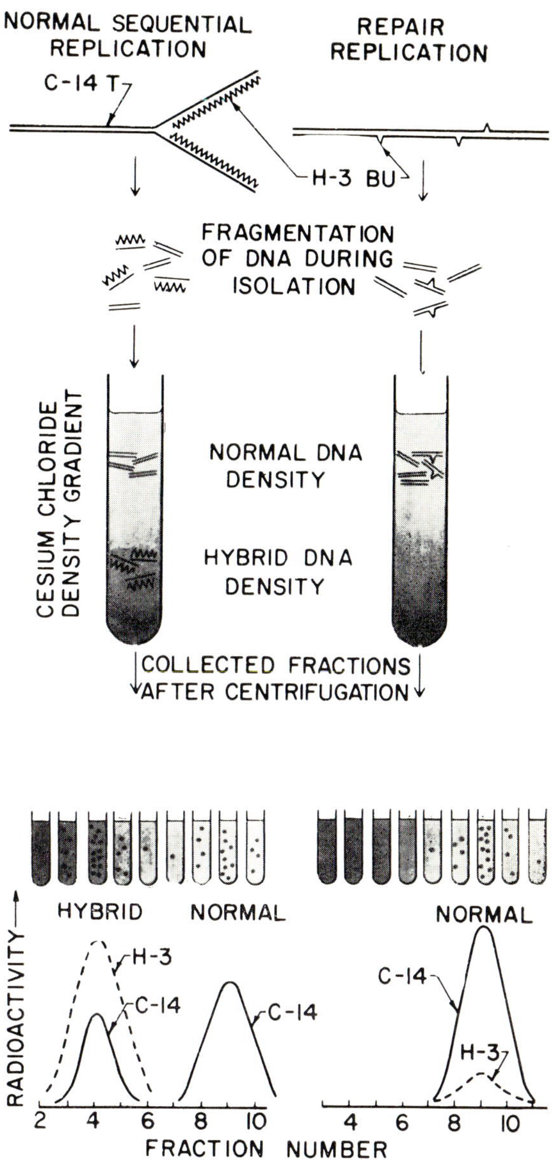

Fig. 2. The protocol for the demonstration of normal replication and/or repair replication of DNA in growing cells. The DNA is first radioactively labeled (*e.g.,* by the incorporation of ^{14}C-thymine as a specific precursor for DNA). After several generations of prelabeling, the cells are harvested by filtration and transferred to a new growth medium that contains both a radioactive label (*e.g.,* ^{3}H) and also a density label (*e.g.,* Br-U to replace thymine). The large DNA molecules are fragmented by shearing during the process of isolation from the cells. Some of these fragments contain only the parental ^{14}C-label. Others that have replicated in the presence of the density label contain both ^{14}C and ^{3}H. The density label has the effect of increasing the density of these fragments to the position in a CsCl gradient called "hybrid". (Such fragments contain roughly as much thymine as Br-U.) After another generation of growth some fragments with Br-U in both strands would be expected and these would appear at a still denser position called "heavy". Parental DNA fragments that contain *short* regions of repair should differ little in density from those that contain no Br-U. Thus, repair would be indicated by the appearance of ^{3}H-label at the normal density position in the gradient. The presence of ^{14}C and ^{3}H in the same sample from the centrifuge tube can be assayed simultaneously in the liquid scintillation counter, which distinguishes the two on the basis of the different energies of their emitted radiation.

A currently favored model for the excision-repair process in the bacterium *Escherichia coli* is illustrated in Fig. 1. The first step must be the recognition of the damage in the DNA. An alteration such as a pyrimidine dimer induced by UV-light distorts the phosphodiester backbone of the DNA since the interchain hydrogen bonds are broken in that region. Even the simple mispairing of bases (*i.e.*, a transition mutation) may cause an adequate distortion of the structure to activate the repair system. The same enzyme that recognizes the damage is probably also responsible for the second step which involves an incision in the damaged strand. Endonuclease activity specific for pyrimidine dimers has been isolated from *Micrococcus lysodeikticus*[4,5] as well as from bacteriophage T4 infected *Escherichia coli*[6] although not yet from uninfected *E. coli*. The recent *in vitro* characterization of the DNA polymerase from *E. coli*[7] strongly implicates this enzyme in both steps, three and four. The enzyme possesses a 5′→3′ exonuclease activity that specifically facilitates the release of thymine dimers (or mismatched base sequences) from DNA as oligonucleotides. Concurrently it polymerizes nucleotides from the 3′-OH end of the initial incision. The logical candidate for the final rejoining step is the polynucleotide ligase[8–10] which restores the intactness of the repaired strand by forming the last 3′—5′ phosphodiester bond. Thus, the excision-repair sequence may involve as few as three enzymes: (1) A damage-specific endonuclease (2) The DNA polymerase with associated 5′-exonuclease activity and (3) The polynucleotide ligase.

STUDIES ON REPAIR REPLICATION IN *E. coli*

The combined labeling of replicating DNA with a density label and a radioactive label is utilized to measure repair replication *in vivo*. The experimental procedure is essentially the same as that used by MESELSON and STAHL[11] to prove the normal semiconservative mode of DNA replication: our adaptation of this method is illustrated in Fig. 2. The use of a density label such as Br-U (5-bromouracil, an analogue of thymine) and a radioactive label permits the determination of the incorporation of this label into short regions of the parental DNA as an indication of repair replication. The experimental protocol has recently been described in detail elsewhere[12]. There are several problems in the *quantitative* estimation of repair by this method, however. It is not possible to ascertain as yet how much of the observed nonconservative replication is *bona fide* repair and how much represents an aberrant form of DNA synthesis occurring perhaps in cells not destined to survive. Also, the dilution of the repair label with endogenous pools of thymine derivatives (some from the excision process itself) presents a problem in the determination of repair. The latter problem has been resolved through the development of a chromatographic analysis to assess the contribution from these endogenous pools[13].

The nonconservative mode of repair replication has been observed at the restrictive temperature in an *E. coli* mutant that is temperature sensitive for normal

DNA replication[14]. This finding suggests that different enzyme systems are involved in the normal replication mode and in the repair mode. It does not imply that two different DNA polymerases are involved, however. Repair replication has also been examined in a number of UV-sensitive mutants of *E. coli*. Repair replication does not occur in an excision-deficient *E. coli* K12 strain AB2500 following UV-irradiation[15]. This mutant is unable to perform the first incision in DNA containing

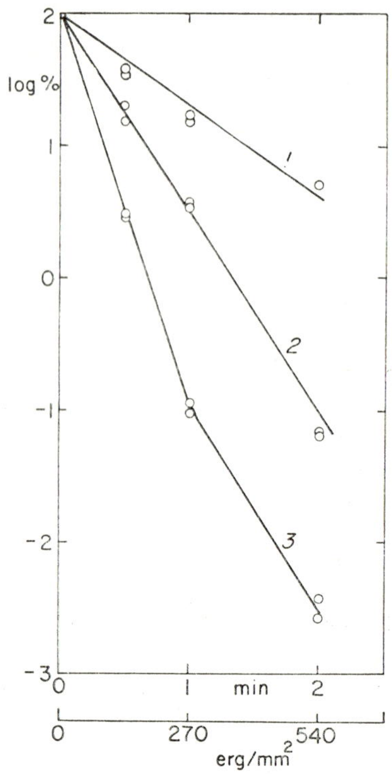

Fig. 3. Survival curves (% survival) for *E. coli* strains following UV-irradiation of exponentially growing cultures with a Pen-Ray Quartz Lamp at 4.5 erg mm^{-2}. . s^{-1}. Colony-forming ability was determined on glucose-salts agar with appropriate supplements (from Ref.[17]). ***1*** strain W3110 *pol*$^+$, ***2*** strain W3110 *polA1*, ***3*** strain AB2500 *uvrA6*.

pyrimidine dimers so this result is as expected. This finding also implies that any aberrant DNA synthesis mode, normally included in the repair assay, must require the first incision step in the repair sequence. Less expected have been the results of repair studies on a bacterial mutant defective in DNA polymerase. The mutant *E. coli* strain W3110 *polA1* was isolated by DE LUCIA and CAIRNS[16] who showed that its extracts contained less than 1% of the DNA polymerase activity of the parent strain W3110 from which it was derived. Furthermore the mutant was found to be UV-sensitive although not as sensitive as the excision-deficient K12 strain AB 2500 — this implies that the mutant has some repair capability[17]. Fig. 3 shows the relative UV-sensitivities of these strains.

The analysis of repair replication in this mutant and its parent gave the surprising result shown in Figs. 4 and 5. Both the parent and the mutant appear to have similar repair capabilities as judged by our operational definition of repair.

Repair can also be assayed by looking at the final rejoining step in the sequence. A comparison of the parent and mutant in this regard at different times after irradiation (Fig. 6) shows that the mutant is less efficient in completing the repair process, as indicated by the lower single-strand molecular weight of its DNA, even 90′ after the irradiation. The DNA from the parent strain has essentially regained the single-strand molecular-weight distribution of the unirradiated control culture within this period.

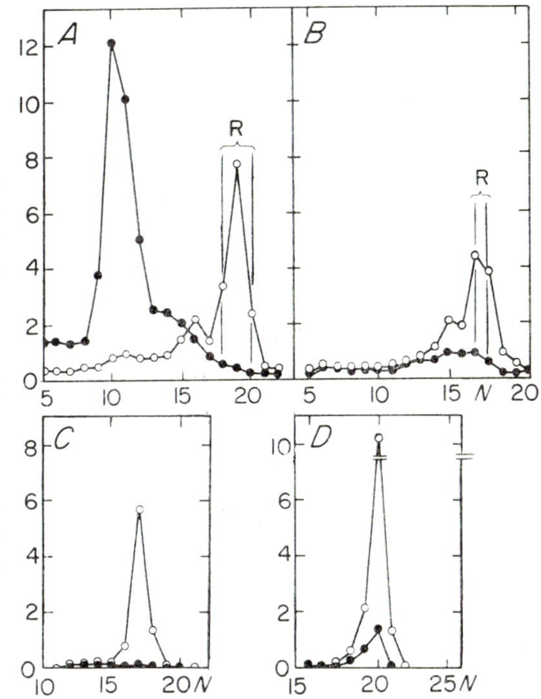

Fig. 4. Repair replication in *E. coli* strain W3110 *pol*$^{+}$. An exponentially growing culture was labeled uniformly with ^{32}P and then subjected to a 1 min dose of UV-light. The cells were transferred to growth medium containing ^{3}H-Br-U for 20 min at 37 °C after irradiation. The density distribution of the labeled DNA was determined by equilibrium sedimentation in CsCl density gradients as described in detail by HANAWALT and COOPER[12]. **A** unirradiated control culture (first banding in CsCl: density increases right to left), **B** irradiated culture (first banding in CsCl), **C** rebanding of fractions shown in *A*, **D** rebanding of fractions shown in *B*; *open circles:* ^{32}P prelabel, c.p.m. $\times$ 10^{-2}; *closed circles:* ^{3}H-Br-U label, c.p.m. $\times$ $\times$ 10^{-2}; N fraction number, R reband.

Loss of assayable DNA polymerase activity and UV-sensitivity have been shown to result from the same mutational step[18]. Based on the data in Fig. 6 the UV-sensitivity seems attributable to a deficiency in dark repair, so that loss of assayable polymerase activity and of dark repair capability can be said to result from the same mutational step. The loss of polymerase activity in extracts from the mutant can mean that either the number of active polymerase molecules present in the cell is very low, or else that the molecules are altered in structure in a way which makes them labile upon extraction, through functioning *in vivo*. The first possibility poses two problems: (a) Since the mutant is unaffected in its ability to synthesize DNA under normal growth conditions this amounts to acceptance of the hypothesis that DNA polymerase is not the enzyme used in normal synthesis or else that very little of it is needed for this function; (b) DNA polymerase has now been shown to excise dimers and repolymerize bases on a single-strand template in a manner entirely consistent with the requirements for the repair enzyme[7]. Assuming, as seems likely, that it is the repair enzyme, then reduction of the enzyme level should correspondingly reduce repair replication in the mutant. Yet this was not observed (Fig. 5). In recent experiments using the chromatographic analysis of repair, the quantitative amount of repair replication in the mutant has been shown to be more than twice as much as that in the parent after an equivalent UV-dose[34]. It is also known that the amount of DNA degradation in the mutant after irradiation is more than that in the parent[19].

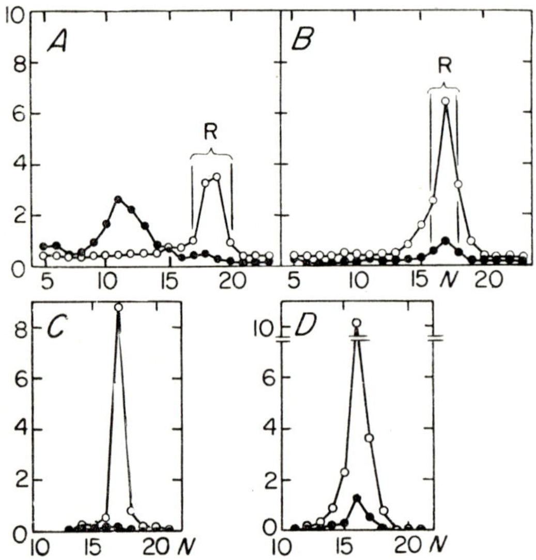

Fig. 5. Repair replication in *E. coli* strain W3110 *polA1*. Procedure was identical to that used for the parent strain (*see* Fig. 4). ***A*** unirradiated control culture (first banding in CsCl: density increases right to left), **B** irradiated culture (first banding in CsCl), **C** rebanding of fractions shown in ***A***, **D** rebanding of fractions shown in ***B***; *open circles:* ^{32}P prelabel, c.p.m. $\times 10^{-2}$; *closed circles:* ^{3}H-Br-U label, c.p.m. $\times 10^{-2}$; *N* fraction number, R reband.

If one assumes that DNA polymerase is actually present and active *in vivo*, but altered in structure, one must further assume that some step in dark repair occurring *after* repolymerization is blocked by this alteration. For example, this might be the joining together of the repaired DNA fragments. It has been shown that UV-sensitivity results when the joining process is defective in ligase-deficient mutants of *E. coli*[20] or bacteriophage T4[21]. A currently-favored model for the excision-repair scheme involves the displacement of the repair polymerase from the DNA by the ligase after repolymerization. In strain W3110 *polA1*, the altered DNA polymerase may bind too tightly to the DNA to be displaced and this may result in abnormally long stretches of replaced nucleotides. Alternatively, it may hinder recognition of single-strand breaks or prevent access to these sites by the ligase. Current studies in our laboratory are designed to test the notion that the extent of repair is responsive to the level of ligase activity in the cell. These studies should also elucidate the role of the DNA polymerase in the repair process.

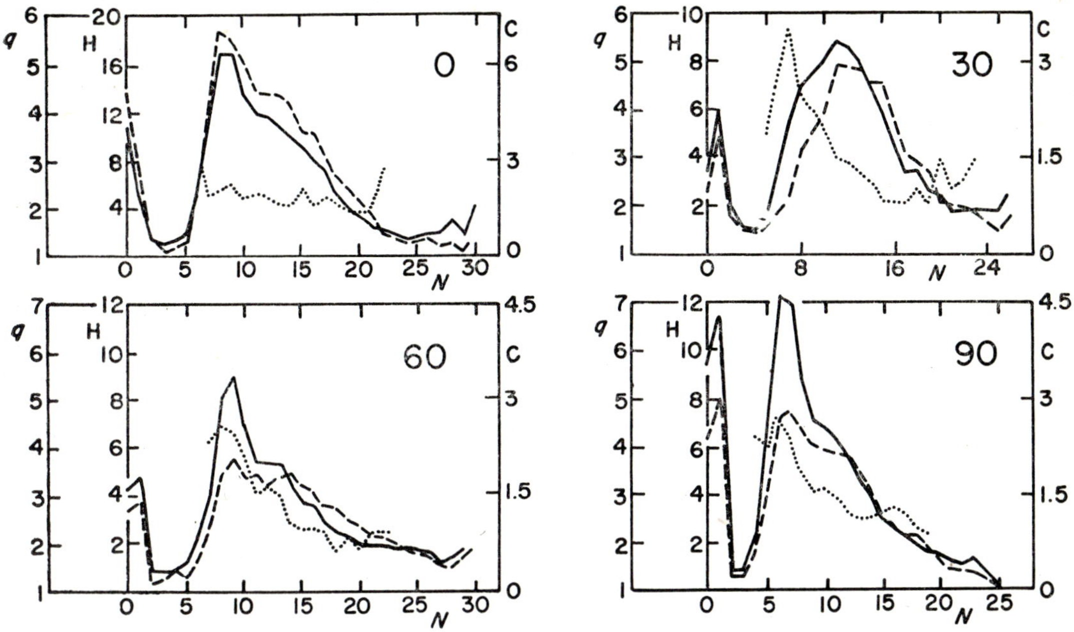

Fig. 6. Alkaline sucrose gradients showing molecular weight distribution of DNA after UV-irradiation and post-irradiation growth. The W3110 *pol*$^+$ strain was labeled with ^{3}H-thymine and the mutant W3110 *polA1* was labeled with ^{14}C-thymine. Each cell suspension received an UV-dose of 270 erg/mm^2. At different times equal portions of the cultures were mixed and the cells were lysed as described elsewhere prior to zone sedimentation at 20 °C and 35,000 r.p.m. for 1 h in linear 5—20% sucrose gradients at pH 12 in the Beckman SW39 rotor. Fractions are numbered from bottom to top of the gradients and fractions "0" is the residue left in the centrifuge tube after collecting the fractions. Graphs consist of line segments joining the data points obtained for each fraction (from Ref.[17]). **0** 0 min after UV, **30** 30 min after UV, **60** 60 min after UV, **90** 90 min after UV; H ^{3}H-thymine incorporated into trichloroacetic acid-precipitable material during prelabeling period by W3110 *pol*$^+$, c.p.m. $\times$ 10^{-2} *(solid lines)*; C ^{14}C-thymine incorporated into trichloroacetic acid-precipitable material during prelabeling period by W3110 *polA1*, c.p.m. $\times$ 10^{-2} *(dashed lines)*; *q* ratio ^{3}H-c.p.m./^{14}C-c.p.m. *(dotted lines)*.

REPAIR REPLICATION IN OTHER SYSTEMS

Although repair replication was first demostrated in bacterial systems and most of our understanding derives from the study of these systems, it has also been shown in a number of other organisms. Of particular interest was the finding that the simplest known living cells, the mycoplasmas, perform repair replication[22]. Since the genome in these cells can only code for about 700 different proteins the demonstration of repair attests to the importance of this process in maintaining genetic stability in even the simplest living cells. In a still simpler system, the T4 bacteriophage, it has been shown that the *v*-gene is responsible for the excision of UV-induced pyrimidine dimers from the phage DNA. The experimental analysis of repair replication in T4 infected bacteria, however, is complicated by the extensive recombination that occurs during the infective cycle[35].

Cellular systems more complex than *E. coli* have also been shown to repair their damaged DNA. The protozoan *Tetrahymena pyriformis* performs repair replication after UV-irradiation[23]. However, an important difference has been noted in comparison with the simpler systems previously studied. A substantially greater amount of Br-U was incorporated into some parental DNA strands and only a small fraction of the parental strands were found to contain appreciable amounts of the Br-U label. The observed density shift in CsCl gradients of those parental strands that contained Br-U led to an estimate of about 5000 nucleotides replaced per repaired lesion in the DNA[24]. It is hard to believe that segments of 5000 nucleotides must be reconstructed to replace a pyrimidine dimer. An alternative model to consider would assume that most of these long stretches of Br-U in these strands came from sister-chromatid exchanges with normally replicated homologous DNA. As with the T4 phage system, these recombinational events would be difficult to distinguish from repair replication. It is possible that the amount of Br-U label incorporated by the recombinational scheme may mask the small amount of DNA turnover due to excision-repair and thus obscure the quantitative estimates of repair. Of course, an observed recombinational mode of repair may be equally important in the ultimate survival of the organism.

The most exciting application of the Br-U labeling method for measuring repair was the demonstration that normal human-skin fibroblasts were capable of repair replication after UV-treatment[25]. Yet, fibroblasts from patients with the rare hereditary skin disease xeroderma pigmentosum exhibited much reduced levels of repair replication. It was concluded that the failure of DNA repair must somehow be related to the fatal skin cancers that these patients generally develop upon exposure to sunlight. This is the first experimental evidence for the possible significance of the excision-repair process in the protection from or recovery from a carcinogenic transformation in human cells. Repair replication has also been shown in other mammalian-cell types in culture[26].

It is evident that many different types of living cells utilize the excision-repair mechanism as at least one means for maintaining genetic stability. As suggested in the introduction the need for such a mechanism may have been significant in the evolution of the *double-stranded* DNA molecule as the principal repository of the precious hereditary information.

Less well defined than the ubiquity of the excision-repair process is the range of defects that can be dealt with effectively.

REPAIRABLE DAMAGE AND MUTAGENESIS

Although repair replication was first demonstrated following UV-irradiation of bacterial cells[3] it has since been shown that a variety of agents which produce structural damage in DNA can lead to the repair replication mode *in vivo*. Thus, the bifunctional alkylating agent, nitrogen mustard, which primarily attacks the 7-nitrogen position of guanine results in repair replication[27]. The most powerful mutagen, nitrosoguanidine, has also been shown to stimulate repair replication[28]. The attempt to detect repair replication after X-ray treatment of bacteria has led to ambiguous results and the difficulty is probably that the extensive degradation of DNA following X-ray damage obscures the repair replication mode by contributing parental label to the precursor pools. Evidence has been presented that mitomycin C[29] and methyl methanesulfonate[30] lead to the excision-repair mode in bacteria. It is now apparent that the excision-repair mechanism can deal with a variety of types of damage that cause cell lethality.

One must distinguish, however, between damage to DNA that leads directly to lethality (*e.g.*, by blocking replication) and mutagenic damage which may or may not be lethal. Certainly UV-light produces damage to DNA that may result in mutagenesis and, in fact, most of the mutagenic effect is attributable to pyrimidine dimers. For low UV-doses the mutation frequency in bacteria may be increased 10^3 to 10^6-fold over the spontaneous mutation level and for high doses nearly all of the survivors may be mutants. Yet lethal mutations really play a very minor role in the UV-inactivation of cells. The role of excision-repair in mutagenesis has been examined by comparing mutant yields in bacterial strains that are deficient in repair. Excision-deficient bacterial mutants are found to be extremely sensitive to UV-mutagenesis[31]. Therefore, more mutations must be due to unrepaired pyrimidine dimers than to the faulty repair of dimers. Excision-deficient mutants do *not* show increased sensitivity to mutagenesis by nitrosoguanidine, suggesting that the mutagenic damage produce by this agent may not be recognizable by the functional excision-repair mechanism[28,31]. WITKIN has recently shown that a tryptophan-requiring derivative of the *polA1* (DNA polymerase deficient) strain does not exhibit any greater frequency of UV-induced mutations to tryptophan independence than does the parental strain[32]. However, COUKELL and YANOFSKY[33] have demonstrated a marked increase in the frequency of spontaneous deletion mutations in the DNA polymerase-deficient strain. Much remains to be done before we will fully understand the role of repair mechanisms in the protection of living systems from mutagenic damage.

It is a pleasure to acknowledge the collaborative efforts of Priscilla Cooper, John Couch, and Lee Kanner in our recent studies on repair replication. John Cairns generously provided the DNA polymerase defective mutant and its parent strain. The research was supported by a contract AT(04-3)326-7 with U.S. Atomic Energy Commission and a grant GM 09901 from the U.S. Public Health Service.

REFERENCES

1. Setlow R. B., Carrier W. L.: Proc. Nat. Acad. Sci. U.S.A. **51**, 226 (1964).
2. Boyce R. P., Howard-Flanders P.: Proc. Nat. Acad. Sci U.S.A. **51**, 293 (1964).
3. Pettijohn D., Hanawalt P.: J. Mol. Biol. **9**, 395 (1964).
4. Takagi Y., Sekiguchi M., Okubo S., Nakayama H., Shimada K., Yasuda S., Nishimoto T., Yoshihara H.: Cold Spring Harbor Symp. Quant. Biol. **33**, 219 (1968).
5. Kaplan J. C., Kushner S. R., Grossman L.: Proc. Nat. Acad. Sci. U.S.A. **63**, 144 (1969).
6. Sekiguchi M., Yasuda S., Okubo S., Nakayama H., Shimada K., Takagi Y.: J. Mol. Biol. **47**, 231 (1970).
7. Kelly R., Atkinson M., Huberman J., Kornberg A.: Nature **224**, 495 (1969).
8. Olivera B. M., Lehman I. R.: Proc. Nat. Acad. Sci. U.S.A. **57**, 1426 (1967).
9. Zimmerman S. B., Little J. W., Oshinsky C. K., Gellert M.: Proc. Nat. Acad. Sci. U.S.A. **57**, 1841 (1967).
10. Gefter M. L., Becker, Hurwitz J.: Proc. Nat. Acad. Sci. U.S.A. **58**, 240 (1967).
11. Meselson M., Stahl F. W.: Proc. Nat. Acad. Sci. U.S.A. **44**, 671 (1968).
12. Hanawalt P., Cooper P.: Methods Enzymol. **21**, D221 (1971).
13. Couch J. L., Hanawalt P. C.: Anal. Biochem. **4**, 51 (1971).
14. Couch J. L., Hanawalt P. C.: Biochem. Biophys. Res. Comm. **29**, 779 (1967).
15. Cooper P., Hanawalt P.: Photochem. Photobiol. **13**, 83 (1971).
16. de Lucia P., Cairns J.: Nature **224**, 1164 (1969).
17. Kanner L., Hanawalt P.: Biochem. Biophys. Res. Comm. **39**, 149 (1970).
18. Gross J., Gross M.: Nature **224**, 1166 (1969).
19. Boyle J. M., Paterson M. C., Setlow R. B.: Nature **226**, 708 (1970).
20. Pauling C., Hamm L.: Proc. Nat. Acad. Sci. U.S.A. **60**, 1495 (1968).
21. Baldy M.: Cold Spring Harbor Symp. Quant. Biol. **33**, 333 (1968).
22. Smith D. W., Hanawalt P. C.: J. Mol. Biol. **46**, 57 (1969).
23. Brunk C., Hanawalt P.: Science **158**, 663 (1967).
24. Brunk C., Hanawalt P.: Radiation Res. **38**, 285 (1969).
25. Cleaver J. E.: Nature **218**, 652 (1968).
26. Painter R. B., Cleaver J. E.: Radiation Res. **38**, 451 (1969).
27. Hanawalt P., Haynes R.: Biochem. Biophys. Res. Comm. **19**, 462 (1965).
28. Cerdá-Olmedo E., Hanawalt P.: Mutation Res. **4**, 369 (1967).
29. Howard-Flanders P.: Ann. Rev. Biochem. **37**, 175 (1968).
30. Strauss B.: Current Topics Microbiol. Immunol. **44**, 1 (1968).
31. Witkin E. M.: Proc. 12th Internat. Congr. Genet. **3**, 225 (1969).
32. Witkin E. M.: Nature, New. Biol. **229**, 81 (1971).
33. Coukell M. B., Yanofsky C.: Nature **228**, 633 (1970).
34. Cooper P.: *Ph. D. Thesis.* Stanford University 1970.
35. Soll J.: *Ph. D. Thesis.* Stanford University 1969.

4 Regulation of Enzyme Synthesis and Modification of Enzyme Activities

Coupling of Energy Conversion and Biosynthetic Processes in Growing Bacteria

HOWARD GEST

Department of Microbiology, Indiana University, Bloomington, Indiana, U.S.A.

For some[1], the "Golden Age" of molecular biology is over; for the rest of us, there remains the prospect of exciting new discoveries, especially in the area of cellular control mechanisms. I think it is safe to say that we still do not have a penetrating knowledge of the means through which the biosynthesis of *any* one of the major classes of biological macromolecules is regulated in growing cells. As one significant example, RNA synthesis may be cited — at present, a number of quite different notions are being pursued and the subject is controversial. Intensive research, especially during the past decade, has revealed numerous elements of biochemical control patterns, but the overall schemes are still quite elusive in many respects. Indeed, the universality of "key" regulatory mechanisms that have been glimpsed is not firmly established, and it is suspected that many variations — and possibly, profound differences — will be found among cells of different types (*e.g.*, prokaryotes *vs.* eukaryotes).

It is likely that much research in the next decade will be directed toward discovering the intricate and sophisticated regulatory "interlocks" among major biosynthetic pathways and other processes. The knowledge already gained suggests the necessity for control networks of great complexity, which will challenge the ingenuity of biochemists and microbiologists. The first breakthrough at this level of understanding of cellular metabolism may well come from further development of current studies on the mechanisms responsible for integration of energy conversion and biosynthesis in microorganisms.

The general outlines of the "mechanics" of ATP regeneration processes are reasonably evident, even though the detailed molecular reactions concerned in phosphorylation associated with electron transport still defy elucidation. With the debatable exception of photophosphorylation, evidence has been gradually accumulating for the existence of "feedback" controls of one kind or another on the velocity of energy conversion sequences. Control of photophosphorylation might at first glance appear to be a special problem in that the energy to be converted comes from outside of the cell, and is not associated with the katabolism of substrates. For reasons to be discussed later, however, it seems probable that fine regulation of photophosphorylation is achieved by mechanisms basically similar, or analogous, to those operating in other kinds of cells. We generally assume that evolutionary pressures resulted in cell types which are very efficient in the sense that they utilize and manage material resources with superb economy. If this is so, we may reasonably also assume that energy conversion processes must normally be under tight control. This is, in fact, indicated in a gross way by the relationship observed, under certain conditions, between yield of microbial cell material and quantity of energy source available.

THE Y_{ATP} VALUE

Studies with anaerobic (or facultatively anaerobic) heterotrophic organisms have shown that the yield of cell material is directly proportional to the quantity of ATP calculated to be generated during katabolism of the organic energy source, when the latter is growth limiting[2]. Moreover, the yield is remarkably constant among the various species that have been examined (*e.g.*, in such diverse organisms as *Saccharomyces cerevisiae, Streptococcus faecalis, Pseudomonas lindneri, Desulfovibrio desulfuricans*, and *Clostridium tetanomorphum*). Expressed as a "growth yield coefficient", the constant (Y_{ATP}) is $\sim$ 10.5 grams dry weight of cells per mole of ATP generated[2,3]. Significant deviations from a value of 10.5 are usually taken as an indication that the *calculated* ATP production is in error and that previously unrecognized ATP-yielding steps probably occur in katabolism of the energy source. More recently, Y_{ATP} studies have been made with aerobic bacteria — this will probably prove to be a valuable avenue of research, but at the moment a number of difficulties are encountered in interpretation of the experimental results, mainly because of inadequate knowledge of actual ATP yields (*see* discussion in Ref.[3]).

If we restrict our considerations, for the moment, to anaerobic organisms, two significant (and related) questions immediately arise:

(1) For a given organism, can Y_{ATP} change significantly under different conditions of growth? In other words, is the "coupling efficiency" between energy conversion and biosynthesis affected by environmental circumstances? The answer seems to be yes[4], but systematic studies are still lacking. The *apparent* invariance of the Y_{ATP} value of $\sim$10.5 for different organisms is a related matter, and it may be that deviations from this average figure reflect differences in coupling efficiency[5]. In this connection, we can also anticipate that mutants may be obtained which differ markedly from wild type in respect to efficiency of ATP utilization (*e.g.*, certain mutations might lead to appreciable "uncoupling").

(2) The other question of import that comes to mind is — can we assume that the *rate* of cell material production is also closely geared to the *rate* of

energy conversion? Another way this question has been asked for heterotrophic organisms is: does the overall biosynthetic rate control the rate of katabolism of the energy source? Older studies on this significant point, which will not be reviewed here, are somewhat contradictory. In this connection, it is worth noting that even though non-growing cells may katabolize an energy source at a high rate, this does not negate the possibility that there is a close interdependence of energy conversion and biosynthetic rates in *growing* cells. In fact, the results of a recent study[6] with *Aerobacter aerogenes* provide quite convincing evidence for this kind of mutual dependence.

For anaerobes, I assume that an intimate temporal coupling between energy conversion and biosynthesis is implied by the constancy of Y_{ATP} under "normal" growth conditions, but it may be that this is influenced by growth rate.

Before proceeding, it is relevant to note the implications of these questions to industrial microbiology. Obviously, the maximum yield of cells obtainable in relation to available quantity of the energy source, and the tightness of coupling between mass increase and energy conversion under different conditions have important practical connotations. A better understanding of the theoretical questions under consideration may well add new dimensions to industrial efforts, such as single-cell protein production.

THE CATALYTIC NATURE OF ATP

BALDWIN[7] has described the catalytic nature of ATP as follows:

"... the amounts of ATP present in most tissues are not very large, and are out of all proportion to the energy turnover of these tissues. It follows, therefore, that ATP must be used and reformed over and over again if the many and diverse activities of the tissues are to be maintained. But, as we know today, new supplies of ~Ⓟ arise in the course of many katabolic transformations and, indeed, it would appear that katabolism as a whole is directed above all to the generation of new energy-rich from energy-poor phosphate radicals. These can be transferred to ADP left after previous breakdown of ATP molecules. It is significant, however, that newly generated ~Ⓟ cannot be utilized directly for the performance of endergonic operations: they can be put to service only through the intermediation of the ADP/ATP system. No major alternative system has so far been brought to light."

As far as is known, ATP is present in essentially catalytic quantities in all types of bacterial cells, autotrophic and heterotrophic. Such cells, moreover, do not store phosphagens such as creatine phosphate. By direct analysis, bacterial cells are ordinarily found to contain of the order of 0.1 mmol of ATP per 10 g dry weight (*e.g.*, *see* Ref.[8]) — in other words, orders of magnitude less than is required for synthesis of a comparable amount of new cell material (*cf.* Y_{ATP}). Another expression of this situation is LEHNINGER's calculation[9] that an *Escherichia coli* cell "contains only 1 million molecules of ATP, which are enough for only a half second of biosynthetic work". Thus, the "instantaneous" pool size of ATP does not give a meaningful indication of the high ATP *flux* during active growth.

INTERMESHING OF ENERGY CONVERSION AND BIOSYNTHESIS; THE "ENERGY CHARGE" CONCEPT

Since ATP is "catalytic" in the sense noted and is also required continuously in considerable total quantity for production of new cell material, it follows that utilization and regeneration of this important nucleotide must be very delicately balanced. A first approximation working hypothesis for explaining how energy conversion and biosynthesis are integrated can be developed on the basis of the assumption that their overall rates are interdependent and are primarily controlled by a common signal system. The adenylate nucleotides, which link biosynthesis and energy conversion, evidently are the most plausible components of a signal system that might serve such a purpose. A quantitative formulation of the control effects exerted by the nucleotides is given by the "Energy Charge" (E.C.) concept advanced by ATKINSON and his colleagues[10]. In their notation, the energy level of a system is represented as a function of the adenylate nucleotide concentration profile by the relation:

$$\text{Energy Charge} = 1/2 \times \frac{2\,\text{ATP} + \text{ADP}}{\text{ATP} + \text{ADP} + \text{AMP}} = \frac{\text{ATP} + 0.5\,\text{ADP}}{\text{ATP} + \text{ADP} + \text{AMP}}$$

Arbitrary use of the factor 1/2 gives a parameter which conveniently varies from 0 (all nucleotide present as AMP) to 1.0 (all present as ATP). In essence, E.C. is a measure of the number of anhydride-bound phosphates per adenosine moiety. The energy level of a system, expressed as E.C., can be likened to the potency of an electrochemical storage battery that can be charged up and discharged. Since the E.C. value depends only on the concentrations of AMP, ADP, and ATP, it is simple to set up any desired E.C. *in vitro* and to observe the responses of enzymes to change of E.C., to obtain indications of their possible responses to *in vivo* energy level alterations. The E.C. idea has become important because ATKINSON *et al.*[11] have shown that the activities of a number of enzymes are affected by the charge as illustrated in Fig. 1.

Two kinds of response have been observed. The **R** type is seen with enzymes involved in ATP regeneration in heterotrophic systems, *e.g.*, phosphofructokinase and isocitrate dehydrogenase (which feeds electrons into the oxidative phosphorylating respiratory chain). **U** refers to biosynthetic enzymes that utilize ATP, such as aspartokinase and ribosephosphate pyrophosphokinase. Responses of this nature are observed *in vitro* when the total adenylate concentration is kept constant and E.C. varied by changing the ratios of the nucleotides. The E.C. region between about 0.7 and 0.9 is of special interest — here, enzyme reaction velocity is a sensitive function of the charge. The mirror-image relationship for ATP-**R**egenerating and **U**tilizing enzymes clearly would be expected to contribute to metabolic stability, and to regulation of the rate of ATP regeneration in response to the cell's momentary needs for ATP (*see* Ref.[11], ATKINSON).

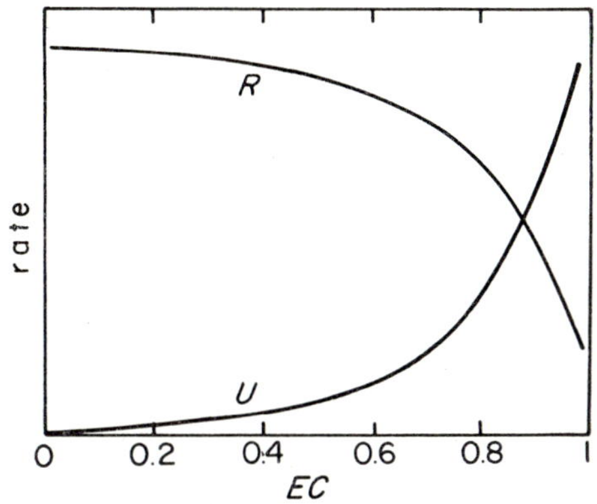

Fig. 1. Generalized representation of enzyme activity responses to the **E**nergy **C**harge EC = = (ATP + 0.5 ADP)/(ATP + ADP + AMP). *R* enzymes involved in ATP regeneration; *U* biosynthetic enzymes that utilize ATP. After ATKINSON[11].

Although it was recognized quite a long time ago that concentration ratios of certain nucleotides (*e.g.*, ATP/ADP) were probably important in determining the velocity of particular reactions and processes, an adequate rationale for understanding the very numerous effects (both stimulatory and inhibitory) of adenylate nucleotides on biosynthetic and other enzymes did not emerge until the late 1960's. The E.C. concept provides a rationale and at the same time gives us a sensible view of how energy-yielding and energy-consuming sequences may be regulated and integrated with each other. In this connection it may be noted that within the past few years the control of the citric-acid cycle, as it is related to the energy state of the cell, has been satisfactorily explained[12] using new knowledge of the effects of adenylate nucleotides on enzymes of the cycle and the E.C. analysis. This general conception also suggests new ways of looking at several unsolved basic problems relating to the dynamics and economy of cell growth (*see* below).

DIRECT DETERMINATION OF THE *in vivo* E.C.

If E.C. functions as a control parameter in the manner suggested, cells in *balanced* exponential growth should have a quite constant adenylate nucleotide profile (at least in the sense that the nucleotide composition affects the numerical value of the charge). Indeed, the capacity of a cell to "buffer" the E.C. may well be the primary requirement for maintaining a state of balanced growth. It appears that the "normal" E.C. in actively growing cells is usually in the neighborhood of 0.8 (Ref.[13]). Clearly, a number of environmental factors might be expected to influence the average E.C., and it is pertinent to ask whether or not determination of the *in vivo* charge under various conditions — by direct analysis — is a worthwhile approach towards testing the predictive value of the hypothesis.

It would seem that the obvious answer is yes, but there are both "theoretical" and practical reasons for giving a qualified answer at this time. For example, it is known from *in vitro* experiments that quite small changes in E.C. may have rather large effects on the activity of certain biosynthetic enzymes (Ref.[11], KLUNGSØYR *et al.*). This state of affairs places a great burden on the reliability of methods for rapidly "freezing" the nucleotide profile preparatory to analysis. Changes occurring during extraction of the nucleotides would, of course, tend to obscure the purpose of the analysis; such experimental artifacts may, in fact, be partly responsible for contradictory observations in the literature on relationships between ATP pool size and growth rate. Significant differences in compartmentalization of nucleotide pools, in respect to growth kinetics, and in biosynthetic patterns of different organisms may also contribute to difficulties in evaluating the meaning of observed changes in E.C. Nevertheless, it is encouraging that an increasing number of investigators are undertaking closer quantitative examination of nucleotide profiles of various microorganisms in connection with studies on regulatory metabolism. These efforts will doubtless pay good dividends.

"DISTURBANCE" OR "RESETTING" OF *in vivo* E.C.; PHOTOSYNTHETIC BACTERIA AS MODEL SYSTEMS

Another major approach for testing the working hypothesis is to examine the consequences of disturbing the E.C. in growing cells. We have been exploring this using photosynthetic bacteria, which are experimentally advantageous in that the activity of their (photochemical) energy conversion system can be manipulated with great ease.

The nonsulfur purple photosynthetic bacteria (*Athiorhodaceae*) are so-called "photoheterotrophs" and, as such, are capable of abundant anaerobic growth on organic carbon sources when suitably illuminated[14]. Under usual circumstances[15], no growth occurs in the absence of light (*i.e.*, anaerobically). Fig. 2 shows the relationship observed between growth rate and incident light intensity for *Rhodopseudomonas capsulata* in a synthetic medium with DL-malate and $(NH_4)_2SO_4$ as the carbon and nitrogen sources, respectively.

The function of light is to provide energy for the phosphorylation of ADP, according to the overall reaction:

$$\mathrm{ADP} + \mathrm{P_i} \xrightarrow[\mathrm{Mg^{2+}}]{\text{light}} \mathrm{ATP} \quad \text{(photophosphorylation)}$$

This kind of energy conversion is *independent* of O_2 and substrates, but resembles oxidative phosphorylation in that the ATP formation is coupled to a series of electron transfers (in this case, light-induced electron transfer). The detailed mechanism is not relevant to this discussion and, in fact, for present purposes we may consider *R. capsulata* and similar organisms to be the equivalent of an *E. coli* specially endowed with the ability to regenerate ATP by a "self-contained" process not associated with metabolism of organic compounds.

Within limits, the growth rate of *R. capsulata* in a given medium can be altered by simply changing the light intensity (*see* Fig. 2). Starting at the light saturation

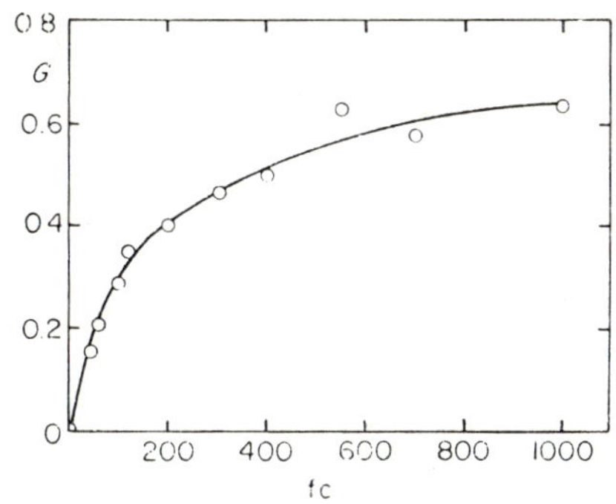

Fig. 2. Relationship of growth rate (G, doublings/h) of *R. capsulata* to incident light intensity (fc). Rates are expressed in terms of mass doublings per hour during logarithmic growth. For experimental details, *see* Ref.[17]; *reprinted by permission of the Proceedings of the National Academy of Sciences of the United States.*

point (about 550 foot-candles (fc)), decrease of illumination intensity leads to progressively slower growth rate. Thus, the net biosynthetic rate can be "set" by "fixing" the activity of the ATP regeneration system, through manipulation of the photon flux supplied. This can also be achieved using saturating light intensity, by intermittently illuminating the cells at certain frequencies[16,17]. For example, cultures exposed to intermittent saturating light in the repeating regimen (30 seconds light: 30 seconds dark) behave as if they were growing in *continuous dim* light of 10-fold lower intensity. Cells growing (slowly) in such regimens seem to "perceive" a disproportionately small fraction of the incident energy. The depressed growth rate and other changes observed amount to the syndrome of energy stress, apparently caused by derangement of the integration between energy conversion and biosynthesis[17].

It is evident that the photosynthetic bacteria offer unique experimental possibilities for attacking problems of "energy metabolism" in relation to control of synthesis of major cell components. In this connection it is of interest to consider the RNA content of bacterial cells. The total RNA (and ribosome) content of heterotrophic bacteria, growing with continuous energy supply at a given temperature, is proportional (within limits) to the growth rate[18]. This kind of proportionality is also observed in *R. capsulata* when continuous illumination is used. With certain intermittent light regimens, however, the usual relationship between RNA content and growth rate is violated[17]. Similarly, pulsing of the energy supply can also cause disturbances in the regulatory mechanisms that govern "balanced" synthesis of the bacteriochlorophyll-containing energy-converting membranes[17]. At this point it is relevant to briefly describe the relation normally observed between growth rate and membrane content. Fig. 3 is from an investigation by SISTROM[19] with *Rhodopseudo-*

monas spheroides; the results may be considered typical for nonsulfur purple bacteria. Bacteriochlorophyll (BChl) content, which can be taken as an index of membrane quantity[20], is inversely proportional to growth rate, when the latter is fixed by continuous light intensity. The production of "excess" energy converting membrane at low intensity reflects operation of a regulatory mechanism that tries to cope with diminishing energy supply.

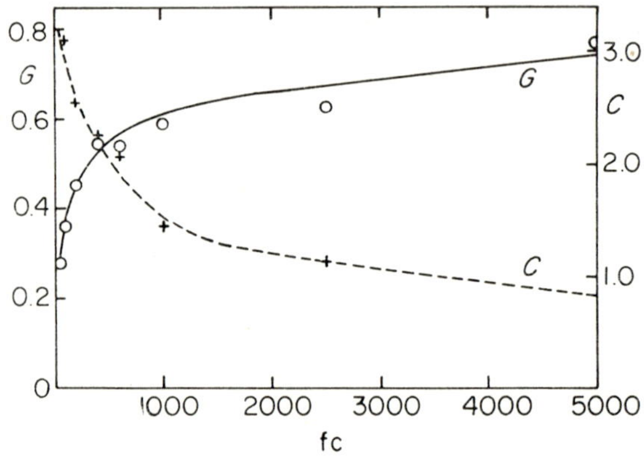

Fig. 3. Growth rates (*G*, doublings/100 min) and specific bacteriochlorophyll contents (*C*, µg chlorophyll/100 µg protein) of *R. spheroides* at various light intensities (fc). Experimental details are given in Ref.[19]; *reprinted by permission of Dr. W. R. Sistrom and the Journal of General Microbiology.*

In our view, the findings made with both continuous and intermittent light can be reasonably reconciled by the interpretation developed below, in which we are concerned only with the "light-limiting" region of the growth response (*i.e.*, below the light saturation point), where intensity fixes the growth rate. For the sake of simplicity, it can be assumed that intermittent illumination in regimens causing pronounced growth inhibition is tantamount to decreasing the energy flux by reducing the continuous light intensity. Starting at zero intensity, it is assumed that for any given quantity of photochemical energy conversion apparatus, increase of the photon flux increases both the absolute rate of ATP regeneration and the E.C. These, in turn, are presumed to be the primary factors governing biosynthetic rates.

REGULATION OF PHOTOPHOSPHORYLATION AND SYNTHESIS OF ENERGY-CONVERTING MEMBRANES

In order to achieve metabolic integration and balance, it follows from the general hypothesis that the rate of ATP regeneration by photophosphorylation must also be regulated in some fashion by the adenylate nucleotides. At least two categories of possible controls can be envisaged — regulation of the actual *activity* of the photophosphorylation system, and regulation of *formation* of the system. In respect to activity, E.C. control would be the analogue of E.C. regulation of enzymes involved in heterotrophic ATP regeneration. There is little evidence as yet for specific "chemical" regulatory control of bacterial photophosphorylation activity, but it is reasonable to suppose that the velocity of the "photoelectron flow" is influenced by the rate at which energy-rich intermediates, or states, are used for the final production of ATP from ADP. The light-dependent flow of electrons in chloroplasts is, in fact, substantially affected by the availability of ADP and P_i (or arsenate)[21]. As in other systems, a protein "coupling factor" is required for the terminal step of bacterial photophosphorylation. The *R. capsulata* coupling factor is a rather labile protein which, suggestively, is stabilized by low concentrations of ATP[22]. Obviously, the coupling factor-dependent reaction would be a likely point for fine control of the photophosphorylation process.

Regulation of the formation of the photophosphorylating membrane complex is a subject of particular interest in the present connection. It was noted earlier that the quantity of membrane formed is inversely proportional to light intensity. Accordingly, we have suggested that high E.C. *inhibits synthesis* of the BChl-membrane complex[17] (supporting observations are reported in Ref.[23]). It might seem that such behavior is contrary to what would be expected for biosyntheses controlled by E.C. Formation of energy converting membrane, however, may be viewed as a special case in that *the biosynthetic enzymes involved give rise to a system with **R**-type function.* Thus, certain of the biosynthetic enzymes active in production of energy converting membranes may behave like typical **R**-type enzymes in respect to E.C. control.

It occurred to us that it might be possible to test the suggested E.C. control of BChl-membrane formation by observing the consequences of artificially raising the E.C. through exposure of growing cells to ATP[24]. In accord with expectations, BChl synthesis and the anaerobic photosynthetic growth of *R. capsulata* were found to be markedly inhibited by exogenous 5 mM-ATP — especially with inoculum cells of low BChl content. The results of suitable control experiments make it very unlikely that the ATP effect is simply due to chelation of Mg^{2+} or some other metal. A particularly interesting control in this connection is the observation that 5 mM-ATP has virtually no effect on the dark aerobic growth rate of *R. capsulata.* It should be noted that the inhibitory effect of exogenous ATP on photosynthetic growth is not permanent and that in the experiments described, conditions were used which required the cells to preferentially synthesize BChl in order to continue growth — thus, we were primarily examining the effect of exogenous ATP on the ability of cells to readjust the rate of synthesis of the BChl-membrane complex. In any event, it appears that sufficient ATP enters the cells or somehow affects synthesis sites on the membrane to mimic the situation and consequences that would be expected if a very high E.C. were established by extraordinarily rapid intracellular photophosphorylation.

COMPETITION IN SYNTHESIS OF MACROMOLECULAR CELL COMPONENTS

Aside from membranous structures, RNA and protein are the most notable major components of bacterial cells. In *R. capsulata* there is a striking reciprocal relation between synthesis of membrane and RNA. Thus, as growth rate increases the RNA content increases and the quantity of membrane decreases, and *vice versa*. This led us to suggest that the syntheses of BChl-membrane and RNA are affected in opposite ways by the ATP regeneration rate and E.C.[17]. Although the mechanisms regulating RNA (and thereby, protein) synthesis in bacterial cells are still not elucidated, recent studies[25] with *E. coli* have implicated the flux of ATP and related nucleotides as important controlling elements.

Presuming that a common signal system does operate in this fashion (*i.e.*, with opposing effects on RNA and BChl-membrane syntheses), we are led to consider the final consequences to cell growth. It seems reasonable to believe that this kind of signal device is designed to help the cell cope with the situation created by a diminishing supply of energy. The demands that production of additional energy converting machinery make on the cell, however, are no doubt quite considerable in themselves — in terms of ATP and "building blocks" — and, consequently, the growth rate does in fact diminish. In other words, it appears that preferential synthesis of the membrane system under energy stress conditions inevitably means decreased formation of other components essential for rapid growth (such as ribosomes[26]). Accordingly, we have suggested[27] that "a regulatory system performing in this fashion should be characterized as a partially compensatory control mechanism rather than as a finely tuned regulation device."

MUTANTS WITH ALTERED "ENERGY METABOLISM" CHARACTERISTICS*

Recently, we undertook experiments aimed at causing disturbances in the *in vivo* E.C. of *R. capsulata* by exposing growing cells to inorganic arsenate (As_i), which competes with P_i and acts as a phosphorylation uncoupler. As_i, equimolar to P_i, proved to be very inhibitory to photosynthetic growth, and As_i-resistant (or tolerant) strains could be readily isolated from surviving cell populations. One of the mutant strains, Z-1, has been studied in some detail[28]. Several kinds of tests failed to show a substantial difference between wild type and Z-1 in respect to As_i uptake. Similarly, photophosphorylating membrane fragments from wild-type and Z-1 showed no difference in K_m for P_i, and in both instances As_i was found to

* A more comprehensive description of the properties of *R.capsulata* Z-1 and discussion of the mechanism of arsenate resistance can be found in Ref.[31].

inhibit photophosphorylation competitively with P_i (with approximately the same K_i). An unexpected difference, however, was found in the maximum velocity (V_{max}) of phosphorylation. In Z-1, the V_{max} is considerably higher.

The following explanation is offered to account for the foregoing observations[29]. As_i can substitute for P_i, and is converted to ADP·arsenate, which is presumed to hydrolyze very rapidly. This fast process would tend to keep the phosphorylation sites saturated with substrates. We assume that As_i stimulates photophosphorylation slightly (<2 fold) in the wild type, but quite significantly (2—4 fold) in Z-1. In analogy with the effects of As_i on basically similar systems, the stimulation would derive from an acceleration of electron flow. With an As_i to P_i ratio of 1, the slightly increased phosphorylation rate in wild type is presumably insufficient to compensate for the decreased ATP regeneration rate occasioned by the As_i *vs.* P_i competition. Accordingly, the cell cannot maintain a physiological E.C., biosynthesis is impaired and growth is consequently inhibited. In Z-1, however, a *marked* stimulation of phosphorylation by As_i (when it is present equimolar to P_i) would enable the cell to maintain the ATP regeneration rate and E.C. within the "normal" range, despite the As_i *vs.* P_i competition.

This explanation implies that we think of *R. capsulata* Z-1 as an "E.C. mutant" of sorts. Such mutants should be helpful in further study of the E.C. regulation hypothesis, and as a by-product, may provide new systems of value for investigating electron flow associated with photophosphorylation. Conceivably, there may also be possibilities for industrial exploitation of mutants of this character.

In connection with "E.C. mutants", note should be made of the interesting recent studies of Brenner *et al.*[30], with tRNA synthetases of *Salmonella typhimurium*. The histidyl-tRNA synthetase responds to E.C., and a mutant enzyme with a significantly altered response pattern was obtainable. These investigators believe that the "properties of the mutant enzyme are consistent with the proposition that the response of the wild-type enzyme to energy charge may be maintained by natural selection." Obviously, we can look forward to a powerful attack on the problems under consideration through genetic approaches.

CONCLUDING REMARKS

The great amplification of knowledge of "intermediary" metabolism, especially in regard to biosynthesis, and of control mechanisms during the past decade has revealed an intricate system of interrelated roles played by the adenylate nucleotides. Energy metabolism requires a continual "turnover" of these nucleotides, which also serve as substrates for macromolecular syntheses in growing cells. Clearly, the net synthesis of nucleotides, which are constantly being drained from the intracellular pools, must be closely regulated and properly geared with the "energy metabolism" turnover. The control network involved must be of enormous sophistication, and it now seems evident that the importance of "energy metabolism" as a primary directive force in cell growth has been underestimated. A major task for the future will be the charting of the complex schematic plan that "determines" the self-regulating value of the *in vivo* E.C.

The concept of "Energy Charge" as a key regulatory element facilitates the construction of a working hypothesis for understanding the integration of the energy conversion and biosynthetic phases of metabolism, and permits the systematization of a large number of observations on nucleotide regulation of enzyme activities. Furthermore, this conceptualization suggests new ways of looking at old problems, and can therefore be expected to stimulate novel approaches to their solution. In applying these notions to such problems, the definition given for Energy Charge should be used with the understanding that it is likely to be only a good first approximation. For certain situations or processes the concentrations of other purine or pyrimidine nucleotides might also be relevant. Perhaps of even greater likelihood is the possibility that a more comprehensive "charge expression" would have to include additional terms for the pyrimidine nucleotides which, like adenylate nucleotides, link energy conversion processes and biosynthesis, and also act as modulating effectors for numerous enzymes.*

Research of the author is supported by Grant GB-7333X from the U.S. National Science Foundation.

REFERENCES

1. STENT G. S.: Science **160**, 390 (1968).
2. BAUCHOP T., ELSDEN S. R.: J. Gen. Microbiol. **23**, 457 (1960); GUNSALUS I. C., SHUSTER C. W., p. 1 in I. C. Gunsalus, R. Y. Stanier (Eds.): *The Bacteria*, Vol. 2. Academic Press, New York 1961; SENEZ J. C.: Bacteriol. Rev. **26**, 95 (1962).
3. PAYNE W. J.: Ann. Rev. Microbiol. **24**, 17 (1970).
4. FORREST W. W., p. 65 in P. Meadow, S. J. Pirt (Eds.): *Microbial Growth*. Cambridge University Press, Cambridge 1969.
5. ROBERTON A. M., WOLFE R. S.: J. Bacteriol. **102**, 43 (1970)**.
6. JENSEN D. E., NEIDHARDT F. C.: J. Bacteriol. **98**, 131 (1969).
7. BALDWIN E., p. 67 in *Dynamic Aspects of Biochemistry*, 3rd Ed. Cambridge University Press, Cambridge 1957.
8. COLE H. A., WIMPENNY J. W. T., HUGHES D. E.: Biochim. Biophys. Acta **143**, 445 (1967).
9. LEHNINGER A. L., p. 198 in *Bioenergetics*. W. A. Benjamin, Inc., New York and Amsterdam 1965.

---

* According to ATKINSON (*see* Ref.[11]), "... we may predict that the charge of the TPNH system, the mole fraction $(TPNH)/[(TPN^+) + (TPNH)]$, will be found to be an important control parameter for those processes whose metabolic function is the regeneration of TPNH."

** Experiments with a strictly anaerobic *Methanobacterium* indicating a very low coupling efficiency.

10. Atkinson D. E., Walton G. M.: J. Biol. Chem. **242**, 3239 (1967); Atkinson D. E., Fall L.: J. Biol. Chem. **242**, 3241 (1967).
11. Atkinson D. E.: Biochemistry **7**, 4030 (1968); Klungsøyr L., Hagemen J. H., Fall L., Atkinson D. E.: Biochemistry **7**, 4035 (1968); Shen L. C., Fall L., Walton G. M., Atkinson D. E.: Biochemistry **7**, 4041 (1968).
12. Atkinson D. E.: Biochem. Soc. Symp. **27**, 23 (1968).
13. Chapman A. G., Fall L., Atkinson D. E., *personal communication.*
14. Gest H., Kamen M. D., p. 568 in W. Ruhland (Ed.): *Encyclopedia of Plant Physiology,* Vol. V/2. Springer Verlag, Berlin 1960*.
15. Uffen R. L., Wolfe R. S.: Bacteriol. Proc. 64 (1970); J. Bacteriol. **104**, 462 (1970)**.
16. Sojka G. A., Din G. A., Gest H.: Nature **216**, 1021 (1967).
17. Sojka G. A., Gest H.: Proc. Nat. Acad. Sci. U.S.A. **61**, 1486 (1968).
18. Schaechter M., Maaløe O., Kjeldgaard N. O.: J. Gen. Microbiol. **19**, 592 (1958); Ecker R. E., Schaechter M.: Ann. N. Y. Acad. Sci. **102**, 549 (1963); Leick V.: Nature **217**, 1153 (1968).
19. Sistrom W. R.: J. Gen. Microbiol. **28**, 607 (1962).
20. Ketchum P. A., Holt S. C.: Biochim. Biophys. Acta **196**, 141 (1970); Steiner S., Sojka G. A., Conti S. F., Gest H., Lester R. L.: Biochim. Biophys. Acta **203**, 571 (1970).
21. Avron M., Jagendorf A. T.: J. Biol. Chem. **234**, 967 (1959); Izawa S., Connolly T. N., Winget G. D., Good N. E.: Brookhaven Symp. Biol. No. 19, 169 (1967).
22. Baccarini-Melandri A., Gest H., San Pietro A.: J. Biol. Chem. **245**, 1224 (1970).
23. Schön G.: Arch. Mikrobiol. **66**, 348 (1969).
24. Sojka G. A., Gest H.: Bacteriol. Proc. 64 (1970).
25. Irr J., Gallant J.: J. Biol. Chem. **244**, 2233 (1969); Gallant J., Harada B.: J. Biol. Chem. **244**, 3125 (1969).
26. Din G. A., Gest H.: J. Bacteriol. **97**, 1518 (1969).
27. Sojka G. A., Baccarini A., Gest H.: Science **166**, 113 (1969).
28. Zilinsky J. W., Gest H.: Bacteriol. Proc. 64 (1970).
29. Zilinsky J. W., Sojka G. A., Gest H.: Biochem. Biophys. Res. Comm. **42**, 955 (1971).
30. Brenner M., DeLorenzo F., Ames B. N.: J. Biol. Chem. **245**, 450 (1970).
31. Lien S., San Pietro A., Gest H.: Proc. Nat. Acad. Sci. U.S.A. **68**, 1912 (1971).

---

* Review of the general physiology and biochemistry of photosynthetic bacteria.

** Authors have recently discovered that if very strict anaerobic procedures (Hungate technique) are used, several nonsulfur purple bacteria can grow anaerobically in darkness, at very slow rates, in certain media.

Regulation of Maximal Enzyme Synthesis*

ARTHUR B. PARDEE

Departments of Biochemistry and Biology, Moffett Laboratory, Princeton University, Princeton, New Jersey, U.S.A.

THE PROBLEM OF GETTING MORE ENZYME FROM A MICROORGANISM

The industrial production of enzymes is becoming increasingly important.[1] To quote a recent review, "Nevertheless, the greatest challenge that industrial microbiologists will face during the next decade will be the development of large scale processes for the production of enzymes, nucleotides, and novel biochemicals"[2]. The obtaining of microorganisms which possess high enzyme activities is of first importance in this endeavor. Searches for a species with optimal properties have in the past been conducted by screening organisms obtained from nature and by random mutation. No doubt these procedures will be most useful for years to come. But an outline of the basic principles of enzyme control which are emerging from molecular genetic studies should be useful to applied microbiologists. After all, methods of mutation and isolation are similar in industrial and basic laboratories; once the organism has been isolated it can be used at any time and on any scale. Also, these problems of optimal enzyme production are of great theoretical interest with regard to the control of gene action.

The main purpose of this article is to describe some ideas and experiments on modes of control of *maximal* enzyme synthesis, and ways in which mutations can change this maximal rate. Both the basic principles of microbial genetics[3] and the genetics of industrial microorganisms[4] have been reviewed recently.

We must first of all distinguish between enzyme amount and enzyme activity, either the activity in an intact microorganism or under cell-free conditions. Enzyme amount means the number of enzyme molecules that are present. The activity of a microbial culture or preparation depends on the enzyme amount, and also on the activity of the individual enzyme molecules. Activity of an extracellular enzyme molecule can vary for two basically different reasons. The first is that the structure of the enzyme itself may be different, for example, in different species or owing to a mutation, so that one kind of molecule is a better catalyst than another. The

* Aided by Grant CAAI-11595 from the U.S. Public Health Service.

second is that the activity of an enzyme molecule depends on the conditions that surround it — pH, substrate concentration, activators and inhibitors. In intact cells, the rates at which substrates can reach the enzymes and the rates at which inhibitors, including reaction products and feedback inhibitors, escape from the cells are of additional great importance in determining rates of reactions. In this article we shall be primarily interested in ways of changing the enzyme amounts, although observations on increases in enzyme activity will be mentioned when they seem relevant.

Two kinds of genes are now recognized. Structural genes determine the structures of proteins. According to the basic statement of BEADLE and TATUM, one gene determines the structure of one enzyme, in the sense that a mutation in this gene either partly or absolutely alters the structure of its enzyme. The genetic code brings this idea to the level of DNA base triplets. One triplet determines one amino acid, and a change in this triplet by a missense mutation can change the amino acid; by a nonsense mutation it can abolish incorporation of any amino acid at this point; or deletion or insertion of one or more bases (a frameshift mutation) alters reading of the code for the whole series of amino acids that follow the mutation point. All of these mutations are most likely to change the catalytic activity of an enzyme (generally reducing it to zero), rather than changing the enzyme amount.

The second type of genes are regulatory. These genes are able to determine the activities of related structural genes. The two gene types remind one of modern regulated machinery which is built with a functional part and also a controlling part which determines the extent of function. Since the machine's function can affect its environment, for example with a home heating furnace, the regulator (thermostat) can make the machine selfregulatory. Similarly, enzyme production can be negatively regulated by the concentrations of reaction products. This is known as repression of enzyme synthesis.

The best studied example of regulation is induction of β-galactosidase of *Escherichia coli*, the enzyme that hydrolyzes lactose into glucose and galactose[5]. In induction the substrate (or a related compound) has a positive effect and permits rapid enzyme synthesis (in contrast to repression mentioned above, in which the product prevents enzyme synthesis). Thus lactose induces a rapid synthesis of this enzyme. It does so by interacting with a system based on three different regulatory genes (Fig. 1). The functions of these genes must be distinguished clearly before we proceed to the rest of this article.

The first step of gene function is called transcription; it is the copying catalyzed by RNA polymerase of the structural genes, *Z*, *Y*, and *A*, to make the corresponding messenger RNA. RNA polymerase initiates its action at a place on the DNA called the promoter gene (*P*), which lies just before the first structural gene (*Z*) of the group (operon), except that another gene called the operator (*O*) lies between. The point of control on the DNA is the operator gene, over which the RNA polymerase must pass. When the *O* gene is combined with a protein named the repressor, the RNA polymerase cannot proceed to copy the structural genes. Thus, the repressor is an inhibitor of enzyme formation.

The *lac* repressor protein (molecular weight 150,000; Ref.[6]) is made according to the information in *its* structural gene (*I*, for inducible). Besides combining with the operator gene, the repressor can also combine with the inducer, lactose. But combination with lactose prevents it from attaching to the *O* site. Thus, the inducer is an inhibitor of the inhibition by the repressor.

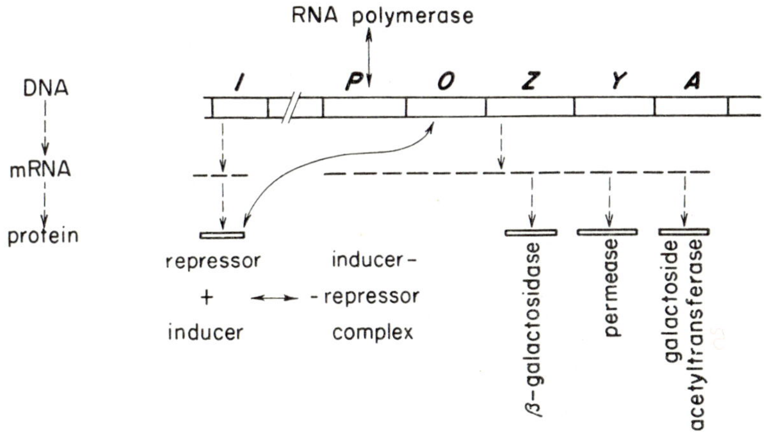

Fig. 1. Control of the lactose operon

Some genes like this set of three (*P*, *O*, and *I*) are thought to be important in regulation of each enzyme synthesis. Clearly, mutation in any one of these genes should modify a corresponding rate of enzyme production.

We can distinguish three basic properties of the rate of enzyme synthesis. One is the rate of enzyme production when the inducer is absent. This is called the basal rate. The second is the dependence of rate on the amount of inducer. The rate increases with increasing inducer concentration. It approaches a maximal value, the property most interesting to us, which is interpreted as the rate of maximal gene function when the repressor is completely saturated with inducer.

Clearly, the effects of inducers and repressors, and similar effects, are of great importance for the practical problem of producing high enzyme amounts[7]. Their actions are limited to the maximal rates of induction. As pointed out some time ago, these maximal rates are by no means the same for different enzymes, since some enzymes are produced maximally in much larger amounts than others[8].

This raises a paradox that is not explained by the mechanism of induction described above. Why does one gene have a different maximal, uninhibited activity than another? Apparently there are "lazy" genes and "active" genes, and those must differ in some chemical way. This distinction is particularly clear when the enzyme is not inducible, that is, when its rate of production is nearly independent of external conditions. Enzymes of this sort are called constitutive. Different constitutive enzymes are produced at rates which can vary by at least 1,000-fold; and yet there is only one gene for each. Constitutive enzymes are extremely common, particularly in higher organisms, including higher microorganisms such as yeast[9].

REPRESSOR AND OPERATOR MUTATIONS

If we eliminate effects of the environment by choosing a system that is naturally constitutive, two main sorts of mechanisms that increase the rate of enzyme production can be imagined. Constitutive rates could be set by a promoter gene, which determines how frequently the RNA polymerase makes copies of the structural gene. This kind of rate control should apply to all enzyme syntheses. Mutations in promoter genes could possibly increase the rate of each enzyme synthesis, up to the maximum that has been observed for any enzyme, which is about 5% of the total protein synthesized. For the second sort of mechanism, a constitutive enzyme synthesis might be held to a low level by a repressor which does not respond to molecules in the environment. β-Galactosidase is synthesized faster if mutation damages or eliminates the repressor or if it modifies its target, the corresponding operator gene. The techniques for distinguishing between all of these possibilities are not yet well worked out, particularly since questions of this sort have not been studied frequently or in depth. Indeed, all the evidence for promoter mutations is very recent[10]. Thus, we have several cases of increased enzyme synthesis for which the genetic basis is not yet clear.

Mutations which modify the maximal rate of β-galactosidase production serve as prototypes for the ways in which other enzymes' rates of production might be changed. We will first discuss the effects of mutations in the *I* gene. These mutations change the repressor; active repressor is not made in the simplest case of I^- mutations. These I^- mutants produce β-galactosidase as about 3% of the total bacterial protein, under the usual growth conditions, independently of the presence or absence of an inducing compound. This is the same level as is found in completely induced I^+ (wild type) bacteria, and it is about 1,000 times higher than the basal level of the wild type. Subtler modifications of *I* can change the affinity of the inducing compound for the repressor, or even the sign of inducer effect on enzyme synthesis[11].

Of particular interest here are mutations which make the repressor insensitive to the inducer (I^s mutants). These bacteria make a very low, basal level of the enzyme under all conditions[12]. Their existence indicates a basis for low level constitutive synthesis of other enzymes, namely, presence of a repressor which does not respond to any inducer.

The rate of enzyme synthesis by these I^s mutants can be increased. One sort eliminates the I^s repressor molecule by a secondary I^- type of mutation; these bacteria make the enzyme at the 3% level[12]. Another kind of second mutation changes the *O* gene, the receptor of the repressor, so that it does not hold the repressor firmly. These double (I^sO^c) mutants make β-galactosidase at an intermediate, constitutive rate which depends on the specific O^c mutation. One example reports a rate about 30× the basal rate[12].

Another well worked out example of constitutive control is seen with synthesis of ornithine transcarbamylase and other enzymes of the arginine biosynthetic pathway[13]. These enzymes are repressible by arginine in *E. coli* strain K12. Their rates of synthesis are high if the repressor-forming gene (*R*) is inactivated. Another mutation of this gene makes enzyme synthesis nearly insensitive to arginine, and this kind of mutation is found naturally in strain B of *E. coli.* Therefore, we have a natural example of low-level constitutive-enzyme synthesis controlled by nonresponsive repressor. As with the lactose I^s mutants, this constitutive, low production of ornithine transcarbamylase *E. coli* strain B is released by further mutation of the *R* gene, or by a mutation of the *O* gene which reduces its affinity for the repressor and increases the rate of enzyme production about 10-fold.

Constitutive mutations of the *O* type can also, of course, be put into wild-type bacteria. For example, *O* mutations on the tryptophan synthesizing system of *E. coli* make up to 12 times the repressed level of several of these enzymes[14]. Derepression conditions give the same enzyme production for both mutant and wild type. The advantage of this sort of constitutivity is that commercial production might favor a medium, such as corn steep liquor, which is cheap and yet which represses production of the desired enzyme. Constitutive mutants would make more enzyme under these conditions.

A number of enzymes are not induced by their substrates, but they are induced by other compounds[15]. Therefore, although a repressor exists, it cannot be detected by the usual nutritional induction or derepression test. Examples are histidinase and urocanase of *Aerobacter aerogenes.* These enzymes act in sequence in the degradation of histidine. They are not induced directly by histidine, but they are induced by urocanic acid. However, in some mutants of *Salmonella typhimurium* urocanic acid is not able to induce them because it cannot enter the cell, and yet the enzymes are inducible by other compounds[16]. Thus, inducibility can be masked by permeability barriers. These results indicate the difficulty of determining whether an apparently constitutive system really possesses a repressor.

PROMOTER MUTATIONS

Mutations in the *P* gene were originally described for the *lac* system[10]. They are thought to change the rate of mRNA synthesis, although changes in the rate at which the mRNA is copied into enzymes are also conceivable[17]. Among other properties, these mutants were able to make maximal levels (on full induction) of

β-galactosidase that were only a few percent of the corresponding levels in the P^+ parental strains. One might suppose that other structural genes normally possess low activity promoters; and these might be mutated to higher activity promoters. However, only a few results are yet available and these have not been well explored. One problem is to determine whether increased constitutive activity appears because of a more active enzyme synthesis or by an increased *activity* of the enzyme molecule owing to a change of structure. Either mutation would be very close to structural gene mutations and would be genetically similar in other ways.

A most interesting example is the dihydrofolate reductase of *Diplococcus pneumoniae*. This enzyme is present at low levels in wild-type bacteria. But in three mutants resistant to the folate analog amethopterin activities of this enzyme are 14, 19, and 120 times the original strain's levels. Furthermore, the number of enzyme molecules could be measured directly by the irreversible binding of amethopterin. These binding studies show that the number of enzyme molecules increase in proportion to the increase in activity[18]. The mutations lie within the structural gene, as demonstrated both by genetic mapping and because the stability of the enzyme to heat is slightly modified, as well as are its properties during purification. These remarkable results suggest that mutation within the structural gene also controls the rate of copying of this gene; or in other terminology the promoter region appears to control some of the enzyme's structure as well as its amount.

As a second example, we have mutations which increase the rate of production of the *lac* repressor protein[6]. This protein is made in very small amounts, about 10 molecules per cell in wild-type *E. coli.* Mutations have been obtained which increase the rate of repressor production 10-fold. They also modify the structure of the repressor, as in the dihydrofolate reductase case. The basis of this effect is unknown, although it is speculated that mutation of a special *I*-gene promoter might be involved, among other possibilities.

Still another case is seen with histidinase and urocanase[16]. The wild-type *S. typhimurium* strain LT-2 cannot grow on histidine as a source of nitrogen. But rare mutants are capable of growth on histidine. These mutants possess about a 10 times higher activity of histidinase and at least 300$\times$ more urocanase activity than does the parent strain. The mutations do not map in the structural gene for either enzyme, but rather are very close, suggesting that a promoter locus is affected.

The activity of a low level promoter has been raised by mutation, with the result that three enzymes of tryptophan synthesis become about 10 times higher[19]. This promoter is in the middle of the tryptophan operon, and it permits a low level of production of these enzymes. Another promoter, part of the regulatory repressible system, controls all five enzyme formations over a wide range. When this latter promoter is destroyed, the secondary constitutive promoter permits only low level function; the mutation of interest here enhances its activity.

A final example is seen in work from my laboratory with nicotinamide deamidase of *E. coli.* This enzyme is present in extremely low concentrations[20], as might be expected of an enzyme that is involved in coenzyme metabolism. It is completely constitutive. Mutations have been obtained, by selection for cells capable of using nicotinamide as a sole nitrogen source. Different mutants produce different (to 50×) activities of the enzyme. (It is not yet known whether this reflects different rates of enzyme formation.) The mutants map very close to the structural gene, and hence, might be promoter mutants.

All of these examples show that one can considerably increase the rates at which constitutive enzymes are made, even of enzymes which are naturally found at very low activities. But in some cases it is not clear whether the increases are merely increases of specific activity. Methods of this sort should be useful, or at least encouraging, for attempts to get higher enzyme production from an organism which has specific commercial advantages.

CHANGES IN STRUCTURAL GENES

Evidently, one can increase the rate of enzyme production by inserting more copies per cell of the gene for this enzyme. Extra genetic material can be included in several ways. One is to obtain gene duplication in the chromosome. An example of this is selection of mutants which grow especially well on very low lactose concentrations. These mutants had 4× more structural genes for β-galactosidase than did the wild type[21]. A more common modification is to insert an extra genetic piece (episome) which carries the structural gene. This can be done with *E. coli* by conjugation. These bacteria have 2 or 3 copies of the structural gene and make several times the usual amount of the enzyme[5]. The most spectacular way is to insert the genetic material into the cell with a lysogenic transducing bacteriophage. When the phage is induced to multiply, its genetic material which includes the structural genes for the enzyme increases many-fold. In this way, both the *lac* repressor and β-galactosidase production are increased 40-fold[6,22].

One way should be mentioned by which a structural gene mutation modifies enzyme production. This depends on the sequences of structural genes called operons (like the *lac* operon shown in Fig. 1). An operon is copied by RNA polymerase starting from the operator end. Certain "polar" mutations in genes copied early will greatly reduce the production of the enzymes further down the sequence[5]. Thus, in the operon for the histidine biosynthetic enzymes, a mutation in an early gene reduces all later gene products by a factor of about three[23]. Possibly, polar mutations of this sort occur in nature; their reversal could increase the rates of expression of genes that follow them.

APPEARANCE OF NEW ENZYME ACTIVITIES

The extreme case of increased enzyme activity is seen when a totally new activity appears. Of course "new" is relative to the sensitivity of the test, since a low basal level might not be detected or considered significant. An example is seen in urocanase of *S. typhimurium*, mentioned above[16]. The structural gene of this enzyme was unchanged, and, hence a new activity was not created, although the increase caused by mutation was spectacular.

A most striking case is the appearance of ability of *A. aerogenes* to utilize xylitol as a carbon source[15]. This ability increased stepwise after each of three mutations. Ribitol dehydrogenase is also able to dehydrogenate xylitol; but it is present at such low levels in xylitol-containing medium (in uninduced wild-type cells) that growth on xylitol is very slow. The first mutation makes ribitol dehydrogenase constitutive. The second mutation increases the ability of the ribitol dehydrogenase to utilize xylitol, and is seen as a change in the structure of the enzyme. The third change is caused by a mutation which increases the transport of xylitol into the cell, thereby concentrating it and improving its rate of utilization. Thus, we are not actually seeing the creation of a completely new gene; what is introduced is built on the preexisting ribitol dehydrogenase system. Yet the overall effect permits an organism which could not grow on a carbon source to achieve growth rates comparable to those found with excellent carbon sources.

As a final case, we have the appearance of what appears to be a new activity, the ability to dehydrogenate 1,2-propanediol[24]. This enzyme is not found in wild-type *E. coli*, but it appears in mutants isolated by their capacity to grow on this compound. The basis for this enzyme's appearance is not yet worked out.

REFERENCES

1. Sci. Res. Council (Britain) Report: *Enzyme Chemistry and Technology*, 1969.
2. Hepner I. L.: Adv. Appl. Microbiol. **11**, 283 (1969).
3. Bradley S. G.: Adv. Appl. Microbiol. **8**, 29 (1966).
4. Hütter R.: Pathol. Microbiol. **34**, 195 (1969).
5. Beckwith J.: Science **156**, 597 (1967).
6. Müller-Hill B., Crapo L., Gilbert W.: Proc. Nat. Acad. Sci. U.S.A. **59**, 1259 (1968).
7. Pardee A. B., p. 3 in D. Perlman (Ed.): *Fermentation Advances*. Academic Press, New York 1969.
8. Pardee A. B., Beckwith J. R., p. 255 in H. J. Vogel, V. Bryson (Eds.): *Informational Macromolecules*. Academic Press, New York 1963.
9. Bussey H., Umbarger H. E.: J. Bacteriol. **98**, 623 (1969).
10. Miller J. H., Ippen K., Scaife J. G., Beckwith J. R.: J. Mol. Biol. **38**, 413 (1968).
11. Myers G. L., Sadler J. R.: J. Mol. Biol., *in press* (1970).
12. Willson C., Perrin D., Cohn M., Jacob F., Monod J.: J. Mol. Biol. **8**, 582 (1964).
13. Jacoby G. A., Gorini L.: J. Mol. Biol. **39**, 73 (1969).
14. Hiraga S.: J. Mol. Biol. **39**, 159 (1969).

15. Wu T. T., Lin E. C. C., Tanaka S.: J. Bacteriol **96**, 447 (1968).
16. Meiss H. K., Brill W. J., Magasanik B.: J. Biol. Chem. **244**, 5382 (1969).
17. Steitz J. A.: Nature **224**, 957 (1969).
18. Sirotnak F. M., Donati G. J., Hutchison D. J.: J. Biol. Chem. **239**, 2677 (1964).
19. Wuesthoff G., Bauerle R. H.: J. Mol. Biol. **49**, 171 (1970).
20. Imsande J.: J. Biol. Chem. **236**, 1494 (1961).
21. Horiuchi T., Tomizawa J., Novick A.: Biochim. Biophys. Acta **55**, 152 (1962).
22. Pfahl M.: Mol. Gen. Genet. **105**, 122 (1969).
23. Ames B. N., Hartman P. E.: Cold Spring Harbor Symp. Quant. Biol. **28**, 349 (1963).
24. Sridhara S., Wu T. T., Chused T. M., Lin E. C. C.: J. Bacteriol. **98**, 87 (1969).

Genetic Modifications in Substrate Specificity of the Amidase Produced by *Pseudomonas aeruginosa*

PATRICIA H. CLARKE

Department of Biochemistry, University College London, London, Great Britain

Microbial enzymes which are inducible are subject to control systems which allow them to be made in large amounts when wanted for cell growth and in small amounts, or perhaps not at all, when not required. Constitutive mutants can be isolated by various methods which synthesise enzymes in high yield in the absence of inducers. Many inducible enzymes are also subject to katabolite repression by various carbon compounds and increased yields of enzyme can be obtained by selecting the least repressing carbon compounds as growth substrates in the medium. It may also be possible to isolate mutants which are relatively resistant to katabolite repression which produce high enzyme yields on most media. This is very important for large scale work since it may make it possible to use cheaper compounds in the growth medium and still get high yields of enzyme.

We have isolated mutants of *Pseudomonas aeruginosa* which produce amidase constitutively and are also resistant to katabolite repression by succinate. One of these mutants is used routinely as a source of amidase and since it produces 5% or more of its total protein as this one enzyme a purification of only twenty-fold is needed to obtain the pure enzyme. This is an example of mutational changes affecting the regulation of synthesis of an enzyme normally produced by the wild-type strain. We have also obtained mutants which produce amidases with altered substrate specificity and are therefore able to grow on amides which are unavailable to the wild type.

The aliphatic amidase produced by the wild-type *P. aeruginosa* PAC has a very limited substrate range and is optimally active on acetamide and propionamide releasing ammonia and the carboxylic acid. The wild-type strain grows well with acetamide or propionamide as sole carbon and nitrogen source for growth. It does not grow on formamide, since it is unable to utilise 1-carbon compounds, but formamide is a substrate, though a poor one, and under certain conditions can be shown to be a weak inducer. The wild-type strain does not grow on the higher aliphatic amides such as butyramide or valeramide or on aromatic amides.

We looked for mutants which could grow on some of the amides which were unavailable to the wild-type strain and we have found that several different types of mutation could achieve this. First we isolated a series of constitutive mutants on plates containing succinate as carbon source and formamide as nitrogen source. On this medium the wild type cannot grow since the enzyme is not induced to any significant extent, whereas constitutive mutants produce large amounts of enzyme and grow well. These mutants were able to grow on formamide because they produced the enzyme constitutively and this is a common type of mutation allowing utilization

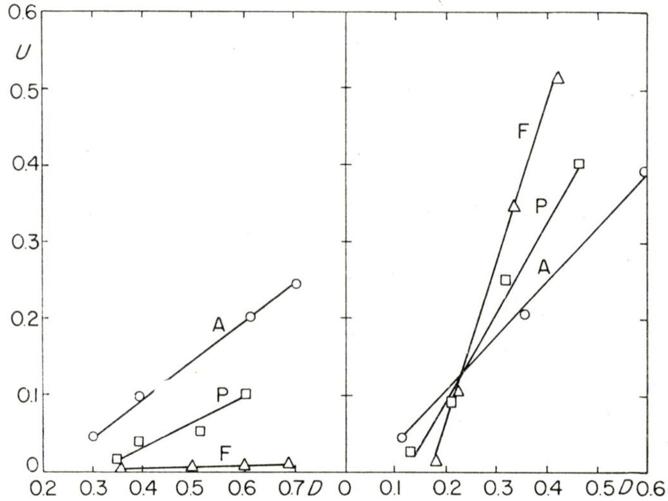

Fig. 1. Differential rates of amidase synthesis by the wild-type strain of *P. aeruginosa* PACI *(left)* and mutant F6 *(right)* induced by formamide (F), acetamide (A), and propionamide (P). *D* bacterial density equiv. (extinction 670 nm), *U* amidase units/ml equiv. (extinction 500 nm).

of a new growth substrate. We also isolated on this medium mutants which grew well but were inducible. These had mutations which conferred altered inducer specificity so that unlike the wild type they were readily induced by formamide. Fig. 1 compares the rates of amidase induction by formamide, acetamide and propionamide for the wild-type strain and for one of the formamide-inducible mutants. In this case we have a regulator mutation giving an alteration in inducibility pattern allowing growth on formamide although, like the constitutive mutants, they produce the same enzyme as the original wild-type strain[1].

One of the constitutive mutants was used to isolate mutants utilizing butyramide. Strain C11 produces the wild-type enzyme, which we term the A amidase, in high yield and was used as the parent strain in the isolation of a series of mutants which grew well on butyramide. Brown, Brown and Clarke[2] showed that the enzyme produced by these strains differed in substrate specificity and in electrophoretic mobility from the wild-type enzyme.

Fig. 2 compares the substrate profiles of washed suspensions of mutant strain C11 and the butyramide-utilizing mutant derived from it, strain B6. The rate of hydrolysis of butyramide by washed suspensions of constitutive strains, or by the fully induced wild-type strain, is about 2% of that for acetamide while strain B6 hydrolyses butyramide at about 30% of the acetamide rate.

In this case the genetic modifications allowing utilization of the new growth substrate involved a mutation in a regulator gene conferring constitutivity and a further mutation in the structural gene of the enzyme giving an alteration in the

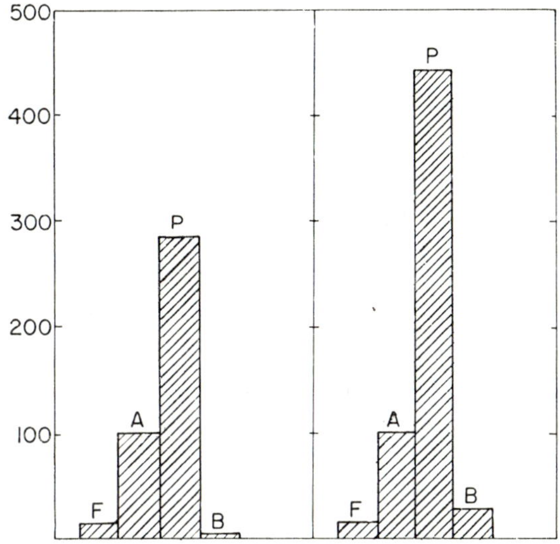

Fig. 2. Relative rates of hydrolysis of formamide (F), acetamide (A), propionamide (P), and butyramide (B) by washed suspensions of mutant strain of *P. aeruginosa* PACI. Constitutive mutant C 11 producing wild--type A amidase *(left)*, constitutive mutant B6 producing mutant B amidase *(right)*.

substrate specificity. The B enzyme was very stable and comparable in specific activity with the wild-type A amidase. It can be considered to be as physiologically efficient as the wild-type enzyme and mutants producing the B amidase grow well on acetamide and propionamide. This was a very straightforward mutational sequence involving mutation in a regulator gene followed by a mutation in a structural gene but we found that there was another way of achieving growth on butyramide.

J. E. Brown isolated a number of mutants from strain C11, and also directly from the wild type, which grew well on butyramide but produced the wild-type A amidase. These were all constitutive. A re-examination of the constitutive strains isolated previously on other media showed that some could grow on butyramide but that others were unable to do so. It was already known that the constitutive strains were potentially capable of producing very large amounts of A amidase. It was also known that the A amidase could hydrolyse butyramide although the activity was very low. The answer to the paradox that some constitutive mutants could not grow on butyramide had to lie in effects exerted by butyramide on the synthesis of the enzyme itself. In fact it was a repressor of the enzyme for which it was the substrate. Amidase synthesis by all the constitutive strains which were unable to grow on butyramide was severely repressed if butyramide was added to the growth medium. Fig. 3 compares the percentage repression produced by various concentrations of butyramide on cultures growing exponentially in succinate medium. Strain C11, which does not grow on butyramide, was almost completely repressed at a very low butyramide concentration. Strain CB4, isolated from C11, and strains CB6 and L10 isolated directly from the wild type, were much less sensitive to butyramide repression, although all four strains were repressed to some extent at the higher concentrations.

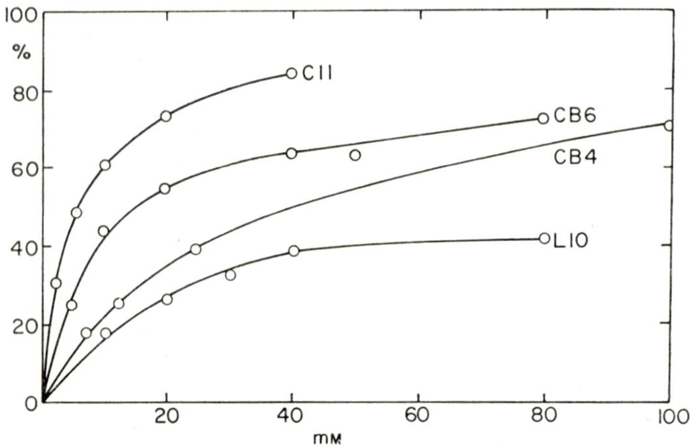

Fig. 3. Repression by butyramide (mM) of amidase synthesis by constitutive mutants of *P. aeruginosa* PACI.

With this enzyme system there is a rather complex regulatory situation determining whether or not growth can occur. The wild-type A enzyme is active enough to allow growth if sufficient enzyme can be synthesised. Earlier studies had shown that butyramide exerted competitive repression of amidase induction in the wild-type strain[3]. These results showed that only certain constitutive mutations would allow growth on butyramide. It was concluded that the cytoplasmic repressor produced by the regulator gene of the constitutive strain C11 still has an amide-binding site and that when butyramide is bound to it the conformation change is such that the repressor is able to bind to the operator and stop amidase synthesis. The constitutive mutants able to grow on butyramide would then be assumed to produce cytoplasmic repressors which had either lost the ability to bind butyramide or, if butyramide was bound, did not undergo a conformation change which increased the probability of binding to the operator. Similar findings of repression of constitutive strains by substrate and inducer analogues have been reported for β-galactosidase[4].

The substrate range of mutant amidases was further extended by the isolation of mutants able to grow on the 5-carbon amide, valeramide and the aromatic amide, phenylacetamide. Strain B6 was used as the parent and a heterogeneous group of mutants were isolated by J. E. Brown on valeramide plates. Some of them had a higher specific activity on butyramide than the parent strain B6, and a few of them hydrolysed butyramide at more than twice the rate at which they hydrolysed acetamide, and some could not grow on acetamide. These strains must have had at least three mutations; the first mutation to constitutivity (C11), the second mutation to produce the B enzyme (B6) and a third mutation which extended the substrate range to include valeramide (V mutations). All the enzymes produced by this group of mutants were unstable in the cell-free state. The differences between the substrate profiles suggests that the mutations were at different sites but that all were deleterious to enzyme stability. However, *in vivo* the enzyme is physiologically efficient and the cultures grow well on valeramide[2].

The mutants which grow on phenylacetamide are again a heterogeneous group and the substrate profiles of independently isolated mutants differ markedly from one another. All of the phenylacetamide-utilizing mutants derived from strain B6 were unable to grow on acetamide but grew on butyramide and valeramide. The enzyme specificity appeared to have shifted in favour of amides with larger side chains. It was interesting that while none of the mutants previously isolated on valeramide plates were able to grow on phenylacetamide, all the phenylacetamide mutants were able to grow on valeramide.

The two groups of mutants, the valeramide group and the phenylacetamide group, are examples of complex mutational situations involving both regulator and structural gene mutations. A rather interesting observation was that higher specific activities towards butyramide were found for some of the mutants isolated from valeramide, and higher activities towards valeramide for mutants isolated on phenylacetamide. This suggests that if a mutant enzyme is required with an activity for a particular substrate it might be an advantage to select for a mutant attacking a larger or smaller substrate to get optimal activity.

Another interesting group of mutants were those which could utilize the N-substituted amide, acetanilide. These were termed AI mutants to distinguish them from mutants attacking the aromatic amides with the substituted group in the side chain. Hydrolysis produces aniline and acetate and the selection medium requires them to use acetate as a carbon source. The substrate specificity of AI mutants differed

TABLE I

Values for Apparent Michaelis Constants (mM) of Three Amidases in *P. aeruginosa* PAC Strains[a]

Substrate	Amidase		
	A	B	AI3
Acetamide	19	15	52.6
Propionamide	55	9.5	208
Butyramide	500	73.5	—

[a] Determinations made using the transferase reaction[2,5].

TABLE II

Peptides from Wild-Type and Mutant Amidase of *P. aeruginosa* PAC Strains

Strain	Amino acid sequence
Wild-type	Ser-Leu-**Thr**-Gly-Glu-Arg
Mutant AI3	Ser-Leu-**Ile**-Gly-Glu-Arg

markedly from those previously described. The enzyme produced by mutant AI3 was found to be less heat stable than both the A and B amidases. This difference in heat stability was not serious enough to interfere with purification and the pure enzyme could be compared in detail with the A and B amidases. Table I shows the apparent K_m values of the three amidase enzymes. The B enzyme has a similar K_m for acetamide as the A enzyme, but the K_m for propionamide is considerably lower and the K_m for butyramide is decreased 10-fold. For technical reasons it was difficult to determine the K_m of the AI enzyme for acetanilide but it can be seen that the K_m values for acetamide and propionamide are significantly altered. Activity of the pure enzyme on butyramide was very low and the AI mutants were unable to grow on butyramide although the parent strain L10 was able to do so.

The peptide maps prepared from a combined trypsin plus chymotrypsin digest of the pure proteins of the A amidase and AI3 amidase showed a difference in the location of one of the peptides. The amino-acid sequence of these peptides was determined and showed a difference in a single amino acid. Table II shows that a threonine residue of the wild-type A amidase had been replaced by an isoleucine residue in the AI3 amidase. This replacement appears to be sufficient to give a protein with significantly different properties so that the substrate range is extended to include the N-substituted amide.

Table III lists some of the mutants which have been described in this paper and shows the differences in utilization of the various amides as growth substrates.

The wild type does not grow on butyramide because it cannot act as inducer; strain C11 does not grow because butyramide represses, while the mutants listed here as the CB group are constitutive, not severely repressed by butyramide, and produce sufficient of the wild-type A enzyme to allow growth. All these strains produce the A amidase and differ in regulatory response to butyramide. The next group of mutants B6, V8, PhB3 were all derived from strain C11 and differ from one another in the amidase proteins which they produce. Their regulatory properties are probably similar to strain C11 in that amidase synthesis is repressed by the same amide analogues. Growth is possible because the enzymes attack the relevant amides. Finally, the AI strains were derived from mutant L10 which was katabolite resistant, and belonged to the group of constitutive strains which, like CB4, vere resistant to butyramide repression. This mutant produces large amounts of the wild-

TABLE III

Amides as Growth Substrates for *P. aeruginosa* PACI and Mutants[a]

Strain	Acetamide	Propionamide	Butyramide	Valeramide	Phenylacetamide	Acetanilide
WT	—	+	—	—	—	—
C11	+	+	—	—	—	—
CB4	+	+	+	—	—	—
B6	+	+	+	—	—	—
V8	+	+	+	+	—	—
PhB3	—	+	+	+	+	—
AI3	+	+	—	—	—	+

[a] Cultures tested on minimal agar plates containing the test amide; + growth, — no growth.

type A enzyme in all media. The mutation of the AI mutants gave them a new enzyme which differing from the wild-type A enzyme only in a single amino acid allowed acetanilide to be used for growth.

REFERENCES

1. BRAMMAR W. J., CLARKE P. H., SKINNER A. J.: J. Gen. Microbiol. **47**, 87 (1967).
2. BROWN J. E., BROWN P. R., CLARKE P. H.: J. Gen. Microbiol. **57**, 273 (1969).
3. KELLY M., CLARKE P. H.: J. Gen. Microbiol. **27**, 305 (1962).
4. JAYARAMAN K., MÜLLER-HILL B., RICKENBERG H. V.: J. Mol. Biol. **18**, 339 (1966).
5. BRAMMAR W. J., CLARKE P. H.: J. Gen. Microbiol. **37**, 307 (1964).

Genetic Control of Biogenesis and Enzyme Formation in Green Algae

RUTH SAGER

Hunter College of the City University of New York, New York, U.S.A.

"The Institute isolated a mutant form of a strain of commonly occurring algae which excreted a broad spectrum antibiotic and was able to fix nitrogen. Because of these two characteristics these algae could be grown in the open in improvised ditches filled with water and they did not require the addition of nitrates as fertilizer. The protein extracted from them had excellent nutritive qualities and a very pleasant taste." (From "*The Voice of the Dolphins*" by Leo SZILARD[1].)

The use of genetically selected strains of algae for industrial production of a protein substitute played a central role in the great prophetic short novel "*The Voice of the Dolphins*" written by Dr. SZILARD several years ago. The attempts now being made in Czechoslovakia to develop mass-scale production of algae are very much in line with SZILARD's prediction. Algal engineering may become of world-wide importance in relation to the world food supply in the foreseeable future. It should be noted that the algal protein postulated by SZILARD was also a contraceptive thus contributing simultaneously to solving the international twin problems of population and food supply.

My purpose is to provide some information on recent developments in basic genetic investigations with algae. In particular I shall discuss the new knowledge which has developed within the last few years concerning the role of autonomous genetic systems located in chloroplasts and mitochondria and their interaction with nuclear genes in regulating growth and development.

The story begins in 1908 with the discovery by the German plant geneticist Carl CORRENS[2] of a gene influencing chloroplast development in the higher plant *Mirabilis* — a gene that did not follow Mendel's laws. In subsequent years numerous investigators described similar patterns of inheritance in a variety of higher plants. By 1950 it was well documented and abundantly clear that many fundamental plant characteristics including chloroplast development and pollen fertility were regulated by so-called non-Mendelian genes[3]. However, no one had managed to establish a readily workable experimental system with which to study these genes in a systematic way.

The problem was too complex to solve with a higher organism. In the history of biology the 1950's represents a landmark decade in which many classical problems of biology were solved with the use of microorganisms. It was the decade in which investigators turned to the simplest possible systems with which to examine fundamental laws of biological organization. So too with non-Mendelian heredity. The correlation of non-Mendelian genes with the DNA of cytoplasmic organelles and the development of cytoplasmic genetics has been carried out exclusively with eukaryotic microorganisms[4,5].

Turning quickly from the past to the present, let us now examine briefly in summary fashion the present status of the problem of "non-Mendelian heredity". The problem has been solved at the genetic level with the development of methods for genetic analysis in algae, yeasts and fungi of systems located in chloroplasts and mitochondria[4]. At the biochemical level the physical identity of non-Mendelian genes has been solved with the discovery of unique sequences of DNA located in chloroplasts and mitochondria[4].

It is now well accepted that eukaryotic cells whether microbial or in higher plants or animals contain organelle genetic systems composed of DNA as well as the classical nuclear genetic systems and that these systems work in synchrony (by mechanisms still unknown) to provide an integrated over-all metabolism. The central role of chloroplasts and mitochondria in energy production and therefore in growth focusses our attention upon the genetics of these organelles and leads us to stress their central importance.

ORGANELLE DNA's IN ALGAE

The first biochemical evidence that chloroplasts might contain a distinctive DNA came from radioisotope labeling studies by IWAMURA[6]. He reported a fast labeling DNA associated with the chloroplast fraction of *Chlorella.* The definitive demonstration of a characteristic or unique DNA in chloroplasts came later with the demonstration that chloroplasts contained a DNA with different average base composition than that of the nuclear component[7,8]. As shown in Table I, the characteristic buoyant densities of nuclear, chloroplast and mitochondrial DNA's have now been demonstrated for a number of algae. The identification of mitochondrial DNA in algae has proven to be very difficult because of the necessity to distinguish it from both nuclear and chloroplast DNA's. This distinction has been achieved best in *Euglena* where colorless mutant strains have been selected which no longer contain either chloroplasts or any detectable chloroplast DNA[9]. In such strains the identification of mitochondrial DNA is considerably easier than in the wild-type material.

In higher plants chloroplast DNA has been identified in a number of organisms, although until recently there was some confusion between mitochondrial and chloroplast DNA's. This problem was resolved by extracting DNA from chloroplasts which had been very carefully purified[10] or treated with DNAase to destroy contaminating nuclear and mitochondrial DNA's[11]. Mitochondrial DNA's of higher plants have

TABLE I

Densities of Organelle DNA's of Algae and Higher Plants

Organism	DNA				Ref.
	Nuclear	Chloroplast	Mitochondrial	Other	
Chlamydomonas	1.724	1.695	(1.715)		8, 34, 35
Chlorella	1.716	1.695			7
	1.717	1.692 1.717	1.712		6
Euglena	1.707	1.685	1.690		9, 36, 37
Acetabularia	(1.702)	1.704	1.714	1.722	38
	(1.702)	1.695		1.724	39
Porphyra tenera	1.720	1.696			40
Tobacco	1.698	1.698			41
	1.690	1.703			42
	1.696	1.706		1.715	43
	1.695	1.697			44
	1.698	1.703			45
	1.697	1.697			10
Spinach	1.695	1.719			7
	1.694	1.696			10
	1.694	1.696			44
Swiss chard	1.689	1.700			47
	1.694	1.696	1.705		44
Mung bean	1.691		1.706		46
(*Phaseolus aureus*)	1.695	1.697			11
Turnip (*Brassica rapa*)	1.692	1.695	1.706	1.700	46
Sweet potato (*Ipomoea batatas*)	1.692		1.706		46
Onion	1.689		1.706	1.718	46
	1.691	1.696	1.706		44
Beet	1.695	1.719			7
Wheat	1.702	1.698			11
Sweet pea	1.695	1.697			11
Lettuce	1.694	1.697			11

been demonstrated most easily in root tissues where mitochondria can be isolated and purified with greater ease than in leaf material. A few characteristic values of bouyant density of organelle DNA's from higher plants are also shown in Table I for comparison with the algal values.

Are organelle DNA's genetically unique? Or do they represent amplified DNA's of nuclear origin? At present the principal evidence that organelle DNA's are unique comes from genetic evidence to be discussed below. However certain DNA–RNA hybridization studies also bear on this problem and they will be discussed in the next section.

Before turning to those data, mention should be made of a new experimental approach to the study of DNA. I refer to physical studies of DNA which measure the kinetics of reannealing of single stranded denatured molecules to reform the double stranded structure. The analyses of reannealing kinetics and its application to biological problems were developed independently by WETMUR and DAVIDSON[12] and by BRITTEN and KOHNE[13].

The principle of the method is straightforward. DNA molecules which have been denatured by heat will reanneal if kept at a suitable temperature (T_m — 25° C.) The rate of reannealing will depend, if other factors are constant, upon the number of copies present. Thus for a given concentration of DNA, the reannealing rate will depend on the genomic complexity: *i.e.* the extent of unique nonrepeating sequences.

Studies of genomic complexity have been carried out with DNA from the chloroplasts of several higher plants[10,14] and from the alga *Chlamydomonas*. The results are similar in indicating that the chloroplasts of both higher plants and of *Chlamydomonas* contain unique DNA sequences of at least 1—2 $\times$ 10^8 daltons. This amount of DNA could code for about 200 proteins of average molecular weight 30,000. What are the functions of the chloroplast DNA?

THE PROTEIN SYNTHESIZING MACHINERY OF ALGAL CHLOROPLASTS

Algal chloroplasts closely resemble the chloroplasts of higher plants with respect to the protein synthesizing machinery contained therein. The presence of ribosomes in algal chloroplasts was first detected in electron micrographs and subsequently positively demonstrated by extraction and characterization[15]. Chloroplast ribosomes are approximately 70S like bacterial ribosomes, whereas the cytoplasmic ribosomes of the same organisms are 80S like those of higher plants and animal cells. A com-

TABLE II

Ribosomes and Ribosomal RNA of Algae

Organism	Source	Sedimentation value		Ref.
		Ribosome	RNA	
Chlamydomonas	cytoplasmic	80	25.1, 17.6	48
	chloroplast	68	22.4, 16.7	48
Euglena	cytoplasmic	86	24, 20	16
	chloroplast	75	23.5, 16.5	16

parison of the properties of cytoplasmic and chloroplast ribosomes in *Chlamydomonas* is shown in Table II in comparison with similar data from *Euglena*.

Hybridization studies have been carried out with RNA's isolated from chroroplast ribosomes and hybridized with the corresponding chloroplast DNA's and nuclear DNA's. In general similar results have been obtained with higher plant systems and with *Euglena*. Comparable studies have not yet been reported with *Chlamydomonas* but are in progress. Chloroplast ribosomal RNA's of *Euglena* hybridize with the corresponding chloroplast DNA's to give a stable DNA—RNA hybrid with a high T_m and a sharp melting profile[16]. A small amount of hybridization has also been reported between chloroplast ribosomal RNA and nuclear DNA's but recent results indicate that the heterologous hybrids are less stable than the homologous chloroplast DNA—chloroplast RNA hybrids[17]. From the available evidence it seems likely that chloroplast ribosomal RNA is transcribed exclusively from chloropast DNA.

Little has yet been reported concerning the tRNA's and corresponding aminoacyl synthetases in chloroplasts, but quite a bit is known about the mitochondrial counterparts in yeast and *Neurospora*. Several tRNA's have been described which are specifically mitochondrial and a few of the corresponding synthetases have been identified as well (reviewed in Ref.[18]).

PROTEIN SYNTHESIS IN ALGAL CHLOROPLASTS

In the previous sections the similarity between chloroplasts of algae and higher plants has been stressed. It seems likely that the nucleic acids and the protein synthesizing machinery are similarly organized in both classes of organisms. However, when we come to considering which proteins are synthesized in the chloroplast, it is possible that greater differences between the algae and higher plants will be evidenced. Thus far the evidence on protein synthesis in isolated chloroplasts comes exclusively from studies with higher plants.

In a series of beautiful experiments, Spencer, Whitfield and their colleagues have examined RNA and protein synthesis in chloroplasts isolated from spinach[19]. Their results show that at least two known proteins, ribulose-diphosphate carboxylase and ferredoxin are synthesized in *in vitro* preparations of isolated chloroplasts. In addition radioactive amino acids are incorporated into a membrane fraction which has not yet been well characterized.

Comparable studies have not been performed as yet with algal chloroplasts primarily because of the difficulties encountered in their preparation. Studies have been carried out with intact algal cells, *Euglena* in particular, utilizing antibiotics in an attempt to distinguish between synthesis on 70S and on 80S ribosomes. In bacterial systems (70S) it is known that the antibiotic chloramphenicol effectively blocks the synthesis of most proteins whereas 80S ribosomes are not affected by this drug. Conversely cycloheximide blocks protein synthesis on 80S ribosomes while having no effect on the 70S class. Despite the clarity of this distinction, studies of the blocking of *in vivo* amino-acid incorporation into protein by antibiotics have not given results that are easy to interpret[20]. The principal reason for the difficulty probably that the cytoplasmic and chloroplast systems are tightly interlocked by regulatory mechanisms, some of which are blocked by antibiotics.

In summary it has been clearly shown that the chloroplasts contain proteins synthesized in the chloroplast as well as proteins synthesized elsewhere in the cell[4]. The complex interactions between the chloroplast and the rest of the cell need to be analyzed in order to further our understanding of cellular organization and growth. One approach to this problem involves the analysis of the chloroplast genetic system *per se* and the next section will deal with that problem.

ORGANELLE GENETIC ANALYSIS

The principal microorganisms in which evidence of non-Mendelian heredity has been documented are the alga *Chlamydomonas*, the yeast *Saccharomyces* and the ascomycete *Neurospora*. With the yeast system, it now appears likely that the cytoplasmic petite mutation leading to loss of respiratory enzymes is the result of alterations in mitochondrial DNA as are mutations to cytoplasmic drug resistance (chloramphenicol and erythromycin). Preliminary evidence of recombination of mitochondrial DNA molecules in yeast has also been reported very recently[21]. It seems likely that mapping of mitochondrial mutations will soon be possible in this genetic system.

Much more extensive evidence about the genetic organization and linkage of a number of mutations has been reported with the alga *Chlamydomonas*[22–26]. The genetic system to be described below probably resides in chloroplast DNA and will be referred to as an "organelle genetic system". Let us turn now to a brief summary of the properties of this system and the methods by which it has been analyzed.

Chlamydomonas is a sexual alga with two mating types determined by a single nuclear gene. The life cycle is very simple: when cells of opposite type are mixed under suitable conditions, they fuse two by two to form the diploid zygote. After several days of maturation the zygote undergoes meiosis and germinates with the liberation of four zoospores which are the four products of meiosis. Two of these are of one mating type and two are the other. All other nuclear genes by which the parents differ also segregate 2 : 2 among these four progeny. This 2 : 2 ratio is a classical Mendelian ratio for a haploid organism.

Non-Mendelian or organelle genes were first identified in *Chlamydomonas* because they did not show this 2 : 2 segragation pattern. On the contrary, they exhibited what we called "maternal inheritance", that is non-Mendelian genes were transmitted from the female or mt^+ parent to all progeny and from the male or mt^- to none. A comparison of these patterns of inheritance is shown in Fig. 1.

This pattern of maternal inheritance is very convenient for distinguishing these organelle gene mutations from the nuclear ones but it has provided a drastic impediment to genetic analysis. Since all such mutations in the female mt^+ parent are transmitted to all the progeny, no recombinants are found and consequently neither

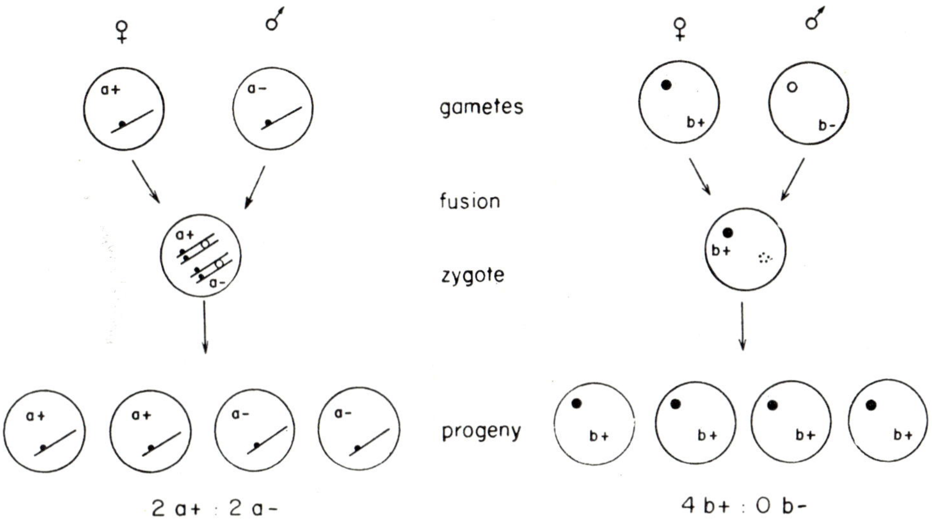

Fig. 1. Chromosomal and non-chromosomal gene segregation in a haploid organism (from Ref.[33]). *Left:* chromosomal gene segregation, *right:* non-chromosomal gene segregation.

recombination analysis nor conventional mapping procedures can be applied. Further genetic analysis required finding a way to circumvent this pattern of maternal inheritance. After some study we found that a low dose of UV-irradiation given to the female parent immediately before mating led to a block in maternal inheritance[24].

The principal feature of biparental inheritance is the transmission of organelle genes from both parents to all of the progeny. Under these conditions we have found that the zoospores (which are haploid for their nuclear complement following meiosis) are diploid and hybrid for their organelle genome having received one complete set from each of the parents. Furthermore, recombination, *i.e.* exchange between the organelle DNA's which came from the two parents, occurs during vegetative growth of the clone issuing from each of the zoospores. The zoospores grow by a regular doubling process, and with the use of suitable markers we have found that genetic recombination of these organelle genes occurs at every doubling. As a result the clones soon develop stable homozygous lines. For example, in a cross of let us say $a+\ b+ \times a—b—$ initially our zoospores are $a+\ b+/a—b—$ but by the time a colony of some 10^6 cells has formed from one zoospore, we find it composed of four cell types: $a+\ b+$, $a+\ b—$, $a+\ b{-}{-}$, and $a—b+$[22].

In subsequent studies, we discovered the existence of linkage between organelle genes[23] and on this basis, developed methods for mapping[24–26]. At present we have placed about 20 genes onto one linkage group or "chromosome", located in organelle DNA, presumably in the chloroplast[27,28].

It is not my purpose in this paper to describe the recombination process or mapping procedures in great detail, but merely to point out that these processes do occur and that they provide the basis for genetic analysis and for the selection of new recombinant types. Thus the chloroplast genetic system is truly a second genetic system in the algal cell to be exploited in the selection of superior strains for industrial purposes.

THE NUCLEAR GENETIC SYSTEM OF *Chlamydomonas*

This article has stressed organelle genetic systems because of their newness on the metabolic scene, but some information should be summarized about the nuclear genetics of *Chlamydomonas*. As noted in Fig. 1, this organism is a simple haploid strain in which the zygote is the only diploid stage in the life cycle under most circumstances. Recently EBERSOLD[29] has described diploid clones which arise with a low frequency after mating. He studied diploids selected with the use of suitable selective markers from a background of zygotes. Diploid strains may be of considerable value in studies of dominance and complementation, and consequently they provide a useful adjunct to the conventional life cycle.

Nuclear genetics of *Chlamydomonas* is classical and very regular. Approximately 100 genes have been mapped into 16 linkage groups[30]. Auxotrophic mutants are rare, probably because most amino acids are not taken up by algae. However arginine-requiring mutants have been described as have mutants requiring *p*-aminobenzoic acid, thiamine, and nicotinic acid. A gene has been mapped that controls the ability of the organism to use nitrate as a sole nitrogen source. A number of non-motile mutants have been studied and genes controlling flagellar motion have been mapped[31].

The largest class of nuclear mutants that have been studied are the so-called "acetate-requiring" strains[32]. These represent a heterogeneous class of mutations, mapping on many of the linkage groups. The majority of the acetate-requiring mutants are able to fix carbon dioxide and the basis of their requirement for acetate is unknown. Mutants which cannot fix CO_2 and are to be considered non-photosynthetic have been studied. Some of them are blocked at particular steps in the photosynthetic electron-transport chain. It is not known whether the mutations have caused changes in components of the electron-transport chain or whether the blocks are the secondary result of primary changes which occurred elsewhere. It seems likely from the large number of acetate-requiring mutants already identified that many genes are involved directly or indirectly in the regulation of photosynthetic activity.

Acetate-requiring mutants are commonly found among the chloroplast gene mutations as well. The position of the mutant block has not been investigated with the chloroplast genetic mutants as yet. Nonetheless the existence of both nuclear and organelle mutations affecting photosynthesis does provide further evidence that chloroplast development is regulated by interactions between these two genetic systems.

CONCLUSION

I wish to stress the central importance and powerful contribution which genetic analysis can play in the industrial development of mass cultures of algae. Genetic methods can uniquely provide information of the following kinds:

(1) Demonstrate whether a particular gene codes for the amino-acid sequence of a particular protein.

(2) Distinguish mutations within the same gene from mutations of different genes affecting the same phenotypic trait.

(3) Identify sets of genes linked into operons.

(4) Identify regulatory sequences such as operators, promoters, initiators, chain terminators.

(5) Distinguish between nuclear and organelle genes.

(6) Determine the number of genetic copies per cell.

(7) Make possible manipulation of the system, *e.g.* to increase the number of copies per cell of particular genes or fragments.

(8) Above all, provide the opportunity by means of selection and recombination to incorporate new gene mutations into existing cell lines.

Mass culture of microorganisms as a source of food has not been popularly received in the past nor has it been extensively developed. In addition to problems of mass culture, yield, nutritional value and balance, an important factor has been the unappetizing flavor. The systematic application of genetic principals has much to offer in the solution of these problems. Microorganisms including algae are remarkably plastic creatures with built-in genetic diversity waiting for the talented investigator. With the proper use of mutagens, selection of promising mutant strains, and recombination by crossing procedures, it should be possible in the future to develop algae with hitherto undreamed of qualities. Why not select not only for high protein and high yield, but also for flavor? Why not select for high sugar content, to get algae that taste like candy? I do not mean to conclude on a frivolous note, but rather to suggest that Leo SZILARD with his nutritive contraceptive algal protein may not have been so fanciful after all.

REFERENCES

1. SZILARD L., in *The Voice of the Dolphins.* Simon & Schuster, New York 1961.
2. CORRENS C.: Z. indukt. Abstamm.-Vererbungslehre **1,** 291 (1908).
3. RHOADES M. M., p. 19 in W. Ruhland (Ed.): *Encyclopedia of Plant Physiology,* Vol. I. Springer-Verlag, Berlin 1955.
4. Austral. Acad. Sci. Symp.: *Autonomy and Biogenesis of Mitochondria and Chloroplasts.* N. K. Boardman, A. W. Linnane, R. M. Smillie (Eds.): North-Holland Publ. Co., Amsterdam 1970.
5. Soc. Exp. Biol. Symp.: *The Control of Organelle Development*, Symp. 24, P. L. Miller (Ed.). Cambridge University Press, London 1970.
6. IWAMURA T., p. 133 in: *Progress in Nucleic Acid Research* Vol. V. Academic Press, New York & London 1966.
7. CHUN E. H. L., VAUGHAN M. H., RICH A.: J. Mol. Biol. **7**, 130 (1963).
8. SAGER R., ISHIDA M. R.: Proc. Nat. Acad. Sci. **50**, 725 (1963).
9. EDELMAN M., EPSTEIN H. T., SCHIFF J. A.: J. Mol. Biol. **17**, 463 (1966).
10. WHITFELD P. R., SPENCER D.: Biochim. Biophys. Acta **157**, 333 (1968).
11. WELLS R., BIRNSTIEL M.: Biochem. J. **112**, 777 (1969).
12. WETMUR J. G., DAVIDSON N.: J. Mol. Biol. **31**, 349 (1968).
13. BRITTEN R. J., KOHNE D. E., p. 78 in: *Yearbook Carnegie Instn.,* 1967.
14. WELLS R. SAGER R.: J. Mol. Biol. **58,** 611 (1971).
15. SAGER R., HAMILTON M. G.: Science **157**, 709 (1967).
16. RAWSON J. R., STUTZ E.: J. Mol. Biol. **33**, 309 (1968).
17. STUTZ E., in N. K. Boardman, A. W. Linnane, R. M. Smillie (Eds.): *Autonomy and Biogenesis of Mitochondria and Chloroplasts.* North-Holland Publ. Co., Amsterdam 1970.
18. Symp. Soc. Gen. Microbiol., Vol. 20.
19. SPENCER D., WHITFELD P. R., BOTTOMLEY W., WHEELER A. M. in N. K. Boardman, A. W. Linnane, R. M. Smillie (Eds.): *Autonomy and Biogenesis of Mitochondria and Chloroplasts.* North-Holland Publ. Co., Amsterdam 1970.
20. SMILLIE R. M., BISHOP D. G., GIBBONS G. C., GRAHAM D., GRIEVE A. M., RALSON J. K., REGER B. J., in N. K. Boardman, A. W. Linnane, R. M. Smillie (Eds.): *Autonomy and Biogenesis of Mitochondria and Chloroplasts.* North-Holland Publ. Co., Amsterdam 1970.
21. COEN D., DEUTSCH J., NETTER P., PETROCHILO E., SLONIMSKI P. P. in *The Control of Organelle Development.* S. E. B. Symp. 24. Cambridge University Press, London 1970.
22. SAGER R., RAMANIS Z.: Proc. Nat. Acad. Sci. U.S.A. **50,** 260 (1963).
23. SAGER R., RAMANIS Z.: Proc. Nat. Acad. Sci. U.S.A. **53**, 1053 (1965).
24. SAGER R., RAMANIS Z.: Proc. Nat. Acad. Sci. U.S.A. **58**, 931 (1967).
25. SAGER R., RAMANIS Z.: Proc. Nat. Acad. Sci. U.S.A. **61**, 324 (1968).
26. SAGER R., RAMANIS Z.: Proc. Nat. Acad. Sci. U.S.A. **65**, 593 (1970).
27. SAGER R., RAMANIS Z., in *The Control of Organelle Development.* S. E. B. Symp. 24. Cambridge University Press, London 1970.
28. SAGER R., RAMANIS Z., in N. K. Boardman, A. W. Linnane, R. M. Smillie (Eds.): *Autonomy and Biogenesis of Mitochondria and Chloroplasts:* North-Holland Publ. Co., Amsterdam 1970.
29. EVERSOLD W. T.: Science **157**, 447 (1967).
30. HASTINGS P. J. *et al.:* Microbial Genet. Bull. **23**, 17 (1965).
31. RANDALL G., WAIR J. R., HOPKIN J. M., MCVITTIE: Nature **203**, 912 (1964).
32. LEVINE R. P.: Science **162**, 768 (1968).
33. SAGER R.: N. Engl. J. Med. **271**, 352 (1964).
34. CHIANG K.-S., SUEOKA N.: J. Cell Physiol. **70**, Suppl. 1., 89 (1967).
35. LANE D., SAGER R., *to be published.*

36. Brawerman G., Eisenstadt J. M.: Biochim. Biophys. Acta **91**, 477 (1964).
37. Ray D. S., Hanawalt P. C.: J. Mol. Biol. **9**, 812 (1964).
38. Green B., Heilporn V., Limbosch S., Boloukhere M., Brachet J.: Proc. Nat. Acad. Sci. U.S.A. **58**, 1351 (1967).
39. Gibor A., p. 321 in T. W. Goodwin (Ed.): *Biochemistry of Chloroplasts,* Vol. II. Academic Press, London 1967.
40. Ishida M., Tadatoshi K., Matsubara T., Hayashi F., Yokomura E.: Res. Reactor Inst. Kyoto Univ. **2**, 73 (1969).
41. Lyttleton J. W., Peterson G. B.: Biochim. Biophys. Acta **80**, 391 (1964).
42. Shipp W. S., Kieras F. J., Haselkorn R.: Proc. Nat. Acad. Sci. U.S.A. **54**, 207 (1965).
43. Green B. R., Gordon M. P.: Science **152**, 1071 (1966).
44. Wells R., Ingle J.: Plant Physiol. **46**, 178 (1970).
45. Tewari K. K., Wildman S. G.: Science **153**, 1269 (1966).
46. Suyama Y., Bonner W. D.: Plant Physiol. **41**, 383 (1966).
47. Kislev N., Swift H., Bogorad L.: J. Cell Biol. **25**, 327 (1965).
48. Hoober J. K., Blobel G.: J. Mol. Biol. **41**, 121 (1969).

5/ Genetic Control of Biosynthetic Pathways

Regulation of Amino-Acid Biosynthesis of the Aspartate Pathway in Different Microorganisms

PRASANTA DATTA, STEPHEN M. DUNGAN, ROSS S. FELDBERG

Department of Biological Chemistry, The University of Michigan, Ann Arbor, Michigan, U.S.A.

The metabolic activity of the living cell largely reflects the delicate balance between two interdependent cellular processes – the synthesis and the utilization of essential metabolites. Since the composition of growing cells changes constantly, and since balanced growth is dependent on continual readjustments to these variations in the cellular environment, it is reasonable to expect that an efficient "coupling" should exist between the rates of synthesis and the rates of utilization. This point of view is generally based on the implicit assumption that the biosynthetic apparatus of the cell is always geared for maximal economy; although, with our present knowledge, we do not know if this assumption is valid. Detailed investigations in the last decade have clearly established that the end products of the biosynthetic sequence are directly implicated in adjusting the rates of their own synthesis. The negative feedback devices, repression of enzyme synthesis and end-product inhibition of enzyme activity, often mutually nonexclusive, are triggered by the accumulation of end products over that required for the normal rate of utilization. Although both these mechanisms are efficient in the regulation of oversynthesis, it appears that control by repression is slow and rather less flexible as a regulatory device as compared to feedback control at the enzyme level. Regardless of the enzyme—end-product combination, only the concentration of enzyme is effectively modulated by end-product repression; it is often argued that most enzymes in the cell are present in excess, and unless the step involved is the rate limiting reaction, a 20 to 50% reduction in the enzyme concentration may not affect the overall synthetic rates.

Regulation of enzyme activity, on the other hand, represents a rapid and more flexible device in which the catalytic activity can be influenced by (a) inhibition of activity, (b) stimulation of activity, (c) compensatory effects of modifiers of opposing influence, (d) synergistic effects of metabolites, and (e) cumulative or additive influence of several effectors. Additionally, the concentration of substrate itself plays an important part in the rate of product formation. In those enzymes where the substrate saturation curves are sigmoidal, a relatively small increase or decrease in the substrate concentration (from that required to half-saturate the enzyme, *i.e.*, $[s]_{0.5}$) will drastically influence the rate of enzyme reaction. It is important to note that in this instance no metabolite other than substrate is involved in the control of enzyme function. The concentration of substrate is also critically important in the modulation of enzyme activity when an end-product inhibitor is kinetically competitive with respect to the substrate.

Nevertheless it is apparent that a combination of end-product repression and feedback-inhibition controls is a powerful tool to ensure a balanced synthesis of metabolites in response to the exacting physiological requirements of the growing cells. Since the genetic expression and metabolic demands in different organisms are not constant, it is not surprising therefore, that even within a given biochemical pathway, considerable variations of these two basic mechanisms are so frequently observed. We have examined the regulation of biosynthesis of amino acids of the aspartate pathway in several Gram-negative bacteria. Elsewhere data are presented on the specific control of other biosynthetic paths in bacteria and in several eukaryotes. Many of these studies have revealed multiple metabolic feedback patterns involving several amino-acid end products which, individually or in various combinations, efficiently control key enzymatic reactions[1].

ASPARTATE PATHWAY

In bacteria, aspartic acid is the common precursor for the synthesis of several important amino acids, namely, lysine, methionine, threonine, and isoleucine (Fig. 1). According to this scheme the first two enzymes, aspartokinase and aspartate-β-semialdehyde dehydrogenase, are common for the synthesis of all four amino acids; the third enzyme, homoserine dehydrogenase, is shared for the synthesis of three amino-acid end products. The rates of synthesis of two branch-point intermediates, aspartate β-semialdehyde and homoserine, are therefore crucial, and these early enzymes required to synthesize these intermediates may be considered as the focal control points in the context of pathway-wide regulation.

Aspartokinase. The first reaction of the sequence involves the phosphorylation of aspartic acid to β-aspartylphosphate by the enzyme aspartokinase; the reaction requires adenosine triphosphate and a divalent cation. In *Rhodopseudomonas capsulata*, the activity of this first enzyme is not influenced by any single amino-acid end product, lysine, methionine, threonine, or isoleucine; however, when lysine and threonine are present simultaneously, the enzyme activity is severely inhibited (Fig. 2). A 50% reduction in enzyme activity is observed at 1 mM concentrations of each of the two amino acids[2]. The inhibition is specific for the L-lysine-plus-L-threonine combination, and a maximum of about 80% of the total enzyme activity is abolished at saturating levels of these amino acids.

Although the aspartokinase activity of *Rhodospirillum rubrum*, a closely related photosynthetic bacterium, is also subject to concerted feedback control by lysine-

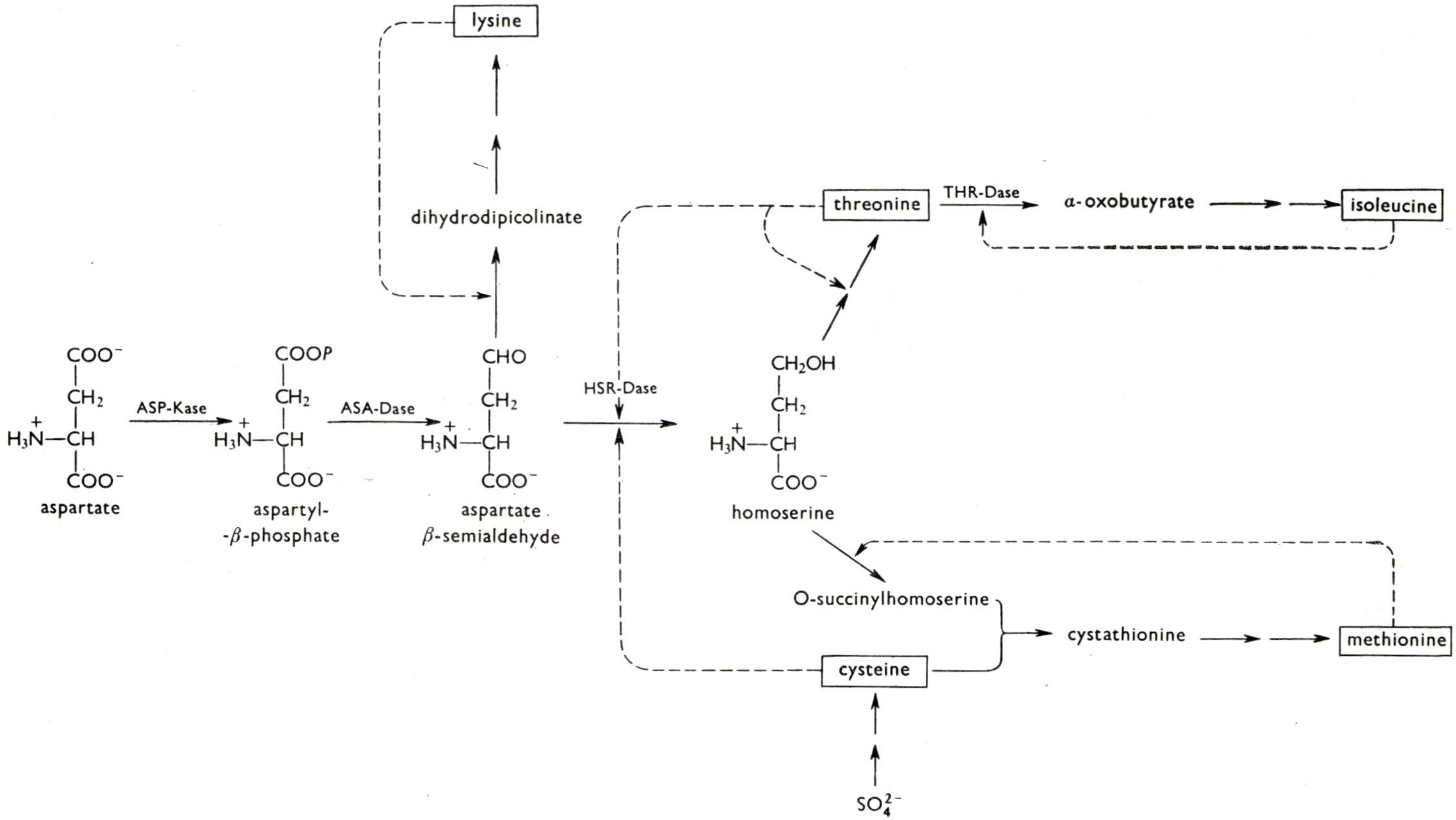

Fig. 1. Summary scheme of aspartate pathway showing regulation of the individual terminal branches and branch-point enzymes. **ASP-Kase** aspartokinase; **ASA-Dase** aspartate-β-semialdehyde dehydrogenase; **HSR-Dase** homoserine dehydrogenase; **THR-Dase** threonine deaminase.

plus-threonine, two important differences are observed in the regulation of aspartokinase in this organism. Firstly, L-threonine by itself is a strong inhibitor of enzyme activity, although the inhibition is not complete even at very high concentrations of threonine; L-lysine alone has no significant effect on the phosphorylation reaction. Secondly, the most important aspect of the concerted feedback is that L-isoleucine and

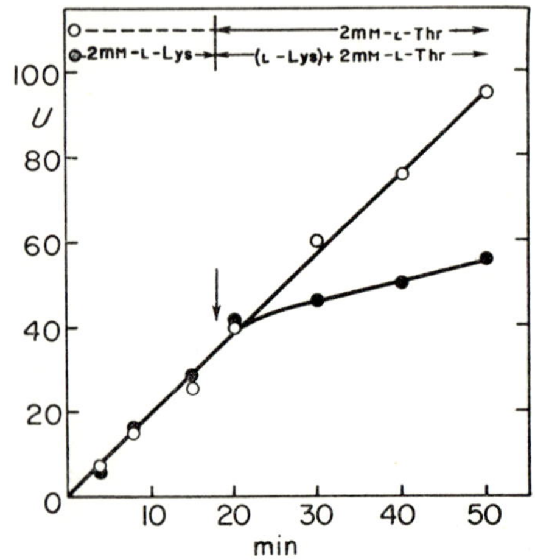

Fig. 2. Effects of L-lysine, L-threonine, and L-lysine + L-threonine on *R. capsulata* aspartokinase activity (units, *U*). Two reaction mixtures were incubated at 26 °C with or without 2 mM L-lysine. At times denoted by the points aliquots were removed and the quantities of aspartohydroxamate estimated. At 18 min *(arrow)*, both solutions were supplemented with 2 mM L-threonine[2].

TABLE I

Concerted Feedback Inhibition of *R. rubrum* Aspartokinase and its Reversal by Isoleucine and Methionine[a]

L-Amino acid added (mM)				Enzyme activity[b]
lysine	threonine	isoleucine	methionine	
—	—	—	—	100
1	—	—	—	100
—	1	—	—	69
—	—	5	—	163
—	—	—	5	165
1	1	—	—	22
1	1	5	—	103
1	1	—	5	83
—	1	5	—	164

[a] The enzyme preparation was obtained by ammonium sulfate fractionation (40—60%) of sonic extract after removal of nucleic acids with protamine sulfate. The precipitate was dissolved in 0.1 M potassium phosphate buffer pH 7.5 containing 0.01 M β-mercaptoethanol and 1 mM each of L-lysine and L-threonine. Just before assay the enzyme solution was passed through a Sephadex G-25 column equilibrated with 0.1 M potassium phosphate—0.01 M β-mercaptoethanol buffer, pH 7.5.

[b] Aspartokinase activity was measured as described[2,30].

L-methionine, individually, can counteract the inhibition due to lysine-plus-threonine or by threonine alone (Table I).

The simultaneous requirement for the lysine-plus-threonine combination for the regulation of aspartokinase is not absolute. In *Pseudomonas putida* and in *P. fluorescens* for example, in addition to concerted feedback by lysine-plus-threonine[3,4],

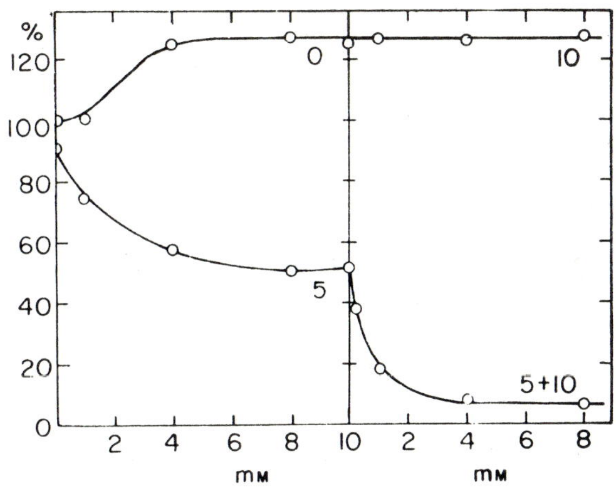

Fig. 3. *Left:* Progressive inhibition of *P. fluorescens* aspartokinase with increasing concentrations of L-methionine (mM) at a constant level of L-threonine (5 mM; 5), or without L-threonine (0). *Right:* Further inhibition of enzyme activity when the concentrations of L-lysine were increased (mM) at a constant level of L-threonine and L-methionine (5 mM and 10 mM, respectively; 5 + 10), and at a constant level of L-methionine (10 mM; 10). No inhibition is observed when L-lysine and L-methionine were present together[4].

TABLE II

Effect of Aspartate β-Semialdehyde on the Aspartokinase Activity of *R. spheroides*[a]

L-Aspartate β-semialdehyde mM	Per cent enzyme activity	
	preparation I[b]	preparation II[c]
0	100	100
1.1	78	81
2.2	67	—
4.4	52	51
6.6	41	—
8.8	32	21

[a] From Ref.[5].

[b] Dialyzed cell-free sonic extract.

[c] 240-fold purified enzyme.

the enzyme activity can also be significantly inhibited at relatively low concentrations of methionine-plus-threonine[4] (Fig. 3). In these organisms threonine alone also inhibits enzyme activity, whereas methionine and lysine separately have a slight stimulatory effect. The fact that methionine does not influence the inhibition due to threonine-plus-lysine suggests that there are independent sites on the protein for the binding of these amino-acid modifiers.

A completely different inhibition pattern of aspartokinase is observed in the bacterium *Rhodopseudomonas spheroides*. The enzyme activity of this organism is strongly inhibited by aspartate β-semialdehyde (Table II); the inhibition is competitive with respect to aspartate and ATP[5]. No other amino acid of the aspartate pathway has any significant effect on the production of β-aspartylphosphate[5,6]. An interesting finding with the aspartokinase of *R. spheroides* is that although the enzyme activity is not affected by lysine, methionine, or threonine, the enzyme molecules can bind these amino acids as judged by the protection of enzyme activity against heat inactivation.

It is clear from these data that the activity of the first enzyme of the aspartate pathway can be influenced by a variety of amino-acid modifiers, and that the mode of inhibition or activation is rather unique depending on the organism in question. In addition, in the *Rhodopseudomonas*[1,7] and *Pseudomonas* species examined thus far[3,8], the rate of enzyme synthesis is repressed by methionine and other amino acids.

Apart from the metabolite control of aspartokinase activity, concentration of substrates would seem to play a minor role in the rate of phosphorylation of aspartic acid. In most cases examined the aspartate and ATP saturation kinetics are of the Michaelis-Menten type, and the feedback inhibitors are kinetically noncompetitive with the substrates; the inhibition of enzyme activity by aspartate β-semialdehyde in *R. spheroides* is, however, competitive with respect to aspartate and ATP, and thus the concentration of this intermediate may not fall below a critical level. It is also noteworthy that when aspartokinase activity is assayed in the absence of hydroxylamine in the reaction mixture[5] both aspartate and ATP show negative cooperativity, that is, the binding of one molecule of the substrate actually decreases the ability of the second substrate molecule to bind on the enzyme[9].

Homoserine dehydrogenase. Considering the biochemical scheme for the synthesis of methionine, threonine, and isoleucine (Fig. 1), the enzyme homoserine dehydrogenase would be expected to be an important step in the context of pathway-wide control. In all bacterial strains and other organisms examined thus far, the activity of homoserine dehydrogenase has been found to be under feedback-inhibition control by L-threonine and L-cysteine, although important species differences exist between the actual mechanisms of modifier effects at the molecular level[1]. The rate of enzyme synthesis can also be repressed by methionine and other end products. The data available on the control of aspartate-β-semialdehyde dehydrogenase, the enzyme preceding the step required for homoserine synthesis, are limited, and therefore the regulatory role of this enzyme cannot be assessed with certainty; in *E. coli* the enzyme is repressible by L-lysine[10], and L-cysteine inhibits its activity in *S. cerevisiae*[11].

A detailed study of the homoserine dehydrogenase of *R. rubrum*[12-14] has indicated several interesting features in regard to its control characteristics. Both in crude extracts and in homogeneous enzyme preparations, L-threonine is a strong inhibitor with a K_i of about 2 mM. The inhibition is nontotal and shows competitive kinetics with respect to homoserine but noncompetitive inhibition in relation to aspartate β-semialdehyde. Under certain conditions the enzyme activity can be stimulated by L-isoleucine or L-methionine; when added simultaneously with the inhibitor L-threonine, the activating amino acids can reverse the threonine inhibition[12]. L-Cysteine is also an inhibitory amino acid[15], and this inhibition by L-cysteine is not reversed by isoleucine or methionine.

The effects of these amino-acid modifiers are highly dependent on the concentrations of certain cations, especially K^+. In buffer solutions containing low K^+, the enzyme is dissociated to fully active half-molecules with concomitant desensitization with respect to allosteric modifiers[14]; addition of salt reassociates the half-molecules, and resensitizes the enzyme. In high K^+-containing buffer, on the other hand, the *R. rubrum* enzyme exists in a different conformational state which exhibits enhanced stimulation of enzyme activity by L-isoleucine and a slightly decreased inhibition by L-threonine; in the presence of high K^+, isoleucine or methionine reversal of threonine inhibition is unaffected. This feature is unique to the homoserine dehydrogenase of *R. rubrum*, and is the basis for the compensatory feedback control described below.

In the context of structural modification of the enzyme in the presence of various ligands, it is sufficient here to point out that the *R. rubrum* enzyme shows a variety of conformational changes and association–dissociation behavior; for example, L-threonine, but not L-cysteine, induces a drastic configurational alteration resulting in the aggregation of the native enzyme — a phenomenon directly involved in the feedback control of this enzyme[14].

In contrast to the *R. rubrum* enzyme the dehydrogenase activities from *R. capsulata* and *R. spheroides* are affected only by L-threonine[6,16] and L-cysteine[15]; threonine inhibition is competitive with respect to both homoserine and aspartate β-semialdehyde with a K_i of about 5 mM. No other amino acid has any inhibitory or stimulatory action on the enzymes. Addition of isoleucine and methionine, individually or together, to the reaction mixture containing L-threonine does not reverse the inhibition by the latter amino acid (Table III). Preliminary experiments indicate the enzyme from *R. capsulata* does not show threonine-dependent association—dissociation reactions, and the enzyme is not desensitized in buffer solutions containing low concentrations of K^+ ions.

TABLE III
Effects of Various Amino Acids on *R. capsulata* Homoserine Dehydrogenase Activity[a]

L-Amino acid added[b]			Enzyme activity	
threonine	isoleucine	methionine	0.06 μM[c]	0.25 μM[c]
—	—	—	100	100
+	—	—	47	58
—	+	—	95	103
—	—	+	105	105
+	+	—	47	61
+	—	+	47	62
—	+	+	100	100

[a] The enzyme preparation was obtained by ammonium sulfate fractionation (31—45%) of a crude cell-free extract in 0.05 M potassium phosphate buffer pH 7.5 containing 1 mM each of EDTA and β-mercaptoethanol. Prior to assay enzyme solution was dialyzed to remove salt. Enzyme activity was assayed at two different substrate concentrations according to the procedure described[12].

[b] All amino acids were 10 mM[16]. The following amino acids, at 10 mM each, had no stimulatory or inhibitory effect: L-valine, L-lysine, L-leucine, L-norleucine, and L-phenylalanine.

[c] L-Aspartate β-semialdehyde concentration.

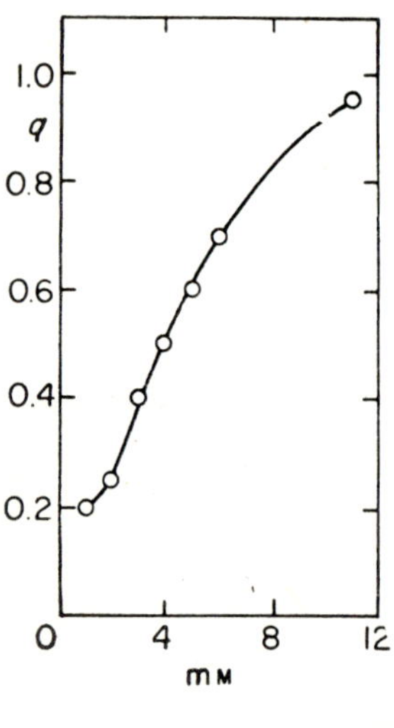

Fig. 4. Loss of L-threonine inhibition of homoserine dehydrogenase of *P. fluorescens* ($V_{Thr}/V_{control}$, q) with increasing concentrations of K^+ (mM). Enzyme activity was measured in the presence and absence of 10 mM L-threonine at various levels of K^+ as specified[17].

Recently we have explored the catalytic and regulatory properties of the homoserine dehydrogenase of the aerobic pseudomonad *P. fluorescens*[17]. As is true with the other bacterial enzymes, the activity of *P. fluorescens* enzyme is also inhibited by L-threonine and L-cysteine. Aspartic acid, lysine, methionine, isoleucine, and valine, at 10 mM each, do not affect enzyme activity[17]. An extensive study by COHEN *et al.*[3,18] on a variety of *Pseudomonas* species has confirmed the same general conclusions.

The most interesting aspect of the threonine inhibition of homoserine dehydrogenase of *P. fluorescens* is its dependence on K^+ ion. At 2 mM K^+, 80% of the enzyme activity is inhibited by 10 mM L-threonine, whereas with increasing K^+ concentrations the inhibitory effect of threonine is progressively decreased until, at 10 mM K^+, in the assay mixture a complete loss of inhibition is observed[17] (Fig. 4). Furthermore, K^+ is also obligatory for the maintenance of the catalytic activity. At 2 mM-K^+, the half-life of the enzyme at 25° and at pH 8.0, is about 10 min; no loss occurs at 25° in 30 min when the buffer solution is supplemented with 20 mM-K^+ ions. These effects of monovalent cation are highly specific for K^+; Rb^+ can replace K^+ to a limited extent. All other cations tested, including Li^+, Na^+, NH_4^+, and Cs^+, are either ineffective or are effective to a very small extent[17]. The *Pseudomonas* enzyme therefore shows regulatory properties which are quite distinct from that of the enzymes from *R. capsulata*, *R. spheroides*, and especially from *R. rubrum* insofar as the effects of cations are concerned.

TABLE IV

L-Isoleucine Inhibition of Threonine Deaminase Activities from *R. spheroides* and *R. rubrum*[a]

Organism	Strain	L-Isoleucine concentration mM	Inhibition[b] %
R. spheroides	wild type[c]	0	0
		0.005	25
		0.01	65
		0.025	95
	mutant[d]	0	0
		5	15
		10	55
		20	90
R. rubrum	wild type[e]	0	0
		5	0
		10	2
		20	5

[a] Adapted from Refs[20-22].

[b] Enzyme activities were measured as described[20] at pH 7.5 with 10 mM L-threonine as the substrate.

[c] Enzyme extracted from the wild-type organism and was partially purified by ammonium sulfate treatment and DEAE-cellulose chromatography[21].

[d] A feedback-resistant enzyme, isolated and purified by salt-fractionation and ion-exchange chromatography[22].

[e] Enzyme purified to homogeneity according to the method described.[20]

Threonine deaminase. The biosynthetic path given in Fig. 1 shows that the enzyme threonine deaminase is directly involved in the synthesis of only one end product of the aspartate family, isoleucine. Nevertheless this enzyme is an important control point since the availability of both threonine and isoleucine for macromolecular synthesis is dependent on the rate of transformation of L-threonine to α-oxobutyrate, an isoleucine precursor. In most bacteria and in higher organisms, the intracellular pool of α-oxobutyrate is precisely adjusted by the concentration of threonine as well as by isoleucine. In general the concentration of L-threonine required to saturate the enzyme is quite high — in the range of 10 to 20 mM; this low substrate affinity for the enzyme may be physiologically significant since the enzymatic step is not reversible and a high rate of α-oxobutyrate formation could lower the threonine pool sufficiently to cause transient slow-down of protein synthesis. Furthermore, threonine saturation curves are either Michaelis-Menten type or sigmoidal depending on the bacterial origin. In most cases, inhibition of enzyme activity by L-isoleucine is kinetically competitive with respect to L-threonine; that is, the extent of inhibition is sensitive to the exact concentrations of L-threonine. This latter feature of threonine deaminase is perhaps designed to ensure no excess build up of threonine which might otherwise repress the synthesis of and/or inhibit the activities of homoserine dehydrogenase and aspartokinase and thus cause lysine and methionine deprivation.

The enzyme from *R. rubrum* is the only example of a biosynthetic threonine deaminase where no rigid feedback control by isoleucine has been detected[19,20]. Table IV shows the comparative aspects of isoleucine inhibition in two species of bacteria. The lack of effective isoleucine control of the *R. rubrum* enzyme is a crucial factor in proposing a novel overall regulatory pattern for the synthesis of the amino acids of the aspartate family discussed below.

MULTIPLE METABOLIC FEEDBACK PATTERNS

Analysis of the experimental data with aspartokinase, homoserine dehydrogenase, and threonine deaminase in a variety of microorganisms reveals several distinct overall control patterns[1]. In the coliform bacteria including the genera *Escherichia, Salmonella, Aerobacter, Serratia, Erwinia*, and *Proteus*, isoenzyme pattern seems to be the principal mode of control[3]. In many other bacteria the concerted action of two end products controls the early enzymes. The aspartokinase of *R. capsulata* is a suitable example to illustrate this control pattern. In this bacterium both lysine and threonine are required to inhibit the enzyme activity; individually they do not. Several variations of this basic scheme of concerted feedback have been observed in a few species of *Bacillus*[23] and *Pseudomonas*[3,4], in *R. rubrum, Brevibacterium flavum*[24] and in a variety of other organisms: (a) in most cases, although the lysine-plus-threonine combination drastically inhibits the aspartokinase activity, these amino acids themselves are also strongly inhibitory when added separately to the reaction mixture, (b) in *R. rubrum* and in *B. flavum*, the concerted action by lysine-plus-threonine can be reversed either by isoleucine, methionine, or valine, and (c) in *P. putida* and *P. fluorescens*, in addition to lysine-plus-threonine feedback, the amino-acid pair methionine-plus-threonine also inhibits the aspartokinase activity (*see* Fig. 3); some inhibition by the same combination of amino acids has also been reported in *B. polymyxa*[25], however, the concentrations required to elicit this effect are too high (40 to 50 mM each) to be considered physiologically significant.

Whatever the minor modifications may be it is well-established that concerted feedback inhibition is an important mode of regulation of the early enzymes of a branched biosynthetic pathway (Fig. 5). In addition to the inhibition control of enzyme activity by more than one metabolite, synthesis of certain biosynthetic enzymes can also be repressed by simultaneous presence of several end products[26]. The dashed lines in Fig. 5 show the multivalent repression by threonine-plus-isoleucine of the aspartokinase, homoserine dehydrogenase, and threonine synthase.

A completely distinct control scheme in which a branch-point intermediate, and not the biosynthetic end products, controls the activity of the first enzyme, aspartokinase, has been reported in *R. spheroides*. In this organism the activities of threonine deaminase, homoserine dehydrogenase, and aspartokinase are inhibited by isoleucine, threonine, and aspartate β-semialdehyde, respectively (Fig. 6). In this case isoleucine, the final end product of the terminal branch, inhibits threonine deaminase causing threonine accumulation which in turn prevents the conversion of aspartate β-semialdehyde to homoserine; the concentration of aspartate β-semialdehyde is thus increased resulting in the slowing down of the aspartokinase reaction. The overall regulatory pattern is a sequential event of progressively lowering the concentration of the earlier metabolites by the negative cascading effect (Fig. 6). This particular pattern of sequential feedback is not restricted to the aspartate pathway, and has been observed in the aromatic-amino-acid biosynthesis of several species of *Bacillus*[27].

Although the aspartokinases and homoserine dehydrogenases of both *R. rubrum* and *R. capsulata* are subject to feedback inhibition by lysine-plus-threonine, and threonine, respectively, an important characteristic that distinguishes these organisms is the reversal of inhibiton by isoleucine or by methionine. In *R. rubrum* concerted feedback of aspartokinase is counteracted by isoleucine or methionine (Table I), and the threonine inhibition of homoserine dehydrogenase activity is also reversed by these amino acids[12]; the inhibition of both aspartokinase and homoserine dehydrogenase activities from *R. capsulata*, on the other hand, is not influenced by isoleucine or by methionine[2] (also *see* Table III). Additionally, threonine deaminase of *R. rubrum* is not a focal control point in isoleucine biosynthesis. In view of these observations the pathway-wide control in *R. rubrum* appears novel and is easily distinguished from that seen in *R. capsulata* (concerted feedback) or in *R. spheroides* (sequential feedback); this scheme, designated as compensatory feedback, involves feedback inhibition by one end product (threonine) or by a combination of end

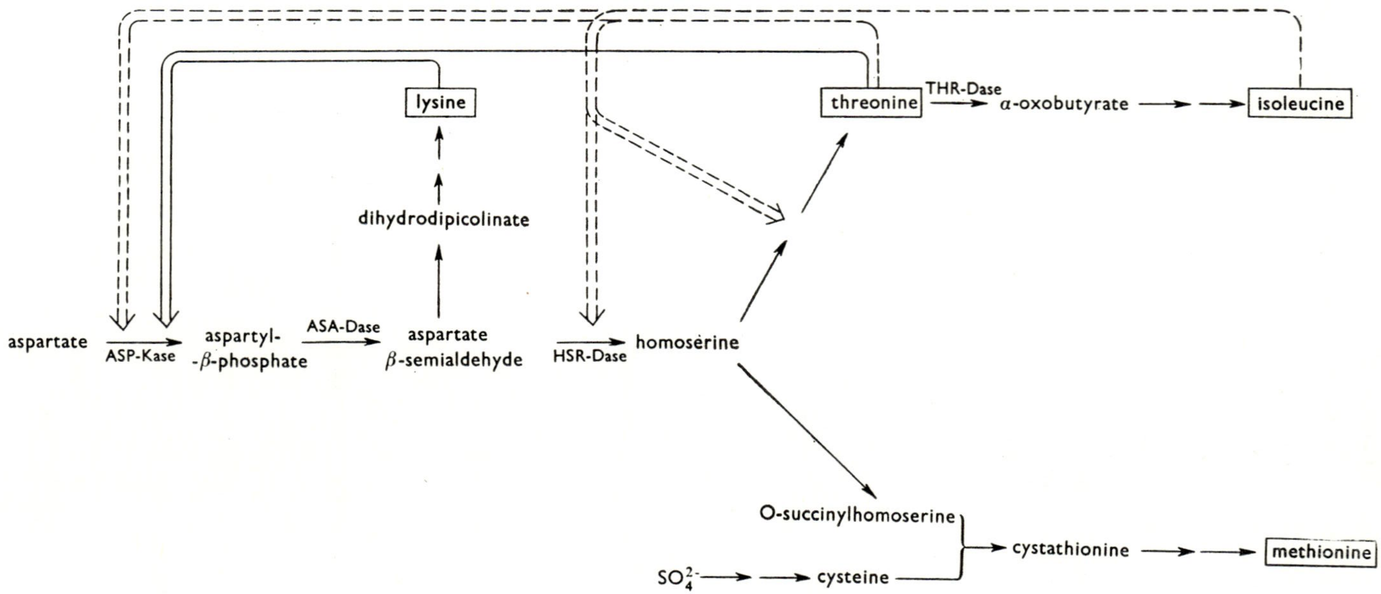

Fig. 5. Summary scheme of aspartate pathway showing concerted feedback inhibition *(solid lines)* and multivalent repression *(dashed lines)*. *See* Fig. 1 for the controls of the individual terminal branches.

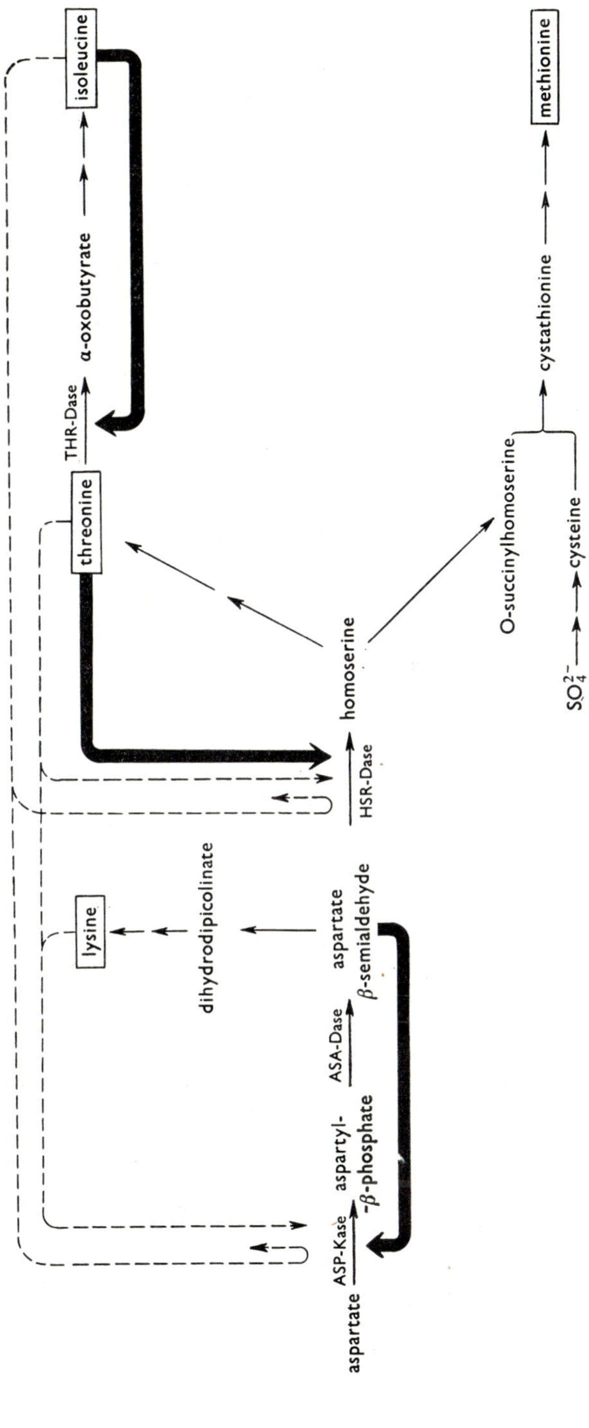

Fig. 6. Summary scheme of aspartate pathway showing sequential feedback *(solid lines)* and compensatory feedback *(broken lines)*. *See* Fig. 1 for the controls of the individual terminal branches.

TABLE V

L-Cysteine Inhibition of Homoserine Dehydrogenase Activity from a Variety of Bacterial Species[a]

Organism	Inhibition by 10 mM L-cysteine %
Escherichia coli[b]	94
Rhodopseudomonas capsulata[b]	78
Rhodopseudomonas spheroides[b]	79
Rhodospirillum rubrum[b]	60
Pseudomonas fluorescens[c]	90
Pseudomonas aeruginosa[c]	100
Pseudomonas putida[c]	100

[a] Enzyme activities were measured as described[15].
[b] For preparation of enzymes *see* Ref.[15].
[c] For preparation of enzymes *see* Ref.[17].

products (lysine-plus-threonine) that can be compensated by a different end product (isoleucine) which shows an opposite effect by stimulating enzyme activity in the absence of the inhibitors[1].

In the aspartate pathway, homoserine is the branch-point precursor for both methionine and threonine; control of homoserine synthesis by these end products is highly desirable. Although threonine is an effective feedback modifier of homoserine dehydrogenase and aspartokinase, in bacteria these enzymes are not subject to inhibition control by methionine; in most organisms methionine represses the synthesis of these enzymes. In view of this observation, the role of L-cysteine as a potent inhibitor of several bacterial homoserine dehydrogenases (Table V) can be rationalized to have especial regulatory significance. The inhibition is stereospecific for L-cysteine; modification of either the -SH or the -COOH, or both, of L-cysteine results in complete abolition of the inhibitory property. Two other intermediates of the methionine branch, cystathionine and homocysteine, have no significant effect on enzyme action[15]. The *in vitro* results strongly suggest that an increase in the cysteine pool in the cell may prevent the overproduction of methionine by curtailing the concentration of homoserine available for the condensation reaction between cysteine and O-succinylhomoserine to form cystathionine. Since cysteine *per se* is not a direct intermediate in bacterial methionine synthesis but rather is a terminal metabolite of a separate series of biochemical reactions, the results indicate a dual regulatory role for this amino acid; on the one hand, cysteine controls the activity of enzymes involved in its own synthesis, and on the other this amino acid functions as a regulatory modifier of a linked, or interconnecting, pathway[1].

PHYSIOLOGICAL RELEVANCE

In any discussion of biological control mechanisms it is highly relevant to ask the simple, yet the most difficult question: is it justified to extrapolate the data obtained *in vitro* to situations as they exist *in vivo*? The rationalization is particularly difficult in those cases where the control of enzyme activity by an end-product metabolite is used as the basis to elucidate the overall metabolic regulation of a series of biosynthetic reactions. The rate of transformation of the substrate to a product may be variously influenced by the concentrations of enzyme and substrates *in situ*, by the levels of various ions, including H^+, by the localized concentrations of the modifying metabolites due to cellular compartmentation, and by other factors such as kinetic peculiarities of the enzyme with respect to substrates and modifiers. Some of these factors can be explored in the reconstituted systems, but many others cannot be tested. For example, at low but not at high concentrations of K^+ the homoserine dehydrogenase of *R. rubrum* dissociates to active half-molecules that are insensitive to feedback inhibition by threonine, whereas the *P. fluorescens* dehydrogenase is maximally sensitive to threonine inhibition only at low K^+ concentrations but becomes completely insensitive to inhibition at 10 mM-K^+. Presumably with changing concentrations of this monovalent cation in the cell the synthesis of homoserine can be affected drastically due to loss or gain of regulatory controls. The essential question in this case is what is the K^+ concentration in the cell, especially at the site of homoserine synthesis? Although the answers to this type of question are not always forthcoming, it is reasonable to postulate that the usual controls by end-product metabolites may undoubtedly be supplemented by one or more of these intricate and subtle devices in modulating the biological efficiency of the regulatory enzymes. If the *in vitro* observations are combined with the results of appropriate physiological experiments it should be possible to arrive at some general conclusions in regard to pathway-wide regulation.

The experimental data at the enzyme level that led to the elucidation of the overall control patterns such as concerted feedback, linked-pathway control and compensatory feedback are compatible with the results obtained by cultivating these bacterial strains under simulated conditions of excess end products by adding preformed amino acids to the basal growth media. If the concerted feedback mechanism by lysine-plus-threonine can indeed operate as regulatory device, it is predicted that, in growing cells, relatively high concentrations of this amino-acid pair will shut-off aspartokinase reaction and will lead to an undersynthesis of methionine; the deficiency will result in an "pseudoauxotrophy" and permit a slow growth or no growth at all. If, however, sufficient methionine were to be available during growth on lysine and threonine, normal growth would ensue. This prediction was borne out when L-lysine-plus-L-threonine, at 1 mM and 0.5 mM, respectively, caused severe reduction in the growth rate of *R. capsulata*, and an additional supplement of 0.5 mM L-methionine along with lysine-plus-threonine, resulted in almost normal growth[7]. The possibility that the combination of lysine-plus-threonine inhibited growth by multivalent repression was ruled out by comparing the rates of synthesis of aspartokinase under various growth conditions. These results strongly implicate concerted feedback as a regulatory device in growing bacterial cells.

The physiological basis of linked-pathway control has also been tested by appropriate growth experiments. Growth of *R. rubrum* is strongly inhibited by addition of 0.6 mM L-cysteine to the basal medium presumably due to interference with methionine and threonine biosynthesis; this inhibition in growth is expected from the *in vitro* results that L-cysteine is a potent feedback inhibitor of homoserine dehydrogenase. A combination of 0.4 mM L-methionine and L-cysteine resulted in a significant increase in the growth rate, and when the basal medium was supplemented with L-cysteine, L-methionine, and L-threonine (0.6 mM, 0.4 mM, and 0.5 mM, respectively) the rate of growth of *R. rubrum* was indistinguishable from that in the basal medium alone without any amino acids[15]. Since either L-threonine or L-methionine, or both, cannot overcome the *in vitro* inhibition of homoserine dehydrogenase activity by L-cysteine, it is reasonable to conclude that in growing culture suppression of growth by L-cysteine is due to depletion of the intracellular homoserine pool resulting in the deprivation of methionine and threonine. If L-cysteine suppressed the growth rate by a mechanism other than that proposed above, it is highly unlikely that addition of methionine and threonine would compensate the cysteine effect. In view of this, and without any experimental evidence on the contrary, it is premature to discard the cysteine effect as "peculiar" and of no physiological consequence[28].

An analogous series of metabolic studies in which the inhibitory effect of high concentrations of L-threonine on growth of *R. rubrum* is shown to be counteracted by L-isoleucine[29] provides the rationale for designating a physiological role of compensatory feedback control for the synthesis of amino acids of the aspartate pathway.

From the above discussion it is clear that metabolite control of biosynthetic enzymes is vastly different from one organism to another. With a given enzyme the catalytic efficiency of the protein is modulated by a variety of allosteric modifiers. In the coliform bacteria isoenzymic pattern seems to be of most common occurrence, whereas concerted feedback is predominantly observed in most species of the genera *Bacillus* and *Pseudomonas*. In *R. spheroides* and in *B. licheniformis*, sequential feedback has been reported. Whatever the overall regulatory pattern may be, it is reasonable to assume that each scheme is adequate for the maintenance of the delicate balance between the synthesis and utilization of small molecular weight metabolites in the cell. It is also reasonable to expect that if a given control pattern is indeed highly efficient, it may be found in a variety of organisms. In fact, all of the multiple metabolic feedback patterns described above are found to exist in the control of many other branched biosynthetic pathways.

It is noteworthy that previous studies with many enzymes subject to metabolic regulation have already established many fundamental concepts that have proven to be highly useful for our understanding of biological control mechanisms. Furthermore, many of these studies have made significant contributions in fermentation technology to synthesize biologically important products for research and industrial applications. It is hoped that continued investigation of the regulation of enzyme synthesis and modification of enzyme activities would contribute substantially to an increasing knowledge of cellular metabolism.

The work performed in the authors' laboratory was supported in part by grants from the National Science Foundation, the National Institutes of Health, and by the Institutional Research Grants to the University of Michigan from the American Cancer Society. S. M. D. was supported by Predoctoral Training Grant, GM-00187 and R. S. F. was the recipient of a Predoctoral Fellowship of the National Institutes of Health, GM 35,576.

REFERENCES

1. DATTA P.: Science **165**, 556 (1969).
2. DATTA P., GEST H.: Proc. Nat. Acad.Sci. U.S.A. **52**, 1004 (1964).
3. COHEN G. N., STANIER R. Y., LE BRAS G.: J. Bacteriol. **99**, 791 (1969).
4. DUNGAN S. M., DATTA P., *unpublished results.*
5. DATTA P., PRAKASH L.: J. Biol. Chem. **241**, 5827 (1966).
6. GIBSON K. D., NEUBERGER A., TAIT G. H.: Biochem. J. **84**, 483 (1962).
7. BURLANT L., DATTA P., GEST H.: Science **148**, 1351 (1965).
8. ROBERT-GERO M., POIRET M., COHEN G. N.: Biochim. Biophys. Acta **206**, 17 (1970).
9. LEVITZKI A., KOSHLAND D. E.: Proc. Nat. Acad. Sci. U.S.A. **62**, 1121 (1969).
10. COHEN G. N., p. 183 in *Current Topics in Cellular Regulation,* Vol. I. Academic Press, New York 1969.
11. DE ROBICHON-SZULMAJSTER H.: Bull. Soc. Chim. biol. **49**, 1431 (1967).
12. DATTA P., GEST H.: J. Biol. Chem. **240**, 3023 (1965).
13. DATTA P.: J. Biol. Chem. **245**, 5779 (1970).
14. DATTA P.: Biochemistry **10**, 402 (1971).
15. DATTA P.: Proc. Nat. Acad. Sci. U.S.A. **58**, 635 (1967).
16. WAN LU L., DATTA P., *unpublished results.*
17. BOTHWELL M., DATTA P.: Biochim. Biophys. Acta **235**, 1 (1971).
18. ROBERT-GERO M., POIRET M., COHEN G. N.: Biochim. Biophys. Acta **206**, 31 (1970).
19. HUGHES M., BRENNEMAN C., GEST H.: J. Bacteriol. **88**, 1201 (1964).
20. FELDBERG R. S.: PhD. Thesis. The University of Michigan, 1970.
21. DATTA P.: J. Biol. Chem. **241**, 5836 (1966).
22. DATTA P., WAN LU L.: J. Biol. Chem. **244**, 850 (1969).
23. PAULUS H., GRAY E.: J. Biol. Chem. **239**, PC4008 (1964).
24. SHIIO I., MIYAJIMA R.: J. Biochem. **65**, 849 (1969).
25. PAULUS H., GRAY E.: J. Biol. Chem. **243**, 1349 (1968).
26. FREUNDLICH M., BURNS R. O., UMBARGER H. E.: Proc. Nat. Acad. Sci. U.S.A. **48**, 1804 (1962).
27. JENSEN R. A., NASSER D. S., NESTER E. W.: J. Bacteriol. **94**, 1582 (1967).
28. UMBARGER H. E.: Ann. Rev. Biochem. **38**, 323 (1969).
29. STURANI E., DATTA P., HUGHES M., GEST H.: Science **141**, 1053 (1963).
30. JOHNSON C., DATTA P., *unpublished results.*

Genetic and Physiological Regulation of Isoleucine, Valine and Leucine Formation in the *Enterobacteriaceae*

H.E. UMBARGER

Department of Biological Sciences, Purdue University, Lafayette, Indiana, U.S.A.

INTRODUCTION

Although it might be appropriate in a book on genetics of industrial microorganisms to confine one's remarks to studies on organisms that have already shown promise in industrial processes, the fact is that the genetics of such organisms is at present very primitive. This is understandable in view of the fact that neither the choice of organisms used by geneticists nor the choice of organisms used by those interested in fermentation processes has been entirely arbitrary. The early attempts to initiate genetic studies in bacteria were, of course, usually done using the common laboratory strains of numerous species; the choice of organisms for any of the successful studies has always been dictated by the empirical observation of amenability to genetic manipulation. As evidence, we can cite the highly empirical first demostrations of transformation[1], conjugation[2] and transduction[3] and the still limited distribution of these modes of genetic recombination in other bacteria.

Even more empirical and accidental are the choice of *Saccharomyces cerevisiae*, *Aspergillus oryzae* and *Penicillium roquefortii* for making better flavors for better living than have our chemist friends at Du Pont. While the entire spectrum of fermentative activities of these organisms is yet too complex to allow biochemical analysis let alone a genetic analysis, it is doubtful that the latter two species would be any more readily studied genetically than are yeasts. The genetic studies with yeasts have been extensive, but the greater number of chromosomes than found in *Neurospora crassa* has made the study difficult and an improvement of brewers' stocks in a rational way is unlikely. Similarly, the production of antibiotics is the result of empiricism and serendipity no matter how rational has been the screening procedure. The industrial production of amino acids that has been so successful in Japan, again, appears to have been the result of patient and effective screening. In some cases, the genetic modification of amino-acid-producing strains has been judiciously employed, but the work has not required, fortunately, a genetic analysis for its success[4].

Since industrial microbiologists and microbial geneticists appear not to be choosing organisms appropriate to the others' interests, is their coming together worthwhile? Or more specifically, can the detailed study of genetic factors affecting the production of amino acids, as will be considered here, make some useful contribution to industrial microbiology? Indeed, I believe there are some useful lessons to be gained from the study of regulation of amino-acid formation in *Escherichia coli* and *Salmonella typhimurium*. Although the genetic analysis of the regulation of biosynthesis of the branched-chain amino acids has not progressed very far, it has been possible to go as far as we have because: (i) we understand the role of the compounds being formed, (ii) we were able to select mutants that affected the formation of the compounds, and (iii) we chose two species that were closely enough related to have homologous genes but appear to have diverged sufficiently at the DNA base-pair level so that nearly unique mutations occur readily. The knowledge that has been gained in the genetic and biochemical analysis of the biosynthesis of the branched-chain amino acids, of the aromatic amino acids and of histidine in several organisms might hopefully suggest ways to improve the production of any metabolite, the role of which in the cells is understood. It is, of course, the latter requirement that might provide the stumbling block in the case of the most useful industrial process. Even in the absence of such information, some of the genetic events that have been found to occur in the more readily understood systems might suggest analogous events that might be sought in more complex systems.

It is with the possibility of analogy in mind that the results of the investigation of the biosynthesis of the branched-chain amino acids in our laboratory and in a few other laboratories are being presented. It is unlikely that I am going to interest our colleagues from the fermentation industries to employ our *E. coli* or *S. typhimurium* mutants even though they convert a significant amount of the available carbon to branched-chain amino acids. However, it is possible that some of our observations will find some application in biochemically quite unrelated systems.

THE USE OF ANALOGUES AS PROBES FOR BIOCHEMICAL AND GENETIC ANALYSIS

One of the first rational, systematic attempts that employed amino-acid analogues to select bacteria that overproduced the natural amino acid was that initiated by ADELBERG[5]. One might today, in retrospect, criticize the rationale on the basis that the approach was indeed a naive one: that resistance to the analogue would result from the overproduction of the natural amino acid. Indeed, it is possible that antagonism by endogenous amino acid would be the basis of resistance. However, today we could visualize mechanisms that might result in overproduction of the amino acid as rather trivial consequences of the actual biochemical changes leading to resistance. Some of these anticipated mechanisms are listed in Table I. Some, but not all, appear to have been found in the case of resistance to analogues of the branched-chain amino acids. It should be pointed out that, in some cases, the me-

TABLE I

Possible Mechanisms of Resistance to Amino-Acid Analogues

Mechanism	Amino-Acid Overproduction
1. End-product-sensitive enzyme no longer inhibited by analogue.	1. Yes, if the end-product-sensitive enzyme also lost sensitivity to the amino acid and the internal pool could not be increased sufficiently to inhibit the resistant enzyme.
2. The end-product-sensitive enzyme (and perhaps other biosynthetic enzymes as well) formed in amounts greatly in excess of wild-type minimal medium levels. a) By loss of a genetic element (operator?) that is necessary for a repression signal to be recognized. b) By alteration of genetic elements needed to generate the repression signal. c) If positive control were involved, alterations that led to a permanently "on" signal either at the site of recognition near the structural gene or at the site of positive control signal generation.	2. Yes, if the end-product sensitivity was not sufficient to permit complete quenching of biosynthesis.
3. Increased discrimination against the analogue at some step in the incorporation of the amino acid into protein. For example, an altered activating enzyme. a) When analogue itself is incorporated into protein. b) When analogue itself merely inhibits incorporation of the natural amino acid.	3. Probably not directly. However, if the increased discrimination was achieved at the expense of a reduced activity toward the natural amino acid and if that step were involved both in protein synthesis and in generation of the repression signal, the condition listed in N° 2 might obtain.
4. Altered uptake of the analogue a) altered binding protein b) altered energetic coupling of transport process and binding protein.	4. Unlikely, although under such conditions the biosynthetic enzyme might be less repressible owing to the limited uptake of amino acid.
5. If "internal" permeation barriers exist ("multiple pools" and "channelling") an increased discrimination might occur in the transfer of the amino acid or its analogue from one pool to another.	5. Unlikely, unless the channelling led to an alteration in transfer of repression signal.
6. Loss of capacity to maintain an amino acid or analogue pool.	6. Possibly, if the internal pool was reduced below the level needed to control metabolic flow.
7. Antagonism of the analogue by overproduction of the natural amino acid by any mechanism given in N° 1 or N° 2.	7. Yes, in this case the overproduction is the essential feature.
8. Enhanced destruction of the analog.	8. No.

chanism of resistance is not well enough understood to decide whether amino-acid overproduction is a cause or an effect of resistance.

Quite unrelated to the dilemma of cause or effect in amino-acid overproduction is the far more interesting question of the molecular mechanism leading to the overproduction. ADELBERG[5] clearly recognized this aspect when he stated "*The study of such mutants promises to provide information on mechanisms of control of biosynthesis and gene—enzyme relationships as well as on drug-resistance*". Much of our current knowledge of regulation testifies to the truth of that prediction. Indeed, a major approach in our laboratory for the last several years has been to employ analogue-resistant mutants to detect altered gene—gene interactions and altered biochemical behavior in the pathways to the branched-chain amino acids.

The approach is not a unique one and I would be the first to admit that it has not been as fruitful in our hands as has the same approach in the case of the histidine pathway. The approach, however, has allowed us to detect relationships between the structural genes for the enzymes forming the branched-chain amino acids and genes that have been found to have a regulatory function. In some cases, the function of the regulatory genes has not been sufficiently clarified to determine whether their regulatory function is their sole role. As will be noted below, it is quite clear that a locus that upon mutation gives rise in *S. typhimurium* to trifluoroleucine resistance is the structural gene for leucyl-tRNA synthetase[6]. In this case, the regulatory role is presumably a minor one.* Again, in *E. coli*, almost parallel experiments with a different leucine analog yielded mutants with a regulatory pattern similar to that in the trifluoroleucine-resistant *S. typhimurium* mutants but which had achieved that resistance through involvement of an entirely different gene[7]. If this gene has any other than its regulatory role, it has not been recognized.

The point that should be emphasized is that the use of analogues has served as a probe not only for biochemical mechanisms underlying regulation of biosynthetic pathways but has also served as a probe for gene—gene interaction (some of them certainly mediated metabolically *via* the cytoplasm but some perhaps more directly as anticipated in the translational or transcriptional control exerted by regulatory elements contiguous to structural genes). Such gene—gene interactions can in some cases be explained in biochemical terms. However, more often than not at this stage in our sophistication, we can only resort to an operational description in very vague terms. Hopefully, this biochemical analysis will be possible someday using the techniques currently being developed that allow direct measurement of genetic messages and proteins formed *in vitro* under control of specific DNA fragments[8]. The operational recognition of these control elements now will perhaps hasten their biochemical definition when techniques are developed.

---

* The possibility, however, must be entertained that the so-called regulatory role is somewho intimately connected to its role in protein synthesis. As yet we have been unable to resolve regulation from protein synthesis in the case of leucine biosynthesis.

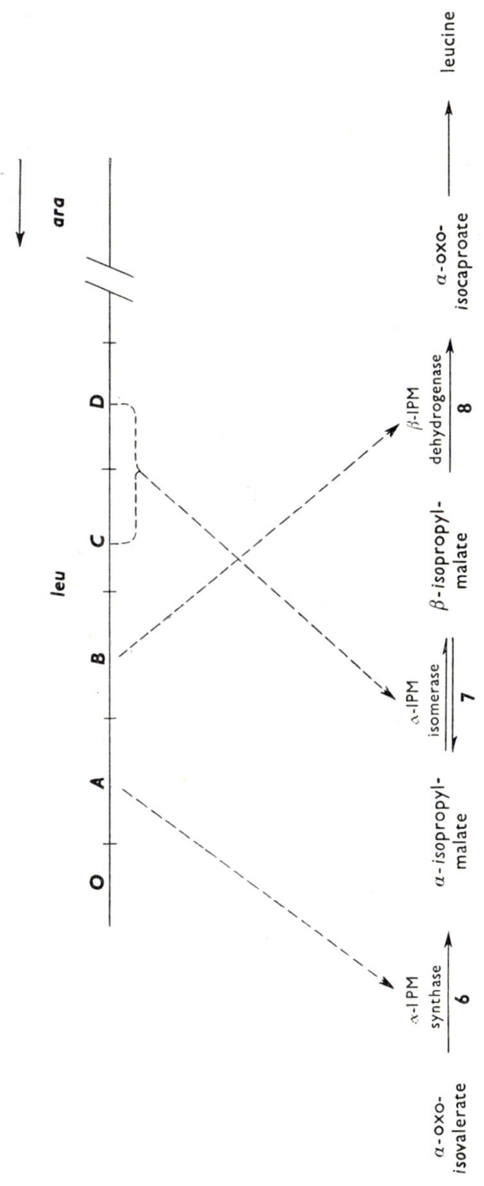

Fig. 1. The relationship between the genes of the *leu* operon and the leucine biosynthetic enzymes.

The following sections will give some idea of the kind of picture that has emerged from a combined biochemical and genetic analysis of mutants resistant to analogues of the branched-chain amino acids. It will be quite clear that the picture is an incomplete one and it will be quite clear what kind of experiments remain to be done. Because the leucine pathway is a little less complex than the pathway to valine and to isoleucine, it will be discussed first.

THE LEUCINE PATHWAY — ENZYMIC STEPS

The pathway of leucine biosynthesis in all microorganisms that form the entire carbon chain of leucine from simple carbon sources is that postulated by STRASSMAN *et al.*[9,10]. That pathway, as shown in Fig. 1, proceeds from α-oxo*iso*valerate, the immediate precursor of valine, *via* the transfer of the acetyl group of acetylcoenzyme A. The reaction is catalyzed by α-*iso*propylmalate (α-IPM) synthase which is the *leuA* gene product. The product, α-IPM, is converted to β-IPM by α-IPM isomerase, an enzyme that appears to have two kinds of subunit as indicated by comparative studies with the yeast enzyme and by the involvement of two genes, *leuC* and *leuD*. The α-oxoacid precursor, α-oxo*iso*caproate, is formed by the oxidation of β-IPM by β-IPM dehydrogenase, a NAD specific enzyme and the product of the *leuB* gene.

The final step in leucine biosynthesis is catalyzed by at least two transaminases distinguished originally by RUDMAN and MEISTER[11]. One, transaminase B, provides the sole mechanism for the conversion of α-oxo-β-methylvalerate to isoleucine and plays a major role in the conversion of α-oxo*iso*valerate to valine (*see* below). The other is transaminase A, which appears to be required for phenylalanine and tyrosine biosynthesis. The role of transaminase A in tyrosine and phenylalanine biosynthesis has not been established unequivocally. However, the enzyme has been shown to respond to repression by tyrosine through a gene (*tyrR*) implicated in the repression of two other tyrosine-specific biosynthetic enzymes[12].

THE LEUCINE PATHWAY — CONTROL OF METABOLITE FLOW

The flow of metabolites along the leucine pathway is controlled through end-product inhibition of α-IPM synthase by leucine. The enzyme in *S. typhimurium* has been studied by KOHLHAW *et al.*[13]. Leucine inhibition is nearly competitive with respect to one of the substrates, acetylcoenzyme A, and noncompetitive with respect to the other, α-oxo*iso*valerate. Pronounced cooperative binding of inhibitor and acetylcoenzyme A was observed. The subunit structure of the enzyme is still under investigation but it appears that the enzyme is a tetramer of approximately 200,000 and that it is dissociated to the monomeric form by the inhibitor, leucine[14].

It was in an effort to study the interaction between leucine and α-IPM synthase that CALVO initiated a study of 5′,5′,5′-trifluoroleucine-resistant mutants of *S. typhimurium* at Cold Spring Harbor several years ago. It was known that the analogue, like the amino acid itself, inhibited α-IPM synthase and it was anticipated that one mechanism of resistance might be loss of sensitivity to leucine and to the analogue. In an extensive selection program for such mutants, however, only one was shown to exhibit an α-IPM synthase that was less sensitive than the wild type to the end-product inhibitor. The majority of the resistant mutants exhibited derepressed levels of the leucine biosynthetic enzymes (*see* below).

As might be expected the trifluoroleucine-resistant mutants were readily recognized as colonies that were surrounded by zones of growth in agar plates seeded with leucine-requiring mutants derived from the same wild-type organism. The lesion is closely linked to the *leuA* gene and is presumably affecting only that gene since no evidence for distinct catalytic and regulatory subunits has been obtained. Physical differences between the end-product-sensitive and end-product-resistant enzymes are currently under study. The inhibitor-induced dissociation, observed with the wild-type enzyme has not been observed with the mutant enzyme.

THE LEUCINE PATHWAY — REGULATION OF ENZYME AMOUNT

The mutants isolated by CALVO in which there were no alterations in end-product sensitivity but in which the enzymes were derepressed are probably of more interest than are those with altered end-product sensitivity. When the leucine biosynthetic enzymes were initially studied in *S. typhimurium*, it was readily demonstrated that they were repressed to about two thirds of the minimal medium level by exogenous leucine[15]. Thus, the leucine biosynthetic enzymes exhibit a pattern of repression intermediate between that shown by the arginine biosynthetic enzymes of *E. coli* (highly derepressed in minimal medium) and the histidine biosynthetic enzymes of *S. typhimurium* (nearly maximally repressed in minimal medium)[16,17]. Some indication of the mode of this repression has been obtained by studying the trifluoroleucine-resistant mutants.

One class of these mutants was shown to contain lesions lying to one side of those lesions that were known to affect *leuA* the structural gene for α-IPM synthase (to the left of *leuA* in Fig. 1)[18]. These mutants, along with the common sense argument based on the direction of polarity (from *leuA* to *leuD*) observed with some *leu* mutants[15], served to define the position of the "operator" controlling the *leu* cluster. The region identified here as an operator should perhaps be more operationally defined as the region concerned with recognition of the repression signal, which is only one of the features of the operator as defined in the Jacob-Monod model[19]. The other features, such as binding of repressor, remain to be demonstrated. Nevertheless, for the present, the term, "operator-constitutive" (O^c) mutation, will be used to describe the lesion in these strains. Furthermore, so far as we now know, the α-IPM synthetase in the O^c mutants is no different from that in the wild-type strain. Thus, we have no evidence that the operator region is not distinct from the *leuA* gene.

When the enzyme levels in the O^c mutants were examined, it was found that they were well above the level found in normal cells grown in minimal medium and, in the presence of excess leucine, were only slightly repressed by leucine[6]. Thus, they tended to be refractory to repression by both endogenous and exogenous leucine. In contrast, the enzymes of the isoleucine and valine biosynthetic pathway were essentially at wild-type levels in minimal medium and exhibited wild-type repressibility when grown in supplemented media.

Another class of trifluoroleucine-resistant mutants that had elevated levels of leucine biosynthetic enzymes bore lesions that were unlinked to the *leu* genes. Those that have been examined have been found to have leucyl-tRNA synthetases with reduced affinity for leucine. The lesions in these mutants have served to define the *leuS* locus, which is presumably the structural gene for leucyl-tRNA synthetase. Such mutants were still somewhat repressible presumably by virtue of the fact that the leucine pool could be raised almost high enough to saturate the synthetase. It is of some interest that the O^c and *leuS* mutants, like the end-product insensitive, were overproducers of leucine. Thus, it would appear that in *S. typhimurium* oversynthesis of leucine is prevented by end-product inhibition when the level of the sensitive enzyme is under normal control by repression but not when the enzyme is derepressed.

A striking demonstration by the CALVO's[20], of the effectiveness of the two kinds of control was achieved by combining in the same strain a *leuS* lesion and the *leuA* lesion that led to loss of end-product sensitivity. Although the strains bearing one lesion or the other were leucine excretors, they appeared to be little the worse for so doing. The strain with both lesions, however, converted as much as 50 per cent of its glucose to leucine, and actually had a reduced rate of growth that was enhanced to that of the wild-type by supplementing the medium with valine. Thus, while approaching the kind of fermentation yield that some of our Japanese colleagues have using *Micrococcus glutamicus*, the overproduction of leucine appeared not to be a drain on carbon and energy but a drain on valine.* One might wonder, nevertheless, how such a strain might compete in other than a laboratory environment with its progeny which might undergo possible compensatory mutations that would abolish the excretion.

The generalization that arises from these studies is that, for leucine to take part in repression of the *leu* gene cluster, the organism must possess an effective leucyl-tRNA synthetase, implying that leucine itself must be activated to repress. What the role of the activated leucine is with respect to generating the repression signal is not clear at present. A leucyl-tRNA species might be involved but there is as yet no evidence for such involvement.

At the present time, we have evidence for only one additional regulatory element that appears to be involved in the generation of the signal, "excess leucine", that leads to repression. This evidence was obtained not from studies on analogue-resistant

---

* It can be readily calculated that leucine (as well as valine) production from glucose is actually an energy yielding, not an energy consuming process.

mutants of *S. typhimurium* but from studies of *E. coli* mutants.* In an attempt to find mutants comparable to the *leuS* mutants of *S. typhimurium*, the analogue, 4-azaleucine[21], which is a more effective leucine analogue than trifluoroleucine in *E. coli*, was used as a probe in a mutant search comparable to that performed by CALVO.

Two kinds of azaleucine-resistant mutants were found, one in which the α-IPM synthase was end-product insensitive and the other bearing an unlinked lesion that led to a derepression of the leucine biosynthetic enzymes[7]. The lesions were not in the *leuS* gene, nor was the leucyl-tRNA synthetase altered in these strains. Examination of the species of leucine acceptor tRNA failed to reveal any difference between that of the mutant and that of the wild-type with respect to either kind or amount. For the present, we can only infer that in these mutants, designated *azl*, some element has been modified that is necessary (along with excess leucine, an effective leucyl-tRNA synthetase, and an intact *leuO* region) for repression of the leucine biosynthetic enzymes. What might be the nature of the *azl* gene product is only for conjecture at this time. Since, as will be shown below, these same elements are also involved in repression of the isoleucine and valine biosynthetic enzymes, a single model for their interaction will be postponed until later.

THE VALINE AND ISOLEUCINE PATHWAYS — THE ENZYMIC STEPS

The pathway to valine has for some time been known to be catalyzed by four enzymes that catalyze four of the five steps in the pathway to isoleucine[22]. While each of these four corresponding steps are catalyzed by the same kind of enzyme, it has never been shown that the same individual enzyme molecule catalyzes the corresponding steps in *both pathways*. Indeed, it is conceivable that the two pathways have distinct enzyme molecules and that a cluster of four enzymes is committed specifically for valine biosynthesis (a "valosome") whereas a nearly analogous cluster of five enzymes is committed to isoleucine biosynthesis (an "isoleusome"). For the present, only a kinetic and genetic argument can be given for such complexes. Thus far, attempts to isolate the enzymes of these pathways as multienzyme complexes have failed. Thus, if they do exist, they must be in a more labile association than has been encountered in some other systems.

Reviewing first the steps in the pathway to valine, the first step is the synthesis of acetolactate, in which an active acetaldehyde group is condensed with pyruvate by acetohydroxyacid (AHA) synthase. In most organisms this enzyme requires

---

* Because the two organisms are biochemically and genetically so similar to each other, it does seem justified to relate the findings in one organism to the other. Indeed, the differences between the two so-called species seem to be no more fundamental than are differences between mutants of either species.

FAD as a cofactor in addition to thiamine pyrophosphate and a divalent cation. The need for FAD is a surprising one for it is not required for another kind of AHA synthase that is thought to play only a katabolic role in the metabolism of some bacteria[23]. The structural gene for AHA synthase has been designated *ilvB* in *E. coli* and, although it has not yet been demonstrated in *S. typhimurium*, a homologous *ilvB* gene is presumably present.

Acetolactate is converted to α, β-dihydroxy*iso*valerate by AHA isomeroreductase in a NADPH specific step that involves both an isomerization (α → β alkyl group migration) and a reduction. The corresponding structural gene is *ilvC*. The dihydroxyacid is, in turn, converted to α-oxo*iso*valerate, a precursor not only of valine but, as mentioned earlier, also of leucine. The enzyme is dihydroxyacid (DHA) dehydratase and is the product of the *ilvD* gene. The final step in valine biosynthesis is catalyzed by *transaminase B*, which was already referred to in the discussion of the pathway to leucine. The corresponding structural gene is *ilvE*. Like leucine, valine can be formed from its α-oxoacid precursor by means of a second transaminase, that can transfer amino groups from and to the carbon chains of alanine, α-aminobutyrate and valine, but not leucine or isoleucine. Its corresponding structural gene has not been identified.

The pathway to isoleucine is exactly analogous except that the active acetaldehyde is transferred by AHA synthase to α-oxobutyrate, a product, in turn, of threonine deaminase, the structure of which is specified by the *ilvA* gene. The condensation product is, in sequence, converted to α-oxo-β-methylvalerate and finally to isoleucine exactly as acetolactate is converted to valine. Unlike valine, however, isoleucine can be formed from its α-oxoacid precursor by only one enzyme, *transaminase B*. Therefore, *ilvE* mutants have an absolute requirement only for isoleucine. However, they are markedly stimulated by valine and some *ilvE* mutants are stimulated by leucine as well.

THE VALINE AND ISOLEUCINE PATHWAYS — CONTROL OF METABOLITE FLOW

The flow of metabolites over the pathway to isoleucine is controlled through end-product inhibition of threonine deaminase by isoleucine. The sensitivity to isoleucine of the *E. coli* enzyme has been studied in several laboratories. The study of the enzyme by CHANGEUX[24] was of considerable importance historically, for his studies stimulated the considerations that led to the allosteric model for regulatory proteins. Although this model was clearly defined by MONOD *et al.*[25] in an essay that has had a tremendous impact on enzymology, it has been widely misinterpreted.

Because the allosteric model has been so uncritically applied to many proteins that may or may not be allosteric, it might be appropriate to reiterate what this writer believes MONOD, WYMAN and CHANGEUX intended when they defined allosteric proteins. First of all, they stated that an allosteric protein is composed of *at least* two identical components that are symmetrically arranged so that each component occupies an equivalent conformation with respect to the other(s). These components were termed protomers to distinguish them from the more general term, subunit. A protomer itself might be composed of subunits that are either non-identical or, if identical, might be arranged within the protomer in a non-equivalent or non-symmetrical fashion. The unit of symmetry is, therefore, the protomer. The model, furthermore, proposed that there are at least two conformational states in which allosteric proteins can exist. In each state, symmetry, and therefore the equivalence of the protomers, is maintained. The transition between the two forms is the allosteric transition. A very important feature of the allosteric model is that the two forms of the protein exhibit a differential affinity for at least one ligand. Such ligands are termed allosteric ligands. Thus, the equilibrium between the two states would be markedly affected by the kind and amount of allosteric ligands present. The allosteric transition itself is, however, spontaneous, *i.e.* independent of allosteric ligand binding. It should be emphasized that the model does not require that the substrate be an allosteric ligand nor that the two forms of the allosteric protein differ in catalytic activity (although these features might be exhibited).

It might be well also to emphasize some features of regulatory proteins that are commonly observed but do not by themselves define an allosteric protein: (i) binding of inhibitor at a site distinct from the catalytic site, (ii) cooperative binding of substrate or inhibitor (sigmoid instead of hyperbolic saturation curves), (iii) differential loss of sensitivity to regulatory effectors and catalytic activity. In many cases, these observations are readily interpreted in terms of the allosteric model. These properties alone are far from sufficient, however, to define an allosteric protein.

Having digressed to consider the allosteric model, let us consider threonine deaminase. While some of the properties of *E. coli* threonine deaminase were used to illustrate the applicability of the allosteric model, all of the early experiments on the *E. coli* enzyme should be re-examined in light of the findings that have been made on the presumably nearly identical enzyme purified from *S. typhimurium* by BURNS and ZARLENGO[26].

Many of the early observations on the isoleucine-sensitive threonine deaminase revealed cooperative binding of substrate. It now appears that cooperative binding is observed only in the presence of at least trace amounts of isoleucine. Valine is a positive effector for the enzyme in that it antagonizes the inhibitory effect of isoleucine. However, at higher concentrations, it too, is inhibitory.

The enzyme has been found to possess only two catalytic sites although it is composed of 4 identical subunits of molecular weight 48,500 (Ref.[27]): The simplest interpretation is that the enzyme is a dimer and that each monomer (protomer, in the terminology of MONOD *et al.*) is formed from the association of two identical subunits in a non-equivalent or nonsymmetrical fashion to generate a single potential catalytic site. Dimers are most readily obtained by removal of pyridoxal phosphate. These apodimers can re-associate to a tetrameric form by incubation with coenzyme to yield a molecule of the same apparent molecular weight and sedimentation properties as the original enzyme[28]. However, this reconstituted enzyme is not catalytically active until it has been incubated with substrate. Actually, pre-incubation with either valine (like substrate, a positive effector) or isoleucine (a negative effector) also results in activation[29]. This activation of the reconstituted tetrameric enzyme has been termed "maturation" and is presumed to be a process that occurs *in vivo* when the enzyme is formed *de novo*. However, the maturation is blocked by the simultaneous presence of both a negative effector (isoleucine) and a positive effector (either valine or threonine). This observation will be considered later when regulation of enzyme activity of the valine and isoleucine biosynthetic enzymes will be considered.

It is of further interest at this point to note that whereas, in some cases, end-product inhibitors cause a dissociation or an aggregation of the sensitive enzymes, the inhibitor of threonine deaminase does neither. Furthermore, it has been shown that the tetrameric form is the catalytically active one[28].

Mutants of *S. typhimurium* in which the threonine deaminase is resistant to isoleucine inhibition have been isolated in our laboratory. These were obtained by selecting strains that were resistant to the isoleucine analogue, 4-thiaisoleucine[30]. The genetic lesion is linked to and probably affects *ilvA*. In contrast, the thiaisoleucine-resistant mutants of *E. coli* are quite different and will be considered in the next section in which the regulation of enzyme activity is considered.

Control of metabolite flow over the pathway to valine cannot be described so explicitly. Because the first step in the pathway to valine is the second step in the pathway to isoleucine, it might be anticipated either that there would be no end-product control of metabolite flow over the valine pathway or that it would be incomplete in the absence of isoleucine to allow for the biosynthesis of that amino acid. (Such a control could be effected if α-oxobutyrate were a positive effector or if the end-product control were "concerted" with both valine and isoleucine being required.) That the control of valine biosynthesis might have been rather ineffective was indicated earlier by an observation of the Carnegie Biophysics Group[31]. They observed with strain B of *E. coli* that the addition of valine to cells growing in a glucose minimal medium did not affect the endogenous synthesis of valine, in contrast to the almost complete quenching of endogenous isoleucine formation upon the addition of that amino acid. Nevertheless, the AHA synthase of that strain is, in fact, inhibited by valine. It is generally found in all organisms that AHA synthase is valine-sensitive although some organisms do have a second, valine-insensitive synthase with a lower pH optimum. *S. typhimurium* may be similar to *E. coli* strain B in this respect since wild-type strains do appear to have a rather ineffective control over biosynthesis even though the AHA synthase is inhibited by valine.

The K-12 strain of *E. coli* offers a sharp contrast in this respect. It has an AHA synthase that is very sensitive to valine[32]. This greater sensitivity contributes in part to the inhibition of growth of the K-12 strain by valine but other factors are also involved which will be considered later. Indeed, because valine is a growth inhibitor and its effect is reversed by isoleucine, valine might be considered one of the analogues that has proven useful as a probe of the genetic and biochemical control of the isoleucine and valine pathways. It is perhaps because the K-12 strain of *E. coli* is valine-sensitive and *S. typhimurium* strains are not that it has been possible to produce regulatory mutations in *E. coli* that have not yet been detected in *S. typhimurium* — an example of the advantage from dealing judiciously with more than one strain or more than one closely related species. Thus, the selection of valine-resistant mutants has yielded, among many types, some that contain valine-insensitive AHA synthases[32]. Such mutants have made it possible to locate the *ilvB* gene in *E. coli*[33]. A similar marker on the *ilvB* gene is not possible with *S. typhimurium* and the gene has not yet been located, presumably because of a second enzyme allowing acetohydroxyacid synthesis.

Owing to the fact that the enzyme in *E. coli* and *S. typhimurium* has proven to be so unstable in extracts, no one has yet succeeded in purifying it. The enzyme appears to be unstable *in vivo* as well, for under conditions of enzyme repression, specific activity obtained in extracts decreases more than can be accounted for by increase in cell mass. The enzyme is quite stable under assay conditions (although large amounts of product tend to hasten inactivation) suggesting that someday a combination of substrate analogues and cofactors might be employed in a stabilizing buffer during its purification.

Until the enzyme is purified, very little can be said about the interaction of valine with the enzyme. However, studies with crude preparations have revealed that valine was competitive with pyruvate with the *E. coli* enzyme but noncompetitive with *S. typhimurium* extracts[34,35]. These differences may not be an indication of real differences in mechanism since the inhibitory ranges studied in the two systems were different.*

THE VALINE AND ISOLEUCINE PATHWAYS — REGULATION OF ENZYME AMOUNT

That a unique kind of control might have affected the levels of the valine and isoleucine biosynthetic enzymes should have been anticipated from the fact that four of the five enzymes were multifunctional with only single genes coding for each. Therefore, the simple pattern of end-product repression found in the leucine pathway could not have been effective in assuring an adequate amount of enzyme for each end product. The pattern of control was only deduced when it was noted, first in *S. typhimurium* leucine- or isoleucine- and valine-requiring mutants, that limita-

---

* It has been noted that the designation "K system" or "V system"[25] in absolute terms may hold only over limited ranges of concentration with regulatory proteins.

tion of any of the branched-chain amino acids resulted in derepression of the isoleucine- and valine-forming enzymes[36]. In other words, repression was multivalent, *i.e.* for repression to occur, all the branched-chain amino acids had to be present in excess.

As of this writing, the mechanism of multivalent repression, like repression in pathways leading to single amino acids, is not understood. Some hints, however, have been revealed from study of the mutants resistant to amino-acid analogues. The first of the hints came from the study of the trifluoroleucine-resistant mutants of *S. typhimurium* isolated by CALVO[6]. As mentioned earlier, of the three classes of leucine-excreting, trifluoroleucine-resistant mutants, one class was found to have a lesion unlinked to the *leu* gene cluster. This class was characterized not only by overproducing leucine but by overproducing and excreting valine and isoleucine as well[6]. Thus, the lesion (*leuS*) that prevented these mutants from generating the "leucine excess" signal for repression of the leucine biosynthetic enzymes also prevented them from generating that signal for multivalent repression of the valine and isoleucine biosynthetic enzymes. That the biochemical lesion in these strains was a faulty leucyl-tRNA synthetase implied that leucine had to be activated to participate in multivalent repression. This conclusion, of course, was of considerable interest in view of the earlier observation of EIDLIC and NEIDHARDT[37], that under conditions of limited valyl-tRNA synthetase activity in *E. coli*, there was a specific derepression of those valine and isoleucine biosynthetic enzymes that they examined. On the basis of the assumption that the findings in the two species can justifiably be considered together, it was concluded that multivalent repression required activation of at least valine and leucine. The obvious question, therefore, was whether the same was true of isoleucine.

The first approach to answer this question was to employ the usual analogue probe, in this case thiaisoleucine. As mentioned earlier, this analogue had yielded with *S. typhimurium* only mutants with isoleucine-insensitive threonine deaminases. However, in *E. coli*, a series of mutants was isolated with three of the five valine and isoleucine biosynthetic enzymes derepressed[38].

Interestingly, the three enzymes that were derepressed in the thiaisoleucine-resistant mutants were those shown earlier by RAMAKRISHNAN and ADELBERG[39] to be products of *ilv* genes *A*, *D* and *E* (Fig. 2). They observed that, in certain valine-resistant mutants of the K-12 strain of *E. coli*, all three of these enzymes were elevated. The lesion was shown to lie to the left (Fig. 2) of the *A*, *D*, *E* gene cluster[33] and was concluded to have affected the region at which the multivalent repression signal was recognized, *i.e.*, an operator (*ilvO*) according to the Jacob-Monod model[19]. Another class of valine-resistant mutant found by RAMAKRISHNAN and ADELBERG[39] was that in which only AHA synthase was derepressed. The lesion in this case was also close to the corresponding structural gene, *ilvB*. Again, it was thought to have affected an operator, which has been designated *ilvP* in the TAYLOR and TROTTER[40] *E. coli* chromosome map. Thus, the *ilv* gene cluster appeared from these genetic studies to consist of at least two operons, one of a single gene, *ilvB*, and the

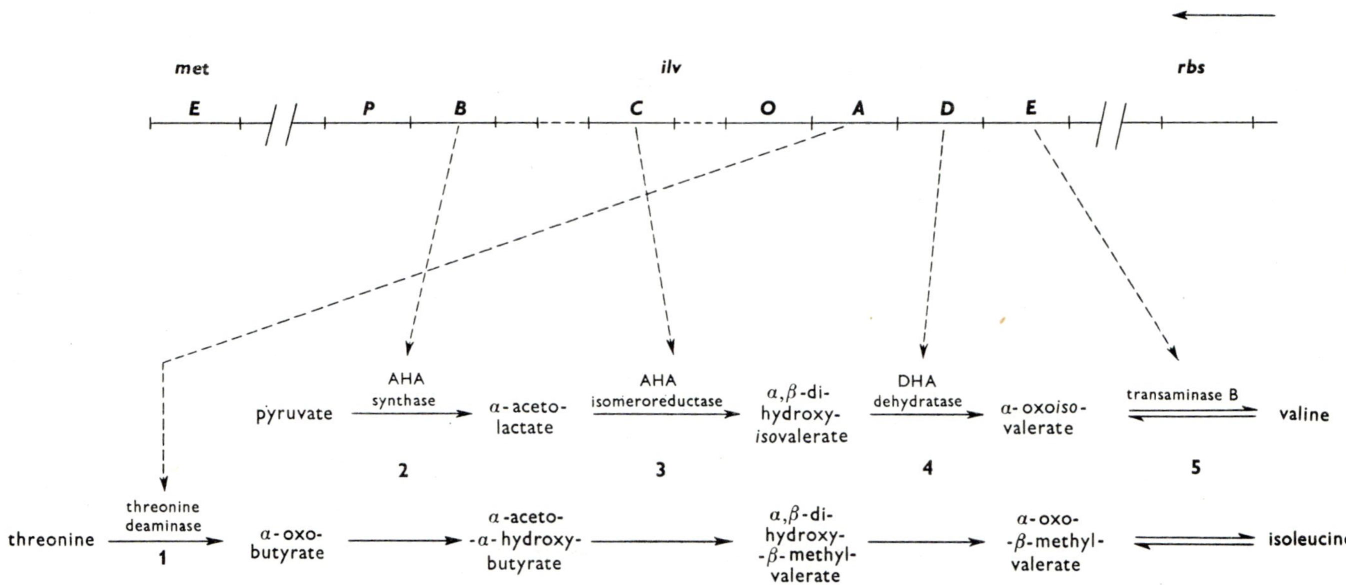

Fig. 2. The relationship between the genes of the *ilv* region and the valine and isoleucine biosynthetic enzymes.

second of three genes, *ilvA*, *D* and *E*. The question of a repressor recognition element for the *ilvC* gene was left unanswered owing to the absence of any mutations affecting its repression. Its regulation *via* an entirely independent mechanism will be considered later.

It was of interest, therefore, to determine whether the thiaisoleucine-resistant mutants were *ilvO*^{c} mutants comparable to the mutants of RAMAKRISHNAN and ADELBERG in which the same enzymes were derepressed. Examination showed, however, that the thiaisoleucine-resistant mutants were quite different. Thus, even with the high levels of the enzymes 1, 4 and 5 (Fig. 2) they were still sensitive to valine[41]. Furthermore, they had low levels of isoleucyl-tRNA synthetase activity and the lesion was unlinked to the *ilv* region[38].

The finding that the isoleucyl-tRNA synthetase activity was low is in accord with the attractive idea that isoleucine, like valine and leucine, would require activation for participation in multivalent repression. However, genetic experiments have revealed that the picture is not that clear, for it has now been found that the thiaisoleucine-resistant mutants have up to three lesions affecting derepression and thiaisoleucine resistance[42]. These have been designated *ilvS*, presumably the structural gene for isoleucyl-tRNA synthase (earlier shown by IACCARINO and BERG[43] to be located between the *thr* and *leu* loci on the *E. coli* chromosome), *ilvT*, also linked by transduction to *thr* but distal to *leu*, and *ilvU*, linked to *proA*.

By itself, the *ilvS* lesion in these strains confers a low level of resistance to thiaisoleucine, but no derepression of enzymes 1, 4 and 5. Its transfer to a wild-type strain results in a level of isoleucyl-tRNA synthetase activity that is intermediate between that in the wild type and that in the thiaisoleucine-resistant mutant. By itself, *ilvU* confers a low level of resistance to thiaisoleucine and low level of derepression of the three enzymes. There appears to be no effect of this lesion on isoleucyl-tRNA synthetase. When the *ilvS* and *ilvU* lesions are both present, analogue resistance and enzyme derepression are higher than that which is observed when either lesion is alone and equivalent to that observed in some of the thiaisoleucine-resistant mutants. In strains doubly transduced for the *ilvS* and *ilvU* markers, the isoleucyl-tRNA synthetase is as low as it is in the original thiaisoleucine-resistant mutants. The introduction of *ilvT* appears to bring about some derepression. By itself, it appears to affect neither resistance nor isoleucyl-tRNA synthetase. It seems to enhance slightly the level of thiaisoleucine resistance of *ilvS* mutants but does not appear to effect the level of synthetase activity of such mutants. It has not been found to be altered in all of the thiaisoleucine-resistant mutants.

A hypothesis currently under test is that *ilvS* is the structural gene for isoleucyl-tRNA synthetase and that the enzyme found in the *ilvS* mutants is a poor catalyst. However, there is a compensatory derepression of the enzyme, which is dependent upon an intact *ilvU* gene. Such a derepression was first observed by NASS and NEIDHARDT[44]. It thus appears that the *ilvS* and *ilvU* lesions act synergistically to bring about a low level of isoleucyl-tRNA synthetase which, in turn, leads to derepression.

The role of the *ilvT* lesion has not been well enough defined to reveal the nature of the interaction with the *ilvS* lesion nor whether there is any interaction with the *ilvU* lesion.

It has been possible to answer the question of involvement of isoleucyl-tRNA synthetase in multivalent repression more directly for *S. typhimurium* than it has been for *E. coli.* Since thiaisoleucine resistance led only to isoleucine-insensitive threonine deaminases, another approach to obtaining *ilvS* mutants was employed. A strain of *S. typhimurium*, exhibiting normal control of the *ilv* pathway and prototrophic for both isoleucine and valine, was infected with an *E. coli* episome containing the *ilv* genes[45]. The resulting strain, which was diploid for the *ilv* region, was mutagenized and isoleucine-requiring strains were isolated. All that have been examined thus far were *ilvS* mutants containing isoleucyl-tRNA synthetases with affinities for isoleucine 1/100 to 1/1000 that of the wild-type enzyme. Such strains were found to have derepressed levels of all five valine- and isoleucine-forming enzymes (in contrast to only three enzymes that were derepressed in the *E. coli* *ilvS,U* double mutants). Transfer of the *ilvS* lesion by transduction demonstrated that derepression, faulty isoleucine activation and isoleucine auxotrophy were simultaneously transferred. No evidence for the presence of an *ilvU* gene in *S. typhimurium* has been obtained.

An explanation of the difference between the number of valine- and isoleucine-forming enzymes derepressed in the two species when the activating enzyme activity is low was found in experiments in which the isoleucine limitation signal was generated in other ways[41]. It was found that, when isoleucine was limiting (or perhaps when some activated derivative of isoleucine was limiting), neither AHA synthase nor AHA isomeroreductase (enzymes 2 and 3, Fig. 2) were derepressed in the K-12 strain of *E. coli.* Delaying for the moment the question of control of the isomeroreductase, this observation is of considerable importance in understanding the mechanism of sensitivity to valine of the K-12 strain. Thus, not only is the AHA synthase of the K-12 strain more sensitive to inhibition by valine but the sensitive enzyme is not derepressed when the signal, "isoleucine-limiting", is generated as a result of the quenching of isoleucine biosynthesis at that step.* The basis of the difference in the responses of operator *ilvP* and of operator *ilvO* to multivalent repression is, of course, not known.

The regulation of the expression of the *ilvC* gene is quite different from that of the other *ilv* genes in that it is induced by either of the substrates of the corresponding enzyme[46]. The induction occurs quite independently of the state of repression or derepression of the *ilvB* or of the *ilvA, D* and *E* genes. This can be readily demonstrated in cultures of *E. coli* or *S. typhimurium* growing in a nutrient medium in which the amino-acid repressible enzymes, such as those in the pathways to valine and isoleucine, are strongly repressed. Such cultures can be readily induced by adding either acetohydroxybutyrate or acetolactate to the medium. Cells may be induced in this way

---

* The synthase is derepressed when either valine or leucine is limiting.

to yield specific activities that would indicate that as much as 6% of the cell protein is isomeroreductase. On the other hand, the isomeroreductase is not induced when the *ilvADE* gene cluster of *E. coli* strain K-12 is derepressed as occurs in O^c type valine-resistant strains, in thiaisoleucine-resistant mutants or during isoleucine starvation. Again, with mutants of *E. coli* in which operator *P* has undergone mutation to the operator constitutive state, the isomeroreductase remains low. The latter observation appears to be readily explained by the previously mentioned fact that the AHA synthase in *E. coli* strain K-12 is extremely sensitive to end-product inhibition by valine, which, in this case, would be in ample supply because of the high level of AHA synthase.

Further support for the role of the sensitivity of AHA synthase in allowing endogenous induction of isomeroreductase in *E. coli* and *S. typhimurium* is obtained by comparing *leuS* mutants in *S. typhimurium* with *azl* mutants in *E. coli.* It was mentioned (*see* above) that the *azl* mutation, like the *leuS* mutation, led to derepression of the leucine-forming enzymes. One class of *azl* mutation leads to derepression of all of the valine- and isoleucine-forming enzymes except the isomeroreductase. In one strain, the AHA synthase was 80-fold derepressed but end-product inhibition of the enzyme by valine was apparently sufficient to prevent any endogenous induction of isomeroreductase over that of the wild type growing in minimal medium. Experiments by WARREN PLEDGER have demonstrated that when such *azl* strains received the *ilvB* gene from a mutant with a valine-insensitive AHA synthase, there was internal induction of the isomeroreductase. Such an internal induction could occur in *E. coli* strains with valine-resistant acetohydroxyacid synthases even when the *ilvC* gene was carried only on an episome bearing an *ilvB* gene for a normal AHA synthase. In *S. typhimurium*, in which the control of acetohydroxyacid formation by valine is less stringent than in *E. coli* strain K-12, the isomeroreductase is induced whenever the AHA synthase itself is derepressed, *i.e.* with limiting leucine, valine or isoleucine, as in *leuS* and *ilvS* mutants grown in minimal medium.

AN OPERATIONAL MODEL FOR CONTROL OF THE LEUCINE, ISOLEUCINE AND VALINE BIOSYNTHETIC ENZYMES

In proposing a working model for regulation of the enzymes that have been of concern in this paper, the principle will be followed that was followed by JACOB and MONOD[19] when their operator–repressor model was first proposed, *viz.*, that only those elements will be incorporated into the model for which genetic or biochemical evidence exists for their involvement. Unlike JACOB and MONOD, no attempt will be made by this writer to define either direct or indirect interactions between known components needed in RNA or protein synthesis and any elements for which only genetic evidence exists. Thus, there have been mutations at loci that might be compared to the repressor genes of the original model but at this stage in our in-

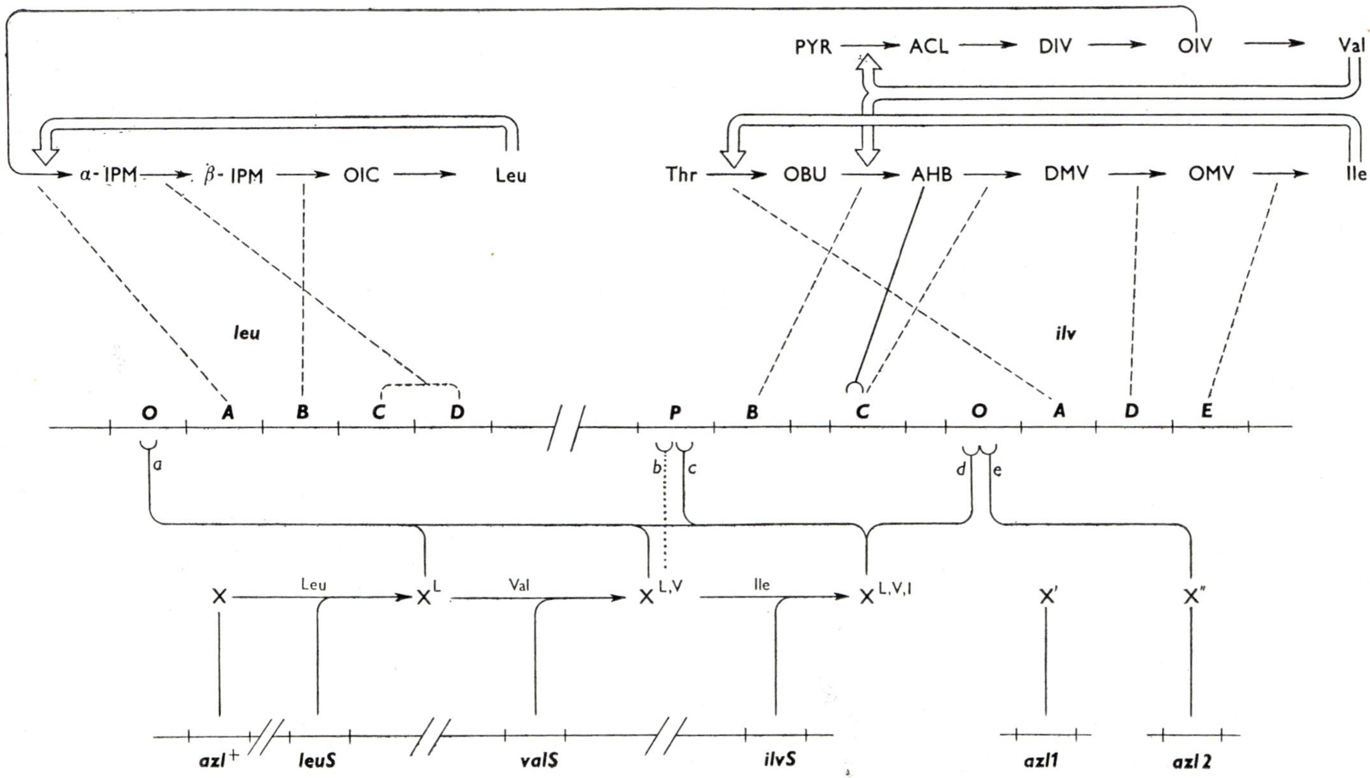

Fig. 3. The regulation of biosynthesis of the branched-chain amino acids. Open arrows represent end-product inhibition of the sensitive enzymes. Solid lines indicate repression or induction of enzyme synthesis. The dotted line indicates repression of the *ilvB* gene in *E. coli* only. The broken lines indicate the correlation between gene regions and enzymic steps as shown in Figs 2 and 3.

ACL α-acetolactate
AHB α-aceto-α-hydroxybutyrate
DIV α,β-dihydroxy*iso*valerate
DMV α,β-dihydroxy-β-methylvalerate
IPM *iso*propylmalate
OBU α-oxobutyrate
OIC α-oxo*iso*caproate
OIV α-oxo*iso*valerate
OMV α-oxo-β-methylvalerate
PYR pyruvate

For details, *see* text.

vestigation they are merely recognized as elements required for interaction between aporepressor and amino acid. Similarly, the model proposed here will not distinguish whether the repression recognition site near the structural genes that it controls functions at the level of transcription or translation.

Fig. 3 contains a diagrammatic representation incorporating most of the elements that have so far been implicated in the regulation of enzymes required for the biosynthesis of the branched-chain amino acids. The simplest features of end-product inhibition are indicated by open arrows. Not shown are the stimulations of the end-product-sensitive enzymes that have been described. The dashed lines interconnect catalytic steps with the corresponding structural genes. The solid lines represent interaction between regulatory genes, amino acids and repression recognition elements to yield repression. The induction of the *ilvC* gene product by substrate is also represented by a solid line.

Considering the leucine cluster first, the scheme in Fig. 3 depicts an interaction (which shall remain undefined) between the product of the *azl* gene (designated X), the product of the *leuS* gene (*i.e.* the leucyl-tRNA synthetase), and an excess of leucine to yield some unspecified component (designated X^L which could be but is not necessarily, the holorepressor) that is required to generate the "excess leucine" signal that is recognized by the *leuO* region or by a product of the *leuO* region. The latter interaction (designated *a* in the diagram) prevents the formation of the leucine biosynthetic enzymes. Whether the transcription of the RNA message corresponding to the *leu* gene cluster is also prevented by this interaction or not has yet to be established. By analogy with experiments performed during repression studies with the valine- and isoleucine-forming enzymes (*see* below), it may be inferred that transcription is prevented during repression.

Since derepression of the valine- and isoleucine-forming enzymes also results from mutation in the *leuS* and *azl* genes, it is assumed in this model that the interaction of the products of both genes needed to repress the leucine-forming enzymes is also needed to repress the valine- and isoleucine-forming enzymes. However, for multivalent repression, this is not sufficient. For the special case of repression of the *ilvB* gene in *E. coli* strain K-12, excess valine and an intact *valS* gene product (*i.e.* valyl-tRNA synthetase) are also needed. The interaction of X^L with valine and valyl-tRNA synthetase is represented by the formation of $X^{L,V}$, which, in the K-12 strain, interacts in some way (dotted line to *b* in Fig. 3), directly or indirectly, with *ilvP*, the repression recognition region for the *ilvB* gene. For interaction in *S. typhimurium* with *ilvP* (*c*) or in either organism with *ilvO* (*d*), the repression recognition region for genes *ilvA*, *D* and *E*, there is the additional requirement for excess isoleucine and an intact *ilvS* gene product (isoleucyl-tRNA synthetase).

In the case of the valine and isoleucine forming enzymes, evidence obtained in our laboratory supports the idea that the derepression signal generated by either limiting valine or limiting isoleucine is accompanied by enhancement of the potential to form the enzymes and that an excess of the branched-chain amino acids prevents the enhancement of that potential. These experiments, modeled after those of FAANES and ROGERS[47], were performed with isoleucine- and valine-requiring mutants that were grown under repressing conditions, transferred to a medium lacking either valine or isoleucine and starved for 30 minutes. When the missing amino acid was added, there was a burst of enzyme synthesis that ceased, after the repression signal so generated took effect. This burst of synthesis was presumably due to the accumulated *ilv* mRNA. Translation of this accumulated message would occur only after the addition of the missing amino acid, which, in turn, leads to the conditions of repression. That the "potential" for enzyme synthesis was, in fact, due to *ilv* mRNA will be known only after hybridization experiments with *ilv* DNA are performed.

The nature of the *azl* gene product is not known, and Fig. 3 is not intended to imply that it becomes bound in any way to the branched-chain amino acids. It may merely be an additional catalytic element required for the sequence of events by which excess leucine is recognized. That it may have a more stoichiometric role may be indicated by one of the two kinds of azaleucine-resistant (*azl*) mutations. The kind that has already been described, which leads to derepression of all the repressible enzymes considered here, is represented in Fig. 3 by *azl1* which yields a product that does not interact with other components needed for repression. The second class, represented by *azl2*, leads to derepression of the leucine biosynthetic enzymes and AHA synthase only. The products of the *ilvA, D, E* cluster are repressed and, even under conditions of limiting leucine, valine or isoleucine, they cannot be derepressed. The *azl2* product (X″), therefore, has been pictured as being sufficient by itself to generate a "repress" signal for *ilvO* (at least none of the other known components appear to be required). Thus, if X does play a catalytic role in repression, that role is probably one of the later ones.

The model in Fig. 3 does not take into account the *ilvT* gene, which plays an as yet ill-defined role in repression of the *ilvADE* cluster, or the *ilvU* gene which appears to be needed for derepression of isoleucyl-tRNA synthetase. The postulated function of the latter could be readily accommodated, however, as could other elements as their presence becomes evident.*

The purpose of a more operational model than that which would result if these observations were interpreted strictly in terms of the repressor—operator model has been to leave open the possibility that the regulation of these pathways may not

---

* Such an observation is the recent one made by HATFIELD and BURNS[29] that the maturation of threonine deaminase from the inactive to the active tetrameric form is impeded by isoleucine and valine whereas either alone hastens maturation. Threonine deaminase, thus, is a protein that has recognition capability for two of the three effectors of multivalent repression. While no binding site for free leucine was found, the immature form (inactive tetramer) was found to bind leucyl-tRNA[48]. They suggest that a protein with such properties might be the aporepressor of the repressor—operator model. The suggestion is extremely interesting, although it is difficult to reconcile such a model with the evidence that the valyl- and isoleucyl-tRNA synthetases play a role in multivalent repression.

be achieved by so simple a scheme. Indeed, it may be that there is no unitary pattern for control of amino-acid biosynthetic pathways. It may be, therefore, that, for a period of time, we should define as clearly and as precisely as we can all the genetically recognized regulatory elements in biochemical terms. Only then shall we be able to recognize in different systems those elements that are unique and those that are common.

ARE THE PATHWAYS TO VALINE AND ISOLEUCINE SEPARATE?

Much of the information that has been reviewed up to this point is quite compatible with the view that the enzymes of valine and isoleucine biosynthesis are, in fact, multifunctional. In other words, the multivalent repression pattern would appear to assure an ample supply of the required enzymes for formation of both amino acids. The question raised earlier, whether there are separate clusters of four and five enzymes that are specifically programmed for valine and isoleucine biosynthesis, has not been considered. In other systems, there is considerable evidence that points to the direction of clusters of biosynthetic enzymes that function as integrated units. WAGNER *et al.*[49] raised the question several years ago in the case of valine and isoleucine biosynthesis in *N. crassa* but they appeared to be considering a single cluster for both functions. The evidence for such integrated multienzyme units in these pathways, however, is not strong. The enzymes are readily separated and, unlike the anthranilate synthase reaction in *E. coli*[50], none of the reactions appear to require two enzyme proteins for the expected catalytic activity.

There are, however, a few observations that are difficult to explain by other than the idea of specific valine- and isoleucine-forming clusters. One is that some mutants of *E. coli* appear to have a non-inducible (*i.e.* constitutively low level) isomeroreductase. Preliminary examination of the enzyme in these strains has revealed no obvious differences between the mutant and wild-type enzyme. Thus, it appears quite likely that some control element required for transcription or translation of the *ilvC* region is missing or is in the permanently "off" state. This element appears to be near the *ilvC* gene but whether separate from it or not or on which side of it is not known.

The intriguing point is that these mutants require only valine for growth.* None of our attempts to spare the function of the limited isomeroreductase activity has made it possible for the strains to form valine. Thus, it appears that the isomeroreductase in these strains is specifically committed to isoleucine biosynthesis. That the postulated isoleusome may have a greater affinity for the first isomeroreductase molecules that are formed or that the mutant isomeroreductase requires

* Those isolated from valine-sensitive strains exhibit valine-dependent growth only at very low levels of valine.

the fifth enzyme of the postulated cluster, threonine deaminase, for its function are possibilities.

Another observation is that under conditions of valine limitation there is strong derepression of the AHA synthase, which functions freely with limiting valine to produce inducing levels of the isomeroreductase substrate. However, if valine is added there is not only an immediate repression of AHA synthase, there is also an immediate halt to the induction of the isomeroreductase. It would thus appear that there is little or no internal pool of the acetohydroxyacids when endogenous induction occurs at a high rate. Such small internal pools would be expected if the enzymes were clustered or if there was "catalytic facilitation" such as has been described to occur between two multienzyme complexes in the pathway to anthranilate in *N. crassa*[51].

Finally, there are some examples of "reverse polarity"[52,53] that might be explained on the basis of protein—protein interactions between some of the individual valine and isoleucine biosynthetic enzymes. It is hoped that the question of a labile complex can be answered in the near future. Indeed, it may be that an understanding of the molecular arrangements of the enzymes may be finally necessary for an understanding of the expression of the *ilv* genes, our current primary goal.

Throughout this paper, I have tried to give proper credit for the development of our current ideas on regulation of the biosynthesis of the three branched-chain amino acids. Lest it appear that I have been personally responsible for all the findings that are discussed here, let me acknowledge and express my deep appreciation to a fairly long list of my past and present colleagues who will recognize so many of their experimental results (and ideas) in this report. It should be mentioned that for the past 19 years, the National Institutes of Health have provided almost all of the support for the research in my laboratory.

REFERENCES

1. GRIFFITH F.: J. Hyg. **27**, 113 (1928).
2. TATUM E. L., LEDERBERG J.: J. Bacteriol. **53**, 673 (1947).
3. ZINDER N. D., LEDERBERG J.: J. Bacteriol. **64**, 679 (1952).
4. YOSHINAGA F., KONISHI S., OKUMURA S., KATSUYA N.: J. Gen. Appl. Microbiol. **12**, 219 (1966).
5. ADELBERG E. A.: J. Bacteriol. **76**, 326 (1958).
6. CALVO J. M., FREUNDLICH M., UMBARGER H. E.: J. Bacteriol. **97**, 1272 (1969).
7. DWYER S.: Ph. D. Thesis, Purdue University, 1969.
8. ZUBAY G., LEDERMAN M.: Proc. Nat. Acad. Sci. U.S.A. **62**, 550 (1969).
9. STRASSMAN M., LOCKE L. A., THOMAS A. J., WEINHOUSE S.: J. Am. Chem. Soc. **78**, 1599 (1956).
10. GROSS S. R., BURNS R. O., UMBARGER H. E.: Biochemistry **2**, 1046 (1963).
11. RUDMAN D., MEISTER A.: J. Biol. Chem. **200**, 591 (1953).
12. WALLACE B. J., PITTARD J.: J. Bacteriol. **97**, 1234 (1969).
13. KOHLHAW G., LEARY T. R., UMBARGER H. E.: J. Biol. Chem. **244**, 2218 (1969).
14. LEARY T. R., KOHLHAW G.: Biochem. Biophys. Res. Comm. **39**, 494 (1970).

15. Burns R. O., Calvo J., Margolin P., Umbarger H. E.: J. Bacteriol. **91**, 1570 (1966).
16. Roth J. R., Anton D. N., Hartman P. E.: J. Mol. Biol. **22**, 305 (1966).
17. Hirshfield I. N., DeKeken R., Horn P. C., Hopwood D. A., Maas W. K.: J. Mol. Biol. **35**, 83 (1968).
18. Calvo J. M., Margolin P., Umbarger H. E.: Genetics **61**, 777 (1969).
19. Jacob F., Monod J.: Cold Spring Harbor Symp. Quant. Biol.**26**, 193 (1961).
20. Calvo R. A., Calvo J. M.: Science **156**, 1107 (1967).
21. Argoudelis A. D., Herr R. R., Mason D. J., Pyke T. R., Zieserl J. F.: Biochemistry **6**, 165 (1967).
22. Leavitt R. I., Umbarger H. E.: J. Biol. Chem. **236**, 2486 (1961).
23. Størmer F. C., Umbarger H. E.: Biochem. Biophys. Res. Comm. **17**, 587 (1964).
24. Changeux J.-P.: Bull. Soc. Chim. biol. **46**, 927, 1115 (1964); **47**, 115 (1965).
25. Monod J., Wyman J., Changeux J.-P.: J. Mol. Biol. **12**, 88 (1965).
26. Burns R. O., Zarlengo M. H.: J. Biol. Chem. **243**, 178 (1968).
27. Zarlengo M. H., Robinson G. W., Burns R. O.: J. Biol. Chem. **243**, 186 (1968).
28. Hatfield G. W., Burns R. O.: J. Biol. Chem. **245**, 787 (1970).
29. Hatfield G. W., Burns R. O.: Science **167**, 75 (1970).
30. McCord T. J., Howell D. C., Tharp D. L., Davis A. L.: J. Medicin. Chem. **8**, 290 (1965).
31. Roberts R. B., Abelson P. H., Cowie D. B., Bolton E. T., Britten R. J.: Carnegie Inst. Wash. Publ. 607 (1955).
32. Leavitt R. I., Umbarger H. E.: J. Bacteriol. **83**, 624 (1962).
33. Ramakrishnan T., Adelberg E. A.: J. Bacteriol. **89**, 661 (1965).
34. Umbarger H. E., Brown B.: J. Biol. Chem. **233**, 1156 (1958).
35. Bauerle R. H., Freundlich M., Størmer F. C., Umbarger H. E.: Biochim. Biophys. Acta **92**, 142 (1964).
36. Freundlich M., Burns R. O., Umbarger H. E.: Proc. Nat. Acad. Sci. U.S.A. **48**, 1804 (1962).
37. Eidlic L., Neidhardt F. C.: Proc. Nat. Acad. Sci. U.S.A. **53**, 539 (1965).
38. Szentirmai A , Szentirmai M., Umbarger H. E.: J. Bacteriol. **95**, 1692 (1968).
39. Ramakrishnan T., Adelberg E. A.: J. Bacteriol. **89**, 654 (1965).
40. Taylor A. L., Trotter M.: Bacteriol. Rev. **31**, 332 (1967).
41. Dwyer S. B., Umbarger H. E.: J. Bacteriol. **95**, 1680 (1968).
42. Coker M., Umbarger H. E.: Bacteriol. Proc. 136 (1970).
43. Iaccarino M., Berg P.: J. Bacteriol. **105**, 527 (1971).
44. Nass G., Neidhardt F. C.: Biochim. Biophys. Acta **134**, 347 (1967).
45. Blatt J., Umbarger H. E.: Bacteriol. Proc. 135 (1970).
46. Arfin S. M., Ratzkin B., Umbarger H. E.: Biochem. Biophys. Res. Comm. **37**, 902 (1969).
47. Faanes R., Rogers P.: J. Bacteriol. **96**, 409 (1968)
48. Hatfield G. W., Burns R. O.: Proc. Nat. Acad. Sci. U.S.A. **66**, 1027 (1970).
49. Wagner R. P., Bergquist A., Barbee T., Kiritani K.: Genetics **49**, 865 (1964).
50. Ito J., Yanofsky C.: J. Biol. Chem. **241**, 4112 (1966).
51. Gaertner F. H.; Ericson M. C., DeMoss J. A.: J. Biol. Chem. **245**, 595 (1970).
52. Elliot C. J., Armstrong F. B.: Genetics **58**, 171 (1968).
53. Wechsler J. A., Adelberg E. A.: J. Bacteriol. **98**, 1179 (1969).

Amino-Acid Production Using Microbial Auxotrophic Mutants

KIYOSHI NAKAYAMA

Tokyo Research Laboratory, Kyowa Hakko Kogyo Co. Ltd., Tokyo, Japan

Industrial production of amino acids started with the discovery of an efficient glutamic acid producer, *Corynebacterium glutamicum* (synonym *Micrococcus glutamicus*), by KINOSHITA *et al.*[1] in 1957. The bacterium was found during the increasing demand for monosodium glutamate as a flavoring agent. However, it was found that most wild strains isolated from nature could not produce industrially significant amounts of other amino acids except a few amino acids (L-glutamic acid, DL-alanine and L-valine). The main cause of this fact is the regulation of cellular metabolism to avoid oversynthesis, the existence of which is now well recognized. An auxotrophic mutant which cannot produce the regulatory effector (usually end product or a derivative of end product) overproduces and excretes the precursor or the related metabolite of blocked reaction when grown in limiting supply of the required nutrient. Simply saying, this is the principle of the application of an auxotrophic mutant to the fermentative production of amino acids. Active attempts to utilize this phenomenon for the industrial production of microbial metabolites were launched up to the 1950s. The first prominent result was obtained by CASIDA and BALDWIN[2] in producing α,ε-diaminopimelic acid using a lysine auxotroph of *Escherichia coli.* Diaminopimelic acid was decarboxylated to produce L-lysine in the second step. Soon afterward, the process for L-lysine production was developed with a homoserine auxotroph of *C. glutamicum*[3]. The process has been operating on a large scale for over 10 years. Nowadays, many processes for various amino acids (generally L-form except for alanine) have been developed, and most of them utilize an auxotrophic mutant. The primary reason for these efforts was the hope to improve the biological value of low-cost vegetable proteins in human and animal nutrition by enrichment with essential amino acids.

During these years, information on the pathway and on the regulation of biosynthesis has accumulated[4–11] and has been utilized as a working hypothesis in obtaining the desired amino-acid producer by mutational techniques. Inversely, the information obtained with the mutant can and did contribute to the elucidation of the biosynthesis mechanism. The success obtained in the production of amino acids extended to the use of microbial mutants for the production of nucleic acid-related substances[12]. This situation seems to persist in other groups of metabolites.

Even in such areas as production of vitamins, antibiotics and other natural products, we are beginning to see the implications of regulation in the process designed for their oversynthesis[13-16].

The present review should bring the field of fermentative production of amino acids up to date and emphasize to a greater degree the regulation mechanism of biosynthesis and the way it is altered for efficient amino-acid production. Furthermore, some of the future aspects will be discussed. Since some excellent reviews[17-19] related to this topic have been published, I shall describe only briefly some of the items reviewed in these papers.

L-LYSINE AND DIAMINOPIMELIC ACID

The biosynthesis of L-lysine by microorganisms has been reviewed[20]. Two distinct pathways, the diaminopimelate pathway found in bacteria and the α-aminoadipate pathway found in yeasts and fungi are shown in Figs 1 and 2.

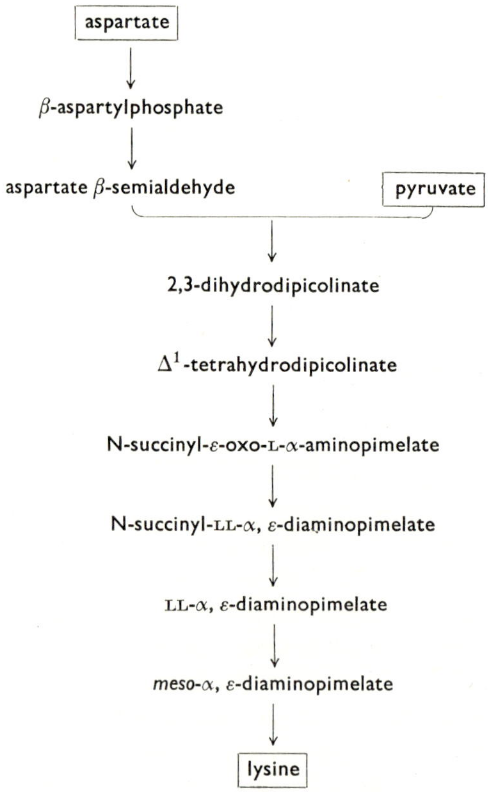

Fig. 1. Diaminopimelate pathway for lysine biosynthesis.

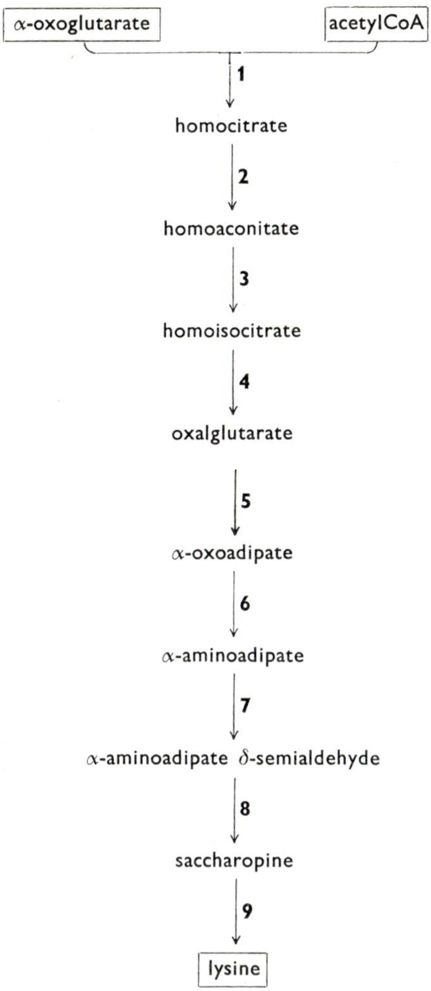

Fig. 2. α-Aminoadipate pathway for lysine biosynthesis.

Besides the direct fermentation method, several methods have been reported. Microbial decarboxylation of α,ε-diaminopimelic acid[1,21] which is itself produced by fermentation[21–23], has been reviewed before[20]. It seems noteworthy that with a double auxotroph of *E. coli* which is deficient in histidine as well as in L-lysine, the final yield of diaminopimelate is controlled not only by the initial concentration of L-lysine but also by that of the second growth factor, *i.e.* L-histidine[23]. Decarboxylation of *meso*-diaminopimelate was applied in the chemical process as a kind of optical resolution[24]. Optically specific hydrolysis of L-α-aminocaprolactam by *Aspergillus niger*[25] functions also as a kind of optical resolution. The conversion of α-aminoadipate or α-oxoadipate by yeasts to L-lysine has been reported[26–28]. The method is evidently more suitable for the preparation of L-lysine-rich yeasts than pure L-lysine, though the lysine concentration reached 20% of the cell weight and the conversion rate of α-aminoadipate to L-lysine reached 83%. The lysine is held in the cell and an extraction process is necessary to release it.

L-Lysine production has been operating on a large scale for more than ten years using a mutant of *C. glutamicum*. L-Lysine was found to be accumulated by various auxotrophs of *C. glutamicum*[29]. These include mutants deficient in threonine, isoleucine, leucine, proline, tryptophan or phenylalanine. However, the most active auxotroph was the one which requires homoserine (or threonine plus methionine). Under the proper conditions, the auxotroph easily produced more than 20 g of L-lysine per liter. This represents a yield of about 30% from glucose on the weight basis. The L-lysine titer reached over 50 g/l with the medium which contained sugar-cane waste molasses and some growth-factor sources. The yield is dependent upon the initial level of both biotin and L-homoserine (or threonine plus methionine). To obtain maximum accumulation of L-lysine, it was necessary to use levels of biotin high enough to inhibit the production of L-glutamic acid and an optimum level of L-homoserine or L-threonine plus methionine. Excess of L-homoserine or L-threonine plus L-methionine inhibited L-lysine production. The effect of excess methionine or L-isoleucine which is required was not so severe that of as L-threonine or L-homoserine. The specific inhibitory effect of L-threonine and L-homoserine was also noted in the formation of L-lysine from glucose and ammonium salt by intact cells. Later, it was found that the activity of aspartokinase was inhibited by the presence of both L-lysine and L-threonine while neither L-lysine nor L-threonine inhibited the enzyme activity[30]. Furthermore, neither lysine nor other amino acids tested exerted an inhibitory effect on the first reaction of the lysine branch of the pathway catalysed by dihydrodipicolinate synthase. Thus, the regulation of the pathway of *C. glutamicum* is in sharp contrast with that of *E. coli* in which three aspartokinases are present[31] and where lysine inhibits the activity of dihydrodipicolinate synthase[32]. The latter effect may be the cause of the low yield of L-lysine[29] by a homoserine auxotroph of *E. coli*. No repression of the lysine-synthesizing system with L-homoserine (1000 μg/ml) was noted in intact cells[29]. Later, some repression was noted in similar experiments with excess L-threonine or L-isoleucine. No prominent repression was noted in the aspartokinase either with L-threonine or L-isoleucine[33].

Thus, in homoserine and threonine auxotrophs of *C. glutamicum*, L-lysine is produced even in the presence of L-lysine unless homoserine or threonine is added externally. The insensitivity of dihydrodipicolinate synthase of *Bacillus subtilis*[34] also explains the relatively high lysine formation by the homoserine or threonine auxotrophs of *B. subtilis*[29]. No decomposition of L-lysine including decarboxylation is also a factor which contributes to the accumulation of L-lysine[35]. L-Lysine is apt to decarboxylate into cadaverine in *E. coli*.

Recently, L-lysine production by homoserine and threonine auxotrophs of *Brevibacterium flavum* was reported[36]. Results of the detailed studies with *B. flavum*[37] coincided with the results with *C. glutamicum*. Furthermore, some interesting phenomena were noted. Simultaneous addition of L-lysine and L-threonine at 1 mM each produced over 90% inhibition on purified aspartokinase while L-lysine or L-threonine at 1 mM gave only 10—20% inhibition[38]. L-Isoleucine and L-valine had an activating effect and recovered the activity from the concerted inhibition caused by L-lysine plus L-threonine. The aspartate kinase reaction was explained as a Michaelis-type rapid equilibrium system with random addition of the substrates (L-aspartate and ATP). The inhibition by L-lysine was competitive with L-aspartate, of mixed type with respect to ATP, while the inhibition by L-threonine was partially competitive with both L-aspartate and ATP. The concerted inhibition was explained by an increased affinity of one inhibitor to the enzyme which was caused by binding another inhibitor. The activation by L-isoleucine was K type. The relation between L-isoleucine and L-lysine was competitive. The reversal of the concerted inhibition by L-isoleucine was explained by a competitive relationship between L-isoleucine and L-lysine. L-Aspartate-β-semialdehyde dehydrogenase was scarecely inhibited by amino acids of the aspartate family, though L-serine inhibited the activity by 40% at a concentration of 10 mM[37]. Dihydrodipicolinate synthase was neither inhibited by L-lysine nor repressed by L-lysine[39]. No prominent repression was noted with aspartokinase and L-aspartate-β-semialdehyde dehydrogenase with the amino acids tested[40].

Recently, SHIIO and SANO[41] isolated leaky-type homoserine-deficient strains by selecting the threonine- or methionine-sensitive cell in the cell suspension of typical homoserine-deficient mutant of *B. flavum* 2247 which received mutagenic treatment. One of these strains, S-20, produced 23 g of L-lysine monohydrochloride from 100 g of glucose. The growth of the strain in a minimal medium was completely inhibited by the addition of 10 μg/ml of L-methionine (or L-threonine) and recovered by the further addition of L-threonine (or L-methionine). It was concluded that the much lower dehydrogenase activity of S-20 than that of 2247 was the apparent cause of the increased sensitivity of the cell growth to inhibition by L-threonine or L-methionine and further production of lysine through lowering of the threonine pool. Some of the auxotrophs derived from such a strain produced much larger amounts of lysine than the simple threonine auxotroph. By-production of L-homoserine was small in contrast with the simple threonine auxotrophs as expected.

Before this observation, several observations have been reported on the competition between threonine and methionine. The effect of methionine—threonine balance on growth, glucose utilization and lysine biosynthesis of a homoserine auxotroph of *C. glutamicum* has also been reported[42]. The growth of a homoserine-deficient mutant strain of *Neurospora crassa* which was originally classified as a threonine-less mutant was inhibited by excess L-methionine. The inhibition was relieved by increasing the threonine concentration[43]. Similar competition between threonine and methionine has been reported and explained by a regulatory mechanism[44–46].

Another recent important feature is the use of an analogue-resistant mutant for L-lysine production. S-(α-Aminoethyl)-L-cysteine (AEC) resistants from *B. flavum* 2247 produced L-lysine in the culture broth[47]. Correlation of the yield of L-lysine in a glucose medium with that in waste molasses medium was poor. Strain FA-315 produced 32 g/l of L-lysine in a glucose (10%) medium. Strain FA1-30 produced 34 g/l of L-lysine in a waste molasses (13% as glucose) medium[48]. Aspartokinases from both strains had the same specific activity as that of the wild strain 2247. The aspartokinase of the mutant was slightly inhibited just as the enzyme from the parent strain. Threonine did not inhibit but stimulated slightly. The concerted inhibition by threonine plus lysine was completely lost and threonine partly released the inhibition by lysine. AEC inhibited both enzymes from parent and mutant to a lesser extent than lysine[49]. The concerted inhibition by AEC plus threonine was not noted while it was found in the parent strain. A DL-*allo*threonine-resistant and a DL-serine-resistant mutant were isolated from a homoserine auxotroph of *C. glutamicum*[50]. They produced L-lysine at a wider range of the concentrations of threonine than that in the parent strain.

Homoserine-deficient mutants of bacteria other than the one mentioned above also produced L-lysine. These include *Brevibacterium* sp.[51], *Brevibacterium ammoniagenes*[52], *Corynebacterium* sp.[53], *Azotobacter suis*[54], *Pseudomonas aeruginosa*[52], *Aerobacter aerogenes*[52] and *Streptomyces griseus*[52]. Furthermore, homoserine-deficient mutants of certain hydrocarbon-assimilating bacteria also produced a considerable amount of L-lysine[55].

Replacement of the carbohydrate material with the cheaper materials produced from petroleum refinery or petroleum chemistry is a tendency of recent fermentation industry. Such an effort has been expressed also in reports which described lysine production. *Mycobacterium tuberculosis* ATCC 15073 produced 2.1 g/l of L-lysine and 0.6 g/l of L-methionine in a medium which contained *n*-decane at 1.5% (totally)[56]. *Streptomyces coroniformis* ATTC 15074 also produced 1.2 g/l of L-lysine and 0.8 g/l of L-methionine on tetrahydroxynaphthalene[56]. L-Lysine production from *n*-paraffins has also been claimed with some mutants of glutamic acid producer (*Arthrobacter paraffineus, Brevibacterium ketoglutamicum, Corynebacterium hydrocarboclastus*)[55]. Accumulation of 10 g/l has been described. Most prominent is the yield of 27 g/l of L-lysine (as monohydrochloride) from the total acetic acid 94 g/l by a homoserine auxotroph of *Corynebacterium acetophilum* nov. sp. A-51[57].

Genetic stability of the microorganism during fermentation is a necessary factor especially for the fermentation using auxotrophic mutants. Typical example has been reported in lysine fermentation with *C. glutamicum*[58]. Certain batches of molasses induced "abnormal" fermentations when little lysine was formed and there was an unsually long lag phase before growth took place. No marked chemical or physical differences between "normal" and "abnormal" molasses were apparent. Most of the bacterial cells from an "abnormal" broth were found to consist of prototrophic revertants. Evidently "abnormal" molasses were able selectively to inhibit growth of auxotrophic mutant, and thus indirectly encourage the proliferation of prototrophic cells which were already present in the inoculum. The occurrence of prototrophic cells limits the duration of effective continuous biosynthesis and is a hazard in continuous fermentation. An example in lysine biosynthesis was described recently[59].

L-THREONINE AND HOMOSERINE

The production of L-threonine in the wild strain was low even with a strain of *Ustilago* species which was selected for its ability to produce threonine[60,61]. The microbial conversion of L-homoserine to L-threonine has been studied first by SHIMURA *et al.*[62–64] and later by other workers[65,66]. Though an excellent direct fermentation procedure for L-homoserine is available[67–71], in view of the advantage of the direct fermentation procedure for L-threonine, the practicability of the method seems to depend on some special conditions. The production of L-threonine by direct fermentation has been investigated using auxotrophic mutants of *E. coli* first by HUANG[72], though the accumulation of threonine by a diaminopimelate (DAP) auxotroph of *E. coli* had been noted by DAVIS[73] previously. The yield of L-threonine reached about 4 g/l using the auxotrophic mutant deficient in both DAP and methionine in the sorbitol medium supplemented with beet molasses or sucrose plus cornsteep liquor. Recently, a triple auxotrophic mutant of *E. coli* W was found to produce much more L-threonine (15—20 g/l)[74]. It required isoleucine in addition to DAP and methionine for growth. The advantage of this strain is that a wider range of carbon sources, such as glucose, fructose, sucrose, molasses and hexitols can be used as raw materials. The by-production of valine was reduced by reverse mutation to leaky type in isoleucine auxotrophy.

When it is assumed that the same regulation mechanism as that in *E. coli* K12 is operative in *E. coli* W[75], the accumulation of L-threonine may be explained reasonably well. In *E. coli*, there exist three separate aspartokinases which are subject to inhibition by L-threonine and L-lysine or repressible by methionine. These may be designated as E_I, E_{III} and E_{II}, respectively. E_{II} is repressible by methionine and E_I is repressible by threonine plus isoleucine. In the DAP auxotroph, E_{III} should continue to synthesize aspartate β-semialdehyde when DAP (and hence L-lysine) is limiting although E_I and E_{II} are made inoperative by the intracellular level of threonine and methionine. In the methionine auxotroph, the repression of homoserine dehydrogenase E_{II} (same complex with aspartokinase E_{II}) by methionine is released by limiting methionine in the growth medium. In the isoleucine auxotroph, the multivalent repression of homoserine dehydrogenase E_I (same complex with aspartokinase E_I) by isoleucine plus threonine is released by limiting isoleucine. The inhibition of homoserine kinase by threonine is not so strong. The feedback inhibition of threonine on homoserine kinase may be rather weak as reported by WORMER and PARDEE[76].

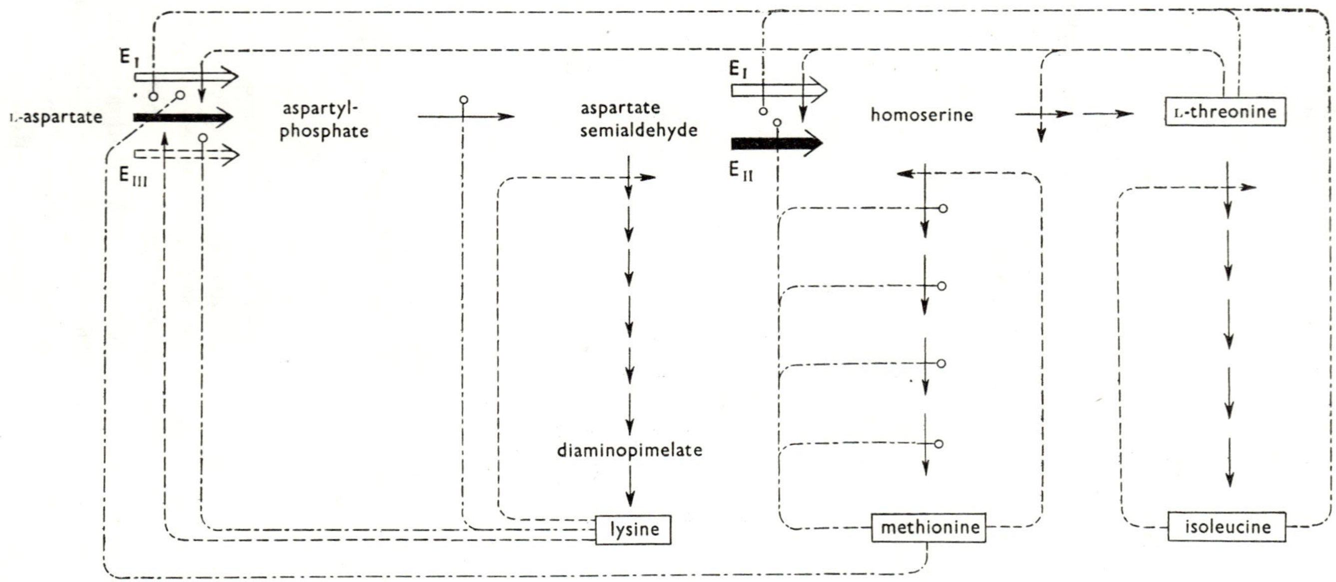

Fig. 3. Feedback inhibition (— — — — — — →) and repression (— · — · — · — ○) in the biosynthesis of amino acids of aspartate family in *E. coli*.

In the screening of the mutants of many species of bacteria having such auxotrophic markers as methionine, isoleucine, DAP and lysine, singly or multiply, some species of *Enterobacteriaceae (E. coli, A. aerogenes, Serratia marcescens)* and an *Arthrobacter* species accumulated L-threonine[74]. This fact is interesting in relation to the existence of isomeric aspartokinases in many members of *Enterobacteriaceae*[75] and the existence of homoserine dehydrogenase, E_{II} in at least one of these species, *Salmonella typhimurium*[77]. In contrast, accumulation of L-threonine by auxotrophs of *C. glutamicum* was rather low. This phenomenon may be explained by the single homoserine dehydrogenase in *C. glutamicum* which is indicated by the fact that homoserine auxotrophs were obtained and reversed each by a one-step mutational event, and the feedback inhibition of homoserine dehydrogenase by threonine. Threonine accumulation up to 4 g/l was found also in a mutant of the yeast *Candida guillermondii* var. *membranaefaciens* which requires isoleucine, methionine and tryptophan[78]. Mutants of *E. coli* which are resistant to the threonine analogue, α-amino-β-hydroxyvaleric acid (AHV), have been isolated and found to excrete threonine[79]. An AHV-resistant mutant, having a nutritional requirement for methionine and isoleucine was found to accumulate 6.2 g/l of L-threonine in a medium containing 30 g glucose per liter[80]. AHV-resistant mutants were isolated from *B. flavum* and *Corynebacterium acetoacidophilum*[81]. The yield of L-threonine by AHV-resistant *B. flavum* reached 13.5 g/l in a medium containing 10% sucrose. In *B. flavum*, L-homoserine has accumulated in addition to L-threonine. Thus, the inhibition of homoserine kinase seems to be stronger than in *E. coli.*

L-Threonine production from *n*-paraffins by a mutant of *A. paraffineus* has been reported recently[82]. The growth of the mutant was promoted by L-isoleucine, α-oxobutyric acid, α-aminobutyric acid, α-hydroxybutyric acid or D-threonine. The accumulation of L-threonine reached a level of 9 g/l with concurrent accumulation of L-valine (9 g/l) in a medium containing 10% (*v*/*v*) *n*-paraffins (C_{13}—C_{18}). The mechanism of accumulation of threonine by the mutant was not explained simply from its complex nutritional requirement though the leakage of threonine deaminase may be its principal cause. The situation resembles the threonine-less *N. crassa* mutant which accumulated threonine[83]. The AHV-resistant mutant of *C. acetoacidophilum* produced 2.7 g/l of L-threonine in a medium containing 50.6 g/l of sodium acetate trihydrate[81]. More recently, the AHV-resistant mutant of *Proteus rettgeri*, having a nutritional requirement for isoleucine, was reported to produce L-threonine in amounts of 8.4 g/l in a medium containing 10% glucose[84]. The amount increased to 12.2 g/l by feeding glucose during fermentation. The feedback inhibition of homoserine dehydrogenase by threonine and somewhat by isoleucine was found to be released greatly in the AHV-resistant *B. flavum*[85].

Prevention of the metabolism of threonine to glycine was a problem in some species to achieve efficient production of L-threonine. Addition of an inhibitor, such as hydroxylamine and hydrazines, was partly effective[86].

BRANCHED-CHAIN AMINO ACIDS

The biosynthetic pathways for isoleucine and valine have been established as homologous biosynthetic pathways, each step (2—5) being catalysed by the same enzyme (Fig. 4).

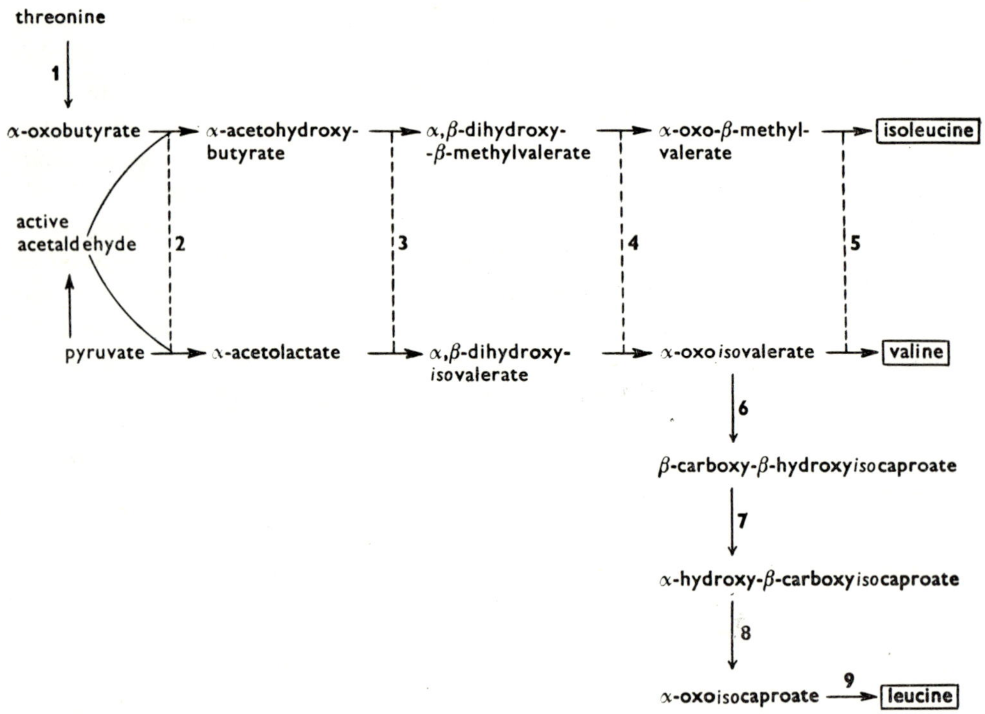

Fig. 4. Pathway of biosynthesis of valine, isoleucine, and leucine.

L-VALINE

The ability to accumulate L-valine in the culture broth is apparently quite common among members of the family *Enterobacteriaceae*. Many papers reported the accumulation of a considerable amount of L-valine. These include various types of auxotrophs of glutamic acid-producing bacteria[87,88], *E. coli*[89,90], chloramphenicol-resistant *A. aerogenes*[91], streptomycin-dependent *E. coli*[92]; the incubation of *A. aerogenes*[93], *B. flavum*[94], *Brevibacterium roseum*[95], *Corynebacterium aminofaciens* nov. sp.[96] and *E. coli*[97] in the presence of aminopterin, arsenite, sodium acrylate, α-oxobutyrate or propionate, inhibitors of protein synthesis or nucleic acid metabolism (chloramphenicol, streptomycin, 6-azauracil, 5-azauracil) and Co^{2+}. Many types of microorganisms have been found to accumulate L-valine in the culture media. These include *Brevibacterium fimbriatum*[98], *Bacterium*[99], *Mycobacterium*[100], *Azotobacter*[101] and many other strains from soil. These observations reflect the ease with which a disturbance of the balance in the intracellular enzyme systems can lead to accumulation of L-valine.

The fact is noteworthy that, unlike with *E. coli*, the acetohydroxyacid synthase from the valine producer studied by UDAKA and KINOSHITA[102] (*Paracolobactrum coliforme* 775 and *B. ammoniagenes* ATCC 6871) is not appreciably sensitive to inhibition by L-valine. Furthermore, the L-valine-producing bacteria do not contain acetolactate decarboxylase. These facts obviously contribute to the high production of L-valine. The prominent stimulation of the formation of valine and derepression of α-acetolactate (ACL) synthase by the exogenously supplied valine, leucine or isoleucine was noted in a wild strain, *Escherichia freundii* var. *ureasophorum*[103–105]. It seemed that valine or leucine contributes to formation of ACL synthase which is insensitive to valine inhibition. Of the three combinations added to the growth medium, valine plus leucine, valine plus isoleucine, leucine plus isoleucine, only valine plus leucine showed a similar effect as valine alone. The participation of the release of multivalent repression in addition to the insensitivity of ACL synthase to valine can be inferred.

Among a series of auxotrophic mutants of *C. glutamicum* which accumulated L-valine, an isoleucine auxotroph and a leucine auxotroph have been studied in some detail[87]. In each case, accumulation of L-valine was optimal at a limiting concentration of the respective growth factor. Excessive levels of the growth factor inhibited L-valine accumulation. In the isoleucine auxotroph, the effect of excess L-isoleucine was shown to be due to repression of enzyme synthesis. Accumulation of L-valine was greatly stimulated by the addition of L-leucine or L-valine to the medium. It was postulated that both L-leucine and L-valine antagonized the repression which L-isoleucine exercised on the synthesis of the enzyme system responsible for L-valine synthesis. L-Valine accumulation was found to occur in a threonine-requiring mutant of *E. coli* W[89] and a homoserine-requiring mutant of *Brevibacterium* sp.[106].

Recently KISUMI *et al.*[107] obtained a L-valine-producing mutant as an α-aminobutyrate-resistant mutant of isoleucine producer, *S. marcescens*. The acetohydroxyacid synthase of the mutant was found to be derepressed while being as sensitive as the parent strain to feedback inhibition by L-valine. The threonine deaminase was also derepressed in the mutant.

L-ISOLEUCINE

Various processes for the production of L-isoleucine were developed using α-aminobutyrate (ABU)[108–114], α-oxobutyrate[115], α-hydroxybutyrate[116,117], threonine[118–123] or homoserine[124] as precursors and various microorganisms such as *Bacillus*, *Pseudomonas*, *Brevibacterium*, *Corynebacterium*, *Microbacterium*, *Arthrobacter*, *Serratia*, *Streptomyces*, *etc.* The yield reported reached 20 g/l. If a considerable amount of ABU remains in the medium when a peak level of L-isoleucine is reached, the recovery of pure L-isoleucine cannot be achieved without careful chromatography on an ion-exchange resin column. The use of α-hydroxybutyrate was claimed to be better for its cost and ease of purification of L-isoleucine. These substances are considered as a kind of precursor. However, the process with *B. subtilis* 14 studied by SHIMURA, UEMURA, HAYASHIBE *et al.*[125–130], is a unique process which is partly microbial conversion and partly direct fermentation. From experiments using ^{14}C-labelled substrates it follows that the carbon skeleton of L-isoleucine was largely derived from DL-ABU during an earlier stage, *i.e.* first and second day of fermentation, while in the later stage it was derived about equally from both DL-ABU and glucose. The analogous role of DL-threonine was found when it replaced DL-ABU. Both ABU and threonine are presumably converted to the key intermediate, α-oxobutyric acid. Although DL-ABU slightly repressed the formation of threonine deaminase, it released the feedback inhibition of the enzyme activity by L-isoleucine. Furthermore, DL-ABU was found to inhibit the activity of the acetohydroxyacid synthase as reflected only in its ability to synthesize ACL from pyruvate, and hence to depress the overall rate of valine biosynthesis[131]. In contrast, ABU stimulated acetohydroxyacid synthase as reflected in its ability to synthesize α-acetohydroxybutyrate (AHB). The formation of AHB-synthesizing activity was repressed by ABU in *B. subtilis* PCI-219 and *E. coli* K12 which do not produce isoleucine while not repressed by ABU in *B. subtilis* 14. The ACL-forming activity was not formed in the minimal medium while it was formed in the presence of high concentration of glucose in *B. subtilis* 14. In *B. subtilis* PCI-219 and *E. coli* K12, the AHB-forming activity and the ACL-forming activity were formed in the minimal medium. Thus strain specificity is involved in isoleucine production. The activity of the ACL-forming enzyme was not influenced by the presence of ABU, valine or isoleucine[132]. The final step (transaminase)[133] and acetohydroxyacid isomeroreductase (3rd step)[132] were not influenced by ABU either in formation or in activity. The mechanism of ABU effect on the acetohydroxyacid synthase step is not known. But it is interesting that the enzyme from *B. subtilis* 14 was inhibited by a SH-reagent in its ACL-forming activity and in AHB-forming activities. The enzyme from *E. coli* was inhibited in both activities[134]. ABU as precursor of L-isoleucine was found to incorporate after transformation into α-oxobutyrate by transamination[134].

In a process using *S. marcescens*[119], ABU was replaced with D-threonine which was a waste product in the chemical synthesis of L-threonine. In this process, D-threonine was converted to α-oxobutyrate by D-threonine deaminase which was induced with D-threonine and was resistant to L-isoleucine while L-threonine was sensitive to L-isoleucine. It was indicated that D-threonine may cause an increase of the intracellular level of L-isoleucine and a decrease of the levels of L-valine and

L-leucine and consequently may cause the derepression of the isoleucine—valine enzymes.

In glutamic acid-producing bacteria (*Arthrobacter aminofaciens* nov. sp. G-56, *Corynebacterium* sp. T-22) and *Corynebacterium amagasakii* nov. sp.[114], the role of ABU is rather different from that in *B. subtilis* 14. With *A. aminofaciens*, L-isoleucine was formed almost solely from ABU *via* α-oxobutyric acid which was formed by L-ABU—pyruvate transaminase[135] or from constitutive D-ABU oxidase. AHB formation from α-oxobutyrate and pyruvate is apparently activated with ABU while practically not inhibited with L-isoleucine. The amino group from ABU seems to contribute as an amino-donor in the reductive amination of α-oxobutyrate to produce glutamic acid which acts as amino-donor in the transamination to oxoisoleucine. Isoleucine formation with intact cells is equally feasible either from glucose plus ABU, or glucose plus ABU plus ammonia. Thus, the amino group of ABU acts efficiently in L-isoleucine synthesis. The activation of AHB synthase by ABU was decreased and no more observed with the progress of purification of the enzyme. In the mixture of the purified AHB synthase and purified ABU—pyruvate transaminase, AHB synthase activity was promoted again by ABU[135].

Recently, KISUMI *et al.*[136,137] isolated an ABU-resistant mutant of *S. marcescens* which was improved in isoleucine production from D-threonine. This mutant produced 15 g/l of L-isoleucine, 1.5 times more than the parent strain. In the mutant, the formation of AHB synthase and L-threonine deaminase was not affected by any branched-chain amino acids, thus both enzymes were found to be genetically derepressed. In contrast, in the parent strain, both enzymes were multivalently repressed by three end products, L-isoleucine, L-valine and L-leucine.

L-LEUCINE

L-Leucine accumulation in amounts of 13 g/l was reported recently[138]. The organism is an isoleucine-requiring (leaky type) mutant derived from ABU-resistant mutant of *S. marcescens* which was described in the section on L-isoleucine. A leucine production process using auxotrophs of *C. glutamicum* requiring one or more of the following amino acids: methionine, phenylalanine, valine and isoleucine was claimed[139].

It seems necessary to re-evaluate the explanations of accumulation of branched-chain amino acids in view of the contamination of effectors in the enzyme preparations studied and interaction among effectors. The present knowledge of the regulation of the biosynthesis of branched-chain amino acids in other microorganisms[140] also points to this necessity.

AROMATIC AMINO ACIDS

The biosynthesis of aromatic amino acids and its regulation has been reviewed extensively[141,142]. The biosynthetic pathway is shown in Fig. 5.

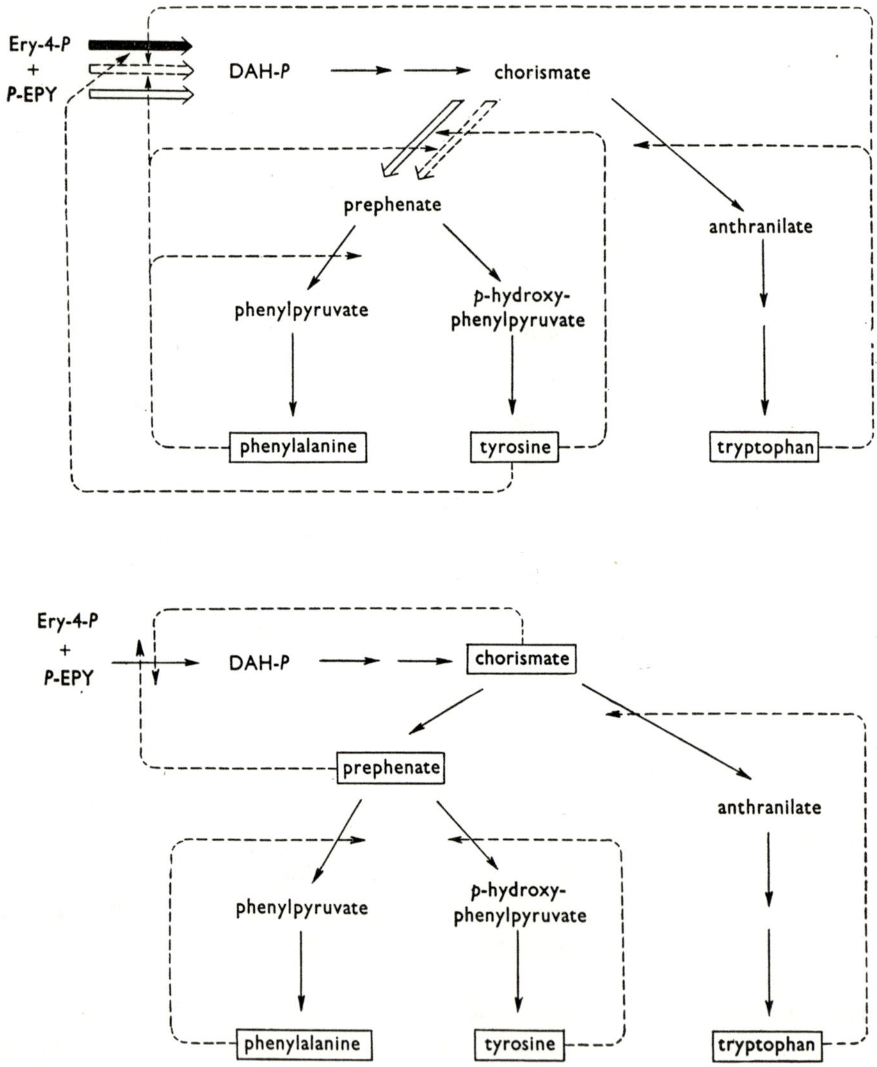

Fig. 5. Two types of feedback inhibition in the aromatic pathway: isoenzyme (*E. coli; above*) and sequential (*B. subtilis; below*). Ery-4-*P* erythrose 4-phosphate, *P*-EPY phosphoenolpyruvate, DAH-*P* 3-deoxy-D-*arabino*heptulosonate 7-phosphate.

L-PHENYLALANINE

A tyrosine auxotroph of *E. coli* accumulated about 2 g/l of L-phenylalanine when grown on glycerol as carbon source[143]. A tyrosine auxotroph of *C. glutamicum* accumulated 1—2 g/l of L-phenylalanine when the level of biotin in the medium was high and the level of L-tyrosine was limiting[144]. The genetic block must lie between prephenate and *p*-hydroxyphenylpyruvate since the addition of shikimate, phenylpyruvate and D-quinate significantly enhanced the yield of L-phenylalanine while *p*-hydroxyphenylpyruvate had no effect on L-phenylalanine accumulation but promoted the formation of L-tyrosine.

L-TYROSINE

A phenylalanine auxotroph of *C. glutamicum* accumulated L-tyrosine[145]. A significant increase in tyrosine production was observed when *p*-hydroxyphenylpyruvate was added to the culture medium while phenylpyruvate was converted to L-phenylalanine. Thus, the block of the pathway of L-phenylalanine biosynthesis must lie before phenylpyruvate. The yield described was low, but it could be greatly improved by recent efforts.

L-TRYPTOPHAN

Processes using some precursors, namely indole[146], indole plus serine[147,148] and anthranilic acid[148−152] have been reported. Microbial conversion of indol-3-ylpyruvic acid to L-tryptophan has also been studied[153]. Amongst these, the process using anthranilic acid seems comparatively economical. The yield of 3.2 g/l of L-tryptophan has been reported using a mutant strain of *Hansenula* sp.[154]. Recently, levels up to 14 g/l of L-tryptophan were attained with *Hansenula anomala* in a jar fermentor in 10 days[155]. Indole and anthranilic acid together with nutrients were fed intermittently. As a means to supply indole gradually to the culture, the use of oil as a reservoir of the lipophilic precursor was proposed.

Direct production of L-tryptophan was reported in the tryptophan analogue-resistant mutant of *E. coli*[156,157] or *Salmonella*[158], but the yield was less than 1 g/l.

AMINO ACIDS OF THE ARGININE PATHWAY

Fig. 6 represents the biosynthetic pathway of arginine and related metabolites.

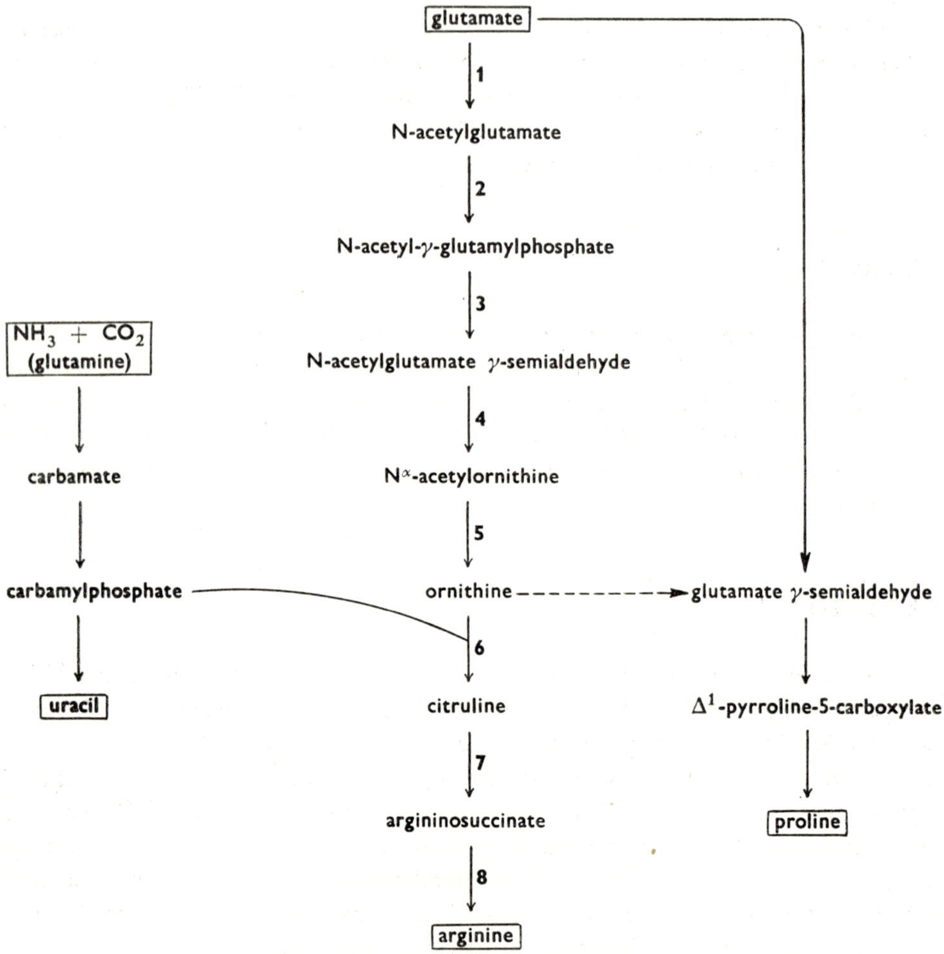

Fig. 6. Pathway of biosynthesis of arginine and proline.

L-ORNITHINE

Production of L-ornithine by a mutant of *C. glutamicum* which requires L-arginine or L-citrulline for growth[159–161] has already been reviewed many times. The process represents the simplest case from the standpoint of regulation of biosynthesis, namely the accumulation of an intermediate of an unbranched pathway. By feeding low levels of arginine, the feedback inhibition (at step **1** in *E. coli* and step **2** in *C. glutamicum*) and repression (generally 8 steps) by arginine is prevented and high levels of ornithine are excreted from the cells. In microorganisms containing constitutive ornithine transcarbamylase, the relation of the pathway to the pathway to proline must be considered.

A mutant of *E. coli* B which had two blocked steps was induced from the mutant which requires arginine or citrulline by transduction, using the p1KC phage grown on an arginine auxotroph of *E. coli* K12 in order to stabilize it against reversion[162]. The mutant produced 18 g/l of L-ornithine in a medium containing 7% glucose. Recently, an ornithine-accumulating mutant of *Streptomyces virginiae* has been reported[163]. This mutant required lysine and was derived from a thiamine auxotroph which did not accumulate L-ornithine. The amount of L-ornithine reached 8—10 g/l and lesser amounts of N^{α}-acetyl-L-ornithine, N^{δ}-acetyl-L-ornithine and α,ε-diaminopimelic acid were produced. Some proline auxotrophs of *C. glutamicum* were found to produce a rather high amount of L-ornithine (11 g/l)[164]. The mechanism of these phenomena awaits to be clarified. Some other processes using arginine auxotrophs of various bacteria including glutamic acid producers (*Brevibacterium lactofermentum*[165], *Arthrobacter citreus*[166], *B. flavum*[167]), *Brevibacterium kawasakii*[168], *B. subtilis*[169], *Bacillus megaterium*[169], *A. aerogenes*[170], *Proteus mirabilis*[170] and *P. rettgeri*[170] have been reported or claimed. Ornithine could be produced also from *n*-paraffins with an arginine auxotroph of *C. hydrocarboclastus*[171] or *A. paraffineus*[172].

L-CITRULLINE

L-Citrulline production using an arginine auxotroph of *B. subtilis*[173,174] or *C. glutamicum*[175] has been reported. The *B. subtilis* mutant produced 16 g/l of L-citrulline in a medium containing 13% glucose. The *C. glutamicum* mutant produced 10.7 g/l in a medium containing 10% glucose. Production of citrulline was also reported by PLACHÝ[176] with an arginine auxotroph of *Corynebacterium* sp. Production of L-citrulline from *n*-paraffins with an arginine auxotroph of *Corynebacterium* sp. reached 8 g/l in a medium containing 10% (*v*/*v*) of *n*-paraffin mixture[177]. The production of citrulline is explained in a way similar to that in the case of ornithine production.

L-ARGININE

Direct production of L-arginine in an industrially significant amount has not been reported though the excretion of arginine by a canavanine-resistant *E. coli*[178] or *Saccharomyces cerevisiae*[179] has been known. L-Arginine can be produced readily from L-ornithine by a synthetic method.

L-HISTIDINE

The pathway of histidine biosynthesis (Fig. 7) has no branches giving rise to other essential metabolites and direct fermentation by an auxotrophic mutant seems

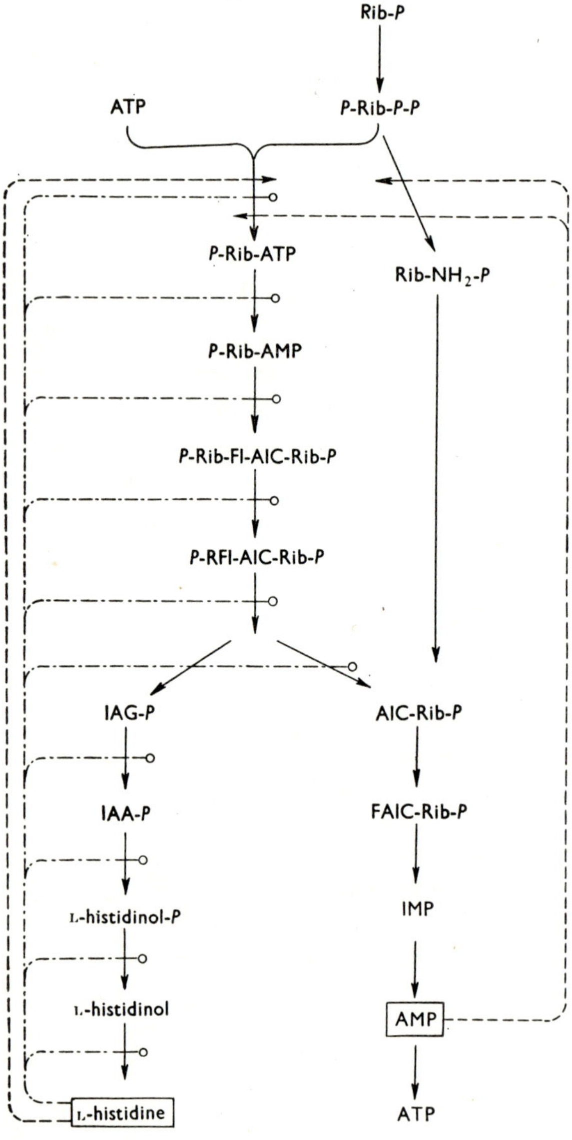

Fig. 7. Pathway of biosynthesis of histidine and regulation in *S. typhimurium* (— — — — → inhibition, — · — · — · — ○ repression).

Rib-*P* ribose 5-phosphate
P-Rib-*P*-*P* 5-phosphoribosylpyrophosphate
P-Rib-ATP 1-(5″-phosphoribosyl)-ATP$^+$
P-Rib-AMP 1-(5″-phosphoribosyl)-AMP$^+$
P-Rib-FI-AIC-Rib-*P* 1-(5′-phosphoribosyl)-N-(5″-phosphoribosylformimino)-5-aminoimidazole-4-carboxamide
P-RFI-AIC-Rib-*P* 1-(5′-phosphoribosyl)-N-(5″-phosphoribulosylformimino)-5-aminoimidazole-4-carboxamide
IAG-*P* 3-(imidazol-5′-yl)glycerol-1-phosphate
IAA-*P* 3-(imidazol-5′-yl)acetol-1-phosphate
Rib-NH_2-*P* ribosylamine-5-phosphate
AIC-Rib-*P* 1-(5′-phosphoribosyl)-5-aminoimidazole-4-carboxamide
FAIC-Rib-*P* 1-(5′-phosphoribosyl)-5-formylaminoimidazole-4-carboxamide

impossible unless some physiological relations (regulatory relations) with other pathways exist. But some intermediates or their derivatives have been found to accumulate in some histidine-auxotrophic mutants. There are reports on the accumulation of histidinol by a histidine auxotroph of *B. flavum*[180] or *C. glutamicum*[181]. The yield of histidinol reached 9.6 g/l (as dihydrochloride) in a medium which contained 10% glucose and 6% ammonium sulfate. Histidinol could be transformed into L-histidine by various bacteria including *B. flavum* during growth in a medium containing 7% glucose and 0.5% histidinol dihydrochloride. The best yield reported was 1.8 g/l[180]. A process providing imidazol-5-ylglycerol has been published[182]. The yield described is about 7 g/l. Recently a direct fermentation process for L-histidine production was developed using a histidine analogue-resistant mutant of *C. glutamicum*[181].

L-PROLINE

L-Proline production by an isoleucine auxotroph of *B. flavum* ATCC 15940 was studied extensively by YOSHINAGA *et al.*[183–188]. Maximum production of L-proline reached 15 g/l[183]. High concentration of nitrogen source (5.5% as ammonium sulfate) and biotin (450 μg/l) were effective for increasing the accumulation of L-proline in a medium which contained 10% glucose. The mutant lacked threonine dehydratase and intracellularly accumulated glycine, methionine, aspartate and threonine. Intracellular valine, leucine and lysine also increased. It was concluded that in strain ATCC 15940, both the higher level of available ATP resulting from the inhibition of aspartate and homoserine kinase, by threonine plus lysine and by threonine, respectively, and accumulation of intracellular L-glutamate promoted by biotin-rich conditions contributed to the high production of L-proline through stimulation of the glutamate kinase reaction[187].

Some auxotrophs of *Brevibacterium* sp. 1201 and 7966 produced L-proline[189]. These include 4 types of auxotrophs requiring histidine, methionine, leucine and isoleucine, respectively. A histidine auxotroph, YS-48, produced most abundantly L-proline. The production of L-proline by YS-48 was maximum with limited supply of L-histidine (50—70 mg/l) and addition of L-isoleucine to the medium increased the proline production to the level of 25 g/l from the level of 20g/l obtained without isoleucine. Activities of glutamate reductase and Δ^1-pyrroline-5-carboxylate reductase from the mutant were higher than those in the parent strain 7966 and they were neither repressed nor inhibited by L-proline. The formation of the enzymes was found to vary depending on the concentration of L-histidine added to the growth medium, being maximal with 50—70 mg/l of L-histidine.

A serine-auxotrophic mutant of *Kurthia catenoforma* produced 1.5 times as much L-proline as its parent[190]. The yield of L-proline reached 30 g/l in a medium which contained high concentration of L-aspartate (2%) in addition to glucose (6%) and others. It seems that glutamate-producing bacteria and also other bacteria such as *P. coliforme*[191], *Bacillus*[192] and *Escherichia*[192] can accumulate L-proline. The markers described above and arginine auxotrophy promote the production of L-proline[193,194].

L-GLUTAMIC ACID

L-Glutamate is produced on a large scale as an intermediate of monosodium glutamate, a flavoring agent. Recently it has been utilized for other purposes, such as the starting material for polyglutamate. Microbial production of L-glutamic acid was reviewed before[18,19]. Here only the recent trends, particularly the production of L-glutamate from noncarbohydrate material, are reviewed.

Since the report by YAMADA *et al.*[194,195] on the isolation of *C. hydrocarboclastus* S1OB1, the production of L-glutamate from hydrocarbon has been studied by many workers attempting to replace carbohydrate with cheaper material, such as hydrocarbons. The strains selected by SHIIO *et al.*[196] also belong to *C. hydrocarboclastus.* However, these methods were far from being practical because of the low yield and long cultivation required for the accumulation of L-glutamate. Afterwards, two prominent phenomena were found to increase the yield of glutamate. First, the process of adding penicillin at an early time of fermentation; this has been established in the production of L-glutamate with molasses medium. The accumulation of L-glutamate by *C. hydrocarboclastus* M-104 increased about 10 times as much as the process without penicillin[197]. This strain does not require biotin for its growth. With the better strain isolated afterwards, the yield of L-glutamate reached about 20 g/l in 48 h by the penicillin-addition process[198]. It corresponds to 72% (*w/w*) of *n*-hexadecane consumed. Similar results were also obtained with *Corynebacterium petrophilum*, nov. sp.[199], *A. paraffineus*[200], *Corynebacterium* sp. 1317A[201]. The accumulation of a trehalose ester of a branched-chain fatty acid[200] or a phospholipid[201] is suggestive of the mechanism of the effect of penicillin. Another prominent phenomenon is the effect of thiamine. TAKAHASHI *et al.*[202] identified the growth factor required by *C. hydrocarboclastus* S1OB1 as thiamine. Limitation with thiamine in the medium to a suboptimal concentration (3—5 μg/l) for growth increased the yield of glutamic acid to about 5 g/l. The effect of thiamine is analogous to that of biotin in the process with carbohydrate. The effect of thiamine was noted in many other bacteria, including *A. paraffineus,* though no optimal concentration was noted in *C. hydrocarboclastus* R7[198].

The reported yields of glutamate were still low, but a much higher yield may be obtained by adjusting the culture condition more precisely to the effective production of L-glutamate.

Another recent trend is the process using acetic acid available from petroleum chemistry. The details have not been published by a glutamate producer. Formation of glutamate from acetic acid by some strains of *Brevibacterium*[203,204] and *Corynebacterium*[205] has been reported. These strains are producers of glutamate from carbohydrate and produce a large amount of glutamate in the more limited supply of biotin than in the case with carbohydrate. *C. acetophilum* nov.sp. isolated by HARADA *et al.*[206] produced 2.71 g of L-glutamic acid from 100 ml culture medium by incubation with an addition of penicillin in a final concentration of 20 units/ml after 24 h fermentation. The growth of the strain was promoted with biotin. The production of L-glutamate by *Microbacterium ammoniaphilum*, a producer of L-glutamate from carbohydrate, formed L-glutamate from acetate with the addition of penicillin[206,207]. A large amount of L-glutamate was produced with the medium containing both acetic acid and sugar. A large amount of L-glutamate was accumulated only when the acetate concentration was maintained low throughout the cultivation. It is assumed that acetate is metabolized by bacteria through the usual pathway which leads to the accumulation of L-glutamate when its residual concentration in the medium is low, and through the glyoxylate cycle or dicarboxylic-acid cycle which leads to complete oxidation when its concentration is high. The yield of glutamate increased with the content of acetic acid in the mixture of acetate and sugar equal to 20%, but the yield declined when the percentage increased to 50%. The yield of glutamic acid reached about 23 g/l, namely 29% (*w*/*w*) of acetic acid supplied in 96 h[208]. In the process, *B. flavum* was incubated in the medium containing 4% acetate as acetic acid initially and with continuous feeding of acetic acid to adjust the pH value of the medium to 7.0—9.0.

OKI *et al.*[209,210] selected *Brevibacterium* sp. B136 as an efficient producer of L-glutamate from ethanol. It produced L-glutamate in the yield of 53 g/l (60% of ethanol supplied). Ethanol was supplied to the medium by the feeding process because of its toxicity toward the organism.

L-GLUTAMINE AND N-ACETYLGLUTAMINE

Since the review of HUANG[18] no prominent progress has been reported on the production of L-glutamine. With a selected strain of *C. glutamicum*, the yield of L-glutamine was equivalent to as much as 20% of the glucose consumed[211]. Low pH and a high concentration of ammonium sulfate was necessary to obtain good yield. The production of N-acetylglutamine as a by-product was reported[212]. To increase the yield of N-acetylglutamine, slightly acid to neutral pH and vigorous aeration were necessary.

ALANINE

The ability to accumulate alanine is widely distributed among microorganisms. Studies with a pseudomonad-like bacterium[213], *Corynebacterium gelatinosum*, *Brevibacterium pentosoaminoacidum* and *Brevibacterium alanicum* were reviewed previously[18]. Many reports described the production of DL-alanine. Some active microorganisms and yields were as follows: *Bacillus coagulans*, a thermophile — 16.5 g/l (25% against glucose supplied)[213]; *Fusarium moniliforme* — 19.3 g/l[214]; a nicotinic acid-auxotrophic mutant of *Ustilago maydis* — 11.5 g/l[215]; *Mycobacterium album* 194

— 20 g/l[216]; mutant 13W from *Brevibacterium* 22 — 30 g/l[217]. The optical forms of alanine accumulated in the medium were usually racemic owing to alanine racemase activity distributed among microorganisms though some exceptions[218–220] were reported.

To obtain L-alanine, the process using enzymic decarboxylation of L-aspartate has been developed[221]. Culture broth of *Pseudomonas dacunhae* converted high concentration of L-aspartate (more than 40%) into L-alanine in the conversion ratio of 90% in 72 h at 37 °C. The reaction was completed in less than 20 h with the addition of a surface-active agent such as polyoxyethylene sorbitan monolaurate. A similar process using *Xanthomonas oryzae* 531 was also reported[222,223]. The reaction was carried out in an acid medium (pH 4—6) to prevent the breakdown and racemization of the L-alanine produced. Members of *Xanthomonas* and *Pseudomonas* were found to be active producers of L-aspartate β-decarboxylase.

L-Alanine production was attempted[224] with a mutant which lost racemase and required D-amino acids. Though L-alanine production by mutants of *B. megaterium*, *Arthrobacter ureafaciens*, *Nocardia globerula*, *B. ammoniagenes*, *Brevibacterium helvolum* and *Streptomyces* sp. was confirmed, the mutants were unstable in their mutated character.

C. hydrocarboclastus S1OB produced DL-alanine in addition to L-glutamate and α-oxoglutarate in a thiamine-deficient medium[225] but the amount was rather small though it increased somewhat in a glucose medium.

L-ASPARTIC ACID

Four groups of investigators reported the conversion of fumarate and ammonia to L-aspartic acid by using crude bacterial preparations. Selected strains of *Enterobacteriaceae* (especially *E. coli*[226,227,229,230], *E. freundii*[229], *S. marcescens*[227]). *B. megaterium*[226], *Bacterium succinicum*[227,228] and *Pseudomonas fluorescens*[229] were found to be suitable. When the culture broth of *E. coli* 2 was employed, almost a theoretical conversion at 5% level of ammonium fumarate could be obtained in the presence of a detergent, such as acetylpyridinium chloride[228]. *B. ammoniagenes* and *Microbacterium* strains which seemed to be poor producers were found to produce L-aspartic acid in high yield by the addition of a cationic surface-active agent such as cetyltrimethylammonium bromide in the fermentation medium[231]. The inhibition of deamination of L-aspartate by fumarate was demonstrated in *E. freundii*[229].

Pseudomonas 141-6 isolated by YAMATODANI *et al.*[232] converted ammonium maleate or ammonium fumarate efficiently to L-aspartate when grown in the medium which contained maleate or L-aspartate. On the other hand, L-maleate replaced a considerable part of L-aspartate in the product when grown on fumarate or glucose. A similar process was independently developed by TAKAMURA *et al.*[233] with *Alcaligenes faecalis* or *Pseudomonas ovalis*. Malonate also was demonstrated to be an inducer of *cis–trans* isomerase as maleate though not utilized as substrate[233,234]. Glucose inhibition of aspartase synthesis was demonstrated in various bacteria [235–237]. Fumaric acid produced by a *Rhizopus* strain was transformed to L-aspartate by mixed culture with *Proteus vulgaris* in the same flask[238,239].

L-METHIONINE, L-SERINE AND GLYCINE

DL-Methionine is being produced at low cost by a synthetic method and the racemic form is satisfactory as a growth factor for animals. This may be one reason for the fact that no efficient microbial process has been developed. Microbial transformation of DL-α-methylmercapto-α-hydroxybutyric acid into L-methionine was reported[240]. O-Succinyl-L-homoserine, an intermediate of methionine biosynthesis, has been found to accumulate in large quantities by an auxotrophic mutant of *A. aerogenes*[241].

Methods utilizing DL-glycerate[242] or glycine[243] as a substrate for the fermentative production of L-serine have been reported. A direct fermentation method preferably with an isoleucine-requiring strain of *Arthrobacter*, *Brevibacterium* or *Corynebacterium* has also been claimed[244]. Because of the ease of chemical synthesis and its lack of optical isomerism, the production of glycine has not been studied in detail though excretion of it is observed sometimes. A nonsporogenic bacillus isolated from soil produced 2.2 g/l of glycine in a medium which contained 10 g/l of ethanolamine[245].

CONCLUSION

On the basis of the above review, some concluding remarks will now be attempted. Some amino acids are produced by simple enzymic processes. The examples are L-aspartic acid and L-alanine. Some amino acids are produced by using precursors, such as L-isoleucine production using α-aminobutyrate or D-threonine. However, most amino acids are produced by direct fermentation. The biosynthetic regulation involved adds complexity to the types of processes described, from simple enzymic reactions to microbial transformation of precursor and direct fermentation. But generally, the cost of the raw material decreases in the same order. Some specific conditions for each amino acid made the process of the first and second types practicable.

From the standpoint of the microorganism, some amino acids are being produced by wild-type strains. Most amino acids are being produced using an auxotrophic mutant, and recently, some are being produced using regulation-less type mutants selected as analogue-resistant. The first type of microorganism, namely the wild-type strain, has been selected for its specific ability from nature, and has been improved for the selected ability and in some cases adapted to some desired condition, including the type of raw material to be used. The selected wild-type strain itself has defects in the economy of biosynthesis of the cell materials under specified environmental conditions. Glutamate producers represented by *C. glutamicum* may be mentioned as an example. The nutritional requirement for biotin and the lack or very low content of α-oxoglutarate dehydrogenase are important factors enabling L-glutamate accumulation. It was indicated that when the permeability mechanism is altered by limiting biotin, glutamate is lost from the cell and is no longer available as an intracellular precursor of the other amino acids.

Limitation of the concentration of an end product (effector) induces the biosynthesis of the precursor (such as ornithine) or the metabolite of the precursor (such as lysine) *via* a branched pathway. Obviously, it is useless to accumulate the end product of an unbranched pathway (such as arginine and histidine). The production of such a metabolite seems to depend on a regulatory mutant. The mutant which lost some biosynthetic regulation can be obtained as analogue-resistant and as a prototrophic revertant[246-249]. L-Lysine production by AEC-resistant mutants and L-threonine production by AHV-resistant mutants may be mentioned as examples. Polyauxotrophy and combination of auxotrophy and regulation deficiency make it possible to control more subtly the fermentation for efficient production of the desired metabolite. Examples are found in the production of lysine, threonine and isoleucine; these have great demands and have been carefully studied. Finer improvement may be attained by introducing some markers not directly related to the biosynthetic pathway. Some examples of metabolic interlock, regulatory interactions taking place between biosynthetic pathways, have been described[250]. Leaky mutants are rather common among auxotrophic mutants. Leakiness may be a useful character to be used in some cases to avoid the by-production of an undesirable metabolite which can accumulate by the introduction of an auxotrophic marker to improve the production of the desired metabolite. An example is found in threonine production by a triple auxotroph of *E. coli*. Improvement is also possible by blocking the further metabolism of the product desired. To reduce the by-production of some undesirable metabolite, a metabolite- (or its analogue)-sensitive mutant can be used. The situation is reversed with the analogue-resistant mutant, though other different mechanisms[251] can also result in the mutation expressed as a sensitive or resistant mutant in addition to the alteration of the regulation of biosynthesis.

The genetic stability of the strain utilized in the fermentation is especially important in continuous fermentation. To obtain a stable mutant, the type of mutagen must be carefully considered but this is beyond the scope of this review. Use of a combination of markers which contribute to the production of the desired substance is a method to stabilize the ability of the strain to keep in long fermentations.

The replacement of agricultural raw material with cheaper materials, such as acetic acid and hydrocarbons from petroleum is another recent trend to be mentioned in addition to continuous fermentation.

REFERENCES

1. KINOSHITA S., TANAKA K., UDAKA S., AKITA S.: Proc. Internat. Symp. Enzyme Chem. **2**, 464 (1957).
2. CASIDA L. E., BALDWIN N. Y.: U.S. Pat. 2,771,396 (1956).
3. KINOSHITA S., NAKAYAMA K., KITADA S.: J. Gen. Appl. Microbiol. **4**, 128 (1958).
4. GREENBERG D. M. (Ed.): *Metabolic Pathways* Vol. III, 3rd Ed. Academic Press, N. York 1969.
5. UMBARGER H. E.: Ann. Rev. Biochem. **38**, 323 (1969).
6. TRUFFA-BACHI P., COHEN G. N.: Ann. Rev. Biochem. **37**, 79 (1968).
7. PARDEE A. B., p. 3 in D. Perlman (Ed.): *Fermentation Advances.* Academic Press, New York 1969.
8. GROSS S. R.: Ann. Rev. Genetics **3**, 395 (1969).
9. SANWALL B. D.: Bacteriol. Rev. **34**, 20 (1970).
10. STADMAN E. R.: Adv. Enzymol. **28**, 41 (1966).
11. COHEN G. N.: *The Regulation of Cell Metabolism.* Holt, Reinhard & Winston Inc., New York 1968.
12. DEMAIN A. L.: Progr. Ind. Microbiol. **8**, 35 (1969).
13. DEMAIN A. L.: Adv. Appl. Microbiol. **8**, 1 (1966).
14. DEMAIN A. L.: Lloydia **31**, 395 (1968).
15. DEMAIN A. L., INAMINE D.: Bacteriol. Rev. **34**, 1 (1970).
16. ELANDER R. P., p. 89 in D. Perlman (Ed.): *Fermentation Advances.* Academic Press, New York 1969.
17. KINOSHITA S.: Adv. Appl. Microbiol. **1**, 201 (1959).
18. HUANG H. T.: Progr. Ind. Microbiol. **5**, 57 (1964).
19. DEMAIN A. L., BIRNBAUM J.: Current Topics Microbiol. **46**, 1 (1968).
20. RODWELL V. W., p. 317 in S. M. Greenberg (Ed.): *Metabolic Pathways,* Vol. III. Academic Press, New York 1969.
21. KITA D. A., HUANG H. T.: U.S. Pat. 2,841,532 (1958).
22. NUBEL R. C.: U.S. Pat. 2,968,594 (1961).
23. HUANG H. T., GRIFFIN J. M., FRIED J. H.: U.S. Pat. 2,955,986 (1960).
24. GROTON B., COKER J. N., BOWDER H. P., DEFIEBRE C. W.: Ind. Eng. Chem. (Product Res. Development) **2**, 308 (1963).
25. SETO T. A.: U.S. Pat. 3,056,729 (1962).
26. SAKISAKA K., SHIMURA K.: J. Agric. Chem. Soc. Japan **31**, 110 (1957).
27. BROQUIST H. P., JEFFEY A. V., ALBRECHT A. M.: Appl. Microbiol. **9**, 1 (1961).
28. LENSEN A. L., SHU P.: Appl. Microbiol. **9**, 12 (1961).
29. NAKAYAMA K., KITADA S., KINOSHITA S.: J. Gen. Appl. Microbiol. **7**, 145 (1960).
30. NAKAYAMA K., TANAKA H., HAGINO H., KINOSHITA S.: Agric. Biol. Chem. **30**, 611 (1966).
31. STADMAN E. R., COHEN G. N., LE-BRAS G., DE ROBICHON-SZULMAJSTER H. D.: J. Biol. Chem. **236**, 2033 (1961).
32. YUGARI Y., GILBARG C.: Biochim. Biophys. Acta **62**, 612 (1962).
33. NAKAYAMA K., Tanaka H., *unpublished results.*
34. CHASSIN L. A., SZULMAJSTER J.: Biochem. Biophys. Res. Comm. **29**, 648 (1967).
35. NAKAYAMA K., KINOSHITA S.: J. Gen. Appl. Microbiol. **7**, 161 (1961).
36. SANO K., SHIIO I.: J. Gen. Appl. Microbiol. **13**, 349 (1967).
37. MIYAJIMA R., OTSUKA S., SHIIO I.: J. Biochem. **63**, 139 (1968).
38. SHIIO I., MIYAJIMA R.: J. Biochem. **65**, 849 (1969).
39. MIYAJIMA R., SHIIO I.: Meet. Agric. Chem. Soc. Japan, Tokyo 1969; Summary, p. 125.
40. MIYAJIMA R., OTSUKA S., SHIIO I.: Meet. Agric. Chem. Soc. Japan, Nagoya 1968; Summary, p. 272.

41. SHIIO I., SANO K.: J. Gen. Appl. Microbiol. **15**, 267 (1969).
42. DAOUST P. R., STOUT T. H.: Developments Ind. Microbiol. **7**, 22 (1966).
43. TEAS H. J., HOROWITZ N. H., FLING M.: J. Biol. Chem. **172**, 651 (1948).
44. STURANI E., DATTA P., HUGHES M., GEST H.: Science **141**, 1053 (1963).
45. GIBSON K. D., NEUBERGER A., TAIT G. H.: Biochem. J. **83**, 550 (1962).
46. DOUDNEY C. O., WAGNER R. P.: Proc. Nat. Acad. Sci. U.S.A. **38**, 196 (1952).
47. SANO K., SHIIO I.: Meet. Agric. Chem. Soc. Japan, Tokyo 1969; Summary, p. 171.
48. SANO K., SHIIO I.: 18th Symp. Assoc. Amino Acid Nucleic Acid, Kyoto 1969; Summary, p. 7.
49. MIYAJIMA R., SANO K., SHIIO I.: Meet. Agric. Chem. Soc. Japan, Fukuoka 1970; Summary, p. 11.
50. MINDLIN S. Z.: Trudy Mosk. obshchest. ispyt. prir., Otd. Biol. **22**, 269 (1966).
51. ARESHKINA, L. Y., BEKER M. E., BUKIN V. N., KARKLINS R., KLYUEVA N. M., KUTSEVA L.S., LIPENS G.: Priklad. biokhim. mikrobiol. **1**, 396 (1965).
52. KINOSHITA S., NAKAYAMA K., KITADA S.: Jap. Pat. 1961-6499 (1961).
53. HEROLD M., NEČÁSEK J., MUSÍLKOVÁ M., ČERKES L., PLACHÝ J.: Česk. farm. **11**, 489 (1962).
54. PETROV D. F., OROBINSKII I. I., SHILOVA M. A., GRABELVA T. I., TARAKANOVA G. E., ZAKIROVA T. F., p. 128 in *Selekciya Mikrobov*. Akad. Nauk SSSR, Sibirsk. Otd., Lab. Tsitol. Rost. i Apomiksia Biol. Inst. 1965.
55. Kyowa Hakko Kogyo Co.: Netherl. Pat. 6,612,504 (1967).
56. DOUROS J. D., RAYMOND R. J.: U.S. Pat. 3,219,543 (1965).
57. SETO K., HARADA T.: Hakko Kogaku Zasshi **47**, 558 (1969).
58. SHIGETO M.: J. Agric. Chem. Soc. Japan **36**, 809, 814, 896 (1962); 899 (1963).
59. SLEZÁK J., SIKYTA B., PLACHÝ J., BŘEČKA A., p. 505 in Proc. 4th Symp. Continuous Cultivation of Microorganisms, Prague 1968. Academic Press, New York 1969.
60. ERICSON L. E., KURZ W. G.: Biotechnol. Bioeng. **4**, 23 (1962).
61. KURZ W. G., ERICSON L. E.: Biotechnol. Bioeng. **4**, 37 (1962).
62. SHIMURA K., WATANABE Y., FUJII Y., KONISHI S.: Amino Acids **1**, 71 (1959).
63. FUJII Y., SHIBUYA S., UJIIE H., KOJIMA M., TAGAWA H., UEMURA T.: Amino Acids **1**, 74 (1959).
64. HAYASHIBE M., SUGIHASHI T., SAEKI N., ICHIBA A., FUJII Y., SHIMURA K., UEMURA T.: Amino Acids **1**, 80 (1959).
65. SUGAWA Y., NOZAKI T., NISHIMURA S.: Amino Acid Nucleic Acid **10**, 68 (1964).
66. FUJITA C., NARA T., SAMEJIMA H., KINOSHITA S.: J. Agric. Chem. Soc. Japan **39**, 216, 222 (1965).
67. SAMEJIMA H., NARA T., FUJITA C., KINOSHITA S.: J. Agric. Chem. Soc. Japan **34**, 750, 754, 824 (1960).
68. SAMEJIMA H., NARA T., FUJITA C., KINOSHITA S.: J. Gen. Appl. Microbiol. **7**, 327 (1961).
69. SAMEJIMA H., NARA T., FUJITA C., ITO M., KINOSHITA S.: Amino Acids **3**, 90 (1961).
70. NARA T., SAMEJIMA H., FUJITA C., ITO M., NAKAYAMA K., KINOSHITA S.: Agric. Biol. Chem. **25**, 532 (1961).
71. NARA T., SAMEJIMA H., FUJITA C., KINOSHITA S.: Amino Acids **4**, 65 (1961).
72. HUANG H. T.: Appl. Microbiol. **9**, 419 (1961).
73. DAVIS B. D.: Nature **169**, 534 (1952).
74. NAKAYAMA K. *et al., unpublished results.*
75. COHEN G. N., p. 183 in B. L. Horecker, E. R. Stadman (Eds.): *Current Topics in Cellular Regulation,* Vol. I. Academic Press, New York 1969.
76. WORMER E., PARDEE A. B.: Arch. Biochem. Biophys. **78**, 416 (1958).
77. FREUNDLICH M., CAFFERATA R. L.: J. Bacteriol. **97**, 193 (1969).
78. TSUKADA Y., SUGIMORI T.: 18th Symp. Assoc. Amino Acid Nucleic Acid, Kyoto 1969; Summary, p. 2.

79. COHEN G. N., PATTE J. C.: Cold Spring Harbor Symp. Quant. Biol. **28**, 513 (1963).
80. SHIIO I., NAKAMORI S.: Agric. Biol. Chem. **33**, 1152 (1969).
81. SHIIO I., NAKAMORI S.: Agric. Biol. Chem. **34**, 448 (1970).
82. TAKAYAMA K., ABE S., KINOSHITA S.: Amino Acid Nucleic Acid **19**, 121 (1969).
83. GARNER H. R., TEAS H. J.: Bacteriol. Proc. 159 (1952).
84. NAKAMURA T., AIDA H.: Meet. Agric. Chem. Soc. Japan, Fukuoka 1970; Summary, p. 9.
85. MIYAJIMA R., NAKAMORI S., SHIIO I.: Meet. Agric. Chem. Soc. Japan, Fukuoka 1970; Summary, p. 8.
86. NAKAYAMA K., TANAKA H.: Jap. Pat. 1970-12273 (1970).
87. NAKAYAMA K., KITADA S., KINOSHITA S.: J. Gen. Appl. Microbiol. **7**, 52 (1961).
88. PLACHÝ J.: Folia Microbiol. **13**, 536 (1968).
89. HUANG H. T.: U.S. Pat. 2,975,105 (1961).
90. MAAS W. K., VOGEL H. J.: J. Bacteriol. **65**, 388 (1953).
91. KATAGIRI H., SUZUKI Y., TOCHIKURA T., SUGIMORI T.: Amino Acids **2**, 86 (1960).
92. BRAGG P. D., POLGLASE W. J.: J. Bacteriol. **84**, 37 (1962).
93. WEBB M.: Biochem. J. **70**, 442 (1958).
94. SHIIO I., MITSUGI K., OOTSUKA S., AOKI R.: Jap. Pat. 1962-1694 (1962).
95. KAWAMURA T., TAKEDA I.: Agric. Biol. Chem. **34**, 335 (1970).
96. KOYAMA Y., SUGISAKI Z.: Amino Acid Nucleic Acid **21**, 82 (1970).
97. ČIHÁK A., ŠKODA J., ŠORM F.: Coll. Czech. Chem. Comm. **28**, 2657 (1963).
98. KAPTEREVA Y. V., ASEEVA I. V.: Mikrobiologiya **36**, 970 (1967).
99. ASEEVA I. V., KAPTEREVA Y. V.: Vestnik Moskov. Univ., Ser. VI, **22**, 88 (1967).
100. KOGUMA T., UEDA K.: Meet. Soc. Ferment. Technol. Japan, Osaka 1964; Summary, p. 3.
101. LEAHY J. J.: J. Bacteriol. **100**, 1194 (1969).
102. UDAKA S., KINOSHITA S.: J. Gen. Appl. Microbiol. **5**, 159 (1960).
103. SUGISAKI Z., KOYAMA H.: 7th Internat. Congr. Biochem. Abstract V, 840 (1967).
104. SUGISAKI Z., KOYAMA Y.: Meet. Agric. Chem. Soc. Japan, Nagoya 1968; Summary, p. 158.
105. SUGISAKI Z., KOYAMA Y.: Meet. Agric. Chem. Soc. Japan, Nagoya 1968; Summary, p. 273.
106. OKUMURA S., YOSHINAGA F., ISHIDA M.: Jap. Pat. 1967-513 (1967).
107. KISUMI M., KOMATSUBARA S., CHIBATA I.: Amino Acid Nucleic Acid **19**, 1 (1969).
108. HAYASHIBE K., WATANABE Y., FUJII Y., SHIMURA K.: Amino Acids **1**, 89 (1959).
109. HAYASHIBE M., ITO M., WATANABE T., UEMURA T.: Amino Acids **2**, 100 (1960).
110. HAYASHIBE M., WATANABE T.: Agric. Biol. Chem. **26**, 82 (1962).
111. HAYASHIBE M., SUGISAKI T., SAKAI T., ANRAKU K.: J. Agric. Chem. Soc. Japan **36**, 437 (1962).
112. CHIBATA I., KISUMI M., ASHIKAGA Y.: J. Biochem. Microbiol. Technol. Eng. **2**, 361 (1960).
113. SUGISAKI Z., YAMAMOTO K.: Amino Acid Nucleic Acid **13**, 112 (1966).
114. OTAWARA S., OSHIMA M., KANDA N., ASANO K.: Amino Acid Nucleic Acid **13**, 78 (1966).
115. OKA K., TAKANASHI Y., SAEKI T., HAGIWARA T.: Jap. Pat. 1969-12720 (1969).
116. WADA H.: Jap. Pat. 1965-11918 (1965).
117. SANKYO SEIYAKU CO.: Brit. Pat. 1,114,964 (1968).
118. SHIDA T., FUJII Y., UEMURA T.: Amino Acids **7**, 16 (1963).
119. KISUMI M.: J. Biochem. **52**, 390 (1962).
120. KISUMI M., KOMATSUBARA S., CHIBATA I.: Amino Acid Nucleic Acid **20**, 178 (1969).
121. KISUMI M., ASHIKAGA Y., KATO J., CHIBATA I.: J. Biochem. **52**, 400 (1962).
122. KISUMI M., KATO J., CHIBATA I.: J. Biochem. **56**, 450 (1964).
123. HORVÁTH I., GADÓ I., SZENTIRMAG A.: J. Bacteriol. **78**, 293 (1959).
124. YOSHINAGA F., SHIBUYA M., YAMADA K., OKUMURA S.: Jap. Pat. 1969-14787 (1969).
125. SHIMURA K., SAITO T.: Amino Acids **2**, 97 (1960).
126. SHIMURA K., SAITO T.: Amino Acids **3**, 101 (1961).

127. SHIMURA K., SAITO T.: Amino Acids **4**, iv (1961).
128. HAYASHIBE M., UEMURA T.: Amino Acids **3**, 108 (1961).
129. HAYASHIBE M., WATANABE T., UEMURA T.: Amino Acids **5**, 92 (1962).
130. HAYASHIBE M., UEMURA T.: Amino Acids **5**, 97 (1962).
131. UZUKA Y., IWAI Y., SHIMURA K.: Amino Acid Nucleic Acid **11**, 128 (1965).
132. UZUKA Y., ABE K., SHIMURA K.: Amino Acids **8**, 74 (1963).
133. SHIMURA K., UZUKA Y.: Meet. Agric. Chem. Soc. Japan, Sendai 1962; Summary, p. 202.
134. UZUKA Y., NAGASAKI Y., SHIMURA K.: Meet. Agric. Chem. Soc. Japan, Kyoto 1966; Summary, p. 223.
135. SUGISAKI Z., KOYAMA Y., YAMAMOTO K.: Meet. Agric. Chem. Soc. Japan, Fukuoka 1970; Summary, p. 7.
136. KISUMI M., KATO J., KOMATSUBARA S., CHIBATA I.: Amino Acid Nucleic Acid **19**, 9 (1969).
137. KISUMI M., KOMATSUBARA S., CHIBATA I.: Amino Acid Nucleic Acid **20**, 70 (1969).
138. KISUMI M., FUKUSHIMA H., KOMATSUBARA S., CHIBATA I.: Meet. Soc. Ferment. Technol. Japan, Osaka 1969; Summary, p. 9.
139. Kyowa Hakko Kogyo Co.: French Pat. 2,000,145 (1969).
140. UMBARGER H. E., p. 57 in B. L. Horecker, E. R. Stadman (Eds.): *Current Topics in Cellular Regulation,* Vol. I. Academic Press, New York 1969.
141. LINGENS F.: Angew. Chem. Internat. Edn. **7**, 350 (1968).
142. GIBSON F., PITTARD J.: Bacteriol. Rev. **32**, 465 (1968).
143. HUANG H. T.: U.S. Pat. 2,973,304 (1961).
144. NAKAYAMA K., SATO Z., KINOSHITA S.: J. Agric. Chem. Soc. Japan **35**, 142 (1961).
145. NAKAYAMA K., SATO Z., KINOSHITA S.: J. Agric. Chem. Soc. Japan **35**, 146 (1961).
146. MALIN B., WESTHEAD J.: J. Biochem. Microbiol. Technol. Eng. **1**, 49 (1959).
147. Chas. Pfizer & Co. Inc.: Brit. Pat. 846,521 (1959).
148. OKAZAKI H.: U.S. Pat. 3,385,762 (1968).
149. TERUI G., ENATSU T., SUEMURA T.: Amino Acids **4**, 74 (1961).
150. TERUI G., ENATSU T.: 8th Internat. Congr. Microbiol. Abstracts, p. 65 (1962).
151. NAKAYAMA K., HAGINO H.: Jap. Pat. 1968-20711 (1968).
152. GALANINA L. A., NIZOVA I. M., AGATOV P. A.: Mikrobiologiya **37**, 58 (1969).
153. AIDA K., ASAI T., IIZUKA K., KAJIWARA T.: Amino Acids **1**, 94 (1959).
154. ENATSU T., KIYOI M., MATSUSHIMA H., TRUI G.: Amino Acid Nucleic Acid **8**, 68 (1963).
155. EBIHARA Y., NIITSU H., TERUI G.: Hakko Kogaku Zasshi **47**, 733 (1969).
156. LIM P. G., MATELS R. I.: J. Bacteriol. **87**, 1051 (1964).
157. MATSUSHIRO A.: J. Mol. Biol. **11**, 54 (1965).
158. LINGENS F., KRAUS H., LIGENS S.: Z. physiol. Chem. **339**, 1 (1964).
159. KINOSHITA S., NAKAYAMA K., UDAKA S.: J. Gen. Appl. Microbiol. **3**, 276 (1957).
160. UDAKA S., KINOSHITA S.: J. Gen. Appl. Microbiol. **4**, 272 (1958).
161. UDAKA S., KINOSHITA S.: J. Gen. Appl. Microbiol. **4**, 283 (1958).
162. MATSUNO T.: Meet. Agric. Chem. Soc. Japan, Tokyo 1967; Summary, p. 178.
163. KONDO E., MITSUGI T., HANAI F.: Meet. Agric. Chem. Soc. Japan, Fukuoka 1970; Summary, p. 8.
164. ARAKI K. *et al., unpublished results.*
165. OKUMURA S., SHIBUYA M., YOSHINAGA F., MAEYASHIKI I., KONISHI S.: Jap. Pat. 1968-8712 (1968).
166. OKUMURA S., SHIBUYA M., YOSHINAGA F., KATSUYA N.: Jap. Pat. 1969-2403 (1969).
167. SANO K., MOMOSE H., SHIIO I., KATSUYA N.: Jap. Pat. 1969-26910 (1969).
168. OKUMURA S., SHIBUYA M., YOSHINAGA F., MAEYASHIKI I., KONISHI S.: Jap. Pat. 1968-8714 (1968).
169. OKUMURA S., SHIBUYA M., KONISHI S., SHIRO T., KATSUYA N.: Jap. Pat. 1965-16756 (1965).

170. ANDO K., OISHI H.: Jap. Pat. 1968-8711 (1968).
171. ISHII R., OTSUKA S., SHIIO Z.: J. Gen. Appl. Microbiol. **13**, 303 (1967).
172. TANAKA K., OSHIMA K., TOKORO Y., KINOSHITA S.: Amino Acid Nucleic Acid **18**, 57 (1968).
173. OKUMURA S., SHIBUYA M., KONISHI S., SHIIO T.: Agric. Biol. Chem. **28**, 742 (1964).
174. SHIBUYA M., ISHIDA M., KONISHI S., SHIIO T., OKUMURA S.: Amino Acid Nucleic Acid **12**, 49 (1965).
175. NAKAYAMA K., HAGINO H.: J. Agric. Chem. Soc. Japan **40**, 377 (1966).
176. PLACHÝ J.: Folia Microbiol. **13**, 536 (1968).
177. TOKORO Y., OSHIMA K., TANAKA K., KINOSHITA S.: Amino Acid Nucleic Acid **19**, 115 (1969).
178. MAAS W. K.: Cold Spring Harbor Symp. Quant. Biol. **26**, 183 (1961).
179. RAMOS F., THURIAUX P., WIAME J. M., BECHET J.: European J. Biochem. **12**, 40 (1970).
180. KUBOTA K., SHIRO T., OKUMURA S.: 18th Symp. Assoc. Amino Acid Nucleic Acid, Kyoto 1969; Summary, p. 5.
181. ARAKI K., NAKAYAMA K.: *unpublished results.*
182. Germ. Pat. 1,915,854 (1969).
183. YOSHINAGA H., KONISHI S., OKUMURA S., KATSUYA N.: J. Gen. Appl. Microbiol. **12**, 219 (1966).
184. YOSHINAGA F., YOSHIHARA Y., OKUMURA S., KATSUYA N.: J. Gen. Appl. Microbiol. **13**, 25 (1967).
185. YOSHINAGA F., TAKEDA Y., OKUMURA S.: Biochem. Biophys. Res. Comm. **27**, 143 (1967).
186. YOSHINAGA F.: J. Agric. Chem. Soc. Japan **42**, 703 (1968).
187. YOSHINAGA F.: J. Gen. Appl. Microbiol. **15**, 387 (1969).
188. YOSHINAGA F.: Meet. Agric. Chem. Soc. Japan, Tokyo 1969; Summary, p. 170.
189. YAMATODANI S., SUZUKI M., NAKAO Y.: Amino Acid Nucleic Acid **16**, 126, 134 (1967).
190. KATO J., YAMADA S., ITO H., ICHIKAWA S.: Appl. Microbiol. **16**, 1200 (1968).
191. NOGUCHI Y., NAKAJIMA J., UNO T., ARAKI K.: Jap. Pat. 1969-1193 (1969).
192. YAMATODANI S., SUZUKI M.: Jap. Pat. 1969-1198 (1969).
193. OKUMURA S., YOSHINAGA F., KONISHI S., KATSUYA N.: U.S. Pat. 3,329,577 (1967).
194. YAMADA K., TAKAHASHI J., KOBAYASHI K., IMADA Y.: Agric. Biol. Chem. **27**, 390 (1963).
195. YAMADA K., TAKAHASHI J., KOBAYASHI K.: Agric. Biol. Chem. **27**, 773 (1963).
196. SHIIO I., OTSUKA S., ISHII R., KATSUYA N.: J. Gen. Appl. Microbiol. **9**, 23 (1963).
197. OTSUKA S., ISHII R., SHIIO T., KATSUYA N.: J. Gen. Appl. Microbiol. **10**, 179 (1964).
198. ISHII R., OTSUKA S., SHIIO I.: J. Gen. Appl. Microbiol. **13**, 217 (1967).
199. IGUCHI T., HAYAKAWA S., TAKEDA I.: Agric. Biol. Chem. **29**, 589 (1965).
200. YAMAGUCHI T., SUZUKI T., KIMURA K., TANAKA K., KINOSHITA S.: Amino Acid Nucleic Acid **20**, 57 (1969).
201. KIKUCHI M., SUGAWARA A., DOI M., SUZUKI M., NAKAO Y.: Meet. Agric. Chem. Soc. Japan, Fukuoka 1969; Summary, p. 184.
202. TAKAHASHI J., KOBAYASHI K., IMADA Y., YAMADA K.: Appl. Microbiol. **13**, 1 (1965).
203. TSUNODA T., SHIIO I., MITSUGI K.: J. Gen. Appl. Microbiol. **7**, 18 (1961).
204. TSUNODA T., SHIIO I., MITSUGI K.: J. Gen. Appl. Microbiol. **7**, 30. (1961).
205. KIMURA K.: J. Gen. Appl. Microbiol. **10**, 23 (1964).
206. HARADA T., SETO K., MUROOKA Y.: Hakko Kogaku Zasshi **46**, 169 (1968).
207. OKABE S., SHIBUKAWA M., OTSUKA T.: J. Agric. Chem. Soc. Japan **42**, 563 (1968).
208. ARIMA S., SHIBUKAWA M., OHUCHI S.: J. Agric. Chem. Soc. Japan **43**, 758 (1969).
209. OKI S., SAYAMA Y., NISHIMURA Y., OZAKI A.: Agric. Biol. Chem. **32**, 119 (1968).
210. OKI S., NISHIMURA Y., SAYAMA Y., KITAI A., OZAKI A.: Amino Acid Nucleic Acid **19**, 82 (1969).
211. YAKISAKA Y., ISHIDA H.: Amino Acids **7**, 59 (1963).

212. Oshima K., Tanaka K., Kinoshita S.: Amino Acids **7**, 73 (1963).
213. Matsumoto T., Yano T., Yamada K.: Agric. Biol. Chem. **31**, 1381 (1967).
214. Carito S. L., Pisano M. A.: Appl. Microbiol. **14**, 39 (1966).
215. Dulaney E. L., James C. A., Dulaney D. D.: Developments Ind. Microbiol. **5**, 242 (1964).
216. Lafitskaya T. N., Aseeva I. V.: Microbiologiya **36**, 568 (1967).
217. Griniva P.: Priklad. biokhim. mikrobiol. **4**, 122 (1968).
218. Samejima H., Nara T., Fujita C., Kinoshita S.: J. Agric. Chem. Soc. Japan **34**, 832, 838 (1960).
219. Perry J. J.: J. Bacteriol. **94**, 1249 (1967).
220. Okumara S., Yamanoi A., Tsunoda T., Kawano K., Sasaki T., Mitsugi N.: Jap. Pat. 1964-16342 (1964).
221. Chibata I., Kakimoto T., Kato J.: Appl. Microbiol. **13**, 638 (1965).
222. Watanabe S., Issiki S., Osawa T.: Hakko Kogaku Zasshi **43**, 95 (1965).
223. Watanabe S., Issiki S., Osawa T., Yawamoto S.: Hakko Kogaku Zasshi **43**, 697 (1965).
224. Hagino H., Nakayama K., *unpublished results*.
225. Imada Y., Yamada K.: Agric. Biol. Chem. **33**, 1326 (1969).
226. Kinoshita S., Nakayama K., Kitada S.: Hakko Kyokaishi **16**, 517 (1958).
227. Kitahara K., Fukui S., Misawa M.: J. Gen. Appl. Microbiol. **5**, 74 (1959).
228. Kitahara K., Fukui S., Misawa M.: J. Agric. Chem. Soc. Japan **34**, 44 (1960).
229. Kisumi M., Ashikaga Y., Chibata I.: Amino Acids **1**, 102 (1959).
230. Kisumi M., Ashikaga Y., Chibata I.: Bull. Agric. Chem. Soc. Japan **24**, 296 (1960).
231. Watanabe S., Osawa Y., Nobukuni T.: J. Agric. Chem. Soc. Japan **38**, 430 (1964).
232. Yamatodani S., Kakinuma J., Kanamaru N., Suzuki M.: Takeda Kenkyujo Nenpo **25**, 32 (1966).
233. Takamura Y., Kitamura I., Iikura M., Kawano K., Ozaki A.: Agric. Biol. Chem. **30**, 338, 345 (1966).
234. Takamura Y., Samejima M., Aoyama T.: Agric. Biol. Chem. **31**, 207 (1967).
235. Epps H. M. R., Gale E. F.: Biochem. J. **36**, 619 (1942).
236. Farley M. A., Lichstein H. C.: Canad. J. Microbiol. **9**, 835 (1963).
237. Leiss K., Mecke D., Holzer H.: Biochem. Z. **346**, 244 (1966).
238. Sasaki T., Takao S., Hotta K.: Meet. Agric. Chem. Soc. Japan, Tokyo 1969; Summary, p. 170.
239. Sasaki T., Takao S., Hotta K.: Meet. Agric. Chem. Soc. Japan, Fukuoka 1970; Summary, p. 292.
240. Takesue S.: U.S. Pat. 3,129,386 (1964).
241. Kase H., Nakayama K., Kinoshita S.: Agric. Biol. Chem. **34**, 274 (1970).
242. Sumitomo Chemicals Co.: Jap. Pat. 1967-17,728 (1967).
243. Kubo K., Shiro T.: Jap. Pat. 1970-11,114 (1970).
244. Nakayama K., Kase H.: Germ. Pat. 1,996,421 (1969).
245. Harada T., Murooka Y.: Hakko Kogaku Zasshi **44**, 192 (1966).
246. Umbarger H. E.: Ann. Rev. Plant. Physiol. **14**, 19 (1963).
247. Changeux J. P.: Cold Spring Harbor Symp. Quant. Biol. **28**, 497 (1963).
248. Cohen G. N., Patte J. C.: Cold Spring Harbor Symp. Quant. Biol. **28**, 513 (1963).
249. Reh M., Schlegel H. G.: Arch. Mikrobiol. **67**, 110 (1969).
250. Jensen R. A.: J. Biol. Chem. **244**, 2816 (1969).
251. Meuris P.: Genetics **63**, 569 (1969).

Lysine and Threonine Production by *Brevibacterium flavum* Mutants and Their Regulatory Mechanisms

ISAMU SHIIO

Central Research Laboratories of Ajinomoto Co., Inc., Kawasaki, Japan

Abbreviations used:

AEC	S-(2-aminoethyl)-L-cysteine
AHV	α-amino-β-hydroxyvalerate
ASA	aspartate β-semialdehyde
DAP	α,ε-diaminopimelate
DDP	2,3-dihydrodipicolinate
HSR	homoserine
HSR-*P*	homoserine-O-phosphate

INTRODUCTION

In general, biosynthesis of cellular constituents is regulated through the feedback inhibition and repression by the end product of enzymes involved in the synthesis. Therefore, overproduction of amino acids is not usually observed in wild-type organisms. In contrast, accumulation of the intermediates is possible simply by blocking the pathway and by limiting the end product required for growth. For example, diaminopimelic acid, a precursor of lysine, as well as homoserine, a precursor of threonine, has been shown to be produced by lysine and threonine auxotrophs, respectively[1,2]. Then, lysine can be produced from diaminopimelic acid by enzymatic decarboxylation[3] and threonine from homoserine by a separate fermentation process[4]. On the other hand, mutants requiring one of the end products in a branched biosynthetic pathway have been often shown to accumulate other end products. Thus, lysine- or diaminopimelic acid-requiring mutants of *Escherichia coli* accumulated threonine in their culture fluid[1,5]. Similarly, threonine plus methionine or homoserine-less mutants of *Micrococcus glutamicus* produced a large amount of lysine[6]. Mechanisms for the accumulation of these amino acids are apparently more complicated than those of the intermediates and remain unclear without detailed knowledge on the regulation mechanisms for the biosynthesis which may be different from species to species of microorganism used. This report deals with some studies on the regulation mechanisms for the lysine and threonine biosynthesis in *Brevibacterium flavum* and derivation of various kinds of lysine- and threonine-producing mutants based on the mechanisms. This bacterium has been selected as the original strain for the mutant derivation because it is an excellent glutamate producer and glutamate has been known to be biochemically closely related to aspartate which is a precursor of lysine and threonine in bacteria.

REGULATION OF LYSINE BIOSYNTHESIS

Aspartate kinase[7,8]. The first enzyme of the lysine biosynthesis is aspartate kinase which catalyses the formation of β-aspartylphosphate from aspartate and ATP. Partially purified aspartate kinase from *B. flavum* shows homotropic cooperativity of substrate, aspartate. The enzyme activity which is very low at lower concentrations of aspartate is promoted by the addition of threonine or inorganic sulfates such as ammonium sulfate and, as a result, kinetically is of the Michaelis type. From a kinetic analysis of substrate interaction and of product inhibition, the reaction mechanism has been assigned as rapid equilibrium random Bi Bi in the presence of ammonium sulfate. Besides the two substrates, the enzyme requires Mg or Mn ion for activity. The optimum pH and molecular weight are 7.5 and 80,000, respectively. Table I shows effects of aspartate-family amino acids and structurally

TABLE I

Effects of Amino Acids on Aspartate Kinase[a]

Addition 1 mM	Relative activity %	Addition 1 mM each	Relative activity %
None	100	L-Lys + L-Thr	6
L-Lys	80	L-Lys + D-Thr	84
D-Lys	96	L-Lys + L-HSR	87
L-Thr	88	L-Lys + L-HSR-*P*	77
D-Thr	104	L-Lys + L-Ser	86
L-Met	112	L-Lys + DL-AHV	70
L-Ile	153	L-Lys + L-Met	98
L-HSR	105	L-Lys + L-Ile	121
L-HSR-*P*	93	L-Thr + D-Lys	65
DL-DAP	103	L-Thr + DAP	89
L-Ser	75	L-Thr + AEC	26
DL-AHV	110	L-Thr + L-Met	91
AEC	88	L-Thr + L-Ile	113
		D-Thr + D-Lys	88

[a] Reaction mixtures contained 10 mM L-aspartate, 10 mM-ATP, 10 mM-$MgSO_4$, 600 mM hydroxylamine, 600 mM-$(NH_4)_2SO_4$, given amino acid(s), and aspartate kinase in 0.1 M Tris-sulfate buffer, pH 7.5.

related amino acids on the enzyme activity. The enzyme is strongly and specifically inhibited by the simultaneous addition of L-lysine and L-threonine at low concentrations where one of these amino acids alone has little effect. This concerted inhibition of aspartate kinase by lysine plus threonine has been first observed with *Rhodopseudomonas capsulata* and with *Bacillus polymyxa* and is quite different from the feedback inhibition of the *E. coli* isoenzymes[9]. Kinetic analysis of the concerted inhibition shows that the rate v of aspartate kinase reaction in the presence of lysine and threonine can be represented by the following equation in which a, b, i, and j are concentrations of aspartate, ATP, lysine, and threonine, respectively, and α, β, V, and K_{1-7} are kinetic constants:

$$\frac{V}{v} = 1 + \left(1 + \alpha \frac{i}{K_1} \frac{j}{K_2}\right) \frac{K_3}{a} + \frac{K_4}{b} + \left(1 + \beta \frac{i}{K_5} \frac{j}{K_6}\right) \frac{K_7}{a} \frac{K_4}{b}$$

The "concerted inhibition coefficients", α and β, represent degrees of the interaction between lysine and threonine on the enzyme and on the enzyme-ATP complex, respectively, and were calculated as 49 and 277, respectively, under standard conditions. At higher concentrations, lysine inhibits aspartate kinase competitively with aspartate even in the absence of threonine, and threonine is a K-type activator or a partially competitive inhibitor, depending on the pH and the presence of ammonium sulfate. The effects of threonine in the presence of ammonium sulfate were attributed to the formation of dimer molecules with different specific activity from the enzyme monomers. A lysine analogue, S-(2-aminoethyl)-L-cysteine, also inhibits the enzyme concertedly with threonine. In the AEC-resistant mutant, strain FA-1-30, which has

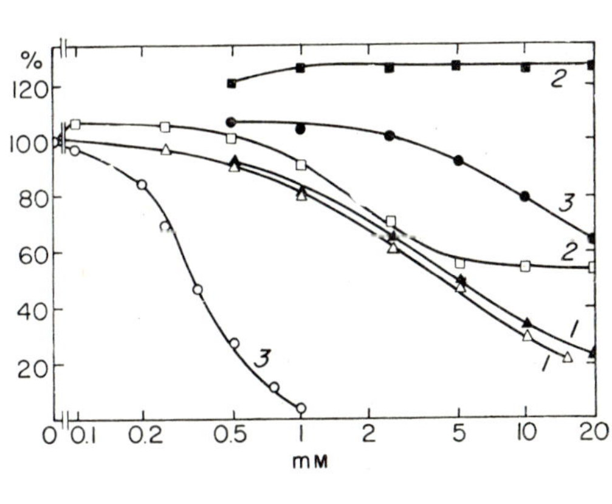

Fig. 1. Effect of lysine and threonine (mM) on aspartate kinase (relative activity, %) from AEC-resistant mutant and the parent strains. *Closed symbols:* mutant FA1-30, *open symbols:* parent strain 2247; *1* Lys, *2* Thr, *3* Lys + Thr. Assay conditions were as given in Table I.

been derived from strain 2247 (wild) and produces a large amount of lysine, aspartate kinase has been genetically desensitized to the concerted inhibition, whereas it is still inhibited by lysine at higher concentrations similar to the parent enzyme. Fig. 1 and 2 show a comparison of these amino-acid effects between the mutant and the parent enzymes. Under standard assay conditions, equimolar concentrations of lysine and threonine giving 50% inhibition are 0.3 and 50 mM in the parent and mutant enzymes, respectively.

It is of interest that isoleucine, one of the end products, not only activates aspartate kinase but also reverses the concerted inhibition at higher concentrations.

No significant repression of aspartate kinase was observed with any of the aspartate-family amino acids.

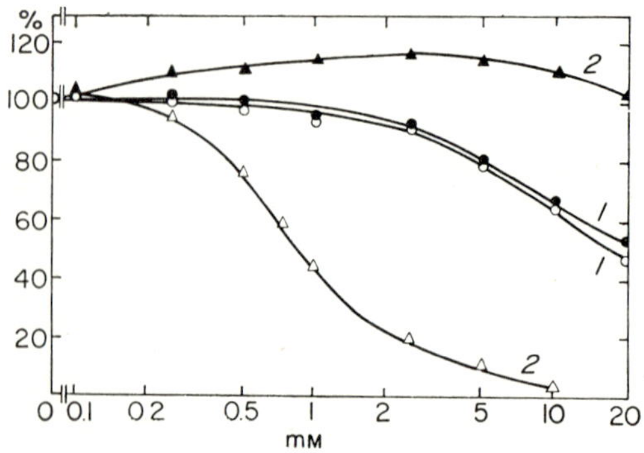

Fig. 2. Effect of AEC and AEC plus threonine (mM) on aspartate kinase (relative activity, %) from AEC-resistant mutant and the parent strains. *Closed symbols:* mutant FA1-30, *open symbols:* parent strain 2247; *1* AEC, *2* AEC + Thr.

Aspartate-β-semialdehyde dehydrogenase, dihydrodipicolinate synthase, and dihydrodipicolinate reductase[10,11]**.** The second enzyme of the lysine biosynthesis, aspartate-β-semialdehyde dehydrogenase, which catalyzes the formation of aspartate β-semialdehyde from β-aspartylphosphate, is specific to NADP and is neither inhibited nor repressed by any of aspartate-family amino acids.

The third enzyme, dihydrodipicolinate synthase, which catalyzes the condensation of ASA and pyruvate to form DDP, has a pH optimum at 8.4. Since this enzyme is the first enzyme in the lysine-specific synthetic pathway and competes with homoserine dehydrogenase for the same substrate, ASA, which exists in the branch point on the threonine and lysine biosynthetic pathway, the K_m for ASA is especially of interest and was calculated as 0.2 mM. The fourth enzyme, DDP reductase, utilizes reduced NAD as well as reduced NADP as cofactor, in contrast with the *E. coli* enzyme which is specific to reduced NADP. Neither significant inhibition nor repression of these two enzymes was observed by the addition of aspartate-family amino acids, whereas DDP synthase of *E. coli* had been known to be strongly inhibited by lysine[12].

The regulation pattern of the lysine biosynthesis in *B. flavum* is quite different from that in *E. coli*. Thus, in *E. coli*, there are three isofunctional aspartate kinases, one of which is inhibited and repressed by lysine. Moreover, DDP synthase of *E. coli* is strongly inhibited by lysine. On the other hand, the only regulatory mechanism observed in the first four enzymes of *B. flavum* is the concerted inhibition of aspartate kinase by lysine plus threonine. Therefore, biosynthesis of lysine in *B. flavum* seems to be controlled through the concerted inhibition of aspartate kinase. This suggestion is strongly supported by the fact that AEC-resistant mutants in which aspartate kinase is genetically desensitized to the concerted inhibition produce large amounts of lysine. However, since the presence of threonine is essential for the inhibition of aspartate kinase by lysine, there must be a mechanism for the preferential synthesis of threonine. This requirement would have been fulfilled by the presence of a much higher level of homoserine dehydrogenase than that of DDP synthase, both of which compete for the common intermediate, ASA, in the lysine and threonine synthesis.

REGULATION OF THREONINE BIOSYNTHESIS

Homoserine dehydrogenase[13,14]. The first and the second enzyme of threonine biosynthesis are the same as those of lysine biosynthesis, the regulation of which has been already described in the preceding chapter. The third enzyme is homoserine dehydrogenase which catalyzes reductive formation of homoserine from ASA. Partially purified homoserine dehydrogenase from *B. flavum* is specific to NADP. The optimum pH and molecular weight are 6.0 and 250,000, respectively. From kinetic analysis of the product inhibition, the reaction mechanism has been described as ordered Bi Bi[15], in which reduced NADP is the first binding substrate and NADP is the last leaving product. The K_m for ASA is 0.19 mM, almost the same as that of DDP synthase which competes with homoserine dehydrogenase for the same substrate, ASA. Effects of aspartate-family amino acids and of their structurally related amino acids on homoserine dehydrogenase activity are shown in Table II. Threonine, one of the metabolic end products, strongly and specifically inhibited the activity, as observed with many microbial homoserine dehydrogenases. The inhibition is kinetically of mixed type for both substrates, ASA and reduced NADP. Isoleucine

TABLE II

Effects of Amino Acids on Homoserine Dehydrogenase[a]

Addition 1 mM	Relative activity %	Addition 1 mM	Relative activity %
None	100	L-Asp	99
L-Thr	0	DL-AHV	86
D-Thr	97	L-HSR	87
L-Ile	91	L-Ser	90
L-Met	108	L-Val	88
L-Lys	101	DL-DAP	95

[a] Reaction mixtures contained 0.4 mM-DL-ASA, 0.1 mM-NADPH, HSR dehydrogenase, and given amino acid in 0.1 M potassium phosphate buffer, pH 7.0.

inhibited the activity competitively for both substrates but much more weakly than threonine. Thus, the concentration of threonine or isoleucine giving 50% inhibition is 0.06 or 4 mM, respectively, under the standard assay conditions. Whereas the enzyme activity itself is affected by pH only slightly, the inhibition by these amino acids depends greatly on pH. For example, 0.06 mM threonine gives 80% inhibition at pH 6.0 but only 5% at pH 8.0. The reaction rate—substrate saturation curve for

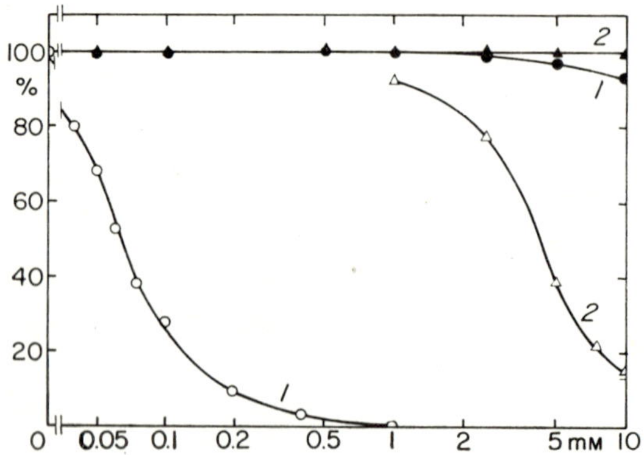

Fig. 3. Effect of threonine or isoleucine (mM) on homoserine dehydrogenase (relative activity, %) from AHV-resistant mutant and the parent strains. *Closed symbols:* mutant BB-69, *open symbols:* parent strain 2247; *1* Thr, *2* Ile. Assay conditions were given in Table II. The enzyme activities without addition were taken as 100.

reduced NADP becomes sigmoidal in the presence of these amino acids, suggesting that the enzyme is allosteric.

A threonine analogue, DL-α-amino-β-hydroxyvaleric acid, competitively inhibits the enzyme, slightly more strongly than isoleucine. Mutants resistant to the analogue produce large amounts of threonine. As shown in Fig. 3, the enzyme from the resistant mutant, strain BB-69, has been almost completely genetically desensitized to the feedback inhibition by threonine or isoleucine. The concentration of threonine giving 50% inhibition is more than 80 mM in the mutant enzyme, more than 1000-fold as much as that in the parent enzyme. Among the aspartate-family amino acids, methionine strongly repressed the formation of homoserine dehydrogenase. Since this enzyme is a common enzyme for the methionine and threonine biosynthesis, the inhibition by threonine or the repression by methionine would cause an inhibition of growth, when either one of these amino acids is added to the culture medium in excess. However, it is not the case in the wild strain, presumably because of the presence of a large amount of the enzyme. In fact, mutants whose growth is inhibited by excess threonine or methionine have been derived and called "*threonine- or methionine-sensitive mutants*". As expected, these mutants have low levels of homoserine dehydrogenase and growth inhibition by threonine (or methionine) is reversed by further addition of methionine (or threonine)[16]. Therefore, the feedback inhibition by threonine and repression by methionine of homoserine dehydrogenase seem to be biologically significant. A typical threonine- or methionine-sensitive mutant, strain S-20, which has only about 2% of the homoserine dehydrogenase activity of the wild strain, gives almost the same growth rate in a minimum medium. This indicates that the enzyme level of the wild strain may be unnecessarily high for the full growth rate and may have been almost completely inhibited by threonine pooled inside the cell, for preventing overproduction of threonine. In this respect, it is of interest that, in the wild strain, the level of homoserine dehydrogenase is about 15-fold higher than that of dihydrodipicolinate synthase, another enzyme in the branch point of the lysine and threonine biosynthetic pathway.

Homoserine kinase and threonine synthase. The fourth and the fifth enzymes of the threonine synthesis are homoserine kinase and threonine synthase which catalyze the formation of homoserine-O-phosphate from homoserine and ATP and the formation of threonine from the former, respectively. Partially purified homoserine kinase from strain 2247 requires Mg or Mn ion besides the two substrates and is activated roughly twice by ammonium sulfate or KCl. The optimum pH and molecular weight are 7.5 and 55,000, respectively. Since homoserine kinase is the first enzyme after the branch point of the threonine and methionine biosynthesis, strong feedback control by threonine of the enzyme could be expected. However, the inhibition by threonine is much weaker than that of homoserine dehydrogenase. Thus, concentrations giving 50% inhibition are 7.5 mM for the kinase and 0.06 mM for the dehydrogenase under their standard assay conditions. Other amino acids tested show no or only slight effect at 10 mM. While homoserine kinase is not repressed either by threonine and isoleucine alone, or by both added together (repression of the *E. coli* enzyme) it is repressed by methionine which is not the metabolic end product, similarly as is homoserine dehydrogenase[17].

Threonine synthase is neither significantly inhibited nor repressed by any of the aspartate-family amino acids[10].

The threonine biosynthesis in *B. flavum* seems to be controlled through the feedback inhibition by threonine of homoserine dehydrogenase, although there may be another possibility, *viz.* control through the concerted inhibition of aspartate kinase by lysine plus threonine. Thus, preferential synthesis of lysine which is required for the latter possible regulation is not likely, as described in the preceding section for the regulation of lysine biosynthesis. Moreover, at least under their respective standard conditions, homoserine dehydrogenase is inhibited by threonine more strongly than aspartate kinase. For instance, the former is inhibited 75% by 0.1 mM threonine which has little effect on the latter enzyme even in the presence of a four-fold concentration (0.4 mM) of lysine. Finally, the above-mentioned suggestion is strongly supported by the fact that α-amino-β-hydroxyvaleric acid-resistant mutants in which homoserine dehydrogenase but not aspartate kinase has been genetically desensitized to the feedback inhibition produce large amounts of threonine, whereas AEC-resistant mutants whose aspartate kinase has been genetically desensitized to the concerted inhibition produce large amounts of lysine but not threonine. Repression of homoserine dehydrogenase by methionine seems to be ineffective *in vivo* for the control of the threonine biosynthesis because the level of the enzyme in the wild strain, which may be controlled by methionine pooled inside the cell, is still very high and about 50 times as high as in the threonine- or methionine-sensitive mutant strain S-20 which gives almost the same growth rate in a minimum medium as the wild strain.

PRODUCTION OF LYSINE

Wild-type strains[17]. Since the lysine biosynthesis in *B. flavum* is controlled through the concerted inhibition of aspartate kinase by lysine plus threonine, any conditions decreasing the threonine level inside the cell may cause an overproduction of lysine. Since threonine biosynthesis is controlled by feedback inhibition of homoserine dehydrogenase, the threonine level in the cell would depend on the level of homoserine dehydrogenase, which, in turn, is affected by the repressive effect of methionine. Therefore, addition of large amounts of methionine to the culture medium would cause an overproduction of lysine even in the wild-type strains. This has been demonstrated with *B. flavum* strain 2247 (ATCC 14067) as shown in Table III. Since the repression of homoserine dehydrogenase by methionine is incomplete, the residual activity is fairly high even in the presence of a large excess of methionine and, accordingly, the maximum amount of lysine produced in this process was only 5 g/l.

Threonine- or methionine-sensitive mutants[16]. To obtain mutants having a genetically low activity of homoserine dehydrogenase, prototrophic revertants derived from

TABLE III

Effect of Methionine on Lysine Production in Wild Strain 2247[a]

Methionine added μg/ml	Growth O.D.$_{562}$ × 26	Homoserine dehydrogenase[b]	L-Lysine-HCl formed g/l
0	1.25	0.177	0.26
250	1.30	—	0.72
500	1.30	—	2.76
1000	1.30	0.053	4.90
2500	1.30	—	5.09

[a] *B. flavum* 2247 was cultured at 30 °C for 48 h in medium A supplemented with methionine which contained 100 g glucose, 40 g $(NH_4)_2SO_4$, 1 g KH_2PO_4, 0.4 g $MgSO_4.7H_2O$, 2 mg Fe^{2+}, 2 mg Mn^{2+}, 300 μg β-biotin, 200 μg thiamine-HCl, 2 ml Mieki (soybean meal acid hydrolyzate solution, 20 g nitrogen/l), 50 g $CaCO_3$ and 1 l distilled water at pH 7.5 after sterilization. L-Lysine was assayed manometrically by using lysine decarboxylase of *Bacterium cadaveris*.

[b] Specific activity.

a homoserine dehydrogenase-defective mutant, strain H-1013, were examined. These mutants were obtained as small-colony-forming revertants and produced lysine. As shown in Fig. 4, the growth of these lysine-producing revertants is inhibited by 10 μg/ml of threonine or methionine which is much less than the concentration required for the maximum growth of threonine or methionine auxotrophs. Therefore, they have been called "*threonine- or methionine-sensitive mutants*". Such growth in-

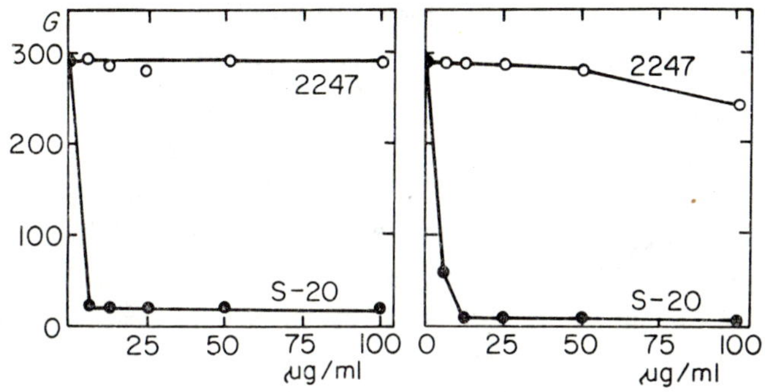

Fig. 4. Growth inhibition by L-methionine *(left)* or L-threonine *(right)* in the sensitive revertant strain S-20. Strain S-20 and 2247 (wild) were inoculated at 4.5×10^7 cells/ml to a minimal medium B-1[16] supplemented with given amino acids (μg/ml). The cell growth (turbidity) was determined at 17.5 h when the nonsupplemented control gave the maximal turbidity.

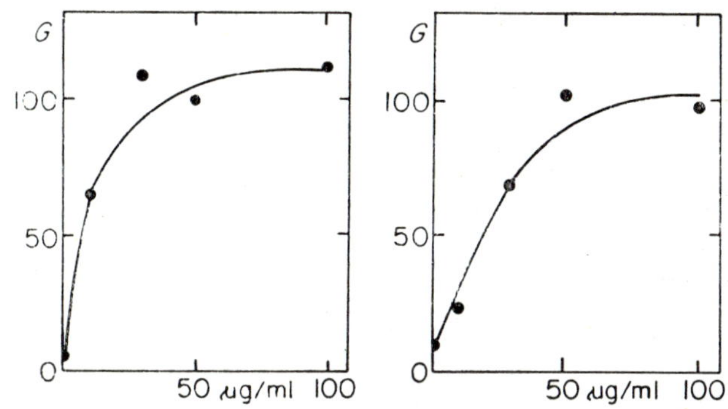

Fig. 5. Restoration by L-methionine (μg/ml; *left*) or L-threonine (μg/ml; *right*) of threonine- or methionine-inhibited growth of a threonine- or methionine-sensitive strain S-20. Conditions were the same as given in Fig. 4. The basal medium was medium B-1 containing 100 μg/ml of L-threonine (*left*) or L-methionine (*right*).

hibition in these revertants seems to be due to their low levels of homoserine dehydrogenase, a common enzyme for the threonine and methionine biosynthesis, because inhibition by threonine or repression by methionine of this enzyme may cause an inhibition of the biosynthesis of methionine or threonine, respectively, and, therefore, an inhibition of growth. In fact, the growth inhibition by threonine (or methionine) is reversed by further addition of methionine (or threonine, respectively), as shown in Fig. 5.

Then, the same kind of mutants were derived from the original wild strain 2247 and selected by their sensitivities or by their abilities to produce lysine which permitted the growth of a lysine auxotroph spread on the same culture plates. In these cases, mutants sensitive to either one of the two amino acids were obtained in addition to the above-mentioned mutants, but did not produce more than 5 g/l of lysine-HCl.

Maximum amounts of L-lysine-HCl produced are 23 g/l in the sensitive revertants and 26 g/l in the sensitive mutants directly derived from the wild strain.

Threonine auxotrophs[18]. To produce lysine with the wild or the sensitive strains, the threonine level inside the cell has diminished in indirect proportion to decreasing the level of homoserine dehydrogenase which is involved in threonine biosynthesis. In the threonine auxotrophs, the cells can be starved simply by limiting the amount of threonine added for growth. Fig. 6 shows the distribution pattern of threonine auxotrophs derived from strain 2247. The auxotrophs are clearly divided into three groups with respect to their abilities for lysine production. The best group corresponds to homoserine dehydrogenase-defective mutants which require methionine in addition to threonine for their growth, while the second and the third groups consist of mutants lacking homoserine kinase and threonine synthase, respectively. It is striking that threonine synthase-defective mutants do not produce lysine more than the wild strain, even under threonine-limiting conditions. These mutants produce small amounts of HSR-*P* in the culture media, which may affect enzymes involved in the lysine biosynthesis.

The difference in the amount of lysine produced between homoserine dehydrogenase- and kinase-defective mutants corresponds to the amount of homoserine which is produced only by the latter mutants in addition to lysine. This suggests that the same threonine-limiting condition in these mutants would give production of the

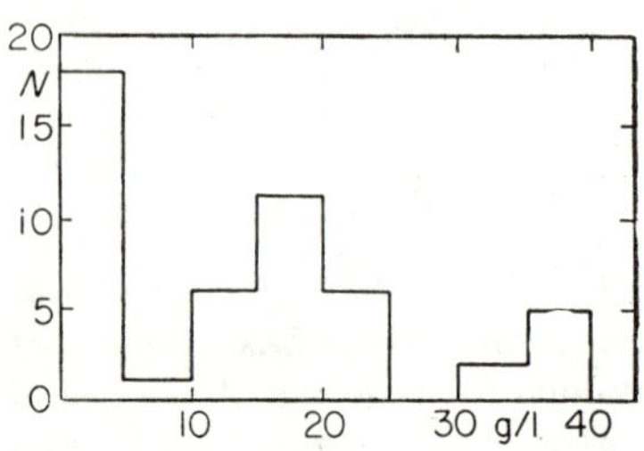

Fig. 6. L-Lysine hydrochloride production (g/l) by threonine auxotrophs. Threonine auxotrophs derived from strain 2247 were cultured at 30 °C for 72 h in medium A supplemented with 600 μg/ml of L-threonine and 200 μg/ml of L-methionine. L-Lysine was determined colorimetrically. *N* number of mutants.

same amount of common intermediates which, in turn, convert to lysine and homoserine with a ratio depending on the level of homoserine dehydrogenase. In this respect, it is of interest that genetic and conditional decrease of the level in the kinase-defective mutants causes significant increase of lysine produced. Thus, addition of methionine to the culture medium causes significant increase of lysine produced as well as repression of homoserine dehydrogenase in homoserine kinase-

TABLE IV

Effect of Methionine on Lysine Production in Homoserine Kinase-Defective Mutant T-36[a]

Methionine added μg/ml	Homoserine dehydrogenase[b]	L-Lysine-HCl produced g/l	Homoserine produced g/l
0	0.429	18.6	23.4
250	0.100	25.5	15.2
500	0.070	28.0	12.1

[a] Strain T-36 was cultured at 30 °C for 48 h in medium A supplemented with 600 μg/ml of L-threonine and given amount of L-methionine.

[b] Specific activity.

-defective mutant, strain T-36, as shown in Table IV. Then, to obtain homoserine kinase-defective mutants having low homoserine dehydrogenase activity, threonine auxotrophs were derived from strain S-20, a prototrophic mutant having low level of homoserine dehydrogenase. As shown in strains ST-4-1 and ST-28 of Table V, homoserine kinase-defective mutants from strain S-20 produced almost the same amount of lysine as homoserine dehydrogenase-defective mutant, strain H-1013, while they produced a much smaller amount of homoserine than the kinase-defective mutants derived from the wild strain. Thus, the total amounts of lysine and homoserine produced become almost the same between homoserine dehydrogenase- and kinase-defective mutants derived from the wild strain and from the mutant having

TABLE V

Production of Lysine and Homoserine by Various Threonine Auxotrophs[a]

Strain	L-Lysine-HCl		Homoserine		Lysine + homoserine
	g/l	mM	g/l	mM	mM
T-36	23.3	128	12.3	103	231
H-1013	34.1	187	0.0	0	187
TM-15	18.5	101	13.5	113	214
ST-4-1	32.7	179	3.1	26	205
ST-28	33.7	185	2.3	19	204
S-20	19.1	105	0.0	0	105
2247[b]	0.4	2	0.0	0	2

[a] The given organism was cultured at 30 °C for 72 h in medium A supplemented with 0.6 mg/ml of L-threonine and 0.2 mg/ml of L-methionine. L-Lysine was assayed manometrically and homoserine by an amino acid autoanalyser.

[b] Wild strain.

lower homoserine dehydrogenase activity, whereas the amounts of homoserine produced are widely different, depending on the level of the dehydrogenase. The results are summarized in Table V.

Isoleucine auxotrophs as well as methionine auxotrophs without threonine requirement did not produce lysine more than the wild strain. Mutants requiring methionine but not homoserine, which were derived from homoserine kinase-defective threonine auxotrophs did not produce lysine more than the parent strains, as shown in strain TM-15 of Table V.

Lysine analogue-resistant mutants[19]. It is likely that mutants where aspartate kinase is insensitive to the concerted inhibition by lysine plus threonine may produce lysine, since this enzyme is the main control site of lysine biosynthesis. Such mutants may be obtained among those resistant to analogues of the end products. AEC inhibited the growth of the wild strain 2247 completely in the presence of threonine, but only partially in its absence. Moreover, the growth inhibition was specifically reversed by a further addition of lysine (Fig. 7). These results suggest that the growth inhibition may be due to a concerted inhibition of aspartate kinase by AEC plus threonine similarly to lysine plus threonine. Then, mutants resistant to AEC were derived from strain 2247 and isolated as large colonies appeared on a minimal-medium plate containing 1 mg/ml of each AEC-HCl and L-threonine. These mutants contained lysine producers in a high ratio. Then 5 excellent lysine producers from among them were further examined for lysine production with a glucose medium (medium A) and with a sugar-cane molasses medium (medium C or C-1). Table VI shows that they produce more than 30 g/l of L-lysine-HCl and that their production largely depends on the cultural conditions. The most remarkable difference

TABLE VI

Production of Lysine by S-(2-Aminoethyl)cysteine-Resistant Mutants[a]

Strain	L-Lysine-HCl produced (g/l) in medium		
	A	C	C-1
FA-1-23	15.6	29.0	33.0
FA-1-24	16.2	25.2	32.6
FA-1-25	15.0	14.9	32.6
FA-1-30	16.3	25.6	30.8
FA-3-115	31.8	3.3	11.5

[a] Mutants were cultured at 30 °C for 72 h in medium A (glucose), for 96 h in medium C, or for 96 h in medium C-1. Medium C contained 255 g cane molasses (equivalent to 130 g sugar), 55 g $(NH_4)_2SO_4$, 1 g TSA-735 (antifoam), 75 g $CaCO_3$, and 1 litre distilled water at pH 8.0. Medium C-1 is medium C supplemented with 50 ml Mieki.

in the lysine production between these AEC-resistant mutants and other prototrophic or auxotrophic mutants as described in the preceding paragraphs is the effect of threonine. Thus, as shown in Fig. 8, 6 mg/ml of L-threonine completely inhibit lysine production by prototrophic threonine- or methionine-sensitive mutant, strain S-20, or by threonine auxotrophs, strains T-36 (homoserine kinase-defective), H-1013 (homoserine dehydrogenase-defective), and ST-4-1 (homoserine kinase-defective and

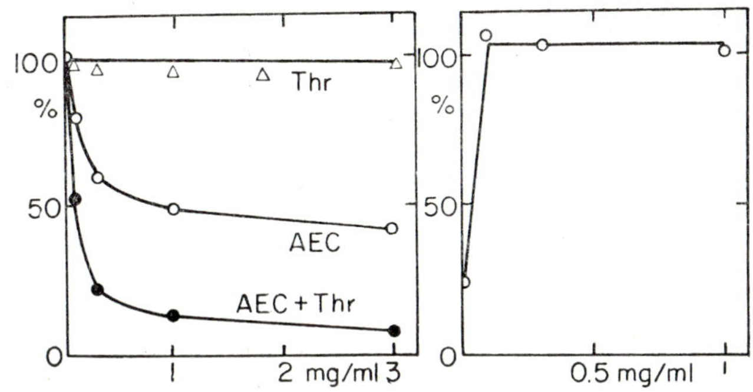

Fig. 7. Growth inhibition (%) by AEC plus threonine and its reversal by L-lysine. Strain 2247 was inoculated at 3×10^7 cells/ml to a minimal medium B-1 supplemented with given amino acids (mg/ml; *left*) or with each 1 mg/ml of AEC-HCl and L-threonine and given concentrations of lysine (mg/ml; *right*). The growth was estimated at 14 h.

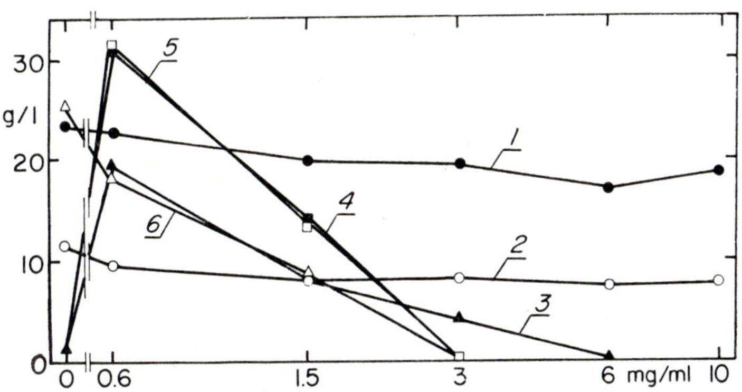

Fig. 8. Effect of L-threonine (mg/ml) on L-lysine production (g/l) in various mutants. Mutants were cultured at 30 °C for 72 h in medium A supplemented with 0.2 mg/ml of L-methionine and given concentration of threonine. L-Lysine was assayed manometrically. *1* FA-3-115, *2* FA-1-30, *3* T-36, *4* ST-4-1, *5* H-1013, *6* S-20.

threonine-sensitive), but not the lysine production by AEC-resistant mutants, strains FA-1-30 and FA-3-115. This suggests that mechanisms for lysine production were quite different in the two groups. In agreement with this suggestion, aspartate kinase of these AEC-resistant mutants has been found to be genetically desensitized to the concerted inhibition by lysine plus threonine, as described previously.

PRODUCTION OF THREONINE

Auxotrophic mutants. Since the main control site of the threonine biosynthesis in *B. flavum* seems to be homoserine dehydrogenase which is inhibited by threonine and repressed by methionine, limitation of methionine in the methionine auxotrophs could cause an overproduction of threonine. However, this is not the case. This is a similar result as in *E. coli* K-12 in which, although enzymes involved in the threonine biosynthesis are repressed by threonine plus isoleucine[20], limitation of isoleucine does not cause threonine accumulation[21]. On the other hand, methionine limitation in methionine auxotrophs of *E. coli* causes a production of threonine[21], presumably because of the presence of methionine-repressible aspartate kinase and homoserine dehydrogenase which are not inhibited by threonine[20]. Thus, the difference in the effect of methionine limitation on the threonine production between the two organisms clearly reflects the difference in the regulation mechanisms. Other auxotrophs of *B. flavum* including lysine and isoleucine auxotrophs did not significantly produce threonine.

Threonine-analogue-resistant mutants[22]. Since the threonine biosynthesis in *B. flavum* seems to be controlled through the feedback inhibition of homoserine dehydrogenase by threonine itself, it is likely that mutants whose homoserine dehydrogenase is insensitive to the inhibition may produce threonine. In this respect, it is of interest that α-amino-β-hydroxyvaleric acid (β-hydroxynorvaline)-resistant mutants of *E. coli* have been shown to produce threonine. The results reexamined show that the maximum amounts of threonine produced are 1.9 g/l in the analogue-resistant mutants of *E. coli* K-12, 4.7 g/l in isoleucine auxotrophs derived from them, and 6.1 g/l in isoleucine and methionine auxotrophs derived from the latter[21]. Similarly as in *E. coli*, the growth of *B. flavum* strain 2247 is inhibited about 50% by 0.25 mg/ml and almost completely by 1 mg/ml of α-amino-β-hydroxyvalerate. Furthermore, this growth inhibition is specifically reversed by threonine. The results are given in Fig. 9. Accordingly, mutants resistant to AHV were derived from strain 2247 and isolated on minimal-medium B-1 agar plates containing 5 mg/ml of DL-AHV. Seventeen strains out of these 23 resistant mutants thus obtained were threonine producers. The best producer, strain B-183, produces 10.5 g/l of L-threonine when cultured at 30 °C for 64 h in medium D. Although this strain is resistant to 5 mg/ml of DL-AHV, the growth is still completely inhibited at 7.5 mg/ml. Therefore, mutants resistant to 8 mg/ml of DL-AHV were derived from strain B-183. Twenty strains of these 200 resistant mutants tested produced more threonine than the parent. The best two mutants, strains BB-82 and BB-69, produce 13.5 and 13.3 g/l of L-threonine in medium D, respectively. Enzymatic studies show that, in the two strains, homoserine dehydrogenase has been genetically desensitized to the feedback inhibition by threonine but neither aspartate kinase nor homoserine kinase has been altered. This result is in good agreement with the concept that the control site of the threonine biosythesis in *B. flavum* is homoserine dehydrogenase.

Since aspartate kinase, the first common enzyme of the lysine and threonine biosynthesis, is still susceptible to the concerted inhibition by lysine plus threonine in these mutants and the inhibition is strong enough to prevent lysine overproduction, threonine overproduction in these AHV-resistant mutants suggests that intracellular levels of lysine in these mutants may be low enough to permit the overproduction. These low lysine levels may be established by preferential utilization of the common

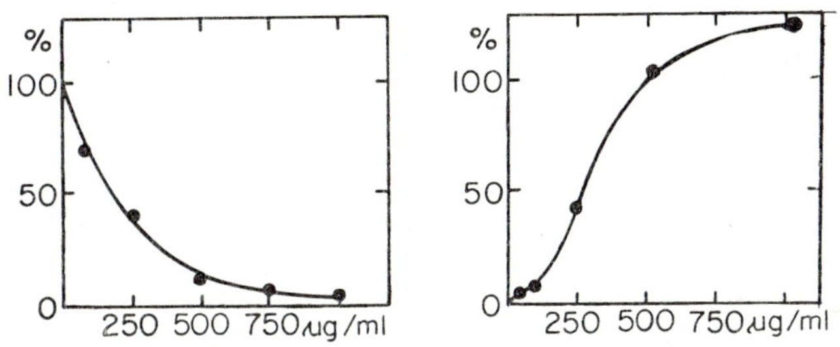

Fig. 9. Growth inhibition (%) by α-amino-β-hydroxyvalerate (µg/ml) and its reversal by threonine (µg/ml). Strain 2247 (wild) was inoculated at 9×10^6 cells/ml to a minimal medium B-1 supplemented with DL-AHV (*left*) or with 1 mg/ml of DL-AHV and given concentration of L-threonine (*right*).

intermediates for the threonine synthesis, through the disproportion of the two enzyme levels at the branch point in the threonine and lysine biosynthetic pathways. As described previously, the level of dihydrodipicolinate synthase at the branch

TABLE VII
Production of Threonine and Other Amino Acids by α-Amino-β-hydroxyvalerate-Resistant Mutants[a]

Strain	Amino acids produced (g/l)						
	L-Thr	L-Val	L-Ile	L-Ala	Gly	HSR	L-Leu
2247[b]	0.02	2.37	0.31	2.40	0.22	—	—
B-183	10.5	1.54	1.40	0.50	0.59	2.46	0.17
BB-82	13.5	0.55	0.70	0.80	0.66	2.06	0.10
BB-69	13.3	1.70	1.38	0.40	—	2.74	—
B-2	6.6	0.68	1.45	1.59	2.03	—	—
BL-4[c]	10.3	1.70	1.26	2.38	2.60	—	—

[a] Mutants were cultured at 30 °C for 64 h (58 h in B-2 and BL-4) in medium D which contained 100 g glucose, 30 g $(NH_4)_2SO_4$, 1.5 g KH_2PO_4, 0.4 g $MgSO_4.7H_2O$, 2 ppm Fe^{2+}, 2 ppm Mn^{2+}, 200 µg β-biotin, 300 µg thiamine-HCl, 4 ml Mieki, 50 g $CaCO_3$, and distilled water in a final volume of 1 litre at pH 7.2. Medium D was supplemented with 500 µg/ml of Lys-HCl for strain BL-4.

[b] Wild strain.

[c] Lysine-requiring strain.

point is only one-fifteenth of that of homoserine dehydrogenase in *B. flavum*. Since the balance between the two enzyme activities is maintained in the wild strain through the strong feedback inhibition by threonine of the latter enzyme, lack of the feedback inhibition in the mutants would cause a disproportion between them and, therefore, give a low lysine level. In fact, inhibitory effect of lysine on the threonine production has been observed with some resistant mutants. Thus, the production by strain B-183 is inhibited about 50% by the addition of 5 mg/ml of L-lysine-HCl. Moreover, lysine-auxotrophic mutant strain BL-4 derived from AHV-resistant mutant strain B-2 produces threonine more than the parent under lysine-limiting conditions, while the addition of excess lysine decreases the amount of threonine produced. Table VII shows that AHV-resistant mutants produce homoserine and isoleucine in addition to threonine, considerably more than the wild strain, homoserine being an intermediate, isoleucine the final product of threonine transformation.

AEC-resistant mutants whose aspartate kinase is insensitive to the concerted inhibition do not produce threonine. This is also in good agreement with the concept that the control site of threonine biosynthesis in *B. flavum* is not aspartate kinase but rather homoserine dehydrogenase.

CONCLUSIONS

In *B. flavum*, the biosynthesis of lysine seems to be controlled through the concerted inhibition of aspartate kinase by lysine plus threonine, while that of threonine would be controlled through the feedback inhibition of homoserine dehydrogenase by threonine. The distribution ratio of the common intermediate at the branch point to these two amino-acid biosynthetic pathways seems to be regulated through the feedback inhibition of the latter. In the absence of the feedback inhibition of homoserine dehydrogenase, the biosynthesis of threonine would be preferred to that of lysine, owing to the much higher level of homoserine dehydrogenase than that of dihydrodipicolinate synthase, the first enzyme of the lysine specific pathway.

In accordance with this regulatory mechanism, S-(2-aminoethyl)cysteine-resistant mutants whose aspartate kinase is insensitive to the concerted inhibition by lysine plus threonine produce about 30 g/l of L-lysine-HCl, while α-amino-β-hydroxyvalerate-resistant mutants whose homoserine dehydrogenase is insensitive to the feedback inhibition by threonine accumulate about 13 g/l of L-threonine.

The concerted inhibition of aspartate kinase which controls the lysine biosynthesis provides another process for the overproduction of lysine. That is, any conditions decreasing the threonine level inside the cell cause accumulation of lysine. The examples tested are repression of homoserine dehydrogenase by excess methionine in the wild strain, low level of this dehydrogenase in the threonine- or methionine-sensitive mutants whose growth is inhibited by threonine (or methionine) and recovered by further addition of methionine (or threonine, respectively), and threonine starvation in the threonine auxotrophs lacking homoserine dehydrogenase or kinase. Homoserine kinase-defective mutants produce homoserine in addition to lysine. The total amounts of lysine and homoserine produced are almost the same as that of lysine produced by homoserine dehydrogenase-defective mutants and the ratio of the amounts produced between lysine and homoserine depends on the level of homoserine dehydrogenase. Among them, lysine producers accumulating more than 30 g/l of L-lysine-HCl are homoserine dehydrogenase-defective mutants and homoserine kinase-defective, threonine-sensitive mutants.

REFERENCES

1. Davis B. D.: Nature **169**, 534 (1952).
2. Kinoshita S., Samejima H., Nakayama K., Nara T., Fujita C.: J. Gen. Appl. Microbiol. **6**, 193, (1960).
3. Casida L. E., Jr.: U.S. Pat. 2,771,396 (1956).
4. Kinoshita S., Samejima H., Fujita C., Nara T.: U.S. Pat. 3,099,604 (1963).
5. Huang H. T.: Appl. Microbiol. **9**, 419 (1961).
6. Kinoshita S., Nakayama K., Kitada S.: J. Gen. Appl. Microbiol. **4**, 128 (1958).
7. Shiio I., Miyajima R.: J. Biochem. **65**, 849 (1969).
8. Shiio J., Miyajima R., Sano K.: J. Biochem. **68**, 701 (1970).
9. Cohen G. N., Stanier R. Y., Le Bras G.: J. Bacteriol. **99**, 791 (1969).
10. Miyajima R., Otsuka S., Shiio I.: J. Biochem. **63**, 139 (1968).
11. Miyajima R., Shiio I.: Agric. Biol. Chem. **34**, 1275 (1970).
12. Ugari Y., Gilvarg C.: Biochim. Biophys. Acta **62**, 612 (1962).
13. Miyajima R., Shiio I.: J. Biochem. **68**, 313 (1970).
14. Shiio I., Miyajima R., Nakamori S.: J. Biochem. **68**, 859 (1970).
15. Cleland W. W.: Biochim. Biophys. Acta **67**, 104 (1963).
16. Shiio I., Sano K.: J. Gen. Appl. Microbiol. **15**, 267 (1969).
17. Miyajima R., Shiio I.: Agric. Biol. Chem. **35**, 424 (1971).
18. Sano K., Shiio I.: J. Gen. Appl. Microbiol. **13**, 349 (1967).
19. Sano K., Shiio I.: J. Gen. Appl. Microbiol. **16**, 373 (1970).
20. Patte J. C., Le Bras G., Cohen G. N.: Biochim. Biophys. Acta **136**, 245 (1967).
21. Shiio I., Nakamori S.: Agric. Biol. Chem. **33**, 1152 (1969).
22. Shiio I., Nakamori S.: Agric. Biol. Chem. **34**, 448 (1970).

Production of Isoleucine, Valine and Leucine by Regulatory Mutants of *Serratia marcescens*

MASAHIKO KISUMI, JOJI KATO, SABURO KOMATSUBARA, ICHIRO CHIBATA

Department of Applied Biochemistry, Chemical Research Laboratory, Tanabe Seiyaku Co., Ltd., Osaka, Japan

INTRODUCTION

Biosynthesis of amino acids is generally regulated by metabolic control mechanisms so as to be economical for the growth of cells. Accordingly, the release of metabolic controls is required for the accumulation of amino acids, although the lack of control of the amino-acid permeability appears to be another important factor. Up to the present, a variety of wild strains and auxotrophs have been employed to industrially produce several amino acids, in the presence or absence of metabolic controllers and precursors[1]. The accumulation of amino acids by wild strains seems to result from the natural lack of control mechanisms. The accumulation by auxotrophs seems to be due to the release of feedback inhibition or repression by the limitation of the required amino acids and due to the interference with the flow of precursors into other branched pathways.

On the other hand, various mutants, such as amino-acid analog-resistant mutants, revertants and temperature-sensitive mutants, have been employed in genetic and biochemical studies on metabolic control mechanisms[2,3]. It has been reported that some of these mutants lacked control mechanisms and excreted a small amount of amino acids[4]. It appears clear from the viewpoint stated above that the use of these mutants is favorable for the accumulation of amino acids. Nevertheless, these mutants had not been used for the production of amino acids. Therefore, we intended to utilize amino-acid analog-resistant mutants in studies on the production of branched-chain amino acids for the reasons described above. Recently, SHIIO and NAKAMORI have also described L-threonine accumulation by the analog-resistant mutants[5].

Biosynthetic pathways for branched-chain amino acids, L-isoleucine, L-valine and L-leucine, in microorganisms are shown in Fig. 1 (Ref.[6]). Regulatory mechanisms for the biosynthesis of these amino acids seem complicated on account of the fact that four of the five isoleucine—valine enzymes are bifunctional for both amino acids, that L-leucine is synthesized from α-oxo*iso*valerate, and that transaminase B participates in the formation of these three amino acids. Regulatory mechanisms in *Escherichia coli* and *Salmonella typhimurium* have been demonstrated as described below. That

IS, L-threonine dehydratase[7,8], acetohydroxyacid (AHA) synthase[9,10] and α-*iso*-propylmalate (α-IPM) synthase[11] are feedback-inhibited by L-isoleucine, L-valine and L-leucine, respectively. Furthermore, the isoleucine—valine enzymes are subject to multivalent repression by these amino acids, while the leucine enzymes are repressed by L-leucine alone[12]. In addition, a variety of regulatory mutants were found to lack either mechanism[8,13-18] and some of these mutants excreted branched-chain amino acids[15,18].

Advantageous methods for L-isoleucine production had been desired for a long time, since isoleucine is composed of four isomers in chemical preparation. About ten years ago, by employing *Serratia marcescens*, we succeeded in L-isoleucine production from D-threonine which was a useless by-product in the chemical synthesis of L-threonine. As shown in Table I, D-threonine was most favorable for L-isoleucine accumulation among several precursors[19]. Thereafter, it was clarified that D-threonine dehydratase was induced by D-threonine and was not feedback-inhibited by L-isoleucine[20]. These phenomena seemed to be the primary cause for L-isoleucine accumu-

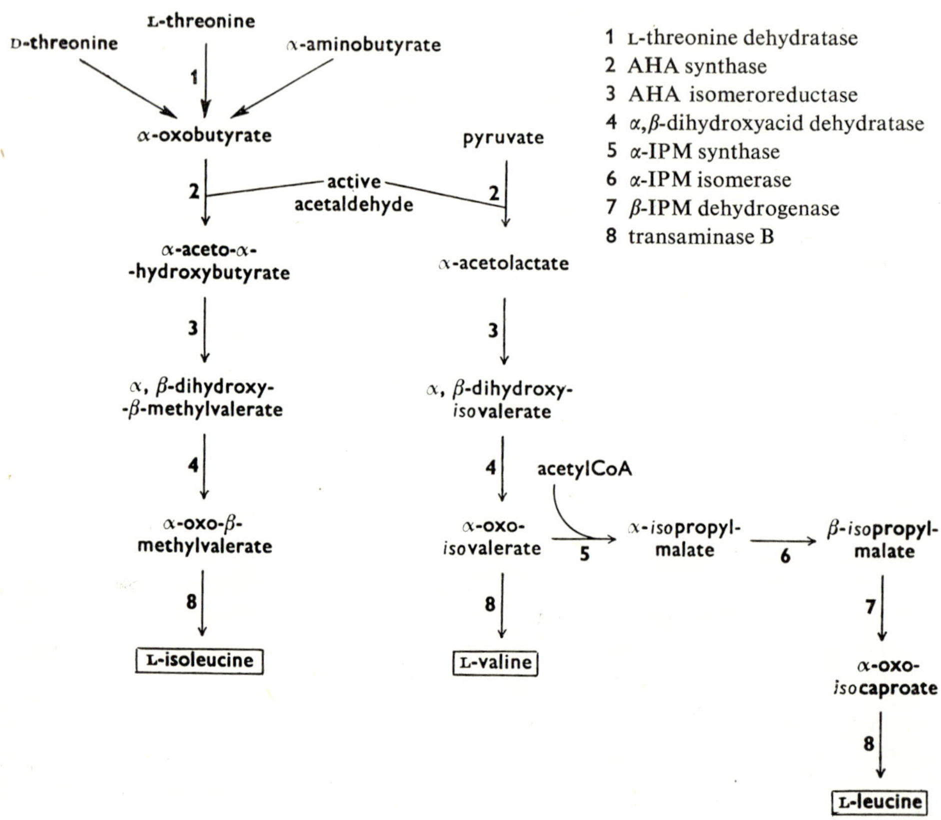

Fig. 1. Biosynthetic pathways of branched-chain amino acids.

TABLE I

Accumulation of L-Isoleucine from Various Precursors by *Serratia marcescens*

Precursor added[a]	Growth O.D.[b]	Remaining precursor mg/ml	L-Isoleucine accumulated mg/ml
None	0.74	—	0.2
L-Aspartic acid	0.65	0	0.2
DL-Homoserine	0.42	2	0.2
α-Oxobutyric acid	0.05	19	0.3
DL-α-Aminobutyric acid	0.75	18	1.0
D-Threonine	0.74	10	7.8
L-Threonine	0.81	12	0.2
DL-Threonine	0.79	8	2.3
DL-*allo*-Threonine	0.74	11	0.2

[a] Precursors were added at the concentration of 2% to a medium containing 10% glucose, 0.5% urea, 0.1% K_2HPO_4, 0.05% $MgSO_4.7H_20$, 0.7% corn steep liquor and 2% $CaCO_3$. The culture was shaken at 30 °C for 40 h.

[b] Optical density at 660 nm, diluting broth to 1 : 10 with 0.1 N-HCl.

lation. It was also demonstrated that D-threonine was incorporated into the carbon skeleton of L-isoleucine[21].

It was, moreover, considered that the production of three branched-chain amino acids would be accomplished by using regulatory mutants of *S. marcescens* on the theoretical basis of metabolic controls. By using regulatory mutants, we have succeeded not only in the improvement of L-isoleucine accumulation but also in the production of L-valine and L-leucine.

INCREASE IN L-ISOLEUCINE ACCUMULATION BY α-AMINOBUTYRIC ACID-RESISTANT (*abu*[R]) MUTANTS[22,23]

In the course of our investigation on L-isoleucine accumulation from D-threonine, it was found that either L-valine or L-leucine inhibited L-isoleucine accumulation and caused α-aminobutyric acid formation[19]. A similar result was obtained when the addition of excess glucose caused a slight formation of L-valine. These facts were considered to be due to feedback control of AHA synthase by these amino acids. In order to relieve the inhibition of accumulation by L-valine and L-leucine, and to increase L-isoleucine accumulation, we intended to release AHA synthase from feedback inhibition and/or repression by using genetic methods.

TABLE II

Antagonism between α-Aminobutyric Acid and Branched-Chain Amino Acids in *Serratia marcescens*

Addition[a]	Growth O.D. in h		
	24	40	48
None	0.51	0.48	0.50
DL-Abu	0	0	0.11
DL-Abu + L-Ile	0	0.19	0.46
DL-Abu + L-Val	0.21	0.45	0.46
DL-Abu + L-Leu	0	0.50	0.49
DL-Abu + L-Val + L-Leu	0.53	0.54	0.56
DL-Abu + L-Ile + L-Val + L-Leu	0.55	0.53	0.55

[a] DL-α-Aminobutyric acid (1 mg/ml) and branched-chain amino acids (each, 100 μg/ml) were added to the minimal medium of Davis and Mingioli.

TABLE III

Effect of L-Valine on the Accumulation of L-Isoleucine, and the Formation of D-, L-Threonine Dehydratases and AHA Synthase in Wild Strain and *abu*[R] Mutant 130-1[a]

Strain	L-Valine added mg/ml	L-Isoleucine accumulated[b] mg/ml	Specific activity[c]		
			D-Threonine dehydratase	L-Threonine dehydratase	AHA synthase
Wild	0	10.4	5.2	26.8	8.1
	1	1.3	4.6	10.0	1.6
130-1 *abu*[R]	0	15.6	5.6	30.4	20.5
	1	14.0	5.1	27.6	23.7

[a] The culture medium (To-3) contained (%): D-threonine, 2; glucose, 2; dextrin, 10; urea, 0.5; K_2HPO_4, 0.1; $MgSO_4.7H_2O$, 0,05; $CaCO_3$, 2. The reaction mixture for enzyme assay contained the following. D- or L-Threonine dehydratase contained in 1 ml (μmol): D- or L-threonine, 20; pyridoxal phosphate, 0.02; Tris-HCl buffer (pH 9.0), 100; cell-free extracts, 0.3 to 1.0 mg of protein. AHA synthase contained in 1 ml (μmol): pyruvate, 40; $MgCl_2$, 10; thiamine pyrophosphate, 0.2; flavine adenine dinucleotide, 0.05; phosphate buffer (pH 8.0), 100; cell-free extracts, 0.3 to 1.0 mg of protein.

[b] L-Isoleucine in the medium cultivated for 72 h.

[c] Enzyme activity (μmol of product per h per mg protein) in cell-free extracts prepared from cells grown for 18 h.

The isolation of valine-resistant (*val*^{R}) mutants was considered advisable for the selection of mutants which have a synthase insensitive to the inhibition by L-valine, in view of the results of LEAVITT and UMBARGER[13,24]. However, the growth of *S. marcescens* was not inhibited by L-valine even at a high concentration, differing from that of *E. coli* K-12 (Ref.[25]). Therefore, *val*^{R} mutants could not be isolated. This fact suggested that AHA synthase of *S. marcescens* might not be strongly

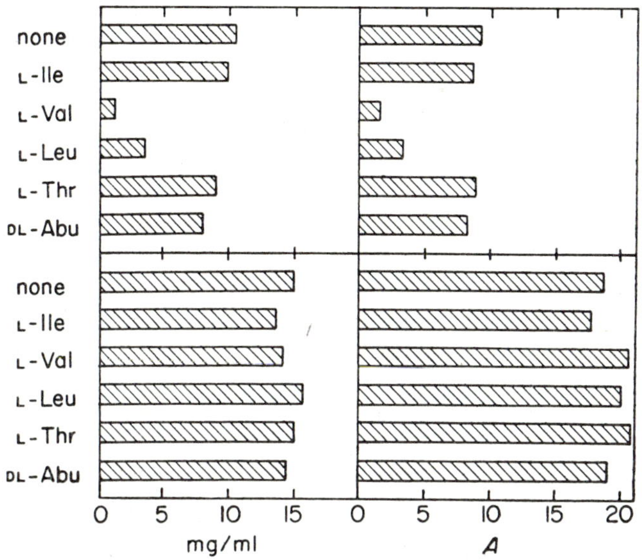

Fig. 2. Effect of amino acids on the accumulation of L-isoleucine (mg/ml) at 72 h and the formation of AHA synthase (specific activity, *A*) at 18 h. Medium To-3 was used. *Above:* wild strain; *below:* strain 130-1 (*abu*^{R}).

inhibited by L-valine. Accordingly, there existed the possibility that the inhibition of L-isoleucine accumulation might be attributed to the repression of this enzyme. Since it was considered that some of the mutants resistant to antagonists of branched-chain amino acids might be derepressed for AHA synthase, the antagonism between branched-chain amino acids and various analogs was examined. As a result, α-aminobutyric acid, norvaline and norleucine were found to be antagonized by branched-chain amino acids. For example, growth inhibition by α-aminobutyric acid was effectively antagonized by three amino acids, especially by L-valine, as shown in Table II. A number of mutants resistant to these antagonists were isolated and tested for L-isoleucine accumulation. Most of *abu*^{R} mutants were found to be superior to the wild strain in L-isoleucine-accumulating activity, while norvaline- and norleucine-resistant (*nva*^{R} and *nle*^{R}) mutants accumulated the same amount of L-isoleucine as the wild strain did. Strain 130-1 shown in Table III was selected as a typical *abu*^{R} mutant which was superior to the wild strain in the activity of L-isoleucine accumulation. Strain 130-1 increased approximately 1.5-fold in L-isoleucine accumulation. Though the accumulation by the wild strain was strongly inhibited by L-valine, the accumulation by strain 130-1 was hardly affected by this amino acid. To elucidate the difference in the accumulation, the levels of three enzymes related to the biosynthesis of L-isoleucine were determined. The levels of D- and L-threonine dehydratases of strain 130-1 were almost similar to those of the wild strain, when both strains were grown in the medium without L-valine. However, L-threonine dehydratase of the wild strain was considerably repressed by L-valine, whereas that of strain 130-1 was hardly repressed. This enzyme of strain 130-1 as well as that of the wild strain was almost completely feedback-inhibited by a low concentration of L-isoleucine. This fact indicates that the increase in L-isoleucine accumulation by strain 130-1 is not due to the formation of this amino acid from glucose *via* L-threonine. Strain 130-1 increased 2- to 3-fold in AHA synthase level as compared with the parent strain. It was also found that AHA synthase of strain 130-1 was hardly repressed by L-valine although that of the parent strain was markedly repressed.

Of the other amino acids related to the biosynthesis of L-isoleucine, L-leucine had the same effect as did L-valine, though to a lesser extent, as may be seen in Fig. 2. As another reason for the insensitivity of L-isoleucine accumulation to L-valine in strain 130-1, the possibility existed that AHA synthase might lack feedback inhibition by L-valine. The synthases of both strains were, however, inhibited about 60% by 10 mM L-valine. From the results described above, it was found that the increase in L-isoleucine accumulation and the release from the inhibition of the accumulation might be mainly due to the derepression of AHA synthase.

ACCUMULATION OF L-VALINE BY *abu*^{R} MUTANTS[22]

In view of the fact that AHA synthase of strain 130-1 grown with D-threonine was not repressed by L-valine, it was anticipated that this *abu*^{R} mutant would accumulate L-valine in a medium devoid of D-threonine. Table IV shows that, when both strains were cultured without D-threonine, strain 130-1 markedly accumulated L-valine and the wild strain accumulated a barely detectable amount of L-valine. Strain 130-1 had a high level of AHA synthase with or without D-threonine. The wild strain, on the contrary, had a lower level of this enzyme when D-threonine was not added to the medium*. In addition, it is of interest that, in the case of strain 130-1, the addition of D-threonine changed L-valine accumulation to L-isoleucine accumulation. This fact indicates that AHA synthase which is a bifunctional enzyme for the formation of both α-acetolactate and α-acetohydroxybutyrate has a higher affinity for α-oxobutyric acid than pyruvic acid.

* It will be described below that D-threonine derepressed the isoleucine—valine enzymes.

TABLE IV

Accumulation of L-Isoleucine and L-Valine by Wild Strain and *abu*^{R} Mutant 130-1[a]

Strain	D-Thr added mg/ml	Amino acid accumulated[b] mg/ml		Specific activity of AHA synthase[c]
		L-Ile	L-Val	
Wild	0	0	0.5	3.5
	5	1.7	0	4.5
	20	11.1	0	8.9
130-1 *abu*^{R}	0	0	7.1	33.5
	5	4.1	6.8	34.2
	20	15.8	0	20.2

[a] Medium To-3.

[b] At 72 h.

[c] At 18 h.

TABLE V

Accumulation of L-Valine and Levels of AHA Synthase in *abu*^{R} Mutants[a]

Strain	Growth O.D. in h		L-Valine accumulated mg/ml in h		Specific activity of AHA synthase[b]
	24	48	24	48	
Parent	0.63	0.77	0.2	0.6	2.6
44	0.61	0.74	0	0.5	2.6
139	0.64	0.76	0	0.9	3.6
117	0.66	0.79	0.6	2.0	9.0
131	0.50	0.70	0.9	2.4	14.4
7	0.57	0.83	2.7	4.2	14.0
106	0.49	0.91	0.9	6.8	26.1
130-1	0.56	0.91	3.5	7.2	31.5
140	0.44	0.75	0.9	8.1	43.9
5	0.61	0.79	3.1	8.9	16.9
9	0.58	0.84	3.8	9.6	24.5

[a] The culture medium (V-1) contained (%): glucose, 2; dextrin, 10; urea, 1; K_2HPO_4, 0.1; $MgSO_4 \cdot 7H_2O$, 0.5; $CaCO_3$, 2.

[b] At 24 h.

TABLE VI

Feedback Inhibition (%) of AHA Synthase in Parent Strain and *abu*^{R} Mutants[a]

Strain	L-Valine added M		
	10^{-3}	10^{-2}	10^{-1}
Wild	42	46	53
Wild[b]	45	63	67
130-1	43	57	70
140	42	53	73
5	16	21	23
9	7	6	8

[a] Cells were grown for 24 h in medium V-1.
[b] Inhibition of AHA synthase in wild strain grown in medium To-3.

Most of *abu*^{R} mutants tested also accumulated L-valine. Table V indicates the L-valine-accumulating activity and the levels of AHA synthase of several *abu*^{R} mutants which are different from each other in *abu*-resistance level. Mutants *abu*^{R} which had a weak resistance, for example strains 44 and 139, accumulated a barely detectable amount of L-valine, and had a low level of AHA synthase as did the wild strain. However, mutants resistant to high concentrations of α-aminobutyric acid, for example strains 106, 140, 5 and 9, accumulated a large amount of L-valine and they had a high enzyme level as did strain 130-1. In other words, a linear correlation was found between resistance level, the amount of L-valine accumulated and AHA synthase level, except in case of strains 5 and 9. It should be noted that strains 5 and 9 accumulated a larger amount of L-valine than strains 130-1 or 140, though AHA synthase of the former mutants was at a lower level than that of the latter mutants. To elucidate the reason for this fact, feedback inhibition of this enzyme was examined. As shown in Table VI, AHA synthase of these mutants was feedback-inhibited by L-valine to a lesser extent than that of most *abu*^{R} mutants, such as strains 130-1 and 140. Most of all, the enzyme of strain 9 was hardly inhibited even by high concentrations of L-valine. It was, therefore, considered that strain 9 was the best accumulator among the *abu*^{R} mutants tested, presumably because of the lack of both control mechanisms.

Mutants *nva*^{R} of *S. marcescens* were found not to accumulate L-valine, though KARLSTRÖM reported[26] that mutants *nva*^{R} of *E. coli* K-12 excreted L-valine. However, certain *nva*^{R} mutants which showed a moderate cross-resistance to α-aminobutyric acid considerably accumulated L-valine. The mutants resistant to norleucine which was moderately antagonized by L-valine did not accumulate any branched-chain amino acid.

MULTIVALENT REPRESSION AND GENETIC DEREPRESSION OF THE ISOLEUCINE—VALINE ENZYMES[27,28]

It has been known that in *E. coli* and *S. typhimurium*, the isoleucine—valine enzymes are multivalently repressed by three end products, L-isoleucine, L-valine and L-leucine[12]. It was considered important to know what mechanisms control the for-

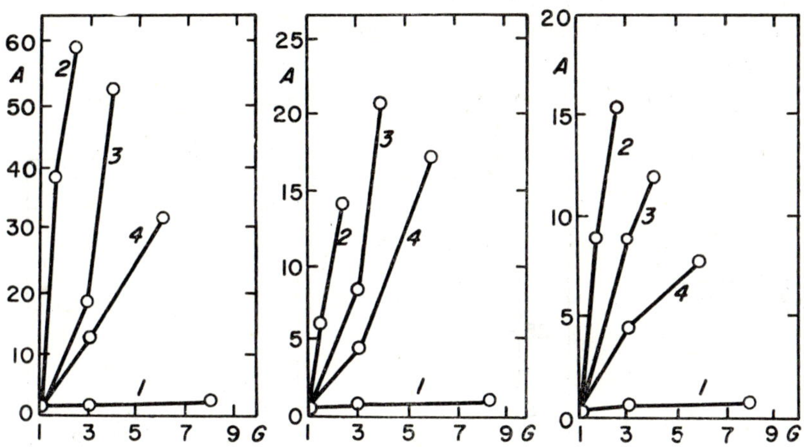

Fig. 3. Multivalent repression of the isoleucine—valine enzymes in an isoleucine—valine auxotroph, strain 102. The minimal medium of Davis and Mingioli containing glycyl-L-valine (10^{-4} M) and Ca-pantothenate (2×10^{-5} M) was supplemented as follows:

Supplement M	Complete *1*	Low Ile 2	Low Val 3	No Leu 4
L-Isoleucine	10^{-3}	10^{-4}	10^{-3}	10^{-3}
L-Valine	10^{-3}	10^{-3}	10^{-4}	10^{-3}
L-Leucine	10^{-3}	10^{-3}	10^{-3}	0

Cells grown in the complete medium were inoculated into the respective medium. The reaction mixture for transaminase B assay contained in 1 ml (μmol) : L-valine, 20, sodium α-oxoglutarate, 20, pyridoxal phosphate, 0.1, phosphate buffer (pH 8.0), 100, cell-free extracts, 0.3 to 1.0 mg of protein. *G* growth (increase in number of cells), *A* specific activity, μmol of product per h per mg of protein. *Left:* L-threonine dehydratase, *middle:* AHA synthase, *right:* transaminase B.

mation of the isoleucine—valine enzymes of *S. marcescens* for the production of three branched-chain amino acids. Fig. 3 shows the results of the experiment employing strain 102 which required L-isoleucine and L-valine, and partially L-leucine and pantothenate. Three isoleucine—valine enzymes, that is, L-threonine dehydratase, AHA synthase, and transaminase B, were repressed when strain 102 was grown in the complete medium. The respective enzymes were found to be significantly derepressed when any single branched-chain amino acid was limiting. In other words, the isoleucine—valine enzymes of *S. marcescens* as well as those of *E. coli* were found to be subject to multivalent repression. However, it has been reported that AHA synthase of *E. coli* K-12 was not derepressed when L-isoleucine was limiting[29]. While pantothenate has been reported to be related to this repression in *E. coli*[30], this compound did not participate in multivalent repression in *S. marcescens*. This fact may indicate that *S. marcescens* partially differs from *E. coli* in control mechanisms for the formation of the isoleucine—valine enzymes.

The levels of three enzymes of the *abu*^{R} mutant 130-1 and the wild strain in the medium with or without branched-chain amino acids were compared in Fig. 4.

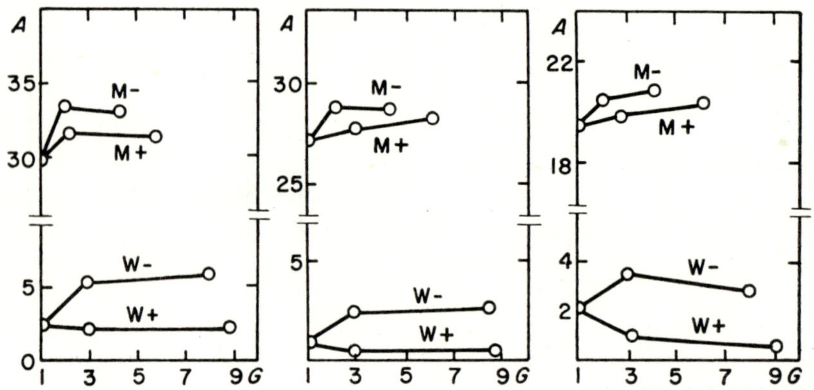

Fig. 4. Genetic derepression of the isoleucine—valine enzymes in *abu*^{R} mutant, strain 130-1. Cells were grown in the minimal medium of Davis and Mingioli with (+) or without (—) the mixture of branched-chain amino acids (each, 10^{-2} M). *G* growth (increase in number of cells), *A* specific activity. *Left:* L-threonine dehydratase, *middle:* AHA synthase, *right:* transaminase B; M mutant 130-1, W wild strain.

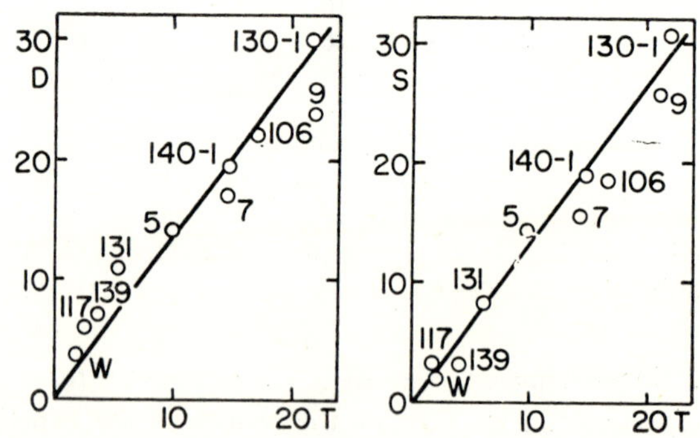

Fig. 5. Coordinate derepression of L-threonine dehydratase (D), AHA synthase (S) and transaminase B (T) in *abu*^{R} mutants (specific activities). Cells were grown in the minimal medium of Davis and Mingioli for 16 h. The numbers refer to the strain numbers of *abu*^{R} mutants; W wild strain.

Strain 130-1 had high levels of the respective enzymes in the presence or absence of the amino acids, while the enzymes of the wild strain were at low levels without them and were repressed by these end products. These data indicated that strain 130-1 was genetically derepressed for the isoleucine—valine enzymes. Fig. 5 shows that *abu*^{R} mutants were coordinately derepressed for three enzymes. Though RAMAKRISHNAN and ADELBERG have reported that an *abu*^{R} mutant of *E. coli* K-12 was genetically derepressed only for AHA synthase[16], *abu*^{R} mutants of *S. marcescens* were derepressed not only for AHA synthase but also for the other two enzymes. This also suggests that *S. marcescens* differs from *E. coli* in control mechanisms.

DEREPRESSION OF THE ISOLEUCINE—VALINE ENZYMES BY D-THREONINE, L-α-AMINOBUTYRIC ACID AND α-OXOBUTYRIC ACID[28,32]

As stated above, in the wild strain the addition of L-valine or L-leucine repressed AHA synthase, the level of which was elevated by the addition of D-threonine. It was, therefore, considered significant to know the role of D-threonine and

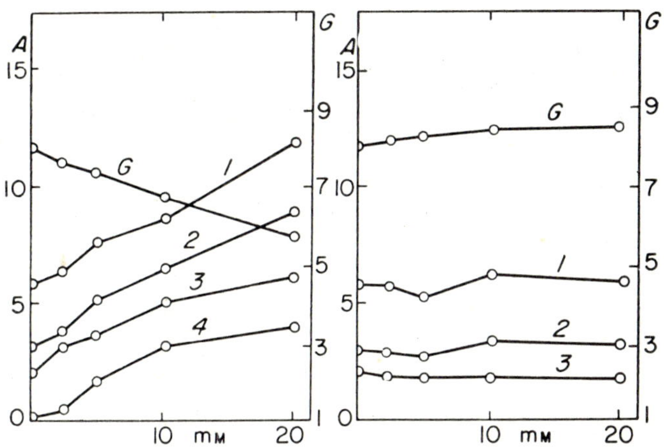

Fig. 6. Effect of D- and L-threonine on the formation of the isoleucine—valine enzymes in wild strain. Cells were grown in the minimal medium of Davis and Mingioli for 4 h. *Left:* D-threonine added, *right:* L-threonine added; *A* specific activity, *G* growth (increase in number of cells); *1* L-threonine dehydratase, *2* transaminase B, *3* AHA synthase, *4* D-threonine dehydratase.

others for elucidating the control mechanisms of the biosynthesis of branched-chain amino acids. As shown in Fig. 6, the three isoleucine—valine enzymes were derepressed by D-threonine but not by L-threonine. D-Threonine-mediated derepression was overcome by either L-valine or L-leucine, most effectively by L-valine plus L-leucine, as shown in Table VII. Fig. 7 gives the results of experiments employing α-aminobutyric acid. L-α-Aminobutyric acid caused not only growth inhibition but also derepression of the enzymes, while the D-form had no effect on the growth and the formation of the enzymes. Table VIII shows that derepression caused by DL-α-amino-butyric acid was overcome by L-valine, L-leucine or their mixture. Results similar to those in Fig. 7 and Table VIII were obtained when α-oxobutyric acid was added to the medium.

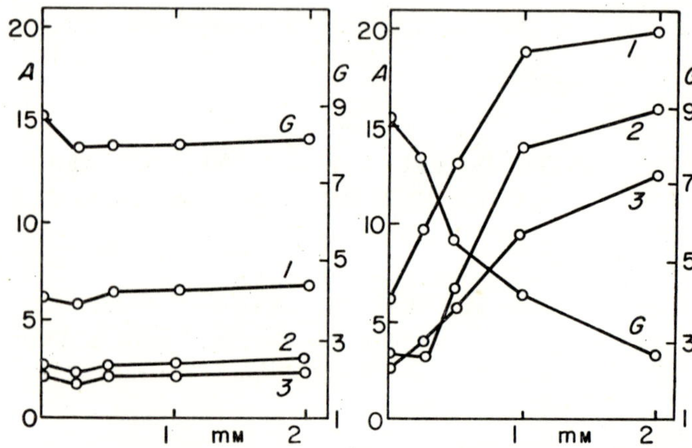

Fig. 7. Effect of D- and L-α-aminobutyric acid on the formation of the isoleucine—valine enzymes in wild strain. Cells were grown in the minimal medium of Davis and Mingioli for 4 h. *Left:* D-Abu added, *right:* L-Abu added; *A* specific activity, *G* growth (increase in number of cells).

TABLE VII

Effect of L-Valine and L-Leucine on the D-Threonine-Mediated Derepression of Isoleucine—Valine Enzymes in Wild Strain[a]

Addition mM			Growth[b]	Specific activity		
D-Thr	L-Val	L-Leu		L-Threonine dehydratase	AHA synthase	Trans-aminase B
0	0	0	8.0	4.5	2.0	3.3
20	0	0	4.8	12.7	7.7	9.2
20	0.25	0	7.3	6.1	3.5	4.8
20	1	0	7.6	5.6	3.0	3.6
20	0	0.25	7.1	6.9	5.9	5.7
20	0	1	7.2	5.7	3.5	4.4
20	0.25	0.25	7.5	5.2	3.6	4.3
20	1	1	8.2	3.2	1.9	2.8

[a] Cells were grown in the minimal medium of Davis and Mingioli for 4 h.
[b] Increase in number of cells.

TABLE VIII

Effect of L-Valine and L-Leucine on the α-Aminobutyric Acid-Mediated Derepression of Isoleucine—Valine Enzymes in Wild Strain[a]

Addition mM			Growth[b]	Specific activity		
DL-Abu	L-Val	L-Leu		L-Threonine dehydratase	AHA synthase	Trans-aminase B
0	0	0	8.2	5.6	2.7	2.8
2	0	0	1.3	21.6	10.6	14.1
2	0.25	0	4.6	18.7	8.3	9.3
2	1	0	6.5	12.7	5.0	6.4
2	0	0.25	2.4	22.4	7.8	13.3
2	0	1	2.8	17.1	6.4	10.9
2	0.1	0.1	5.6	19.8	8.3	11.1
2	0.25	0.25	6.3	9.8	5.8	7.2
2	1	1	8.2	1.0	0.7	2.1

[a] Cells were grown in the minimal medium of Davis and Mingioli for 4 h.
[b] Increase in number of cells.

FREUNDLICH and CLARKE have reported[33,34] that in *E. coli* W growth inhibition and derepression of the isoleucine—valine enzymes caused by α-aminobutyric acid might be attributed to inhibition of AHA synthase or of valyl-tRNA synthetase. AHA synthase of *S. marcescens* was not inhibited by α-aminobutyric acid at concentrations where growth inhibition was caused. In addition, α-aminobutyric acid considerably derepressed the enzymes in strain 102, which is an isoleucine—valine auxotroph, when branched-chain amino acids were supplemented at concentrations where a strong repression occurred. These facts suggest that the physiological effects of α-aminobutyric acid on *S. marcescens* may be due to inhibition of valyl-tRNA synthetase.

The extent of the derepression by D-threonine was less than that observed with L-α-aminobutyric acid even at high concentrations. D-Threonine, in addition, had little effect on enzyme levels in strain 102 under a strongly repressed, partially repressed, or derepressed condition. Therefore, it is assumed that D-threonine differs from α-aminobutyric acid in derepression mechanisms. AHA synthase is a bifunctional enzyme which forms both α-acetolactate and α-acetohydroxybutyrate, and appears to have a stronger affinity for α-oxobutyric acid than for pyruvic acid. Accordingly, α-acetolactate formation is probably inhibited by α-oxobutyric acid formed from D-threonine without any feedback control. This seems to result in excess formation of L-isoleucine and in inhibition of the formation of L-valine and L-leucine. α-Oxobutyric acid, in fact, caused a strong growth inhibition and derepression at low concentrations. L-Isoleucine was also found to cause a slight derepression. Therefore, these facts indicate the possibility that D-threonine-mediated derepression may be attributed to the release of multivalent repression caused by the unbalance of levels of three amino acids in cells.

ACCUMULATION OF L-ISOLEUCINE FROM α-AMINOBUTYRIC ACID AND α-OXOBUTYRIC ACID[35]

From the results described above, it was considered to be possible for the wild strain to accumulate L-isoleucine from α-aminobutyric acid and α-oxobutyric acid, if appropriately high enzyme levels are maintained by the addition of L-valine and L-leucine at a suboptimum concentration where growth inhibition is reversed. It was, in addition, expected that abu^R mutants would have a higher accumulating activity than the wild strain. Table IX shows that L-isoleucine accumulation from α-aminobutyric acid by the wild strain was accelerated by the suboptimum addition of L-valine and L-leucine. Excess L-valine and L-leucine, however, inhibited the accumulation owing to the repression of the isoleucine—valine enzymes. On the other hand, strain 130-1 accumulated a larger amount of L-isoleucine than the wild strain. When D- and L-forms of α-aminobutyric acid were employed, the wild strain accumulated L-isoleucine from the L-form but not from the D-form. Strain 130-1,

TABLE IX

Accumulation of L-Isoleucine from DL-α-Aminobutyric Acid by Wild Strain and abu^R Mutant 130-1[a]

Strain	DL-Abu added mg/ml	L-Val, L-Leu added mg/ml	Growth O.D. in h		L-Ile accumulated mg/ml in h	
			48	96	48	96
Wild	0	0	0.71	0.81	0.1	0.1
	20	0	0	0.25	0	1.5
		0.1	0.31	0.43	1.8	4.8
		1	0.51	0.70	0.1	0.1
130-1 abu^R	0	0	0.74	0.89	0.1	0.2
	20	0	0.35	0.57	2.6	7.3
		0.1	0.35	0.67	3.2	7.8
		1	0.44	0.69	3.1	8.5

[a] Medium To-3 devoid of D-threonine was used. The mixture of L-valine and L-leucine was added at the concentrations indicated.

TABLE X

Accumulation of L-Isoleucine from α-Oxobutyric Acid by Wild Strain and *abu*^{R} Mutant 130-1[a]

Strain	Obu[b] added mg/ml	L-Val, L-Leu added mg/ml	Growth O.D. in h		L-Ile accumulated mg/ml in h	
			48	96	48	96
Wild	0	0	0.72	0.88	0.1	0.1
	10	0	0.07	0.09	0.5	0.2
		0.5	0.12	0.32	0.5	3.0
		1	0.22	0.50	0.2	1.8
130-1 *abu*^{R}	0	0	0.72	0.88	0.1	0.1
	10	0	0.12	0.61	1.2	6.0
		0.1	0.33	0.71	2.8	6.6
		1	0.66	0.73	4.1	5.9

[a] Medium To-3 devoid of D-threonine was used. The mixture of L-valine and L-leucine was added at the concentration indicated.

[b] Sodium α-oxobutyrate.

however, accumulated L-isoleucine from the D-form as well, though to a lesser extent than from the L-form.

When α-oxobutyric acid was replaced with α-aminobutyric acid, as shown in Table X, a slight accumulation of L-isoleucine by the wild strain was observed because growth inhibition was stronger than that by α-aminobutyric acid. On the other hand, strain 130-1 showed a considerable accumulation of L-isoleucine with or without L-valine and L-leucine. From these results it became clear that L-isoleucine accumulation is possible not only from D-threonine but also from α-aminobutyric acid and α-oxobutyric acid.

L-LEUCINE ACCUMULATION BY LEAKY ISOLEUCINE REVERTANTS RESISTANT TO α-AMINOBUTYRIC ACID[36]

L-Leucine is formed from α-oxo*iso*valerate by the metabolic pathway shown in Fig. 1. In view of the facts that simultaneous addition of L-valine and L-leucine was required for the complete reversion of growth inhibition caused by α-aminobutyric acid, and that *abu*^{R} mutants formed a small amount of L-leucine as a by-product in L-valine accumulation, it would be expected that the leucine enzymes of *abu*^{R} mutants are also genetically derepressed. If so, mutants accumulating L-leucine might be obtained by mutating *abu*^{R} mutants. On the other hand, from the results of the studies on L-isoleucine accumulation from D-threonine, it was anticipated that L-isoleucine would be accumulated from L-threonine by mutants which lack feedback inhibition of L-threonine dehydratase. For this reason, it was of interest that some prototrophic revertants isolated from an isoleucine auxotroph had an altered L-threonine dehydratase which was not sensitive to the inhibition by L-isoleucine[8,14].

For selecting the mutants which would accumulate L-isoleucine from L-threonine, several isoleucine auxotrophs were isolated from the *abu*^{R} mutant, strain 130-1. When some of the auxotrophs isolated were cultured in the medium devoid of L-isoleucine, considerable growth was observed after 72 h. Of these auxotrophs, strain 149 accumulated neither L-isoleucine nor L-valine but, unexpectedly, L-leucine, as indicated in

TABLE XI

Accumulation of L-Leucine and L-Valine by Mutants 130-1 and 149[a]

Strain	Growth O.D. in h		Amino acid accumulated mg/ml in h			
			L-Leucine		L-Valine	
	48	96	48	96	48	96
Wild	0.79	0.80	0	0	0.5	0.5
130-1 *abu*^{R}	0.79	0.82	0	0.5	8.5	8.5
149 *abu*^{R} *ile*	0.04	0.77	0	7.5	0	0

[a] The culture medium (Vo-1) contained (%): glucose, 2; dextrin, 10; urea, 1; K_2HPO_4, 0.1; $MgSO_4.7H_2O$, 0.05; $FeSO_4.7H_2O$, 0.0001; $CaCO_3$, 2.

TABLE XII

Effect of Isoleucine Precursors on the Accumulation of L-Leucine and L-Valine by Strain 149[a]

Precursor added[b]	Amino acid accumulated mg/ml in h			
	L-Leucine		L-Valine	
	48	96	48	96
None	0.5	9.5	0	0.5
D-Threonine	0.5	0.5	7	6.5
L-Threonine	1	6.5	0	0.5
DL-α-Aminobutyrate	0.5	1.5	8	8.5
α-Oxobutyrate	0	0.5	7.5	7.5
L-Isoleucine	1	0.5	6.5	8

[a] Medium Vo-1.
[b] 1 mg/ml.

TABLE XIII
Accumulation of L-Leucine by Revertants Derived from Strain 149[a]

Strain	Amino acid accumulated[b] mg/ml in h			
	Vo-1		Vo-1 + L-Ile	
	L-Leucine	L-Valine	L-Leucine	L-Valine
149	7.5[c]	0[c]	0.5	7
S-3	9.5	0	6	0
S-5	8.5	0	4.5	0.5
S-6	11	0	6	0
S-8	10	0	9.5	0.5
S-10	11.5	0	11	0.5
S-11	14	0	12	0

[a] Medium Vo-1 with or without L-isoleucine (1 mg/ml).
[b] At 48 h.
[c] At 96 h.

Table XI. When L-isoleucine and its precursors were added to the medium, strain 149 accumulated L-valine instead of L-leucine, except in the case of the addition of L-threonine, as shown in Table XII. Prototrophic revertants were, subsequently, isolated by nitrosoguanidine treatment, in order to select the strains which accumu-

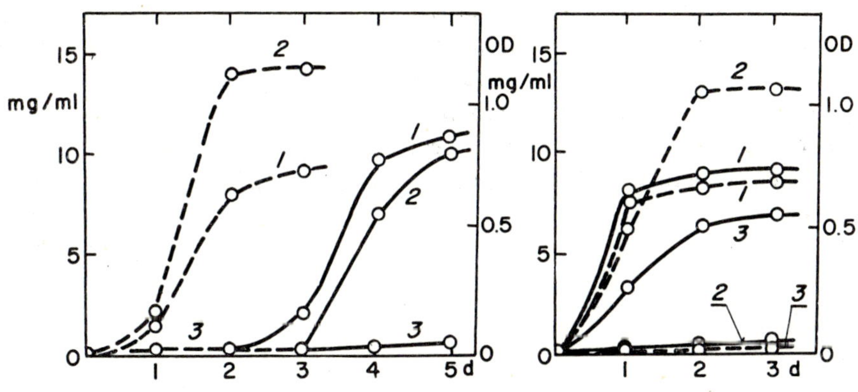

Fig. 8. Accumulation of L-leucine and L-valine by strains 149 *(solid lines)* and S-11 *(dashed lines)*. Media Vo-1 *(left)* and Vo-1 plus L-isoleucine (1 mg/ml; *right*) were used; *1* growth, *2* L-leucine, *3* L-valine.

late L-leucine in a short culture. Some revertants accumulated L-valine as did strain 130-1 and the others in Table XIII accumulated L-leucine even in the medium supplemented with L-isoleucine. The accumulation by the latter revertants was not affected by the addition of isoleucine precursors. Strain S-11 was selected as the best stable accumulator. In Fig. 8, the growth and the accumulation pattern of S-11 were compared with those of strain 149. Without L-isoleucine, strain S-11 grew faster than strain 149 but more slowly than strain 130-1. This observation, together with the other growth expreriment in Davis-Mingioli minimal medium, indicates that S-11 is a leaky isoleucine revertant resistant to α-aminobutyric acid.

It was of interest why L-leucine accumulation by strain 149 was changed to L-valine accumulation by the addition of L-isoleucine. To answer this question, spontaneous revertants were isolated and tested. Complete revertants accumulated L-valine like strain 130-1 and leaky revertants showed the same cultural pattern as did S-11. It was observed that S-11-type revertants were obtained at a high mutation frequency (over 100-fold as compared with that of the 130-1-type). These results probably indicate that L-leucine accumulation by strain 149 is based on the preferential growth of these S-11-type revertants which are present in the inoculum, and that L-valine accumulation on adding L-isoleucine is caused by the dominant growth of auxotrophic cells. Other leaky revertants accumulating L-leucine were spontaneously obtained from other isoleucine auxotrophs.

We studied then the mechanism of L-leucine accumulation by abu^R, leaky isoleucine revertants. It was found that the abu^R mutation caused a constitutive synthesis of α-IPM synthase and that the partial reversion of isoleucine auxotrophy was accompanied by the desensitization of this enzyme. It was, therefore, concluded that L-leucine accumulation was due to lack of both feedback inhibition and repression.

Recently, we succeeded in establishing L-isoleucine production without addition of any isoleucine precursors, as stated below. Mutants resistant to isoleucine hydroxamate, which was found to be a new isoleucine antagonist, had an altered L-threonine dehydratase insensitive to feedback inhibition, and accumulated L-isoleucine in a medium containing L-threonine[41,42]. Double mutants resistant to isoleucine hydroxamate and α-aminobutyric acid lacked both feedback inhibition and repression, and accumulated large amounts of L-isoleucine in a medium without any precursors[43].

DISCUSSION

As described above, we have succeeded in the production of three branched-chain amino acids by using a wild strain and some regulatory mutants of *S. marcescens*. These results are summarized in Fig. 9. The accumulation of these amino acids has been carried out by wild strains or auxotrophs of several bacteria[1]. It should be noted that all three amino acids were accumulated by abu^R mutants and leaky isoleucine revertants. It may be concluded that the increase in L-isoleucine accumulation from D-threonine, α-aminobutyric acid, or α-oxobutyric acid, and L-valine accumulation depends on genetic derepression of AHA synthase.

The causes for L-leucine accumulation by *abu*^{R}, leaky isoleucine revertants are still unknown. Derepression of AHA synthase may cause the increase in the formation of α-oxo*iso*valerate, a substrate for α-IPM synthase. This is considered as an important factor required for L-leucine accumulation. Some observations described above indicate that *abu*^{R} mutants may be derepressed for the leucine enzymes as well as the isoleucine—valine enzymes. It is, however, unknown whether

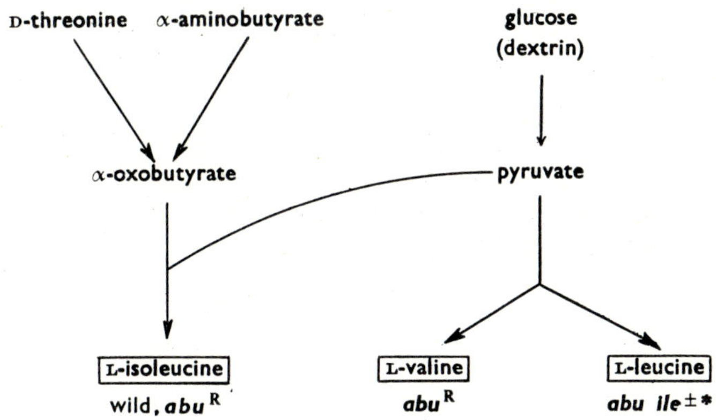

Fig. 9. Accumulation of branched-chain amino acids by *S. marcescens*.

a partial requirement for L-isoleucine is related to the release from either metabolic control, though a close correlation was observed between L-leucine accumulation and the partial requirement for L-isoleucine. Since it was reported that α-oxobutyric acid inhibited α-IPM synthase of *S. typhimurium*[37,38], there was the possibility that a limiting level of intracellular α-oxobutyric acid in leaky isoleucine revertants might relieve the inhibition of α-IPM synthase. However, the addition of α-oxobutyric acid did not prevent L-leucine accumulation. This indicated that a limiting level of α-oxobutyric acid might not play an important role for the accumulation of L-leucine. The investigation of feedback inhibition of α-IPM synthase will lead to the elucidation of mechanisms for L-leucine accumulation.

It is of interest to discuss the genetic aspects of the *abu*^{R} mutation. An *abu*^{R} mutant of *E. coli* K-12 was found to be derepressed for AHA synthase alone[16]. It has been also reported that *val*^{R} mutants were coordinately derepressed for L-threonine dehydratase and two other enzymes[15]. On the other hand, *abu*^{R} mutants of *S. marcescens* were almost coordinately derepressed for L-threonine dehydratase, AHA synthase and transaminase B. These facts suggest that *S. marcescens* differs from *E. coli* K-12 either in the operators controlling structural genes which govern these enzymes or in the *abu*-resistance locus. In view of the report that *val*- and *abu*-resistance resulted from alteration in operator loci[15,16], it is considered that the *abu*^{R} mutation in *S.*

* A leaky isoleucine revertant resistant to α-aminobutyric acid.

marcescens may be the alteration of a common operator controlling the synthesis of these enzymes. As state above, there is the possibility that *abu*^{R} mutants of *S. marcescens* are derepressed for the leucine enzymes as well. This, together with the observation of CALVO *et al.*[39], does not exclude the possibility that *abu*-resistance is due to the alteration of a regulatory gene which controls the synthesis of all the enzymes forming three branched-chain amino acids. On the other hand, FREUNDLICH and CLARKE reported[34] that inhibition of valyl-tRNA synthetase by α-amino-butyric acid might be one of the reasons for growth inhibition of *E. coli* W. It has been also known that amino-acid-analog-resistant mutants which had altered aminoacyl-tRNA synthetases were derepressed for the enzymes involved in the biosynthesis of amino acids[40]. These informations do not rule out the possibility that *abu*^{R} mutation is the alteration of the structural gene for valyl-tRNA synthetase. Further genetic analysis will be necessary for determining the locus where the mutation causes the *abu*-resistance of *S. marcescens.*

CONCLUSION

This paper presented the application of analog-resistant mutants and prototrophic revertants to the production of branched-chain amino acids. Especially, the use of analog-resistant mutants is considered to be advantageous for the industrial production of amino acids, as well as other useful metabolites, for the reasons described below.

1. The possibility is great that the resistant mutants are released from metabolic control.
2. The resistant mutants may be readily isolated from parent strains.
3. In contrast to amino-acid production by auxotrophs which is influenced by the concentration of amino acids required for the growth, the addition of such supplements is not necessary for the production by resistant mutants.
4. The preservation of resistant mutants is easy because back-mutation is prevented by the addition of corresponding analogs into the medium.

Moreover, the introduction of genetic knowledge and techniques, such as recombination, transformation and transduction, into the improvement of strains will contribute to the development of the industrial production of amino acids and other metabolites, together with the progress of the studies on regulatory mechanisms. Conversely, research of the production is expected to provide new knowledge of regulatory mechanisms and to contribute to genetics and biochemistry.

We are grateful to Mr. T. Takayanagi, Manager of Development Division, and Dr. K. Fujii, former director of this laboratory for their encouragement in the course of this investigation.

REFERENCES

1. HUANG H. T.: Progr. Ind. Microbiol. **5**, 55 (1964).
2. TRUFFA-BACHI P., COHEN G. N.: Ann. Rev. Biochem. **37**, 79 (1968).
3. UMBARGER H. E.: Ann. Rev. Biochem. **38**, 323 (1969).
4. FOWDEN L., LEWIS D., TRISTRAM H.: Adv. Enzymol. **29**, 89 (1969).
5. SHIIO I., NAKAMORI S.: Agric. Biol. Chem. **33**, 1152 (1968).
6. MEISTER A., p. 729 in *Biochemistry of the Amino Acids*. Academic Press, New York 1965.
7. UMBARGER H. E.: Science **123**, 848 (1956).
8. FREUNDLICH M., UMBARGER H. E.: Cold Spring Harbor Symp. Quant. Biol. **28**, 505 (1963).
9. UMBARGER H. E., BROWN B.: J. Biol. Chem. **233**, 1156 (1958).
10. BAUERLE R. H., FREUNDLICH M., STØRMER F. C., UMBARGER H. E.: Biochim. Biophys. Acta **92**, 142 (1964).
11. JUNGWIRTH C., UMBARGER H. E.: Fed. Proc. **21**, 10 (1962).
12. FREUNDLICH M., BURNS R. O., UMBARGER H. E.: Proc. Nat. Acad. Sci. U.S.A. **48**, 1804 (1962).
13. LEAVITT R. I., UMBARGER H. E.: J. Biol. Chem. **236**, 2486 (1961).
14. CHANGEUX J.-P.: Cold Spring Harbor Symp. Quant. Biol. **28**, 497 (1963).
15. RAMAKRISHNAN T., ADELBERG E. A.: J. Bacteriol. **87**, 566 (1964).
16. RAMAKRISHNAN T., ADELBERG E. A.: J. Bacteriol. **89**, 661 (1965).
17. BURNS R. O., CALVO J., MARGOLIN P., UMBARGER H. E.: J. Bacteriol. **91**, 1570 (1966).
18. CALVO R. A., CALVO J. M.: Science **156**, 1107 (1967).
19. KISUMI M.: J. Biochem. **52**, 390 (1962).
20. KISUMI M., ASHIKAGA Y., KATO J., CHIBATA I.: J. Biochem. **52**, 400 (1962).
21. KISUMI M., KATO J., CHIBATA I.: J. Biochem. **56**, 450 (1964).
22. KISUMI M., KOMATSUBARA S., CHIBATA I.: J. Bacteriol. **106**, 493 (1971).
23. KISUMI M., KATO J., KOMATSUBARA S., CHIBATA I.: Appl. Microbiol. **21**, 569 (1971).
24. LEAVITT R. I., UMBARGER H. E.: J. Bacteriol. **83**, 624 (1962).
25. UMBARGER H. E., BROWN B.: J. Bacteriol. **70**, 241 (1955).
26. KARLSTRÖM O.: Biotechnol. Bioeng. **7**, 245 (1965).
27. KISUMI M., KOMATSUBARA S., CHIBATA I.: J. Bacteriol. **107**, 824 (1971).
28. KISUMI M., KOMATSUBARA S., CHIBATA I.: Paper at 41st Ann. Meet. Genet. Soc. Jap.; Jap. J. Genet. **44**, 394 (1969).
29. UMBARGER H. E., FREUNDLICH M.: Biochem. Biophys. Res. Comm. **18**, 889 (1965).
30. FREUNDLICH M., UMBARGER H. E.: Bacteriol. Proc. 126 (1963).
31. RAMAKRISHNAN T., ADELBERG E. A.: J. Bacteriol. **89**, 654 (1965).
32. KISUMI M., KOMATSUBARA S., CHIBATA I.: Amino Acid Nucleic Acid **20**, 78 (1969).
33. FREUNDLICH M.: Science **157**, 823 (1967).
34. FREUNDLICH M., CLARKE L. P.: Biochim. Biophys. Acta **170**, 271 (1968).
35. KISUMI M., KOMATSUBARA S., CHIBATA I.: Paper at 17th Symp. Amino Acid, 1969; *to be published.*
36. KISUMI M., FUKUSHIMA H., KOMATSUBARA S., CHIBATA I.: Paper at Ann. Meet. Soc. Ferment. Technol., 1969; *to be published.*
37. KOHLHAW G. B., UMBARGER H. E.: Fed. Proc. **24**, 356 (1965).
38. KOHLHAW G., LEARY T. R., UMBARGER H. E.: J. Biol. Chem. **244**, 2218 (1969).
39. CALVO J. M., FREUNDLICH M., UMBARGER H. E.: J. Bacteriol. **97**, 1272 (1969).
40. NEIDHARDT F. C.: Bacteriol. Rev. **30**, 701 (1966).
41. KISUMI M., KOMATSUBARA S., SUGIURA M., CHIBATA I.: J. Bacteriol. **107**, 741 (1971).
42. KISUMI M. *et al.*: J. Gen. Microbiol. **69**, 291 (1971).
43. KISUMI M. *et al.*: J. Bacteriol. **110**, *in press* (1972).

6/ Cell Walls and Membranes

Substitution of Unusual Amino Acids for Lysine or Diaminopimelic Acid in the Cell-Wall Mucopeptide of Gram-Positive Bacteria

H.R. PERKINS

National Institute for Medical Research, Mill Hill, London, Great Britain

Abbreviations used: UDP-MurNAc, UDP-N-acetylmuramic acid
Dap, diaminopimelic acid

The key precursor of the peptidoglycan (mucopeptide) of bacterial cell walls is the UDP-N-acetylmuramylpentapeptide, the first example of which was isolated from penicillin-treated *Staphylococcus aureus*[1]. The peptide chain has been shown subsequently to consist of L-alanyl—D-*iso*glutamyl—L-lysyl—D-alanyl—D-alanine and all known peptidoglycan precursors are of this pattern. For many years it appeared that the N-terminal amino acid attached to the muramic acid was invariably L-alanine whatever the source of the peptidoglycan[2]. More recently it has been found that this position may be occupied by L-serine, as in *Butyribacterium rettgeri*[3], or glycine, as in *Corynebacterium poinsettiae*[4] and *Microbacterium lacticum*[5]. Probably, however, the exact amino acid that is inserted in a given organism is under strict genetic control, because no evidence of variability has been reported so far in cells grown under normal conditions. Incubation of *S. aureus* Copenhagen in the presence of 0.5 M glycine led to the accumulation of amino-sugar nucleotides, chief among which was UDP-MurNAc. In two minor components, glycine replaced L-alanine, giving rise to UDP-MurNAc—Gly—D-Glu—L-Lys and UDP-MurNAc—Gly—D-Glu—L-Lys—D--Ala—D-Ala[6]. Evidently in extreme circumstances the enzyme that normally attached L-alanine was able to transfer glycine, but at the same time growth of the cells was halted by lack of completed wall synthesis. In general, then, the first amino acid of the peptide chain in peptidoglycan is specific, and most commonly it is L-alanine.

$$\text{(1) UDP-MurNAc} + \text{L-alanine} \xrightarrow[\text{Mn}^{2+}]{\text{ATP}} \text{UDP-MurNAc—L-Ala}$$

$$\text{(2) UDP-MurNAc—L-Ala} + \text{D-glutamic acid} \xrightarrow[\text{Mn}^{2+}]{\text{ATP}} \text{UDP-MurNAc—L-Ala—D-Glu}$$

$$\text{(3) UDP-MurNAc—L-Ala—D-Glu} + \text{L-lysine} \xrightarrow[\text{Mn}^{2+}]{\text{ATP}} \text{UDP-MurNAc—L-Ala—D-Glu—L-Lys}$$

$$\text{(4) UDP-MurNAc—L-Ala—D-Glu—L-Lys} + \text{D-Ala—D-Ala} \longrightarrow \text{UDP-MurNAc—L-Ala—D-Glu—L-Lys—D-Ala—D-Ala}$$

SCHEME I. Biosynthesis of the peptidoglycan precursor of *S. aureus*[8,9].

The biosynthesis of the pentapeptide takes place by sequential addition of the first three amino acids, followed by a dipeptide of D-alanine[7-9] (Scheme I). In the limited number of species so far examined, each of the enzymes has been specific for its relevant substrate, that is, for the next lower member of the peptide series extending from the carboxyl group of UDP-MurNAc. The enzymes adding respectively L-alanine, D-glutamic acid and L-lysine were prepared and partially purified from sonically treated *S. aureus* harvested during exponential growth[8,9]. A related enzyme adding *meso*-diaminopimelic acid instead of lysine was prepared from *Corynebacterium xerosis*[7] and the final enzyme adding D-alanyl–D-alanine has been prepared from *Escherichia coli* K235[7,10] and *Streptococcus faecalis* R[11]. The serial addition of amino acids by separate enzymes recalls the serial addition of sugars in the formation of the O-antigens (lipopolysaccharides) of enteric bacteria. The genetic control of the latter process is fairly well understood[12], but much of it has been elucidated because of the isolation of mutants in which some key enzyme of carbohydrate metabolism is missing, leading to production of a lipopolysaccharide devoid of the part of its polysaccharide on the nonreducing side of the affected sugar residue. It is unlikely that the genetic control of peptidoglycan synthesis can be worked out by a similar method, because any interference with the addition of amino acids must inevitably eliminate the last added, namely the D-alanyl—D-alanine dipeptide. This particular sequence is the key to the vital process of peptidoglycan cross-linking[13] and anything that prevents its addition, such as the antibiotic D-cycloserine, leads to cessation of wall replication and inhibition of bacterial growth.

A preliminary approach to the genetics of peptidoglycan synthesis in *E. coli* K-12 has been made[14]. A strain of *E. coli* derived from K-12, bearing several auxotrophic markers, was treated with N-methyl-N′-nitro-N-nitrosoguanidine and mutants able to grow at 30 °C but not at 42 °C were selected. From these strains were chosen those which, when grown in liquid culture at 30 °C, lysed on being warmed to 42 °C. Chloramphenicol-resistant incorporation of ^{14}C-alanine (into peptidoglycan) was inhibited at the higher temperature and in some strains labelled nucleotides accumulated, either UDP-MurNAc-pentapeptide or the same molecule lacking the C-terminal di- or tripeptide. In the strain accumulating UDP-MurNAc—L-Ala—D-Glu, synthesis of *meso*-diaminopimelic acid was impaired at 42 °C, and addition of diaminopimelic acid to the medium completely restored growth. Similarly, growth of the strain that accumulated UDP-MurNAc-tripeptide was restored by a high concentration of added D-alanine. Four of the ts-genes are mapped.

The D-alanyl—D-alanine synthetase itself is not in fact specific for this particular D-amino acid. If it is supplied with D-serine, or D-*allo*threonine or D-α-aminobutyric acid as well as D-alanine, then mixed dipeptides in which alanine is N-terminal are also formed[15]. The dipeptide of D-α-aminobutyric acid was also produced to some extent. There is no evidence, however, that in the living cell these analogue dipeptides could be incorporated into the peptidoglycan or its nucleotide precursor, so that the incorporation of D-alanyl—D-alanine is probably effectively specific, in spite of some nonspecificity of one of the enzyme-binding sites.

The second residue in the primary peptide chain is D-glutamic acid, linked by its γ-carboxyl group to the next amino acid. So far no peptidoglycan precursor has been shown to contain any other amino acid in this position, although in the completed peptidoglycan of *M. lacticum* grown in a plentiful oxygen supply the D-glutamic acid residue became converted to *threo*-3-hydroxy-D-glutamic acid[5].

The greatest variability of composition occurs in the third position. Until 1964 it was thought that all peptidoglycans contained L-lysine, *meso*-diaminopimelic acid or LL-diaminopimelic acid, although there were some reports of the occasional occurrence of DD-diaminopimelic acid[2,13,16]. Then it was discovered that these amino acids could be entirely lacking, peptidoglycans from phytopathogenic corynebacteria containing 2,4-diaminobutyric acid or D-ornithine[17], while *Micrococcus radiodurans* contained L-ornithine[18]. At first it seemed a natural assumption that these diamino acids had replaced lysine or diaminopimelic acid in the peptidoglycan, but in fact this proved to be true only for some species. It transpired that the UDP-MurNAc--pentapeptide precursor accumulating in *C. poinsettiae* inhibited by vancomycin contained no diamino acid, the molar proportions being alanine, 2; glycine, 1; glutamic acid, 1 and homoserine, 1[19,20]. The peptidoglycan, however, also contained D-ornithine and a structural study led to the formulation of a new type of cross-linking, in which the diamino acid was joined by its ω-amino group to the α-carboxyl group of D-glutamic acid in one primary chain and by its α-amino group to the carboxyl group of the D-alanine residue in fourth position of another chain[4]. This type of cross-linking has been found in other organisms such as *B. rettgeri*[3] and the resulting peptidoglycans have been classified by GHUYSEN as chemotype IV[21].

As a result of the extensive study of peptidoglycans from many species, notably by GHUYSEN and KANDLER and their colleagues, it has become clear that the third position in the primary peptide chains can be filled by an L-amino acid (lysine, ornithine, 2,4-diaminobutyric acid, homoserine) or the L-centre of a diamino dicarboxylic acid (*meso*- or LL-diaminopimelic acid)[22]. In multifunctional amino acids both the amino group and the carboxyl group linked in the primary chain belong to the same centre of L-configuration[23-26].

Hitherto the configuration of the homoserine found in some peptidoglycans has remained unknown, because on acid hydrolysis homoserine readily lactonizes and becomes racemized. It has now proved possible[27] to establish the configuration by the following route. The UDP-MurNAc—Gly—D-Glu—homoserine—D-Ala—-D-Ala from *C. poinsettiae* was brought to pH 8.3 and oxidized by oxygen gas in the presence of reduced platinum catalyst. By this method the homoserine was converted to aspartic acid. After hydrolysis, the aspartic acid was recovered from a paper electropherogram and converted to its trinitrophenyl (TNP) derivative, which was purified by TLC. The eluted TNP-aspartic acid was dissolved in 1% $NaHCO_3$ and its optical rotatory dispersion (ORD) was compared with the curves given by authentic TNP-L- and TNP-D-aspartic acid. The substance derived from the bacterial nucleotide corresponded to the L-compound and hence the homoserine in the peptidoglycan precursor was also of the L-configuration.

Another variation in the amino acid in the third position occurs in the *Actinomycetales*. A slow-moving component was originally observed on chromatograms of hydrolysates in 1957[28] and a similar substance was later found in many strains, particularly those of the *Actinoplanaceae*[29,30]. A sample of this component was isolated from the cell walls of *Ampullariella regularis* and shown by comparison with synthetic material[31] to be 2,6-diamino-3-hydroxypimelic acid, isomer B[32]. The proportion of the hydroxy acid relative to diaminopimelic acid varied from species to species and from strain to strain[33,34]. In six species studied in detail the predominant hydroxydiaminopimelic acid found was always isomer B, although some of it had been converted to isomer C during the acid hydrolysis[33]. Diaminopimelic acid always occurred both as its *meso*- and LL-isomer, the latter being in smaller amount[33,34]. Isomer B of hydroxydiaminopimelic acid has the *threo* configuration for the vicinal hydroxyl and amino groups and the two amino groups are of opposite configuration (*meso*)[31]. By the successive action of periodate and borohydride it proved possible to show that the isomer B occurring in the peptidoglycan of *A. regularis* was hydroxylated on the carbon atom next to the end with the D-α-amino-acid configuration. Furthermore, the L-centre was shown to be the one involved in the primary chain[33].

The simultaneous presence in the *Actinoplanaceae* peptidoglycans of *meso*- and LL-diaminopimelic acid on the one hand, and 3-hydroxyderivatives of the same compounds on the other, suggested that the hydroxylation might occur at some stage after the formation of the UDP-MurNAc-pentapeptide. Although some attempt has been made to relate cell-wall composition and taxonomy in *Actinoplanaceae*[34] it is probably premature to seek reliable information until the influence of oxygen supply upon the proportion of hydroxydiaminopimelic acid in the peptidoglycan has been more thoroughly studied. Similarly, if the hydroxylation occurs as a separate step, then the precise stage of growth may well influence the results. Once it is established which species have the capability to produce peptidoglycan containing hydroxydiaminopimelic acid, it will perhaps become possible to relate this to the phenotypic expression of a specific enzyme.

Hydroxylation of a proportion of the lysine in the peptidoglycan of *Streptococcus pyogenes* Group A, Type 14, has also been observed[35]. Again, although about 10% of the total diamino acid was present as hydroxylysine, it is not known whether this was an invariable proportion, nor whether the hydroxylation took place at an early or late stage in biosynthesis. In *S. faecalis* deprived of lysine, wall synthesis can continue by utilizing exogenous hydroxylysine[36]. Here, evidently, the hydroxyamino acid could pass through the whole biosynthetic process to the completed, cross-linked peptidoglycan. The same was apparently true of an *E. coli* auxotroph requiring diaminopimelic acid for growth, in which hydroxydiaminopimelic acid, isomer D, would serve as a substitute. Here, too, the walls of the organisms grown in this way contained hydroxydiaminopimelic acid in place of the *meso*-diaminopimelic acid normally present[37]. It is not known which isomer occurred in the final wall.

An unusual type of substitution of the diamino acid occurs in *Bacillus sphaericus*, of which the vegetative cells have lysine in their cell walls, whereas the spore coats contain *meso*-diaminopimelic acid[38-40]. Apparently, therefore, this organism in different growth phases can introduce either lysine or diaminopimelic acid into its peptidoglycan. At present it is not known how this change is related to the rest of the metabolic turmoil that accompanies sporulation.

Substitution of one diamino acid for another in the precursor nucleotide has been shown to have a profound effect on peptidoglycan synthesis in a cell-free system from *E. coli* B[41]. When UDP-MurNAc—L-Ala—^{14}C-D-Glu—*meso*-Dap—D-Ala—D-Ala was replaced by UDP-MurNAc—L-Ala—^{14}C-D-Glu—L-Lys—D-Ala—D-Ala, the radioactivity incorporated into peptidoglycan fell by 57%. Nevertheless, such a proportion of incorporation would lead to peptidoglycan of mixed composition unless the specificity were exercised at another stage. It seems that, in *E. coli* at least, the specificity of the enzyme adding *meso*-diaminopimelic acid to the growing peptide chain of the nucleotide precursor controls the diamino acid composition of the final peptidoglycan. This in turn implies that genetic control must be exercised by influencing the production and specificity of this enzyme.

The phytopathogenic organisms at present included in the genus *Corynebacterium* form a fairly closely related group, where differences in amino acid residue 3 of the primary peptidoglycan chain occur. Taxonomic studies on the Adansonian principle based on characters not including cell-wall composition placed them in a cluster of high similarity[10,42]. Of the species included in this group by DA SILVA and HOLT[42], *C. flaccumfaciens*, *C. poinsettiae* are known to have homoserine in their peptidoglycan[20], *C. tritici*, *C. insidiosum* and *C. sepedonicum* have 2,4-diaminobutyric acid[17,43] and although *M. lacticum* has lysine, its cross-linking is of the same general type as in *C. poinsettiae*, involving a peptide bridge between the α-carboxyl group of the D-glutamic acid and the D-alanine at position 4 of another chain[4,5], namely GHUYSEN's Type IV[21]. These organisms suggest a comparatively simple genetic relationship controlling the inclusion of L-homoserine instead of L-2,4-diaminobutyric acid in the primary chain of the peptidoglycan. Since these amino acids are respectively the 4-hydroxy- and 4-amino-derivatives of 2-amino-*n*-butyric acid they are close analogues and their substitution one for another might seem reasonable. At first sight one might think that the 4-amino-compound offered a facility for cross-linking that the 4-hydroxy-compound did not, but in fact, in *C. insidiosum* and *C. sepedonicum* at least, the nucleotide precursor is UDP-MurNAc—Gly—D-Glu—4-N-acetyl-L-diaminobutyryl—D-Ala—D-Ala and cross-linking is mediated by a molecule of D-2,4-diaminobutyric acid[43]. These plant pathogenic organisms seem to offer an excellent prospect for studying the biological control over the nature of the amino acid introduced at position 3 in the primary peptide chain of peptidoglycan.

REFERENCES

1. Park J. T.: J. Biol. Chem. **194**, 897 (1952).
2. Salton M. R. J.: *The Bacterial Cell Wall.* Elsevier, Amsterdam 1964.
3. Guinand M., Ghuysen J.-M., Schleifer K. H., Kandler O.: Biochemistry **8**, 200 (1969).
4. Perkins H. R.: Biochem. J. **102**, 29C (1967).
5. Schleifer K. H., Plapp R., Kandler O.: Biochim. Biophys. Acta **154**, 573 (1968).
6. Strominger J. L., Birge C. H.: J. Bacteriol. **89**, 1124 (1965).
7. Ito E., Nathenson S. G., Dietzler D. N., Anderson J. S., Strominger J. L.: Methods Enzymol. **8**, 324 (1966).
8. Ito E., Strominger J. L.: J. Biol. Chem. **237**, 2689, 2696 (1962).
9. Ito E., Strominger J. L.: J. Biol. Chem. **239**, 210 (1964).
10. Harrington B. J.: J. Gen. Microbiol. **45**, 31 (1965).
11. Neuhaus F. C.: J. Biol. Chem. **237**, 778 (1962).
12. Lüderitz O., Staub A. M., Westphal O.: Bacteriol. Rev. **30**, 192 (1966).
13. Rogers H. J., Perkins H. R., p. 285 in: *Cell Walls and Membranes.* Spon, London 1968.
14. Matsusawa H., Matsuhashi M., Oka A., Sugino Y.: Biochem. Biophys. Res. Comm. **36**, 682 (1969).
15. Neuhaus F. C.: J. Biol. Chem. **237**, 3128 (1962).
16. Work E.: Nature **179**, 841 (1957).
17. Perkins H. R., Cummins C. S.: Nature **201**, 1105 (1964).
18. Work E.: Nature **201**, 1107 (1964).
19. Chatterjee A. N., Perkins H. R.: Biochem. J. **100**, 32P (1966).
20. Perkins H. R.: Biochem. J. **97**, 3C (1965).
21. Ghuysen J.-M.: Bacteriol. Rev. **32**, 425 (1968).
22. Perkins H. R.: Adv. Pharmacol. Chemotherapy **7**, 283 (1969).
23. Bricas E., Ghuysen J.-M., Dezélée P.: Biochemistry **6**, 2598 (1967).
24. Dezélée P., Bricas E.: Biochemistry **9**, 823 (1970).
25. Diringer H., Jusic D.: Z. Naturforsch. **21b**, 603 (1966).
26. van Heijenoort J., Elbaz L., Dezélée F., Petit J.-F., Bricas E., Ghuysen J.-M.: Biochemistry **8**, 207 (1969).
27. Perkins H. R : Biochem. J. **121**, 417, (1971).
28. Hoare D. S., Work E.: Biochem. J. **65**, 441 (1957).
29. Becker B., Lechevalier M. D., Lechevalier H. A.: Appl. Microbiol. **13**, 236 (1965).
30. Yamaguchi T.: J. Bacteriol. **89**, 444 (1965).
31. Stewart J. M.: J. Amer. Chem. Soc. **83**, 435 (1961).
32. Perkins H. R.: Nature **208**, 872 (1965).
33. Perkins H. R.: Biochem. J. **115**, 797 (1969).
34. Szaniszlo P. J., Gooder H.: J. Bacteriol. **94**, 2037 (1967).
35. Muñoz E., Ghuysen J.-M., Leyh-Bouille M., Petit J.-F., Heymann H., Bricas E., Lefrancier P.: Biochemistry **5**, 3748 (1966).
36. Shockman G. D., Thompson J. S., Conover M. J.: J. Bacteriol. **90**, 575 (1965).
37. Sundharadas G., Gilvarg C.: J. Biol. Chem. **241**, 3276 (1966).
38. Hungerer K. D., Tipper D. J.: Biochemistry **8**, 3577 (1969).
39. Powell J. F., Strange R. E.: Biochem. J. **65**, 700 (1957).
40. Tipper D. J.: Bacteriol. Proc. **69**, 24 (1969).
41. Araki Y., Shirai R., Shimada A., Ishimoto N., Ito E.: Biochem. Biophys. Res. Comm. **23**, 466 (1966).
42. da Silva G. A., Holt J. G.: J. Bacteriol. **90**, 921 (1965).
43. Perkins H. R.: Biochem. J. **110**, 47P (1968).

Bacterial Peptidoglycans in Relation to the Membrane and the Mechanism of Action of Penicillin

Jean-Marie Ghuysen, Mélina Leyh-Bouille, Jacques Coyette

Service de Microbiologie, Université de Liège, Belgium

and

Harold R. Perkins, Manuel Nieto

National Institute for Medical Research, Mill Hill, Great Britain

This paper is dedicated to Professor Maurice Welsch on his 60th birthday in acknowledgement of his pioneering work* on the bacteriolytic enzymes from *Streptomyces* sp. and of his continuing stimulating support to the research group of the University of Liège.

INTRODUCTION

The framework of the bacterial envelope is a peptidoglycan network (the terms mucopeptide, glycopeptide or murein, used by some authors, are all synonymous with peptidoglycan) composed of glycan strands that are interconnected through peptide chains (Fig. 1). Study of the topology of the layers within the bacterial envelope[1] has revealed that the peptidoglycan is sandwiched between the underlying plasma membrane and one or several superimposed outermost layers. In Gram-positive bacteria, at least, the peptidoglycan contributes to the cell-surface properties by providing the chemical groups to which the superficial polymers are covalently bound. In *Staphylococcus aureus*, for example, the N-acetylglucosamine residues of the teichoic acid are known to be essential for phage fixation but, in order to be operative, these groups must possess a definite orientation that is imparted by the binding of the teichoic acid to the supporting peptidoglycan[2]. The main function of the peptidoglycan, however, is the preservation of the integrity of the cell within its fragile cytoplasmic membrane which contains high osmotic pressure. In this respect, the obvious advantage of a net-like structure (Fig. 1) is that its mechanical strength, or perhaps better its elastic restraining structure[3], is not impaired by the hydrolysis of a limited number of linkages. The safe enlargement of the net during cell expansion and division can thus result from a strict coordination between the creation in the peptidoglycan of new receptor sites by hydrolysis and the insertion of newly synthes-

* Welsch M.: Rev. Belge Pathol. Méd. Exp. **28**, Suppl. 2, 1 (1947); Welsch M.: Pathol. Microbiol. **29**, 571 (1966).

ized building blocks into the gaps thus formed. In this process, the membrane plays roles of prime importance. It provides an important part of the biosynthetic machinery *(see below)* and at the same time, it may be involved in the protection of this machinery, by means of specialized enzymes, against the deleterious effects of some chemicals, such as penicillins, that can occur in the cell environment[4]. As the title suggests, this paper will be devoted to some properties of both the peptidoglycan and the membrane. Our intention is to rationalize these properties in terms of the functions played by each of these layers within the complex bacterial envelope.

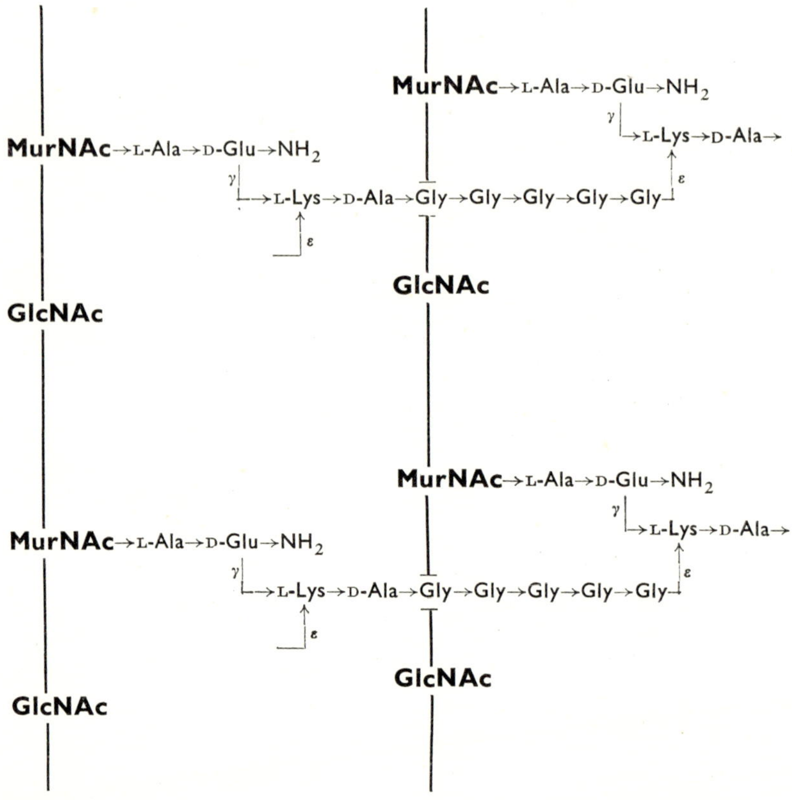

Fig. 1. The peptidoglycan net in *S. aureus* strain Copenhagen. The peptidoglycans are always composed of:

1) *glycan chains* consisting of β-1,4-linked N-acetylglucosamine (**GlcNAc**) and N-acetylmuramic acid (**MurNAc**) (*see* Fig. 2);

2) *tetrapeptide units* (in *S. aureus* L-alanyl-γ-D-*iso*glutaminyl-L-lysyl-D-alanine);

3) *peptide bridges* (in *S. aureus* pentaglycines).

The arrows indicate the CO → NH direction of linkage. α-Peptide linkages are represented by horizontal arrows.

PRIMARY STRUCTURES OF THE PEPTIDOGLYCANS

The structural features of the bacterial peptidoglycans have been recently reviewed[5]. The main purpose of this section is to emphasize the remarkable consistency of structure exhibited by the peptidoglycans both in the glycan and in the peptide moieties.

THE GLYCAN MOIETY

The glycan moiety is essentially a chitin-like structure formed by linear strands of β-1,4-linked N-acetylglucosamine pyranoside residues except that every other sugar is substituted on C_3 by a lactyl group which has the D-configuration (*i.e.* an alternating sequence of N-acetylglucosamine and N-acetylmuramic acid; Fig. 2). For a long time the β-anomery of the linkage between N-acetylmuramic acid and N-acetylglucosamine rested upon the assumption that the *endo*-N-acetylmuramidases which hydrolyze these linkages within the glycan, had a strict β-specificity. Recently[6], the β-anomery has been directly demonstrated, in *S. aureus* at least, by NMR spectroscopy of the N-acetylmuramyl-N-acetylglucosamine disaccharide isolated after ap-

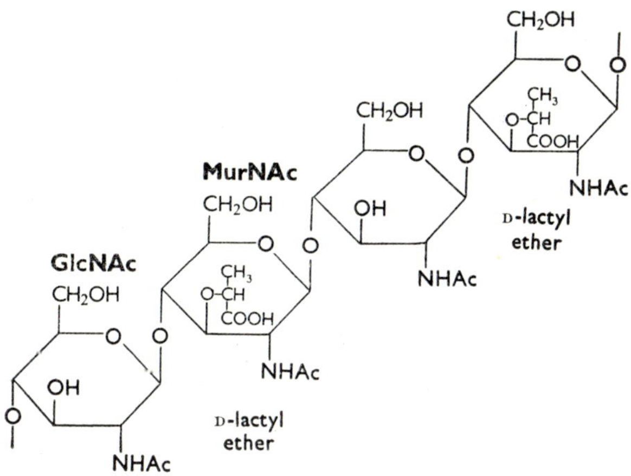

Fig. 2. A portion of a glycan strand. In the intact peptidoglycans, the tetrapeptide units substitute through their N-termini (Fig. 3), the D-lactyl groups of the glycan strands.

propriate enzymatic degradation of the walls. Glucosamine, after acid hydrolysis, has been routinely identified not only by chromatography retention time but also by means of the specific D-glucosamine-6-phosphate N-acetyltransferase. Moreover, a recent survey including a large number of Gram-negative and Gram-positive bacteria from a wide variety of taxonomic groups has revealed that only *gluco*-muramic acid occurs naturally[7,8]. No *galacto*-muramic acid could be identified. It thus appears that the linear arrangement of β-1,4-linked 2-deoxy-D-glucose derivatives confers to the glycan backbone a conformation which is probably essential for the function of the polymer so that any mutation which would alter this conformation would probably be lethal (*galacto*-muramic acid could readily result, in principle, from the action of a 4′-epimerase upon the nucleotide precursor UDP—*gluco*-muramic acid). Alterations on C_2, however, that do not modify the basic conformation of the polymer may possibly occur. The presence of very small amounts of *manno*-muramic acid in walls of *Micrococcus lysodeikticus* has been briefly reported[9].

Alterations in the glycan moiety may affect the nature of the substituent on the C_2-amino group or they may consist of the presence of substituents on the C_6-hydroxyl group of muramic acid. In many Gram-positive bacteria phosphodiester bridges extending from the C_6 of some muramic acids are involved in the binding of the outermost polymers to the wall peptidoglycan matrix[5]. In *S. aureus*, in some strains of *M. lysodeikticus* and in *Lactobacillus acidophilus*[10], 60 to 70% of the muramic acid residues occur as N,6-O-diacetylmuramic acid. In *Mycobacteria* and *Nocardia*, N-acetylmuramic acid is replaced by N-glycolylmuramic acid[11]. In the spore peptidoglycan of *Bacilli*, about half of the muramic acid has been reported to occur in the form of a lactam derivative[12], a compound which must be under the control of a "sporulation-specific" gene since it is unique to sporulation.

Unlike chitin or cellulose, the bacterial glycan is a soluble polymer composed of chains of short length, which is unsuitable for wall function unless its strands are interlinked through peptide chains. Insolubility and mechanical strength are properties solely of the intact peptidoglycan and a loss of integrity resulting from the breakdown of either the glycan or the peptide moieties brings about the solubilization of the whole complex.

THE PEPTIDE MOIETY

The consistency of structure found in the glycan is reflected in the peptide moiety although at first sight this unity is hidden by a wide variation in structural details.

1) The tetrapeptide units which substitute through their N-termini the D-lactic acid of the muramic acid residues in the glycan or at least of some of them, have always the general sequence L-alanyl (or sometimes L-seryl or glycyl)—γ-D-glutamyl——L-R_3—D-alanine (Fig. 3). The glutamyl linkage is always γ. The other linkages are always α. Depending upon the bacterial species, the L-R_3 residue may be L-homoserine, or a diamino acid such as L-diaminobutyric acid, L-ornithine, L-lysine, LL-diaminopimelic acid or *meso*-diaminopimelic acid. In the latter case, both the amino group linked to D-glutamic acid and the carboxyl group linked to D-alanine are located on the same asymmetric carbon, that which has the L-configuration[13].

2) The cross-linking between two tetrapeptides of adjacent glycan chains always involves the C-terminal D-alanine residue of one tetrapeptide and either the amino

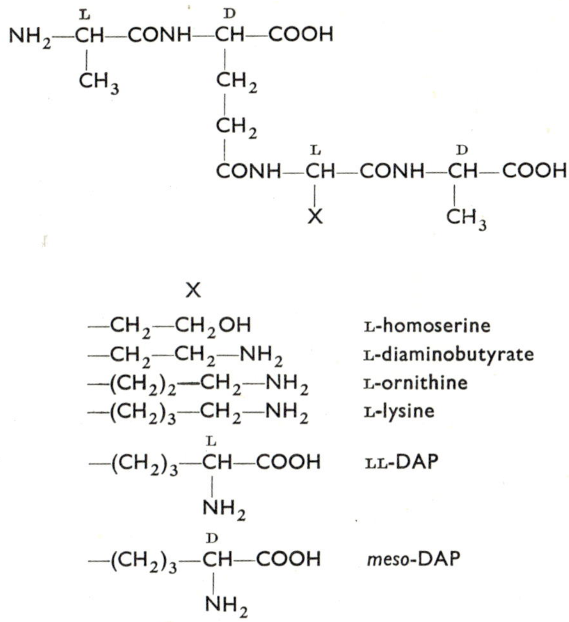

Fig. 3. Structure of the tetrapeptide units. The main variations reside in the structure of the side chain of the L-R_3 residue. Sometimes the N-terminal amino acid is L-serine or glycine (*see* Fig. 7).

group at the end of the side chain of the L-R_3-diamino acid (chemotypes I, II and III; Figs 4—6) or the α-carboxyl group of D-glutamic acid (chemotype IV; Fig. 7), of another tetrapeptide. In chemotype I, the bridge consists of a direct N^{ω}-(D-alanyl—L-R_3) linkage. In chemotype II, one additional amino acid or an intervening short peptide extends between the two tetrapeptides. In chemotype III, a variation of chemotype II, the bridge is made up from one or several peptides each having the same amino acid sequence as the peptide subunit. In chemotype IV, one additional

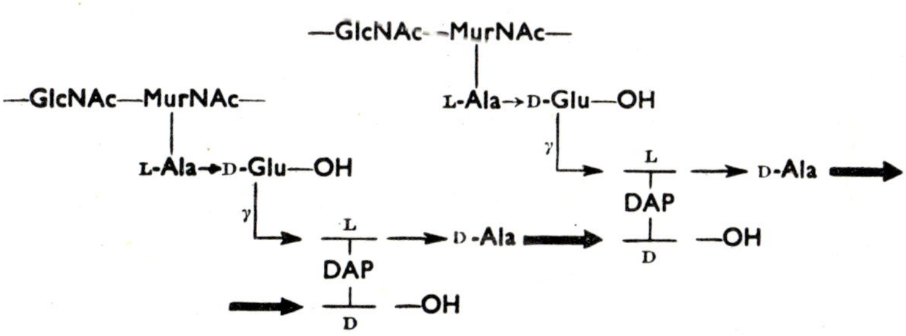

Fig. 4. Peptidoglycan of chemotype I (*E. coli*). Note that the cross-linking bridge is a direct D-alanyl-(D)-*meso*-diaminopimelic acid linkage in a C-terminal position (*heavy arrows*). In other bacteria (*i.e. Bacilli*) the COOH groups not engaged in peptide bonding, or some of them, are amidated.

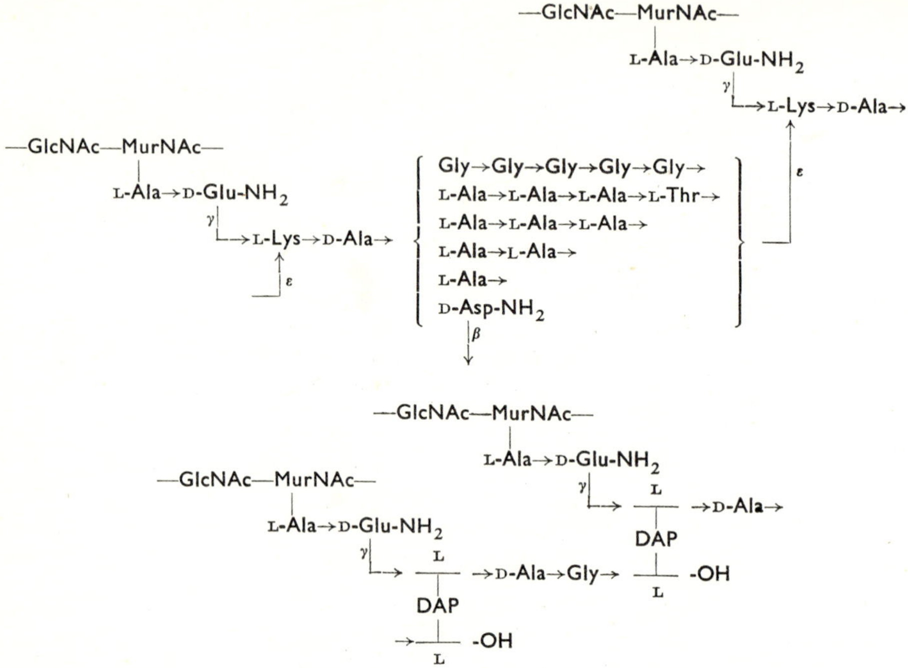

Fig. 5. Peptidoglycan of chemotype II. *Upper* part, *see* Ref.[5]. The pentaglycine bridge is the one found in *S. aureus* (Fig. 1). *Lower* part, peptidoglycan in *Streptomyces* sp.[14].

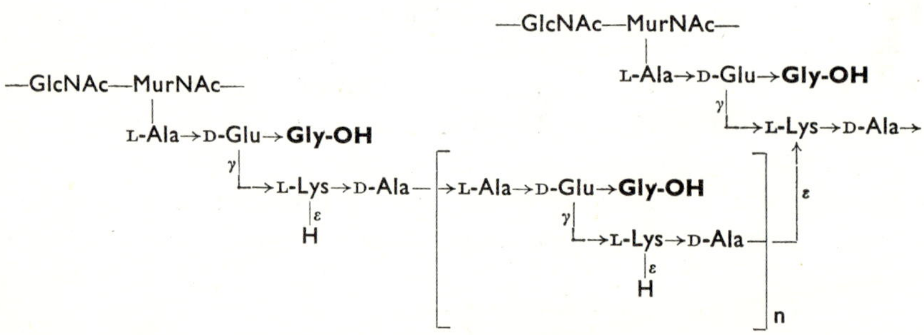

Fig. 6. Peptidoglycan of chemotype III (*M. lysodeikticus* and related *Micrococcaceae*[5,15]). Note that the α-carboxyl group of D-glutamic acid is substituted by a glycine residue (*heavy type*).

diamino acid such as D-diaminobutyric acid, D-ornithine or D-lysine or a short peptide containing a diamino acid, extends from the C-terminal D-alanine to the α-carboxyl group of D-glutamic acid.

3) Depending upon the bacterial species, the α-carboxyl group of D-glutamic acid when not engaged in peptide cross-linking, can be either free or amidated or substituted by an additional amino acid such as glycine. Similarly, the carboxyl groups of the diaminopimelic acid residues which are not engaged in peptide bonding are either free or amidated.

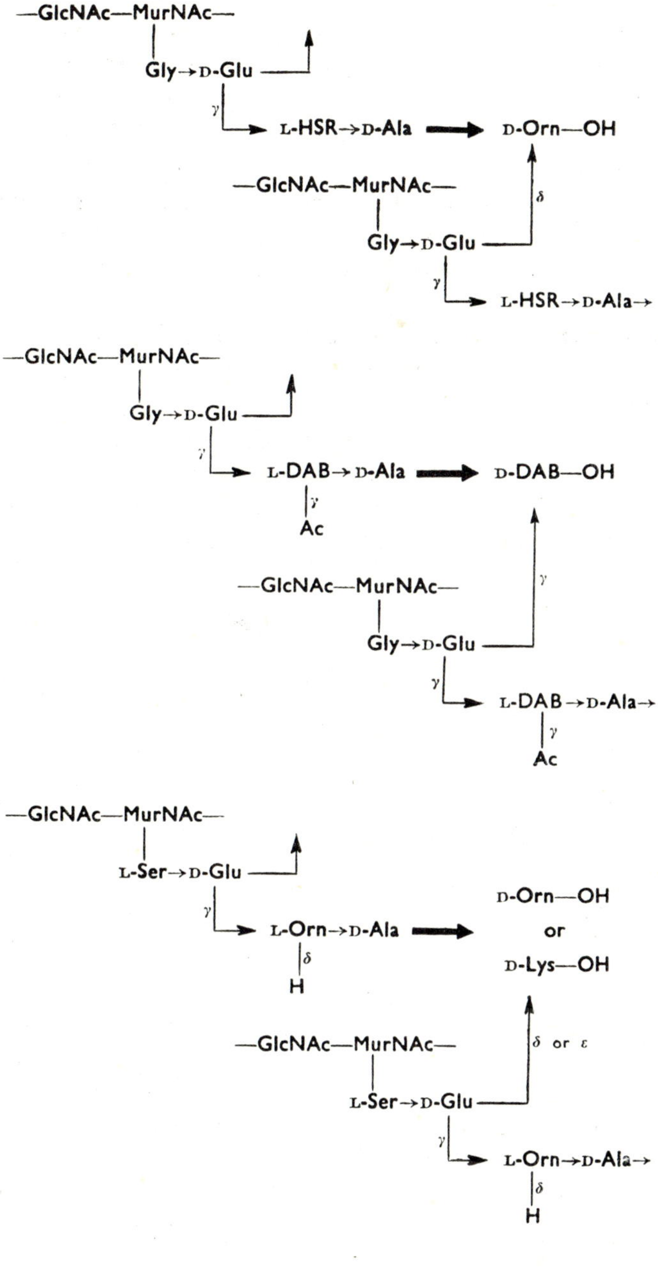

Fig. 7. Peptidoglycan of chemotype IV. From *top* to *bottom:* *Corynebacterium poinsettiae*[16], *Corynebacterium insidiosum*[17] and *Butyribacterium rettgeri*[18]. Note that the cross-linking bridges are mediated *via* D-alanyl-D-linkages in C-terminal position (*heavy arrows*).

THE NET

The tightness of the peptidoglycan net depends upon the length of the glycan strands, the frequency with which the glycan strands are substituted and the frequency with which the tetrapeptides are cross-linked. Depending upon the bacteria, the chain length of the glycan strands averages from 20 to 140 hexosamine residues, the percentage of peptide-substituted muramic acid varies from 50 to 100 and the average size of the peptide moieties is between 2 and 10 cross-linked peptide subunits. Evidently, many terminal groups are present in both the glycan and the peptide moieties of the net. They reflect, at least in part, the dynamics of bacterial growth *(see below)*.

No study has yet been made of the secondary structures of the peptidoglycans. Recently, however, models of each of the four chemotypes described above have been constructed[19]. They suggest that extensive hydrogen bonding giving rise to a regular net-like arrangement is possible only if the polysaccharide chains with the preferred conformation of chitin are run head to tail and if the peptides are extended in the pleated sheet or β-configuration. Of particular interest is the observation that these structures require the D-glutamic acid residue γ-linked to the next amino acid in the sequence of the tetrapeptides.

BIOSYNTHESIS OF THE PEPTIDOGLYCANS

Many important steps of the biosynthesis of the wall peptidoglycan[20] are carried out on or in the close vicinity of the plasma membrane from the two following presynthesized nucleotide precursors: uridine-5'-pyrophosphoryl—N-acetylglucosamine (UDP-GlcNAc) and uridine-5'-pyrophosphoryl—N-acetylmuramyl—L-alanyl (or L-seryl, or glycyl)—γ-D-glutamyl—L-R_3—D-alanyl—D-alanine (UDP-MurNAc-pentapeptide). Note that the peptide moiety of the precursor differs from the tetrapeptide unit in the completed peptidoglycan (Fig. 3) in that the precursor ends in a D-alanyl——D-alanine sequence. The machinery involved is complex and has to carry out: 1) The assembly of the activated precursors into β-1,4-N-acetylglucosaminyl—N-acetylmuramyl peptide units and, when necessary, appropriate alterations of the peptide moiety *(see below)*; 2) The creation of receptor sites in the growing wall peptidoglycan; 3) The transport *via* an intermediate carrier, of the disaccharide-peptide units from the inner surface of the membrane to the receptor sites on the outside of the membrane; 4) The polymerization of the newly inserted peptidoglycan material into an insoluble net-like structure. The carrier involved in the process has been identified as a C_{55}-polyisoprenoid alcohol phosphate. This lipid is clearly relevant to problems of orientation and transport. It is not unique to wall-peptidoglycan synthesis[21].

SYNTHESIS AND TRANSPORT OF DISACCHARIDE–PEPTIDE UNITS

The sequence of the reactions can be visualized as follows[20] (Fig. 8). 1) The MurNAc (pentapeptide)-monophosphate residue is transferred from UDP-MurNAc pentapeptide to the P-C_{55} lipid, resulting in the formation of UMP and of MurNAc-

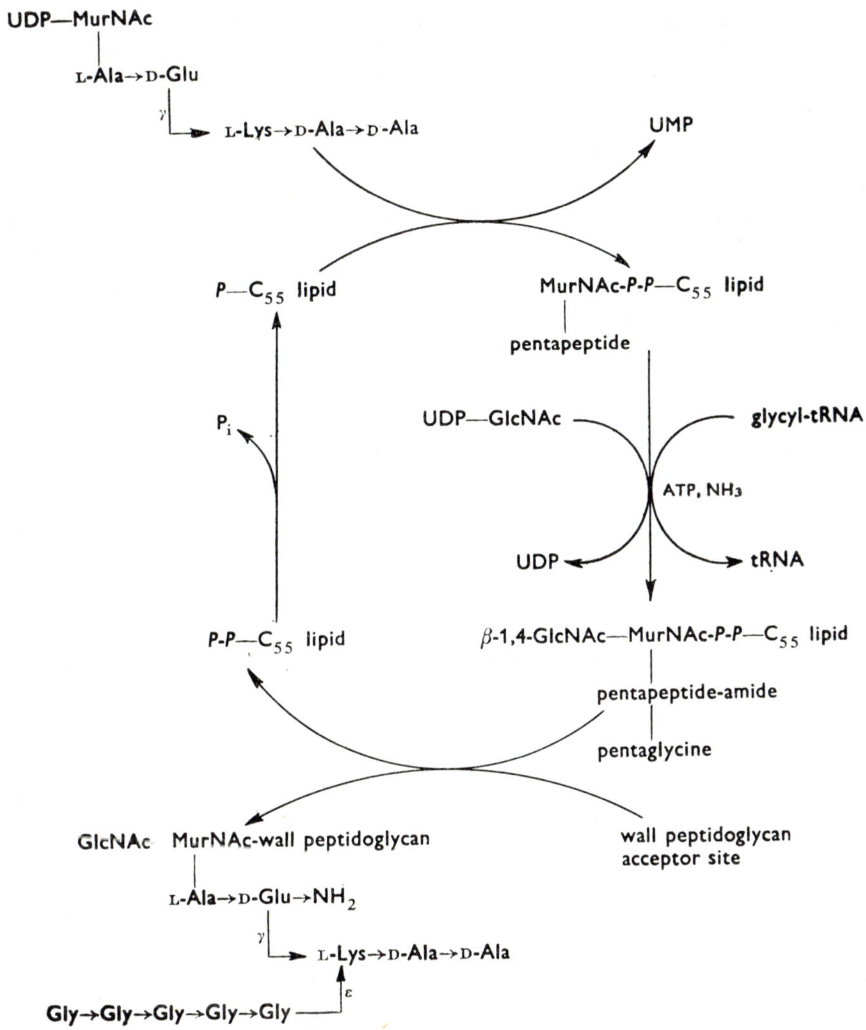

Fig. 8. The lipid cycle in *S. aureus*[20] with formation of disaccharide-peptide units followed by their transport through the plasma membrane and their insertion into the growing wall peptidoglycan (*compare* with Figs 1 and 5). In *E. coli* the same lipid cycle occurs with the exception that D-glutamic acid is not amidated and that no additional amino acid is incorporated into the pentapeptide moiety. Note that at the end of the cyclic reaction, the incorporated disaccharide-peptide units are uncross-linked.

(pentapeptide)-*P*-*P*-C_{55} lipid; 2) GlcNAc is transferred by transglycosylation from UDP-GlcNAc with the liberation of UDP and the formation of β-1,4-GlcNAc-MurNAc(pentapeptide)-*P*-*P*-C_{55} lipid; 3) Often, but not always, the pentapeptides undergo modifications such as the amidation of some carboxyl groups or, more important, such as the incorporation on the side chain of the R_3-residue (chemotypes II and III, Figs 5 and 6) or on the α-carboxyl group of glutamic acid (chemotype IV, Fig. 7) of those amino-acid residues which in the completed peptidoglycan will function as "specialized" peptide cross-linking bridges. Sometimes, but not always, this incorporation is tRNA-dependent and the acceptor is the disaccharide-pentapeptide-lipid intermediate; 4) The disaccharide pentapeptide (that has been modified, if necessary) is finally transferred from the lipid intermediate to the wall-peptidoglycan receptor sites. The liberated *P*-*P*-C_{55} lipid is dephosphorylated so that the *P*-C_{55} carrier can begin in a new cycle. Evidently this cyclic reaction through which pre-synthesized disaccharide-peptide units are transported through the plasma membrane to the extracellular sites of incorporation, is responsible for the alternating sequence of N-acetylglucosamine and N-acetylmuramic acid residues found in the completed wall-glycan strands.

THE WALL-PEPTIDOGLYCAN RECEPTOR SITES

From the above proposed mechanism of disaccharide-peptide transport, the appropriate receptor sites in the expanding peptidoglycan should be nonreducing N-acetylglucosamine termini, *i.e.* termini that result from the action of autolytic *endo*-N-acetylmuramidases which specifically hydrolyze β-1,4-N-acetylmuramyl-N-acetylglucosamine linkages in *endo* position in the glycan strands. Actually, *endo*-N-acetylmuramidases have been identified in many autolytic systems and shown to be involved in wall biosynthesis. In *Streptococcus faecalis*[22-24], the site of action of the *endo*-N-acetylmuramidase autolysin has been localized exactly at the septum formation. Moreover, it has been proved that the wall-peptidoglycan material first to be dissolved through its action is the most newly synthesized part of the peptidoglycan so that the cell equator appears to be the region of new wall synthesis and the region where active autolysin is localized. In *E. coli*[25], inhibition of wall synthesis by penicillin was shown to induce bulge formation on the cell envelope exactly at the sites where normally new cell wall is formed in the process of cell division. The interpretation was that more bonds in the wall peptidoglycan were locally hydrolyzed than could be closed by the inhibited biosynthetic apparatus. With *L. acidophilus* strain 63 AM Gasser[26], the autolytic activity of the cell population progressively disappears during the stationary phase. The walls isolated from both log-phase cells and stationary-phase cells, however, undergo rapid solubilization through the action of a powerful *endo*-N-acetylmuramidase. It is possible that during log phase the enzyme, or at least some of it, is not in the wall, but can be transported to it and that in the transition to stationary phase this ability is rapidly lost. It thus appears that simultaneously with the ability to divide, the cells lose the capability of using the muramidase autolysin which, somehow, is prevented from reaching the wall-peptidoglycan substrate. It may be hypothesized that "*the signal which releases the* endo-*N-acetylmuramidase autolysin is transmitted* via *the plasma membrane to which the replicating DNA and the autolysin would be structurally bound*"[25].

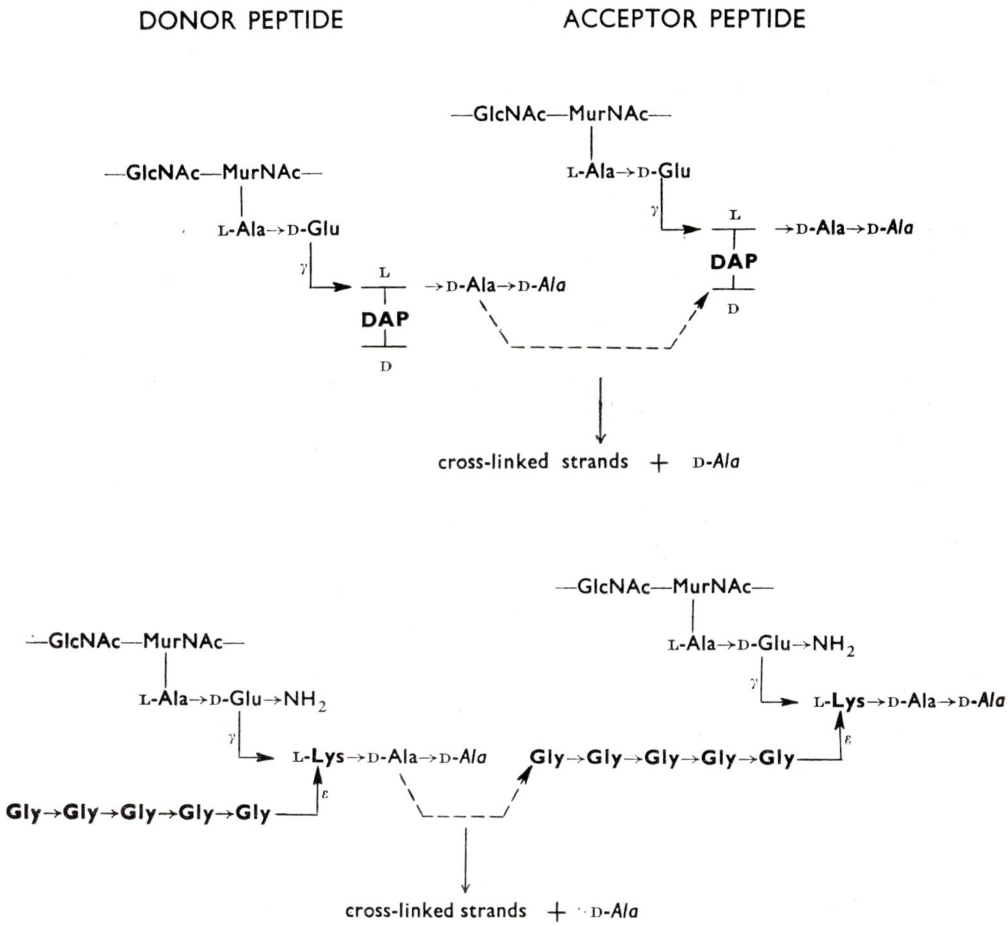

Fig. 9. The transpeptidation reaction with formation of cross-linked peptidoglycan strands and the transformation of the pentapeptide units into tetrapeptides. *Upper* part: transpeptidation in *E. coli*. Note that the residue at the R_3 position is *meso*-diaminopimelic acid (*heavy type*), *i.e.* that found in the nucleotide precursor. *Lower* part: transpeptidation in *S. aureus*. Note that the residue at the R_3 position is N^{ε}-pentaglycyl-L-lysine (*heavy type*).

POLYMERIZATION OF THE NASCENT PEPTIDOGLYCAN

The insertion into the wall peptidoglycan of newly synthesized disaccharide-peptide units, whatsoever its precise mechanism, must be followed by the closure of the peptide bridges if the process is to yield an insoluble network. This last reaction which occurs at sites where ATP is not available, is believed to be introduced by transpeptidation[20]. The mechanism of this reaction is such that the penultimate C-terminal D-alanine residue of a *donor* peptide is transferred to the amino group of an *acceptor* peptide (Fig. 9). Interpeptide bonds are formed and equivalent amounts of D-alanine residues are released from the donor peptides.

REGULATION OF THE SIZE OF THE PEPTIDE MOIETY

Three mechanisms, working alone or in conjunction, may be involved in the regulation of the size of the peptide moiety of the wall peptidoglycans: the efficiency of the transpeptidation reaction, the control of the number of peptides that undergo transpeptidation, the hydrolysis of the peptide cross-linkages in the completed wall peptidoglycan.

1) The transpeptidation reaction does not involve large changes in free energy and probably never reaches completion. The efficiency of the transpeptidation in a given organism should thus be reflected, at least in principle, by the extent of peptide cross-linking in the completed wall. Such relationship is valid only for those bacteria such as *S. aureus* and *L. acidophilus* 63 AM Gasser (Fig. 10) in which the residual not cross-linked C-termini of the wall-peptide moieties have retained the D-alanyl—D-alanine sequence found in the nucleotide precursor, thus demonstrating that peptide hydrolases are not involved in the regulation of the size of the peptides. The extent of peptide cross-linking indicates that the transpeptidation reaction is most efficient in *S. aureus* (80%)[27] whereas it is poorly efficient in *L. acidophilus* (30%)[10].

2) In contrast to *S. aureus* and *L. acidophilus*, the C-termini of the peptides in the wall peptidoglycans of most bacteria have not retained the D-alanyl—D-alanine sequence of the nucleotide precursors. This structural feature indicates, of course, the active presence in these bacteria of D-alanyl—D-alanine carboxypeptidases. Such enzymes were actually isolated[20,28,29] and were shown to hydrolyze specifically the C-terminal D-alanyl—D-alanine linkage in nucleotide precursors UDP-MurNAc-pentapeptide, in β-1,4-N-acetylglucosaminyl—N-acetylmuramyl—pentapeptide units and in a number of related peptides. Evidently, the size of the peptide moiety in these bacteria does not depend solely upon the efficiency of the transpeptidation reaction. It also depends upon the number of peptide units ending in D-alanyl—D-alanine that are allowed to undergo transpeptidation.

3) Finally, the interpeptide bonds in some wall peptidoglycans are mediated through D-alanyl—D-linkages in C-terminal position (Fig. 4 and 7) which, because of this peculiarity, are susceptible to be hydrolyzed by the DD-carboxypeptidases[28,29]. It thus follows that the extent of peptide cross-linking in these bacteria may also be controlled through the "*endo*peptidase" activity of the DD-carboxypeptidases upon the completed wall peptidoglycan.

If the biosynthesis of the bacterial peptidoglycan is understood at the molecular level, our concept of the mode of growth, replication and regulation of the wall at the cellular level is still largely obscure. Further progress depends upon a better

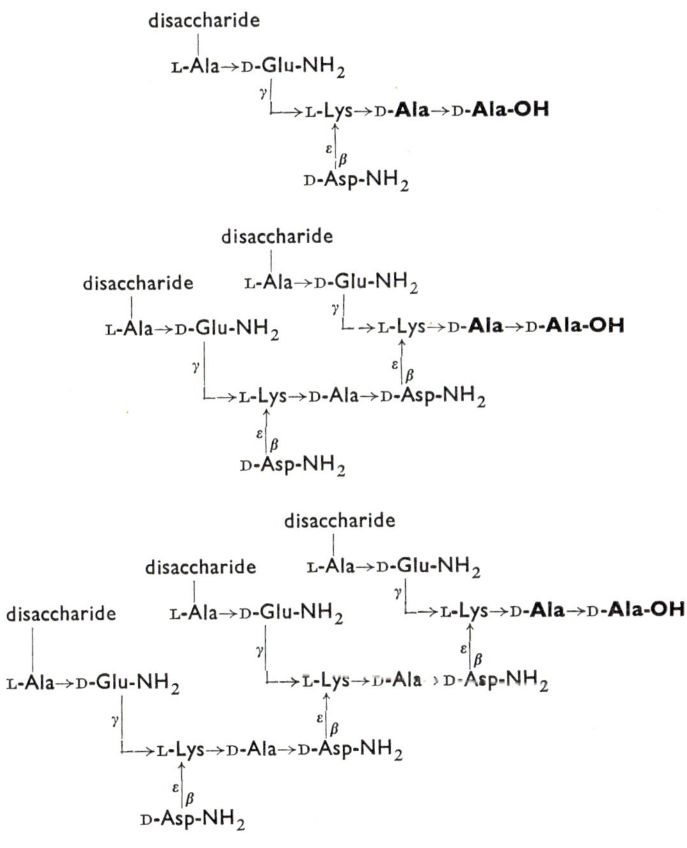

Fig. 10. The peptide moiety in *L. acidophilus* 63 AM Gasser. Disaccharide-peptide monomer (*above*), bis-disaccharide-peptide dimer (*middle*) and tris-disaccharide-peptide trimer (*below*) were isolated after solubilization of the walls (from log-phase cells) through the action of the *endo*-N-acetylmuramidase autolysin. In log-phase walls, 10% of the peptide subunits occur as monomers, 37% as dimers and 30% as trimers. Note that the C-termini of the peptides have retained the D-alanyl-D-alanine sequence found in the nucleotide precursor (*heavy type*). disaccharide = β-1,4-GlcNAc—MurNAc; some of the N-acetylmuramic acid have an additional acetyl group on C_6.

understanding of the precise architecture of the membrane and of the "vectorial" enzymology that is involved in membrane transport. It also depends upon technical improvements in the isolation of the relevant "particulate" enzymes from their sites of attachment in the membrane. Progress in this direction has been reported[21].

MECHANISM OF ACTION OF PENICILLIN*

Penicillin when added at sublethal-dose levels to growing Gram-positive bacteria, was shown to reduce the extent of wall peptide cross-linking, *i.e.*, it decreased the efficiency of the transpeptidation reaction[30]. This conclusion is further supported by the demonstration that penicillin was able to block the transpeptidase activity of a particulate cell-free system prepared from *E. coli*[20]. Among the numerous enzymes involved in peptidoglycan synthesis, the transpeptidase is thus the target of the penicillin molecule. Two mechanisms, *i.e.* penicilloylation and/or competitive inhibition, possibly working in conjunction, may be invoked to explain the molecular basis of penicillin action.

THE PENICILLOYLATION HYPOTHESIS

Analyses of the effect of penicillin on the *E. coli* cell-free system indicated that the transpeptidase was inactivated, rather than inhibited, *via* acylation through the highly reactive CO-N bond in the lactam ring of the penicillin molecule, with the formation of an inactive penicilloyl—protein complex. More recently, a particulate D-alanyl—D-alanine carboxypeptidase from *B. subtilis* (*i.e.* the "uncoupled" transpeptidase; *see below*) was also shown to be inactivated by penicillin in a similar fashion[20]. Further treatment of the penicillin-inhibited DD-carboxypeptidase with hydroxylamine or ethylmercaptan reversed the binding. Penicilloylhydroxamate or ethylthiopenicilloate were formed and, concomitantly, the carboxypeptidase activity was restored[20]. It is not known, however, whether penicilloylation occurs on the enzyme itself or on membrane sites localized in its vicinity. The acylating activity of penicillin is not specific for the membrane-bound transpeptidase or the particulate DD-carboxypeptidase. It is also responsible for the formation of penicilloyl-allergen through acylation of serum protein. Finally, 6-aminopenicillanic acid has no antibacterial activity although its β-lactam ring is intact. It thus appears probable that acylation is at least not the essential part of the mechanism of penicillin action.

---

* For more recent views of the mechanism of action of penicillin from these laboratories, *see* Refs[35,36].

THE STRUCTURAL ANALOGY HYPOTHESIS

Further progress in the elucidation of the exact mechanism of action of penicillin would require the isolation from the membrane of the relevant transpeptidase protein followed by its purification and characterization. The problem is made difficult because it is not known how much of the *in situ* structural integrity must be retained by both the enzyme and the substrate in order to be operative *in vitro*. Another approach[28,29,31] has been recently undertaken which rested upon the observations that DD-carboxypeptidases are spontaneously excreted in the culture media by strains of *Streptomyces* and that there is a remarkable similarity between the substrate requirements of these DD-carboxypeptidases and those of the transpeptidase involved in the peptidoglycan-bridge closure reaction. Studies carried out with small, well defined peptides demonstrated that there is not only a strict requirement for the C-terminal residue of the peptides to have a D-configuration and for the penultimate C-terminal residue to be solely D-alanine, but also that the length, the structure and the polarity of the side-chain of the L-amino acid preceding the C-terminal D-alanyl—D-sequence (*i.e.* the R_3-side-chain) are of prime importance for the enzyme activity (Tables I and II). From the scanty information so far available[20,33], the same observation can be made with the DD-carboxypeptidase solubilized from *E. coli* by physical disruption of the cells. Though the structural features of the R_3-side-chains that fulfill the requirements of the DD-carboxypeptidases vary according to the bacteria (Table I and footnote), it appears that those structures occurring in the wall peptides before cross-linking, as deduced from the structures of the final peptidoglycans, are compatible with the observed specificity of the DD-carboxypeptidases. In other words, peptide cross-linking does not occur unless the R_3-side-chain is such that the C-terminal D-alanine can be removed by the DD-carboxypeptidases. This conclusion gave rise to the idea[29] that the DD-carboxypeptidases are, in fact, the membrane-bound transpeptidases that have undergone solubilization (*i.e.* uncoupled transpeptidases) and that the difference in effective function of the enzyme could be, at least partially, a question of the availability of water. In an aqueous environment, attack of the peptide—enzyme complex, after elimination of the terminal D-alanine, by OH^- would lead to simple hydrolysis. In the hydrophobic environment of the membrane, attack by a recognizable $—NH_2$ would lead to transpeptidation:

R—D-Ala—D-Ala + enzyme (*E*) → R—D-Ala—D-Ala—*E*

R—D-Ala—D-Ala—*E* → R—D-Ala—*E* + D-Ala

R—D-Ala—*E* + OH^- → R—D-AlaOH + *E* (carboxypeptidase activity)

R—D-Ala—*E* + R′-NH_2 → R—D-Ala—NH-R′ + *E* (transpeptidase activity)

TABLE I

Efficiency of Hydrolysis by the DD-Carboxypeptidases from *S. albus* G and from *Streptomyces* R 61[a]

Substrates[b]	*S. albus* G[29,31,c]			*Streptomyces* R 61[32,d]		
	K_m	V	E	K_m	V	E
Ac → L-Ala → γ-D-Glu → L-Lys (↑ε Ac) → D-Ala → D-Ala	1.80	100	56	10	215	21
Ac → L-Lys (↑ε Ac) → D-Ala → D-Ala	0.33	100	300	12	890	72
Ac → L-Lys (↑ε Ac) → D-Ala → D-Lys (↑ε H)	0.80	85	106	13	90	7
Ac → L-Lys (↑ε Ac) → D-Ala → D-Leu	0.33	33	100	10	50	5
Ac → L-Lys (↑ε Ac) → D-Ala → Gly	2.50	60	24	36	200	6
Ac → L-Lys (↑ε Ac) → D-Ala → L-Ala[e]	no hydrolysis			virtually no hydrolysis		
Ac → L-Lys (↑ε Ac) → Gly → D-Ala	15.0	107	7	15.5	1.7	0.1
Ac → L-Lys (↑ε Ac) → D-Leu → D-Ala[e]	no hydrolysis			10	10	1
Ac → L-Lys (↑ε Ac) → L-Ala → D-Ala[e]	no hydrolysis			virtually no hydrolysis		

Ac → L-Lys → D-Ala → D-Ala ↑ε H	6.0	20	3	15	4	0.3
UDP-MurNAc → L-Ala → γ-D-Glu → (L) -*meso*-DAP-(L) → D-Ala → D-Ala	0.4	10	25	11	3	0.3
Nα-(GlcNAc-MurNAc → L-Ala → γ-D-Gln) → Lys → D-Ala → D-Ala ↑ε H-(Gly)5	0.28	9	32	14	800	57

[a] *Upper* and *middle* parts: influence exerted by the C-terminal dipeptide sequence; *lower* part: influence exerted by the preceding residue at the R_3-position. K_m values are expressed in mM; V values in μmol per mg of enzyme per h; efficiency (E) in V/K_m. These values were obtained at 37 °C in 0.01 M Tris-HCl buffer pH 7.5 (R 61 enzyme) or in 0.02 M Tris-HCl buffer pH 7.5 supplemented with 2 mM-$MgCl_2$ (*S. albus* G enzyme).

[b] DAP = diaminopimelic acid, GlcNAc-MurNAc = β-1,4-N-acetyl-glucosaminyl-N-acetylmuramic acid.

[c] Penicillin-resistant.

[d] Penicillin-sensitive.

[e] These peptides are not inhibitors.

Footnote to Table I.

1) Note that in *Streptomyces* sp., the amino group involved in transpeptidation is a glycyl residue which substitutes the LL-diaminopimelic acid at the R_3 position (Fig. 5, *bottom*). In *E. coli*, this amino group is that located on the D-carbon of *meso*-diaminopimelic acid (Fig. 9, *upper part*).

2) The nature of the residue at the N^{α}-terminus of the LDD-tripeptides has no great influence on the activity of the *Streptomyces* enzymes.

3) Both *Streptomyces* enzymes have a strict requirement for a C-terminal D-alanyl—D-linkage. The penicillin-sensitive R 61 enzyme has a more strict requirement for a D-alanine at the C-terminal position.

4) With the penicillin-sensitive *Streptomyces* R 61 enzyme, there is a strict requirement for an amino group on the L-residue (here an N^{ε}-L-lysine group) to be either acylated (here an acetyl group) or transformed into an α-amino group by its substitution by a pentaglycine. With the penicillin-resistant *S. albus* G enzyme, the transformation of the ε-amino group of L-lysine to an α-amino group by introduction of a carboxyl group in α-position (*i.e.* the replacement of L-lysine by diaminopimelic acid) is sufficient to induce a high efficiency of hydrolysis.

5) By comparison with the penicillin-resistant *S. albus* G enzyme, the penicillin-sensitive R 61 enzyme exhibits more strict substrate requirements, has a mediocre fit with all the substrates that were used (high K_m values) but has better groups to promote the hydrolysis of at least some of them (high V values).

6) Note that no activity at all of the *E. coli* carboxypeptidase was demonstrable when UDP-MurNAc-pentapeptide containing L-lysine rather than *meso*-diaminopimelic acid was used[33].

TABLE II
Influence of the Length of the Side-Chain of the Residue that Precedes C-Terminal D-Alanyl—D-alanine on Carboxypeptidase Activity[31,32]

Substrate	Specific activity[a] of enzyme from	
	S. albus G	*Streptomyces* R 61
Ac → L-Ala ⟶ D-Ala → D-Ala	0.3	0.65
Ac → L-DAB[b] → D-Ala → D-Ala ↑γ Ac	22	4
Ac → L-Orn ⟶ D-Ala → D-Ala ↑δ Ac	22	21
Ac → L-Lys ⟶ D-Ala → D-Ala ↑ε Ac	40	47

[a] Expressed in microequivalent of D-Ala → D-Ala linkage hydrolyzed per mg of enzyme per h. Conditions of incubation: 15 nmol of peptide were incubated at 37 °C in the presence of enzyme in a final volume of 35 μl of 0.01 M Tris-HCl buffer pH 7.5 (enzyme from *Streptomyces* R 61) or of 0.02 M Tris-HCl buffer pH 7.5 supplemented with 2 mM-$MgCl_2$ (enzyme from *S. albus* G).
[b] Diaminobutyric acid.

The postulated identity between the DD-carboxypeptidases and the transpeptidases is further supported by the fact that the DD-carboxypeptidase solubilized from *E. coli*[20] is sensitive to penicillin and so, too, is the transpeptidase of this organism. Similarly, the DD-carboxypeptidase excreted by the penicillin-sensitive *Streptomyces* strain R 61 is, also, very sensitive to penicillin[32]. In marked contrast to both the particulate transpeptidase from *E. coli* and the particulate DD-carboxypeptidase from *B. subtilis* which are inactivated by penicillin apparently through acylation, both soluble DD-carboxypeptidases from *E. coli* and from *Streptomyces* R 61 are inhibited by penicillin G in a true competitive fashion ($K_i = 1.6 \times 10^{-8}$ M and 7.5×10^{-8} M, respectively), demonstrating that penicillin G resembles the substrate sufficiently so as to be bound in its stead on the active site. A structural analogy between penicillin and the peptides involved in transpeptidation is thus likely to be relevant to the antibacterial action of penicillin. Both parts of the penicillin molecule, *i.e.* the acyl group and the 6-aminopenicillanic acid moiety, must be involved in the analogy. Indeed, it has been shown[33] that penicilloic acid and 6-aminopenicillanic acid are also competitive for the substrate of the *E. coli* DD-carboxypeptidase but only at concentrations which are 100-fold higher than the concentrations required of penicillin G. Actually, one edge of the penicillin molecule, *i.e.* the 6-aminopenicillanic acid part of it, has a conformation that resembles one of the conformations of the C-terminal D-alanyl—D-alanine backbone[34], *i.e.* the *donor* site in the transpeptidation. A structural analogy with the *acceptor* site involved in transpeptidation, however, is not apparent in the constructed models of the penicillin molecule.

The penicillin-resistant *Streptomyces albus* strain G secretes a DD-carboxypeptidase which is not inhibited by penicillin. The enzyme was isolated and studied[28,29,31]. The K_m and V values were found to be quantitatively different from those exhibited by the R 61 penicillin-sensitive enzyme[32] (Table I and Footnote). Again, however, the values demonstrated that both the C-terminal D-alanyl—D-sequence and the side chain of the preceding L-amino acid in the peptides are essential for the enzyme activity. The isolation of such an enzyme which hydrolyses peptides ending in D-alanyl—D-alanine, but which is not inhibited by penicillin, seems to be at variance with the structural analogy hypothesis or, at least, it proves that this analogy is not universal among bacteria. One explanation is that both the penicillin-sensitive and penicillin-resistant "DD-carboxypeptidase—transpeptidase" systems would recognize one specific configuration in the substrates and that penicillin would be structurally analogous only to the configuration recognized by the penicillin-sensitive enzyme. The molecular basis for such a penicillin resistance that does not involve the enzymatic degradation of the antibiotic, must reside in a peculiar structure of the enzyme that, essentially, does not modify its catalytic activities but prevents it from being attacked by the penicillin molecule*. Evidently, this resistance must result from one or several mutations in the corresponding structural gene. Hence penicillin-sensitive and penicillin-resistant carboxypeptidases could be used as markers in a genetic approach to the exact molecular basis of the antibacterial action of penicillin. Similarly, the comparative biochemistry of isolated and purified DD-carboxypeptidases that differ from each other by their capabilities of recognizing penicillin, should also contribute to the solution of this fascinating problem.

The investigations were supported in part by Grant UR-E4-10-2 from the U.S. Department of Agriculture under Public Law 480 and by Grants N° 515 and N° 1000 from the Fonds de la Recherche Fondamentale Collective, Belgium (to J.-M. G.).

REFERENCES**

1. GLAUERT A. M., THORNLEY M. J.: Ann. Rev. Microbiol. **23**, 159 (1969).
2. COYETTE J., GHUYSEN J.-M.: Biochemistry **7**, 2385 (1968).

---

* The penicillin-sensitive DD-carboxypeptidase from *Streptomyces* R 61 has a poorer fit (higher K_m) than the penicillin-resistant enzyme from *S. albus* G (Table I). Hence, at least in this case, all the difference could be explained as a narrowing of the type of molecules accepted by the active site of the latter so that it will no longer accept penicillin. This explanation, however, may not be universal since the DD-carboxypeptidase from *E. coli* has an excellent fit[33] (low K_m, Table I) for its substrate and, at the same time, it is competitively inhibited by very low concentrations of penicillins.

** A more complete list of references can be found in the Refs[5,20].

3. Li-Tse Ou, Marquis R. E.: J. Bacteriol. **101**, 92 (1970).
4. Coles N. W., Gross R.: J. Bacteriol. **98**, 659 (1969).
5. Ghuysen J.-M.: Bacteriol. Rev. **32**, 425 (1968).
6. Tipper D. J.: Internat. J. System. Bacteriol. **20**, 361 (1970).
7. Wheat R. W., Kulkarni S., Cosmatos A., Scheer E. R., Steele R. S.: J. Biol. Chem. **244**, 4921 (1969).
8. Wheat R. W., Ghuysen J.-M.: J. Bacteriol. **105**, 1219 (1971).
9. Sinay P., Jeanloz R. W., Gross P. H.: 156th Nat. Meet. Am. Chem. Soc. Carbohydrate Chem., 1968.
10. Coyette J., Ghuysen J.-M.: Biochemistry **9**, 2935 (1970).
11 Azuma I., Thomas D. W., Adam A., Ghuysen J.-M., Bonaly R., Petit J. F., Lederer E.: Biochim. Biophys. Acta **208**, 444 (1970).
12. Warth A. D., Strominger J. L.: Proc. Nat. Acad. Sci. U.S.A. **64**, 428 (1969).
13. van Heijenoort J., Elbaz L., Dezélée P., Petit J. F., Bricas E., Ghuysen J.-M.: Biochemistry **8**, 207 (1969).
14. Leyh-Bouille M., Bonaly R., Ghuysen J.-M., Tinelli R., Tipper D.: Biochemistry **9**, 2944 (1970).
15. Campbell J. N., Leyh-Bouille M., Ghuysen J.-M.: Biochemistry **8**, 193 (1969).
16. Perkins H. R.: Biochem. J. **102**, 29c (1967).
17. Perkins H. R.: Biochem. J. **110**, 47p (1969).
18. Guinand M., Ghuysen J.-M., Schleifer K. H., Kandler O.: Biochemistry **8**, 200 (1969).
19. Kelemen M. V., Rogers H.J.: Proc. Nat. Acad. Sci. U.S.A. **68**, 992 (1971).
20. Strominger J. L., p. 187 in T. Bücher, H. Sies (Eds.): *Inhibitors — Tools in Cell Research.* 20th Colloq. Ges. Biol. Chem., Springer-Verlag, Berlin 1969.
21. Ellar D. J., p. 167 in H. P. Charles, B. C. J. G. Knight (Eds.): *Organization and Control in Prokaryotic and Eukaryotic Cells.* 20th Symp. Soc. Gen. Microbiol., 1970. University Press, Cambridge 1970.
22. Shockman G. D., Pooley H. M., Thompson J. S.: J. Bacteriol. **94**, 1525 (1967).
23. Shockman G. D., Martin J. T.: J. Bacteriol. **96**, 1803 (1968).
24. Pooley H. M., Shockman G. D.: J. Bacteriol. **100**, 617 (1969).
25. Schwarz U., Asmus A., Frank H.: J. Mol. Biol. **41**, 419 (1969).
26. Coyette J., Ghuysen J.-M.: Biochemistry **9**, 2952 (1970).
27. Tipper D. J., Berman M. F.: Biochemistry **8**, 2183 (1969).
28. Ghuysen J.-M., Leyh-Bouille M., Bonaly R., Nieto M., Perkins H. R., Schleifer K. H., Kandler O.: Biochemistry **9**, 2955 (1970).
29. Leyh-Bouille M., Ghuysen J.-M., Bonaly R., Nieto M., Perkins H. R., Schleifer K. H., Kandler O.: Biochemistry **9**, 2961 (1970).
30. Tipper D. J., Strominger J. L.: Biochemistry **243**, 3169 (1968).
31. Leyh-Bouille M., Ghuysen J.-M., Nieto M., Perkins H. R., Schleifer K. H., Kandler O.: Biochemistry **9**, 2971 (1970).
32. Leyh-Bouille M., Coyette J., Ghuysen J.-M., Idczak J., Perkins H. R., Nieto M.: Biochemistry **10**, 2163 (1971)
33. Izaki K., Strominger J. L.: J. Biol. Chem. **243**, 3193 (1968).
34. Tipper D. J., Strominger J. L.: Proc. Nat. Acad. Sci. U.S.A. **54**, 1133 (1965).
35. Leyh-Bouille M., Nakel M., Frère J. M., Johnson K., Ghuysen J.-M., Nieto M., Perkins H. R.: Biochemistry **11**, *in press* (1972).
36. Pollock J. J., Ghuysen J.-M., Linder R., Salton M. R. J., Perkins H. R., Nieto M., Leyh-Bouille M., Frère J. M., Johnson K.: Proc. Nat. Acad. Sci. U.S.A., *in press* (1972).

7/ Genetic Control of Sporulation

Early Events in Sporulating *Bacillus subtilis*

ERNST FREESE, KIN-ICHI SUGAE*, WALTHER KLOFAT**

Laboratory of Molecular Biology,
National Institute of Neurological Diseases and Stroke,
Bethesda, Maryland, U.S.A.

COMMITMENT OF SPORULATION WITH RESPECT TO GLUCOSE

Most cells of *Bacillus subtilis* that reach the end of growth in a nutrient sporulation medium (NSMP) begin a differentiation process which produces heat- and chemically-resistant spores 7—8 hours later. This sporulation process can be suppressed when glucose, or any other readily metabolizable carbon source, is added to the medium with the inoculum (Fig. 1) or at any time during exponential growth and is maintained thereafter. However, when glucose is added at any time after growth has declined from its exponential rate, it no longer suppresses sporulation and actually leads to a slightly increased spore titer[1]. This can be seen in Fig. 1, in which glucose had been added 1 hour (= T_1) after the growth curve had declined from the exponential rate. The cells are then committed, with respect to glucose, to sporulate. The OD_{600} as well as the total cell titer still increase slowly in the presence of glucose, apparently because a few cells have not yet been committed and can continue to grow. Growth is also indicated by the presence of a few motile cells among the majority of nonmotile ones: during the process of sporulation cells are nonmotile.

The commitment of sporulation with respect to glucose occurs earlier than that with respect to resuspension of cells in fresh NSMP. When cells are diluted in NSMP soon after T_0 cells just continue to grow, but even at T_2—T_3, when cells have a forespore septum but not yet a complete forespore double membrane, can most cells resume growth: Cell-wall material is laid down between the two membranes of the forespore septum; in *B. cereus* both the mother cell and later the forespore cell grow and divide[2], whereas in *B. subtilis* only the mother cell grows but the development of the forespore cell is arrested[3]. The commitment of most cells to develop complete spores after resuspension in fresh medium is observed only upon completion of the forespore double membrane. Since the active transport mechanisms of the two forespore membranes should operate in opposite directions, leaving extracytoplasmic space between them, only passive diffusion or facilitated transport but no active

* Present address: Nara Women's University, Nara, Japan.
** Present address: Deutsche Forschungsgemeinschaft, Bonn, G.F.R.

transport should occur through this layer. But conceivably the two membranes could transiently form tight junctions, as observed between cells of multicellular organisms, permitting passage of larger molecules. In any case the closure of the forespore membrane must severely affect the metabolism inside the forespore which in turn apparently can influence the mother cell, rendering it incapable of further division[3]. Nevertheless, even after a complete forespore membrane has been formed, up to about 2 hours before spores are normally completed, can sporulation (measured at the time at which it usually occurs) be suppressed (delayed) by high concentrations of casein hydrolysate (10 mg/ml)[4]. It is therefore clear that the word "commitment" has to be specified with respect to media and other conditions under which sporulation is measured.

In order to determine whether any enzyme changes might be responsible for the commitment with respect to glucose, the glycolytic enzymes were assayed

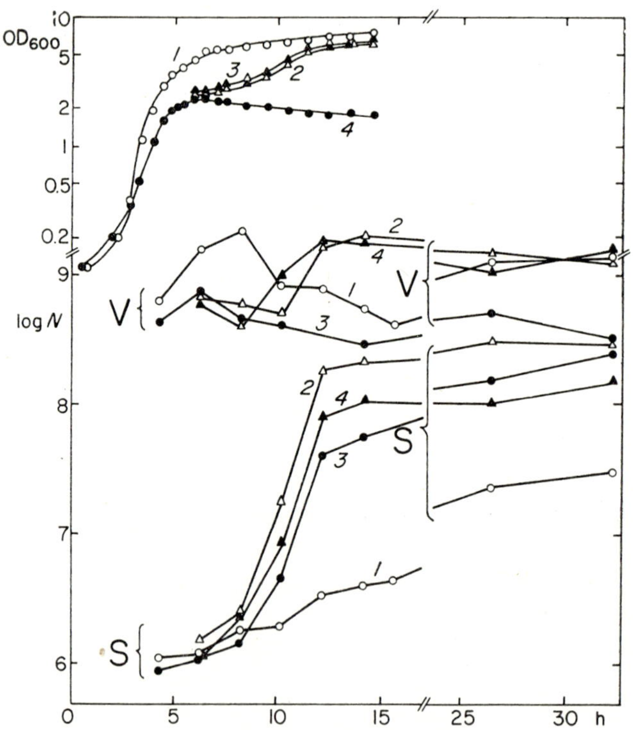

Fig. 1. Commitment of sporulation with respect to glucose addition. The standard strain 60015 was grown in nutrient sporulation medium (NSMP) plus L-tryptophan (25 μg/ml) and 0.1 or 2% glucose were added either at the beginning or at T_1 (1 hour after the growth curve deviated from the exponential increase). Subsequent to addition the glucose concentration was maintained constant. V viable cell titer, S spore titer, OD_{600} optical density; *1* 0.1% Glc continuously, *2* 0.1% Glc at T_1, *3* 2% Glc at T_1, *4* no Glc. (For details *see* Ref.[1])

during growth and sporulation. The glycolytic path is the major path of glucose utilization, because a phosphoglucose isomerase mutant (61158) grows in glucose minimal medium at a greatly reduced rate (doubling time 6.5 hours) compared to its parent, our standard strain (60015) (doubling time 1.9 hours). Toward the end of exponential growth in NSMP no significant increase or decrease in the specific

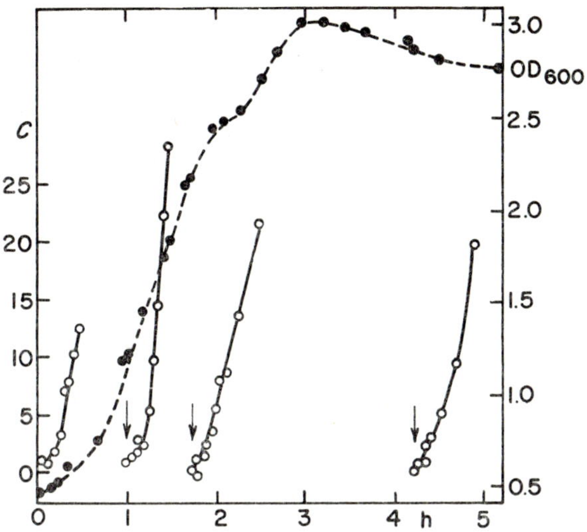

Fig. 2. Induction of U-^{14}C-glucose uptake into the standard strain. Strain 60015 was grown in NSMP and aliquots were transferred at different times to prewarmed Erlenmeyer flasks containing ^{14}C-glucose to give the final concentration of 5 mM and 50 nCi/ml. 0.5 ml samples were withdrawn at different times, placed on membrane filters, washed 7 times with 5 ml warm buffer (14 g K_2HPO_4, 6 g KH_2PO_4, 0.25 g $MgSO_4 . 7H_2O$ per liter), and counted. C carbon (from glucose), nmol/ml of cells (*solid lines*); OD_{600} optical density (*dashed line*).

Fig. 3. ^{14}C-Glucose uptake at different times of growth in nutrient sporulation medium. Strain 60015 was grown in NSMP and 2 ml aliquots were transferred at different times to prewarmed (25 mm tubes containing U-^{14}C-glucose to give the final concentration of 1 mM and 100 nCi/ml. Incorporation into $\frac{1}{2}$ ml aliquots was determined after 15 (15) and 30 (30) min as in Fig. 2. C c.p.m. $\times$ 10^{-3} (*solid lines*); OD_{600} optical density (*dashed line*).

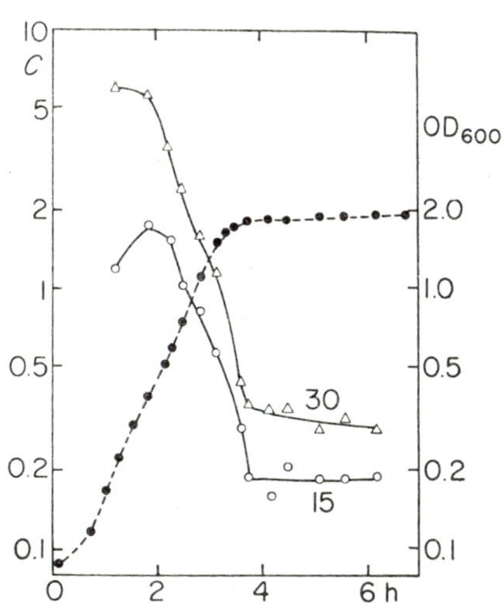

activity of any glycolytic enzyme was observed that could explain the resistance to glucose.

A significant change, however, was found for the uptake of glucose, as can be seen in Fig. 2. In contrast to Gram-negative bacteria the uptake of glucose is inducible in *B. subtilis*. This inducibility declines toward the end of growth, as can be seen better in Fig. 3, which shows at different times of growth the incorporation of

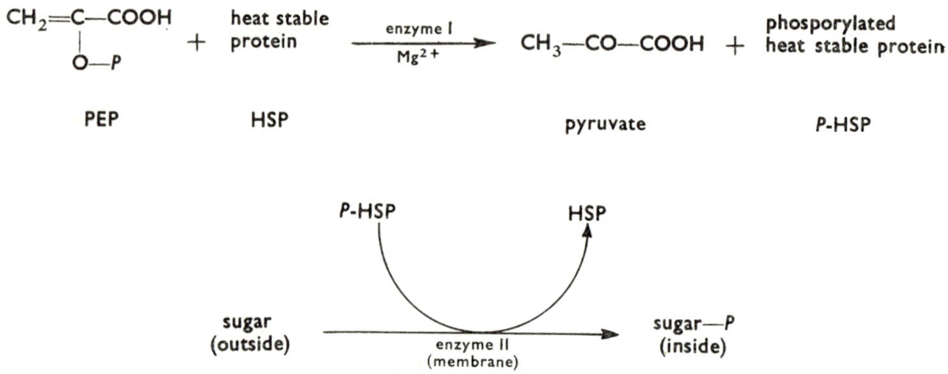

Fig. 4. Reactions leading to uptake and phosphorylation of glucose by PEP—glucose phosphotransferase.

glucose measured 15 and 30 minutes after glucose addition. A similar decline of uptake has been observed for methyl α-glucoside and 2-deoxyglucose, neither of which can be used as carbon sources (within the period of the experiment) or induce their

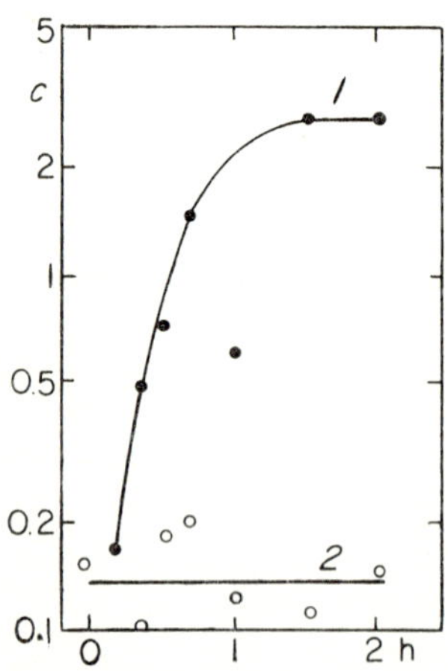

Fig. 5. Phosphorylation of glucose by PEP—glucose phosphotransferase isolated at different times after addition of glucose to cells growing exponentially in NSMP; *c* PEP—glucose phosphotransferase, nmol/min per mg protein; *1* glucose added, *2* no glucose.

own uptake. Glucose is taken up by the phosphoenolpyruvate (PEP)—glucose phosphotransferase system, whose reaction scheme is shown in Fig. 4. In *B. subtilis* this system is inducible by glucose (Fig. 5), whereas glucokinase (using ATP as phosphate donor), which can phosphorylate only internal glucose, has a constant high activity, even without glucose in the medium. The PEP–glucose phosphotransferase phosphorylates also methyl α-glucoside, which cannot be phosphorylated by glucokinase. This substrate is therefore preferable in the assay of PEP–glucose phosphotransferase. Methyl α-glucoside, however, does not induce its own uptake system. Similar to the uptake mechanism of glucose, the inducibility of PEP–glucose phosphotransferase decreases drastically toward the end of exponential growth

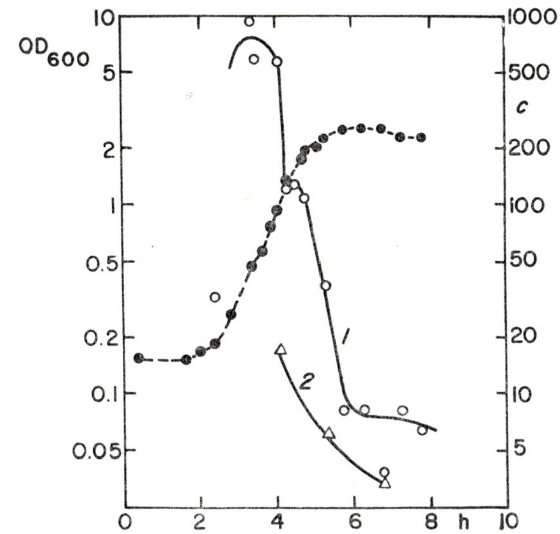

Fig. 6. Decline of PEP–glucose phosphotransferase inducibility during growth in nutrient sporulation medium; *c* PEP–glucose phosphotransferase, pmol/min per mg protein (*solid lines*); *1* cells harvested after growing 20 min in the presence of glucose (0.2%), *2* cells grown without glucose; OD_{600} optical density (*dashed line*).

(Fig. 6). Since the inducible part of PEP–glucose phosphotransferase resides in enzyme II which is membrane-bound, and since the spontaneous activity, observed without glucose addition, also decreases towards the end of growth, an alteration of the bacterial membrane may be responsible for the decrease. The decrease is avoided in the presence of excess glucose; but it occurs also when a limiting amount of glucose has been used up during exponential growth, a condition which allows subsequent sporulation. The lack of PEP–glucose phosphotransferase induction during the developmental period is sufficient to explain the early commitment of sporulation with respect to glucose since sporulation is not suppressed by glucose in a mutant (61310) which lacks PEP–glucose phosphotransferase activity. (This mutant probably is deficient in enzyme I or HSP of the PEP–glucose phosphotransferase system, because it also does not grow on several other carbon sources.)

Shutting off the uptake mechanism with respect to specific metabolites seems to be a simple device by differentiating cells to avoid the disturbance of the differentiation process by external compounds. It will be interesting to see if this device has been used in other differentiating systems to maintain a commitment with respect to external compounds.

THE EFFECT OF GLYCINE ON SPORULATION

In order to determine the minimal conditions under which sporulation can occur, one can transfer growing cells to synthetic media with different supplements. Some microorganisms, such as *B. megaterium*, can sporulate when they are transferred to distilled water. Internal carbon and other nutrient resources, *e.g.* poly-β-hydroxybutyrate in *B. megaterium*, are utilized for this endotrophic sporulation. *B. subtilis* has the experimental advantage that it cannot accumulate poly-β-hydroxybutyrate and can sporulate only if some external carbon sources are present. It sporulates when cells are transferred at times of growth from a rich medium to a buffer (Sl) containing slowly metabolizable carbon sources, such as L-glutamate or L-lactate. Most cells enter sporulation because they could start growing only after prolonged adaptation to the carbon source. (Owing to this sporulation it is difficult to measure the growth rates of sporulating strains in these carbon sources, whereas such rates can be easily determined with asporogenous mutants.) Since these carbon sources support growth, all possible metabolites can be made from them and it would be difficult to determine which of them are actually required for sporulation.

An exceptional situation arises when cells are transferred to a buffer (Sl) containing acetate as sole carbon source. This compound cannot support growth of our strains of *B. subtilis*, which lack both enzymes of the glyoxylate bypath. Acetate alone does not support sporulation. This result is surprising, because previous experiments[5,6] had shown that during sporulation in nutrient sporulation medium, cells utilize acetylCoA as a carbon source for ATP synthesis in the citric-acid cycle; this acetylCoA is not derived from accumulated pyruvate, because that is used up by the end of growth[7], but from some other carbon source such as acetate. The argument[7] was this: In a sporulating strain (60015), ATP is maintained at an approximately constant concentration throughout growth and development in NSMP. In any citric-acid-cycle mutant the concentration of ATP (or the incorporation of uracil) severely decreases during the stationary phase. The decrease can be (transiently) counteracted by any carbon source that enters the citric-acid cycle after the block, the citrate synthase being considered as the start of the cycle. In particular, in an aconitase or isocitrate dehydrogenase mutant ATP synthesis is restored by the addition of glutamate, malate, succinate, and other compounds including pyruvate, but it does not respond to acetate or citrate. In an α-oxoglutarate dehydrogenase mutant acetate, citrate, or glutamate (α-oxoglutarate) have no effect, whereas all other carbon sources restore ATP synthesis. And so on it goes around the cycle. From this finding it was concluded that the ATP synthesis in the normal strain must be supported during the developmental period by a compound which enters the cycle but is unable to be utilized in any citric-acid-cycle mutant. This compound must be acetylCoA. How acetylCoA is normally made is not known, but it can be derived from acetate, which accumulates towards the end of growth in NSMP. The enzymes necessary for the uptake of acetate are present (*see* later Fig. 11, *right*)

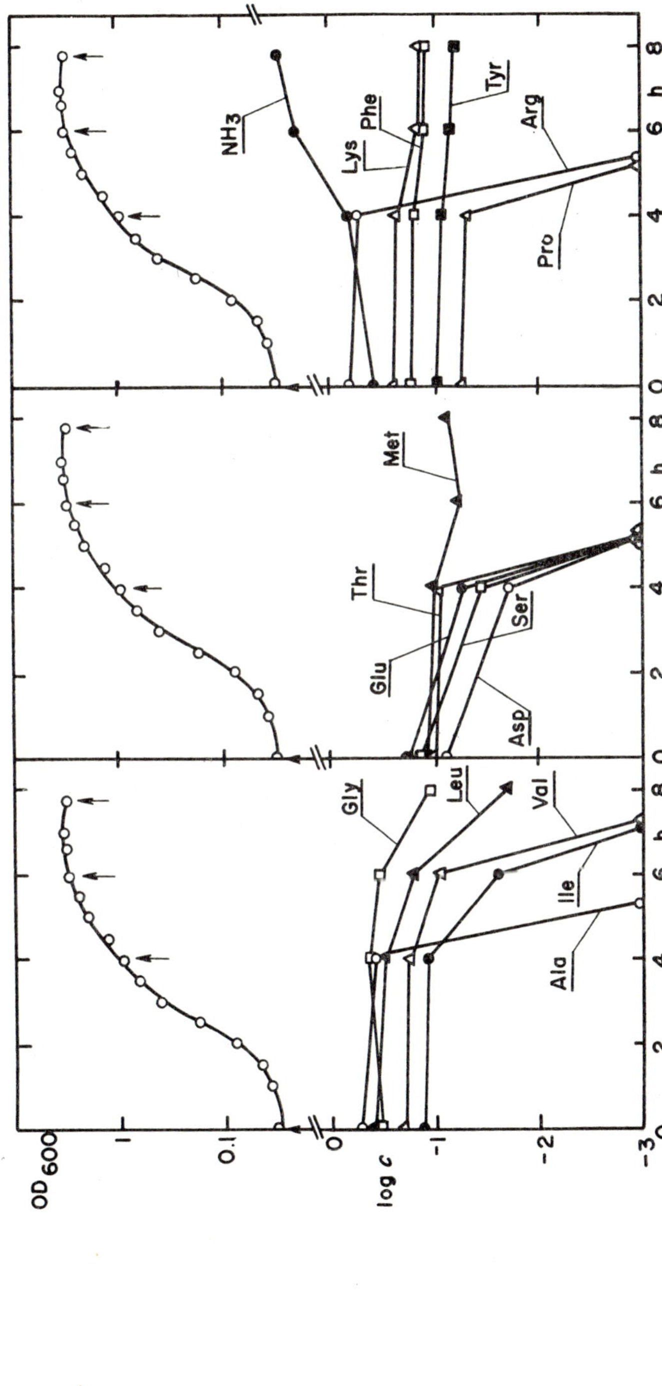

Fig. 7. Change of amino-acid concentration (*c*, mM) in nutrient sporulation medium during growth of *B. subtilis* 60015. Aliquots of the culture of 60015 were harvested at the points indicated by arrows and the amino-acid concentrations of a Millipore filtrate were determined. (For details *see* Ref.[9])

and a number of mutants have been found which sporulate only in the presence of acetate[7,8]. Actually, after transfer of cells (60015) to Sl + acetate the concentration of ATP is maintained whereas it decreases without acetate[9].

To determine which other compound is needed for sporulation in nutrient sporulation medium in addition to acetylCoA, we have examined the amino-acid composition of th. . medium at different times of growth and sporulation[9].

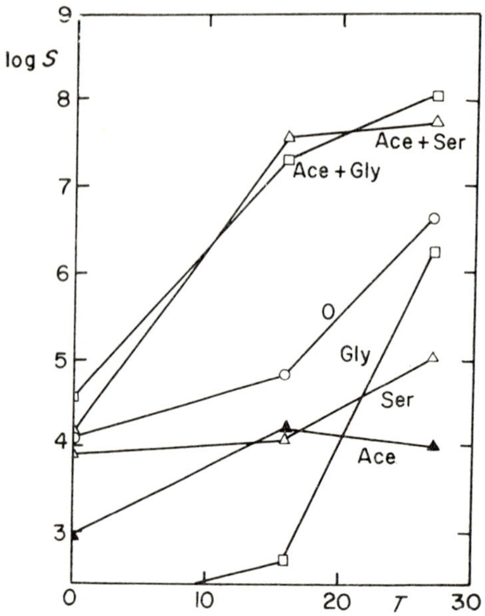

Fig. 8. Sporulation in buffer Sl plus differen additions. *B. subtilis* 60015 was grown in nutrient sporulation medium until the maximum OD_{600} had been reached. The cells were centrifuged at room temperature, washed with warm (37 °C) Sl + 50 mM acetate + 10 mM of the compounds stated (neutral potassium salts). T time after transfer, h, S spore titer; 0 nothing added, Ace acetate.

Fig. 7 shows that most amino acids are used up at the end of growth and only glycine, leucine, and a few other amino acids are still present. In fact, glycine added to buffer plus acetate could support sporulation, whereas none of the other remaining amino acids was effective (Fig. 8). Fig. 8 shows that glycine can be replaced by serine, but neither serine nor glycine alone supports significant sporulation. (The limited sporulation observed 27 hours after transfer to buffer Sl without any addition or with glycine alone is due to extensive cell lysis which supplies nutrients for some of the unlysed cells. Acetate prevents this lysis.) The optimal concentration of both glycine and serine was determined as 10^{-2} M. Neither glycine nor serine could support growth alone or in combination with acetate as carbon sources, so that the role of glycine cannot be merely one of supplying a carbon fragment for general growth. In fact, sporulation occurs, after transfer to acetate plus glycine, only when the cells have undergone a certain development in nutrient sporulation medium, as can be seen in Fig. 9. This result is quite different from the response to glutamate or any of the other slowly metabolizable carbon sources which allow sporulation at any time after transfer, as can be seen in Fig. 10 for glutamate. We do not know which developmental reaction is necessary to allow sporulation in acetate plus glycine.

One could imagine that this reaction involves one of the enzymes between acetate and α-oxoglutarate, all of which are needed for the utilization of acetate to regenerate ATP *via* the citric-acid cycle. Four of these enzymes, acetokinase, phosphate acetyltransferase, citrate synthase (Fig. 11, *left*), and isocitrate dehydrogenase (Fig. 11, *right*) show already during exponential growth a significant specific activity which increases by less than a factor of two toward the end of growth. The only enzyme showing

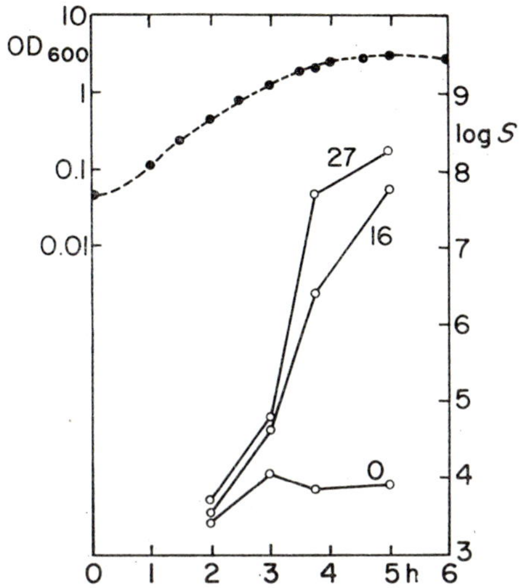

Fig. 9. Sporulation of standard strain 60015 grown for the indicated time in nutrient sporulation medium and then transferred to buffer Sl + + 0.05 M potassium acetate + 0.01 M glycine. S spore titer 0 (0), 16 (16), 27 (27) h after transfer (*solid lines*); OD_{600} optical density (*dashed line*).

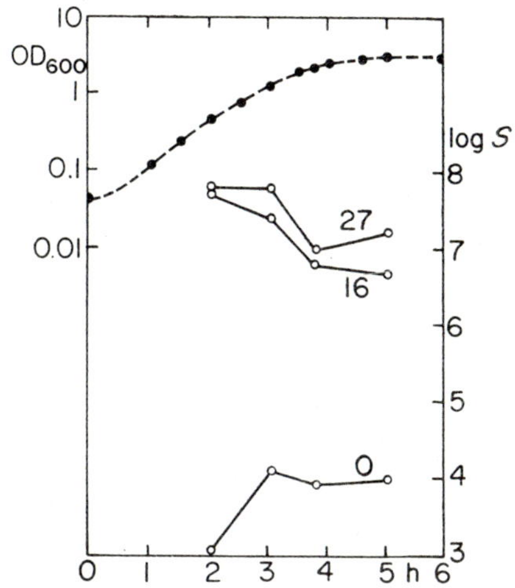

Fig. 10. Sporulation after transfer to Sl + 0.01 M potassium L-glutamate. S spore titer 0 (0), 16 (16), 27 (27) h after transfer (*solid lines*); OD_{600} optical density (*dashed line*).

a significant late increase is aconitase (Fig. 11, *right*), which therefore might represent the developmental reaction. However, we have found another nutrient medium (from Baltimore Biological Laboratories instead of Difco), which is slightly deficient in tryptophan. When NSMP is prepared from this medium, sporulation does not occur at any time of transfer to the acetate-plus-glycine buffer medium (containing tryptophan) unless tryptophan is added already to the NSMP. In the absence of added tryptophan, the above developmental reaction does therefore not seem to occur, because the cells stop growing just a little too early. Nevertheless, the specific activity of aconitase (and isocitrate dehydrogenase) increases in this medium late in growth equally well with tryptophan addition as without it. Some other reaction will therefore have to be found. We also do not know the biochemical role of glycine in the process of sporulation, but we have eliminated several trivial possibilities, such as the conversion of glycine to any one of the known metabolic intermediates, including glyoxylate, *etc.* Since the concentration of glycine is rather high, it may be used as an inhibitor of some metabolic reaction. Such a possibility is indicated by the finding of TURIAN that conidiation of *Neurospora crassa* in the presence of fructose is enhanced by the addition of glycine, which reduces the rate of sugar consumption[10,11].

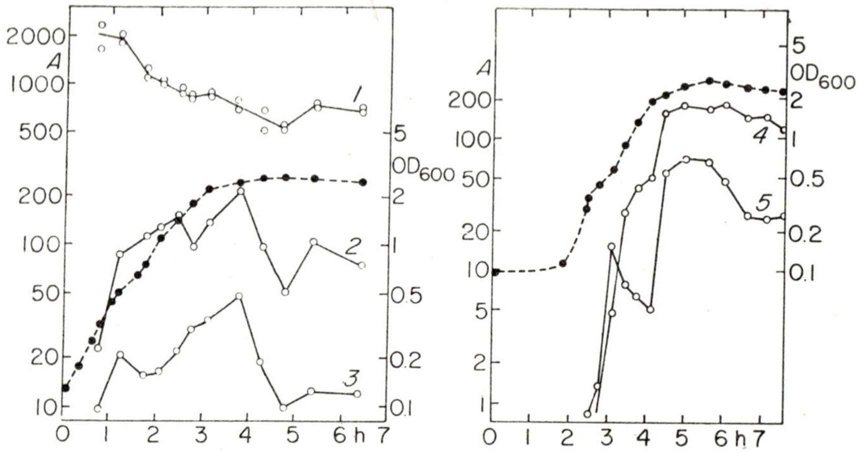

Fig. 11. Change of some enzyme activities during growth and sporulation in nutrient sporulation medium. Extracts were prepared and citrate synthase was assayed as described by FORTNAGEL and FREESE[5]. *A* specific activity, nmol/min per mg protein (*solid lines*), OD_{600} optical density (*dashed lines*); *1* phosphate acetyltransferase, *2* acetokinase, *3* citrate synthase, *4* isocitrate dehydrogenase, *5* aconitase.

We have similarly found that *B. subtilis*, when transferred during exponential growth from NSMP to buffer plus glucose, can sporulate only when the medium also contains glycine.

These findings render it now possible to examine in biochemical terms some of the early reactions leading to sporulation.

POSSIBLE INDUSTRIAL APPLICATIONS

Often without realizing it, industry has used for a long time certain developmental states of microorganisms to isolate antibiotics, metabolites, or enzymes in large quantities. The antibiotic activity of microorganisms increases at the end of growth and in sporulating organisms is closely related to sporulation, if not necessary for it[10,12]. Perhaps owing to this relationship, the assumption seems to have developed, with respect to some antibiotics (*e.g.*, penicillin), that antibiotic-producing organisms must be able to sporulate. Since the antibiotic activity is produced early in the development toward spores, mutants can be isolated which are blocked late in this development and still produce the antibiotic; some mutants may even accumulate it in larger quantity than sporulating cells. Alternatively, higher quantities of antibiotics may be obtained in mutants that sporulate rather slowly. Since glycine stimulates the process of sporulation under certain conditions, it may be worthwhile to see whether it also has a specific effect on antibiotic production.

The production of metabolites is also related to certain stages of growth and sporulation, as can be seen by the fact that pyruvate and acetate are excreted by

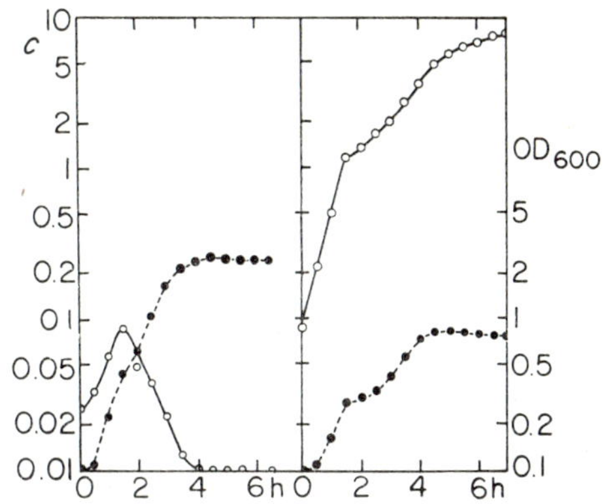

Fig. 12. Accumulation of pyruvate during growth in nutrient sporulation medium of the standard strain 60015 (*left*) and the pyruvate dehydrogenase-deficient mutant 61141 (*right*); c pyruvate, mM (*solid lines*), OD_{600} optical density (*dashed lines*).

sporulating cells and are later reutilized during the process of sporulation (Fig. 12). An asporogenous mutant may allow the accumulation of much larger quantities of a desirable metabolite. Fig. 12 shows, for example, that an acetate-requiring mutant blocked in pyruvate dehydrogenase accumulates, under conditions under which it cannot sporulate, high quantities of pyruvate[7].

The developmental state of the cells is also important for the isolation of some enzymes which are only produced during exponential growth and later destroyed or which appear only during the developmental period. Both this fact and the finding that specific uptake mechanisms of external compounds change during growth and sporulation have to be considered if one wishes to convert a compound into another one by microorganisms.

It seems clear that a detailed enzymic analysis of development will open up new tools for the industrial use of microorganisms.

REFERENCES

1. FREESE E., KLOFAT W., GALLIERS E.: Biochim. Biophys. Acta **222**, 265 (1970).
2. FITZ-JAMES P., YOUNG E., p. 39 in G. W. Gould, A. Hurst (Eds.): *The Bacterial Spore.* Academic Press, London and New York 1969.
3. FREHEL C., RYTER A.: Ann. Inst. Pasteur **117**, 297 (1969).
4. MANDELSTAM J., STERLINI J. M., p. 180 in L. L. Campbell (Ed.): *Spores,* Vol. IV. Am. Soc. Microbiol., Bethesda 1969.
5. FORTNAGEL P., FREESE E.: J. Bacteriol. **95**, 1431 (1968).

6. KLOFAT W., PICCIOLO G., CHAPPELLE E. W., FREESE E.: J. Biol. Chem. **244**, 3270 (1969).
7. FREESE E., FORTNAGEL U.: J. Bacteriol. **99**, 745 (1969).
8. SCHMITT R., FREESE E.: J. Bacteriol. **96**, 1255 (1968).
9. SUGAE K., FREESE E.: J. Bacteriol. **104**, 1074 (1970).
10. TURIAN G.: Nature **202**, 1240 (1964).
11. OULEVEY-MATIKIAN N., TURIAN G.: Arch. Mikrobiol. **60**, 35 (1968).
12. SCHAEFFER P., IONESCO H., RYTER A., BALASSA G.: Colloq. Internat. Centre Nat. Rech. Sci. **124**, 553 (1963).

Regulations of Sporogenesis in *Bacillus subtilis*

J. SZULMAJSTER, P. KERJAN, J.C.C. MAIA*

Laboratoire d'Enzymologie du C.N.R.S., Gif-sur-Yvette, France

„Garde-toi aussi de la sainte simplicité"
(Ainsi parlait Zarathoustra)
Nietzsche

INTRODUCTION

Bacterial sporogenesis has become attractive to molecular biologists as a subject for studies of cellular differentiated structures. Vegetative cells of spore-forming bacteria are, at the end of growth, converted to spores in a relatively synchronous fashion, through a time-ordered sequence of morphogenetic and biochemical steps. The transition from the vegetative to the dormant form constitutes an example of unicellular differentiation, unique in the biological world. Each of these forms can be maintained indefinitely. The emergence of a vegetative cell from the spore under favourable physiological conditions takes place within a few minutes in response to specific chemical agents.

Bacterial sporogenesis thus represents a biological problem whose regulation is intermediate in complexity between the regulation of the production and activity of bacterial enzymes on the one hand and the differentiation of specialized tissues in higher organisms on the other hand.

LIFE CYCLE OF SPORULATING BACTERIUM

At the end of exponential phase of growth, cells of many bacterial species may become committed to sporulation providing they have the genetic capacity and the appropriate environmental conditions. Instead of dividing, each cell will then produce a single spore over a period of 8—10 h.

The morphology of spore formation is now well established in many sporogenic bacteria and will not be described here[1−4]. Fig. 1 gives a schematic presention of the life cycle of a sporogenic bacterium with the seven characteristic and distinctive structural steps recognisable on the electron microscope during growth cycle.

The main subject of this paper is to present some experimental data obtained

* Fellow of the Affaires Etrangères, France, and of the Fundaçao de Amparo a Pesquisa do Estado de Sao Paulo, Brasil.

in our laboratory in the last two years, which, in addition to the important work carried out by others, might contribute in some way to a better understanding of the complex mechanism of regulation involved in bacterial sporogenesis.

The work which will be discussed in the present paper was carried out on *B. subtilis*, an organism which is particularly appropriate for genetic and biochemical studies of sporulation. Mutations in more than 20 different genes of *B. subtilis* were reported, which prevent the development of spores without affecting vegetative growth[5-9].

The *biochemical events* required for the formation of a mature spore could therefore be under the control of genes or operon-like groups of genes. The idea now generally accepted is that the spore genome is repressed during vegetative growth and derepressed at the end of logarithmic growth, although the exact nature of this repression still remains unknown[10-13].

A number of physical and chemical agents are known to affect bacterial sporogenesis, but until now no single gene or gene product has been identified as being responsible of the sporulation process. Genetic studies from different laboratories[5-9] lead to the conclusion that spore genes or groups of genes are randomly distributed along the *B. subtilis* chromosome. The main questions, however, which remain to be answered are to know if there exists a primary locus for the initiation of the overall complex sporulation process and if there are regulatory genes controlling the sequential biochemical reactions catalyzing the synthesis of spore constituents.

Our present knowledge of the genetic, cytological and biochemical events associated with bacterial sporogenesis does not provide precise information regarding the primary initiation step or the nature and number of indispensible functions leading to the formation of a mature spore.

Such information could be obtained if conditional sporulation mutants preventing the expression of these functions were available for studies.

Temperature-sensitive sporulation mutants (t_s-mutants). Temperature-sensitive mutants have already been used effectively to gain information about indispensable gene functions and have proved to be useful tools for the understanding of macromolecular synthesis and its regulation in prokaryotic organisms[14,15]. We have therefore focussed our attention on the isolation of temperature-sensitive sporulation mutants. The method of isolation of such mutants has been described elsewhere[16].

General properties of these mutants. Table I shows some of the properties of the temperature-sensitive mutants grown in nutrient-broth medium at 30 and 42 °C. It can be seen that all the mutants listed (except the t_s-93) have a low capacity to sporulate at the nonpermissive temperature although no change in vegetative growth was observed at either 30 or 42 °C.

Electron-microscopic studies of cell populations of these mutants have shown that at the restrictive temperature all of them are morphologically blocked at stage zero of spore development.

TABLE I

Some Properties of Thermosensitive Sporulation Mutants of *B. subtilis*[a]

Mutant	S/C $\times 10^3$		Protease activity								RNAase		Anti-biotic	
			azo-albumin		casein		gelatin		albumin (denatured)					
	30°	42°	30°	42°	30°	42°	30°	42°	30°	42°	30°	42°	30°	42°
t_s-2	420	1.5	+	+	+	+	+	−	+	+	−	+	+	+
t_s-3	320	3	+	+	+	+	+	−	+	+	−	+	+	+
t_s-4A	560	0.3	+	+	0	0	0	0	0	0	0	0	+	−
t_s-7	550	31	+	+	+	+	+	−	+	+	+	+	+	+
t_s-8	1,000	3	+	+	+	+	+	−	+	+	+	+	+	+
t_s-11	680	65	+	+	+	+	+	−	+	+	+	+	+	+
t_s-27	320	1	+	+	+	+	+	−	+	+	+	+	+	+
t_s-32	230	7	+	+	+	+	+	−	+	+	+	+	+	+
t_s-40	420	2	+	+	+	+	+	−	+	+	+	+	+	+
t_s-42	230	4	+	+	+	+	+	−	+	+	+	+	+	+
t_s-77	1,000	40	+	+	+	+	+	−	+	+	+	+	+	+
t_s-82	400	1	+	+	+	+	+	−	+	+	+	+	+	+
t_s-93	60	260	+	+	+	+	+	−	+	−	+	+	+	+

[a] C viable cells plated at T_2, S viable spores plated after heating the cell suspension 10 min at 80 °C at T_{22}; + halo formation on agar plates, — no halo formation, 0 not tested. The sensitive strain used for the antibiotic tests was *B. subtilis* H.

EXOENZYMES AND ANTIBIOTICS

A correlation between the excretion of exoenzymes and antibiotics in the culture filtrates of spore-forming bacteria at the end of exponential growth and spore formation is now generally admitted, although by no means proved. These traits have been used for general characterization of early blocked sporulation mutants[6,8]. The results summarized in Table I show that with the exception of the t_s-4A strain, all the mutants tested produce, at the two temperatures, the antibiotic active on *B. subtilis* H. All the mutants seem also to produce proteases active on azo-albumin, casein and denatured albumin*. However, using gelatin as substrate, proteolytic activity was produced only at the permissive temperature and not at the restrictive one. The exocellular RNAase activity was also tested on agar plates using yeast RNA as substrate. With the exception of mutants t_s-2 and t_s-3, all of them show RNAase activity at the two temperatures.

* The t_s-93 shows no activity on denatured albumin when tested at 42 °C but this mutant seems to differ from the others by its higher sporulation rate at 42 than at 30 °C.

These results would suggest that at the restrictive temperature, although all the mutants are morphologically blocked at stage zero in sporulation process, there is a production of proteolytic activities at least with some of the substrates tested. It has also to be noticed that the sensitivity of the plate tests used will not permit the detection of quantitative differences in the level of enzyme or antibiotic produced. On the other hand, mutants of *B. subtilis* deficient in antibiotic production which show protease activity (*abu prt*$^+$) have also been found[7,17]. Furthermore, the studies with *B. megaterium* would suggest that none of the proteases are essential for sporulation[18].

The t_s-4A mutant has been chosen for more extensive biochemical studies and the results to be described are relative to this mutant.

Experiments of a biochemical nature further strengthen the belief that the thermosensitive sporulation mutants we have isolated are deficient in an initial step of sporulation process when grown at the nonpermissive temperature.

An important effort has been made in the last ten years to correlate the biochemical and the morphological events during spore development. Although a considerable number of biochemical changes are generally observed during the period separating the end of growth and the appearance of a mature spore, only a limited number of spore-specific components have been well characterized.

This is also true for the enzymes specifically involved in the synthesis of the spore constituents. There is no doubt that many of the biochemical changes observed during sporulation, like the appearance and disappearance of some enzymatic activities, are very often coincident but such changes are without direct relation to the sporulation process, and rather due to some secondary effects. Other changes are functionally associated with the sporulation phenomenon[19-26].

RNA AND PROTEIN TURNOVER

One of the earliest biochemical events which can generally be observed at the end of exponential growth (T_0) in a spore-forming bacterium is an important increase in RNA and protein turnover, as compared to the early blocked asporogenic mutants[27,28]. Fig. 2 shows a similar difference in RNA and protein turnover in the t_s-4 mutants when grown at the permissive and nonpermissive temperatures. In the parent strain or in a revertant strain of t_s-4 (t_s-4-R1) no such difference was observed.

Commitment in the t_s-4 mutant. Experiments were performed to determine the time of commitment to either the irreversible loss of the sporulating capacity or to its recovery when the t_s-4 mutant culture was shifted "down" from 42 °C to 30 °C and "up" from 30 °C to 42 °C, respectively. Typical shift "down" and "up" experiments were carried out with the t_s-4 mutant and the results are shown on Figs 3 and 4.

When the cultures were grown at 42 °C and shifted at 30 °C before T_1 the cells were unable to regain sporulation capacity. The sporulation rate increased rapidly if the cultures were shifted after this time. In a similar experiment, the cultures were grown at 30 °C for different periods of time and shifted to 42 °C and incubated

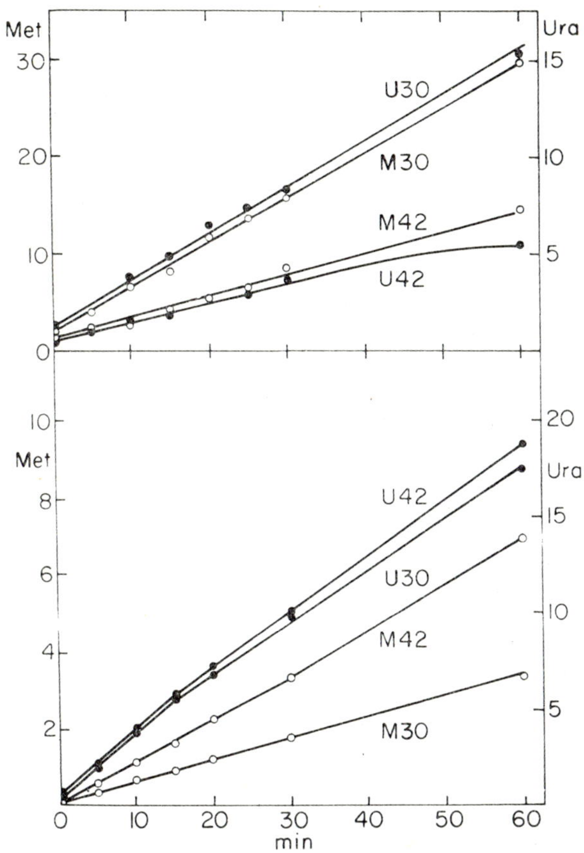

Fig. 2. Turnover of RNA and protein in t_s-4 mutant. Cultures (50 ml) of the t_s-4 mutant (*above*) and of the parent strain (*below*) were grown at 30 and at 42 °C. At T_3 (different for the two cultures), each of the cultures was divided into two equal volumes. To one, 0.2 μCi of ^{14}C-methionine per ml was added; to the other, 1 μCi of ^{14}C-uracil per ml was added. Samples of 1 ml were taken at the times indicated on the figure and added to test tubes containing 1 ml of cold 5% trichloroacetic acid. The acid-insoluble material was filtered and washed on membrane filters (Millipore Corp.), and the radioactivity incorporated was counted in a scintillation counter. Met ^{14}C-methionine, c.p.m./ml × 10^{-3}, Ura ^{14}C-uracil, c.p.m./ml × 10^{-3}; M methionine, U uracil; 30 30 °C, 42 42 °C. (From Ref.[16])

for 22 h. In this case if the shift takes place before T_2 a very small percentage of the bacterial population was able to form spores. This percentage increased rapidly if the shift took place after this time. The difference in the critical limit of the shifting time observed at 30 °C and 42 °C is probably due to the delay of the initiation of the sporulation process at the lower temperature.

All the above results based on experiments of different nature, strongly suggested that the temperature-sensitive event in the mutant analyzed occurs at a very early stage of spore development.

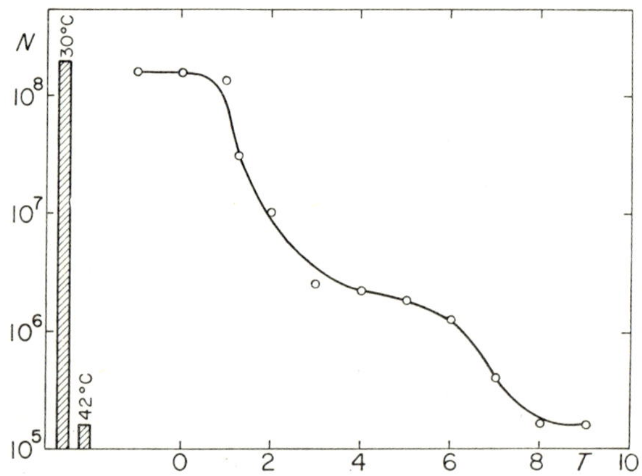

Fig. 3. Sporulation of the t_s-4 mutant cultures "shifted down" from 42 to 30 °C at different times during the growth cycle. T_0 designates the end of exponential growth, and T_1, T_2, *etc.*, designate the time (in hours) after exponential growth stopped. This designation is valid for all the figures where T is on the abscissa. Spores were counted by plating the cell suspensions at T_{22} (22 h after the end of exponential growth) after heating for 10 min at 80 °C. T time of shift from 42 to 30 °C (hours); N number of spores at T_{22}. (From Ref.[16])

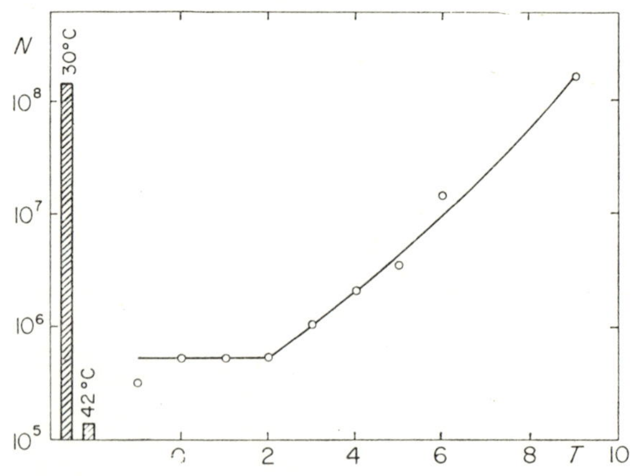

Fig. 4. Sporulation of the t_s-4 mutant cultures "shifted up" from 30 to 42 °C at different times during the growth cycle. Spores were counted at T_{22}, as described in the footnote of Fig. 3. T time of shift from 30 to 42 °C (hours); N number of spores at T_{22}. (From Ref.[16])

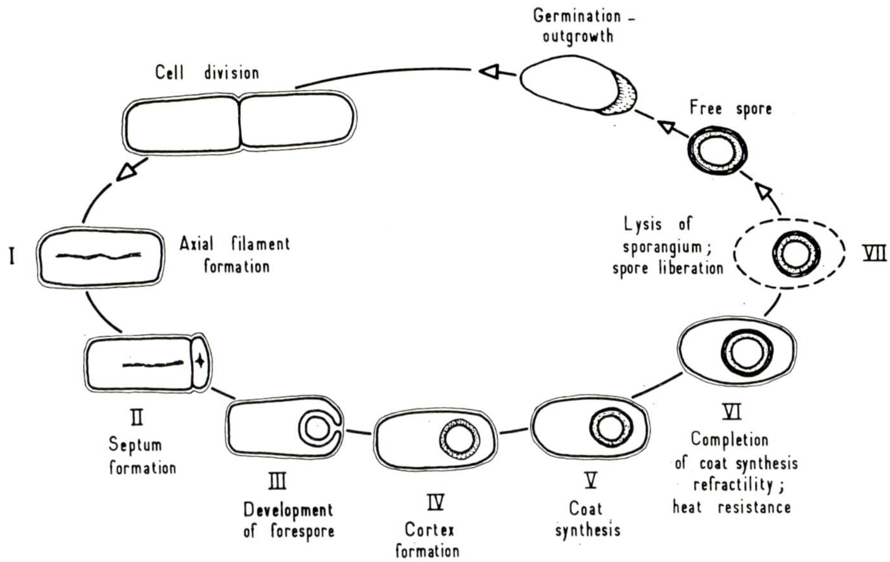

Fig. 1. Diagrammatic representation of the life cycle of a sporulating bacterium.

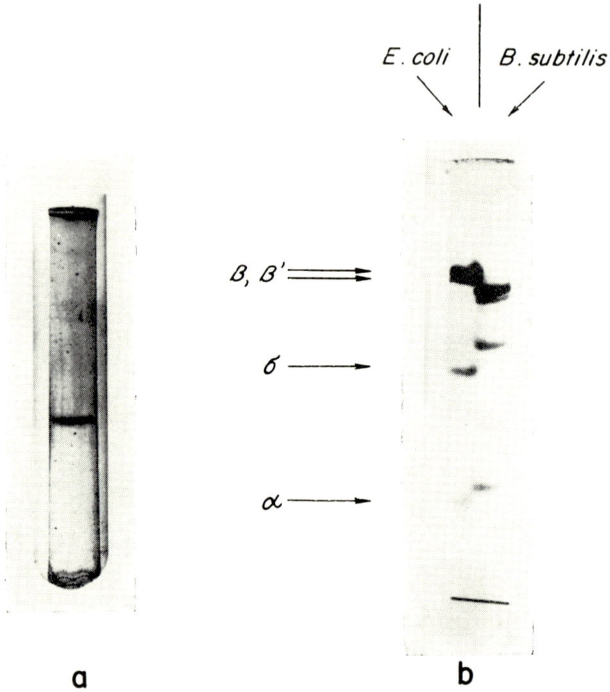

Fig. 14. Polyacrylamide gel electrophoresis patterns of different RNA polymerase preparations. **a** — 15 μg of purified *B. subtilis* enzyme (after agarose-chromatography step) was layered on the gel: 5% of acrylamide in Tris—glycine buffer, pH 8.4. **b** — 0.1% SDS-gel (5%) in 0.1M sodium phosphate buffer, pH 7.2 was run according to SHAPIRO *et al.*[47]. The "split-gel" was used to compare the subunits from *E. coli* RNA polymerase (*left*) to those obtained from *B. subtilis* (*right*). The amounts of enzyme layered on the gel were 5 and 8 μg, respectively. In both cases (**a** and **b**) the gels were stained by 0.2% coomassie brillant blue in methanol—acetic acid—water (5 : 1 : 5, *v*/*v*/*v*) for about 1 h and destained electrophoretically in 7.5% acetic acid.

One of the main characteristics of spore development like those of any other differentiating cell, is that the whole process is a succession of physiological and biochemical events which are regulated in time and in space. These events are under a set of controls responsible for the expression of the phenotypic character.

LATE ENZYMES IN SPORULATION PROCESS

Since the thermosensitive mutant described here seems to be defective in the initiation step of sporulation when growth occurred at the nonpermissive temperature, one would expect that it would also be defective in the spore-specific enzymes involved in the late process of sporulation.

It is desirable at this point to consider in some detail the enzymic system involved in the biosynthesis of dipicolinic acid (DPA) as an example of late events associated to the sporulation process.

DPA is produced in bacteria exclusively during sporulation and is normally never found in vegetative cells. In the form of its calcium salt DPA is a major component of mature spores, representing 5 to 15% of the dry weight. The DPA content

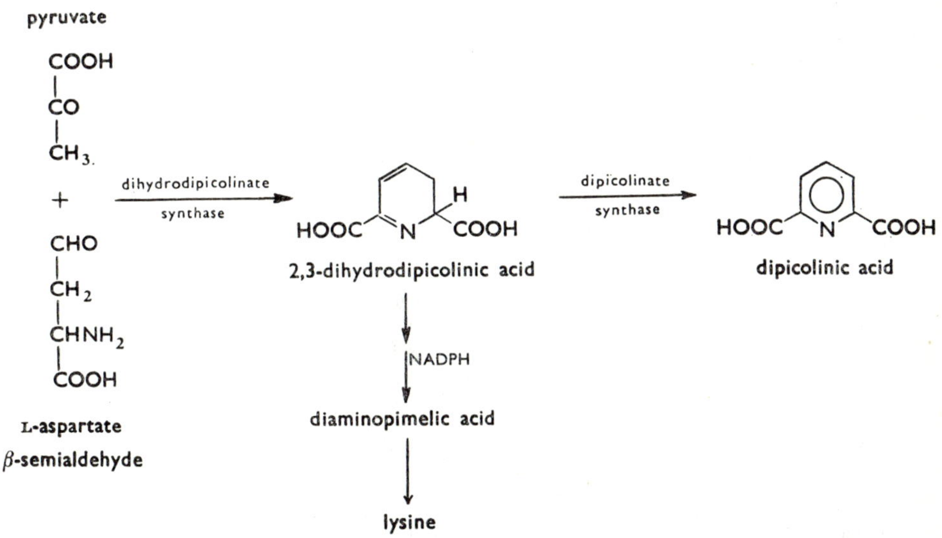

Fig. 5. Pathway of dipicolinic acid synthesis.

has been shown to be associated with the resistant properties of bacterial spores, although its mode of action in this and possibly other functions still remains unknown.

Furthermore, mutants which have lost the capacity to synthesize dipicolinic acid have either completely lost the capacity to make spores or are producing defective "spores" unable to resist heat treatment[29].

The biosynthetic pathway of dipicolinic acid in *B. subtilis*[26,30] and *B. megaterium*[31] has recently been established and is shown in Fig. 5. The demonstration of dihydrodipicolinic acid as an intermediate in lysine biosynthesis in *E. coli*, suggested that this compound is a likely precursor of DPA[30,32].

Suporting evidence for the involvement of this pathway in the synthesis of DPA has been provided by the *in vitro* formation of DPA from aspartate semialdehyde

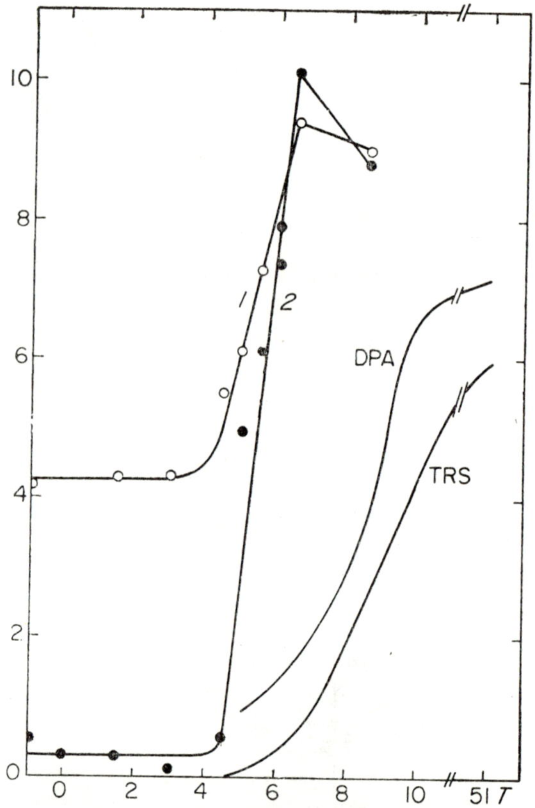

Fig. 6. Kinetics of dihydrodipicolinate synthase and DPA synthase production. Cells were grown in peptone medium at 37 °C. The curves indicating the DPA content of the culture (DPA) and the relative number of thermally resistant spores (TRS) were taken from a separate but similar experiment. *T* time after the end of exponential growth (hours); *1* dihydrodipicolinate synthase, units/mg × 10^{-2}, *2* DPA synthase, units/mg × 4. Both enzyme activities were measured in detergent-treated cells. (From Ref.[26])

(ASA) and pyruvic acid, catalyzed by enzyme systems derived from sporulating *B. megaterium* and *B. subtilis*[26,31].

Kinetics of late events during sporulation. Strong evidence for the physiological significance of the two enzymes, dihydrodipicolinate synthase (condensing enzyme) and dipicolinate synthase, for spore development was based on the kinetic studies of the events occurring during sporulation. Fig. 6 shows the appearance of the enzyme activities of the dihydrodipicolinate synthase and DPA synthase. The specific activity of the first enzyme which is present at a constant level during exponential growth increases about three-fold around T_5 (that is, 5 h after the end of exponential growth). At about the same time, DPA synthase activity rises sharply from a level not significantly different from zero. The increase in both of the activities precedes both the production and accumulation of DPA and the development of thermal resistance in the culture. These results support strongly the idea that the production of dipicolinic acid in sporulating cells is dependent on the presence of DPA synthase.

Regulation of the two DPA enzymes. Although the two enzymes studied are involved in the same biosynthetic pathway, they pose two different regulatory problems to the sporulating cell. Dihydrodipicolinate synthase serves a dual purpose in *B. subtilis.* In exponentially growing cells, dihydrodipicolinate synthase which catalyzes the penultimate step in DPA synthesis, is the first enzyme specific for the metabolic pathway leading to diaminopimelic acid and lysine. It is important to emphasize that, in contrast to the strong inhibition exerted by lysine on the *E. coli* enzyme[33], no feedback inhibition of *B. subtilis* dihydrodipicolinate synthase was demonstrable by L-lysine (10^{-2} M) or DPA (10^{-2} M). The lack of inhibition by lysine insures the continued function of the enzyme (in *B. subtilis*) during sporulation, when the intracellular pool of lysine may reach a high level due to protein turnover and to the presence of exogenous lysine in the complex media usually employed for sporulation of *B. subtilis* cultures.

We have further demonstrated, by the use of chloramphenicol and actinomycin D, that the increase in activity of the two DPA enzymes during sporulation is dependent on *de novo* protein synthesis, and so probably represents enzyme synthesis rather than activation of pre-existing enzyme protein[26].

Recent results obtained in this laboratory have shown that in *B. subtilis* the dihydrodipicolinate synthase isolated from the sporulating cells is identical, by different physico-chemical criteria, with the enzyme isolated from the exponential-growth cells[34]. This seems to be also the case in *B. megaterium*, since a mutant which is deficient in the dihydrodipicolinate synthase is unable to synthesize diaminopimelic acid and lysine and is also unable to make dipicolinic acid and cannot sporulate[31].

The above results imply that the regulation of the appearance of the two DPA enzyme activities during sporulation occurs at the level of either DNA transcription or translation of preformed mRNA. This type of control seems not to be the case with the two DPA enzymes studied in *B. subtilis*, since the addition of actinomycin D prior to the increase in activity prevents all subsequent accumulation of either enzyme. Dihydrodipicolinate synthase production is immediately stopped by this drug, even when the drug is added during the period of increased enzyme synthesis (Fig. 7, *curves 1*).

DPA synthase is also completely inhibited if actinomycin is added at any time prior to the appearance of the enzyme (Fig. 7, *curves* 2). The synthesis of the mRNA responsible for these two enzymes must therefore be concomitant with, or very closely precede, the increase in enzyme synthesis. On the other hand, a con-

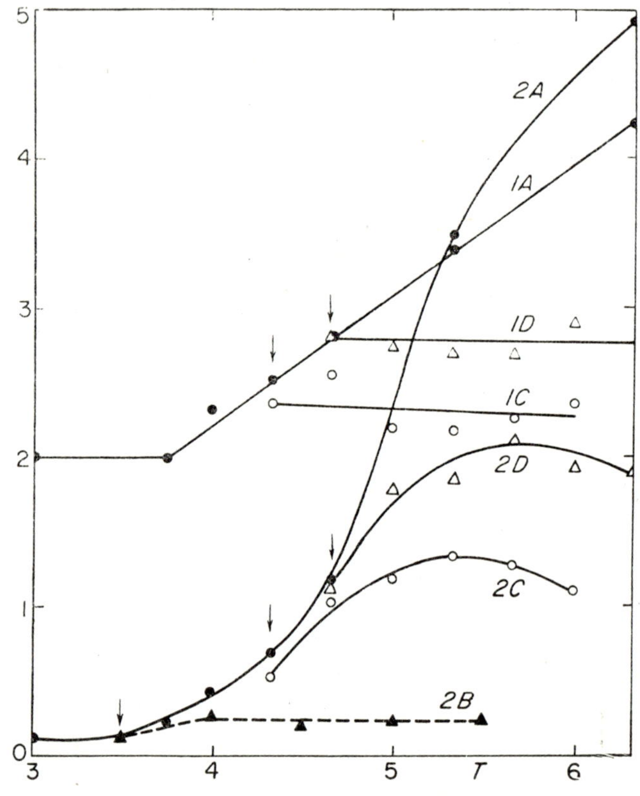

Fig. 7. Effect of actinomycin D on dihydrodipicolinate synthase and DPA synthase formation. Actinomycin (10 μg/ml) was added as indicated by the arrows. Enzyme activities in detergent-treated cells were measured. *T* time after the end of exponential growth (hours); *1* dihydrodipicolinate synthase, units/mg $\times$ 10^{-2}, *2* DPA synthase, units/mg $\times$ 2; *A* no actinomycin, *B* actinomycin at $T_{3.5}$, *C* actinomycin at $T_{4.3}$, *D* actinomycin at $T_{4.7}$. (From Ref.[26])

siderable amount of DPA synthase accumulates when actinomycin is added after enzyme synthesis has been initiated (Fig. 7). Our experiments have shown[26] that the accumulation of DPA synthase is several-fold greater after actinomycin addition than after the addition of chloramphenicol and would suggest that the mRNA for this enzyme may be relatively stable (a half-life of about 15 min) as compared with the stability of general pulse-labeled RNA or the mRNA for histidinase in *B. subtilis*, with a half-life of 1.0 and 2.4 min, respectively[35].

DPA enzymes in the t_s-4 mutant. For the preceding obvious reasons, it was important to test the activities of the DPA enzymes in the temperature-sensitive mutant. Fig. 8 shows that the kinetics of activity of the dihydrodipicolinate synthase in the t_s-4 mutant grown at 30 °C are similar to those observed in the parent strain

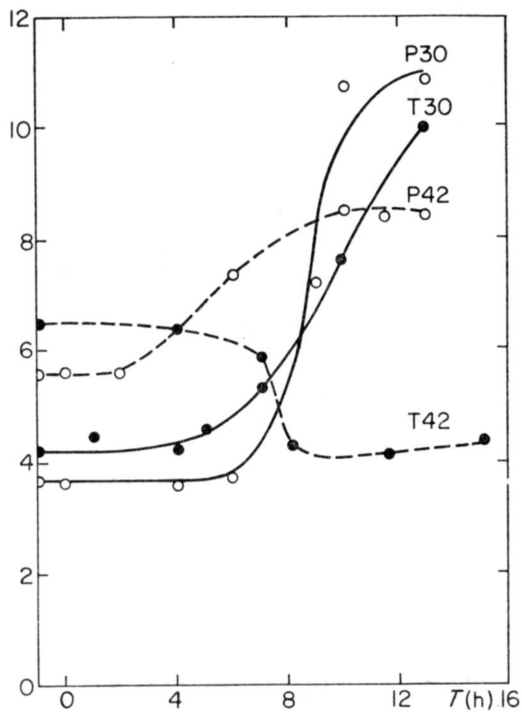

Fig. 8. Activity of the dihydrodipicolinate synthase (units/mg $\times$ 10^{-2}) of the t_s-4 thermosensitive mutant and of the parent strain grown at 30 and 42 °C. T t_s-4 mutant, P parent strain; 30 30 °C, 42 42 °C. (From Ref.[16])

or as shown in Fig. 6, relative to the wild type of *B. subtilis*. Under these conditions there was a net increase in the specific activity of the enzyme during sporulation as compared to that present in exponentially growing cells. A similar increase in activity was also observed in the parent strain when grown at 42 °C. However, when the t_s-4 mutant was grown at the high temperature, not only was such an increase never observed, but generally rather a decrease of about 30 to 40% of the basal level around T_7 to T_8.

Fig. 9 shows the kinetics of DPA synthase activity of the mutant and the parent strain grown at 30 and 42 °C. Here again, an increase can be observed in the specific activity during sporulation in the culture of the t_s-4 mutant when grown at 30 °C and practically no enzyme was formed in the culture at the nonpermissive temperature. In the parent strain, the normal increase in enzyme activity during sporulation took place in the cultures independent of the temperature they were grown, 30 or 42 °C.

Temperature-shift experiments. Experiments were also performed on the effects of temperature shifts, both "up" from 30 to 42 °C, and "down" from 42 to 30 °C, on the activities of the two DPA enzymes during the bacterial life-cycle. In the first series of experiments, cultures were grown at 42 °C for different periods of time

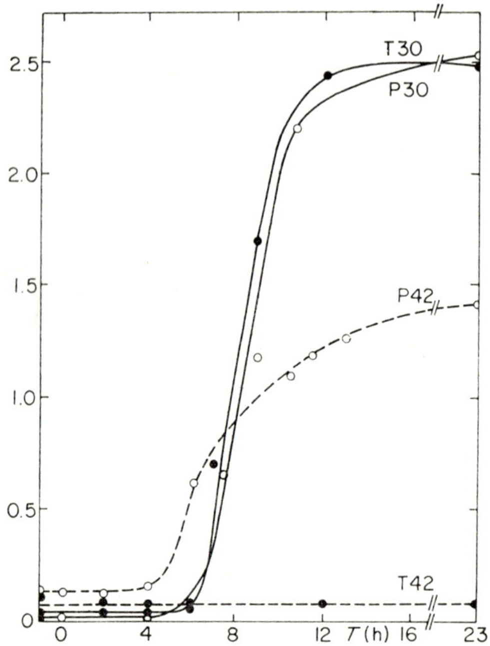

Fig. 9. Activity of the dipicolinate synthase (units/mg) of the t_s-4 mutant and of the parent strain grown at 30 and 42 °C. T t_s-4 mutant, P parent strain; 30 30 °C, 42 42 °C. (From Ref.[16])

and then shifted "down" to 30 °C. The incubations were continued until T_{22}. Thereafter the cultures were analyzed for enzyme activities in comparison to two control cultures maintained continuously, one at 30 °C, the other at 42 °C.

In a similar series of experiments, cultures were shifted up from 30 to 42 °C and the enzyme activities followed as above. The profiles of the curves corresponding to the two enzymatic activities of the DPA enzymes coincide practically with those observed for spore formation under identical shift experiments (*see* Figs 3 and 4).

t_s-4 as a pleiotropic mutant. It is necessary at this point to draw a first conclusion from the overall studies described above, on the thermosensitive sporulation mutant. It is clear that the t_s-4 mutant is an early blocked sporulation mutant. These types of mutants are perfectly capable of performing vegetative functions, but are unable to carry out the biochemical reactions unique to the spore formation. In general, these early blocked mutants exhibit a pleiotropic phenotype proper for regulatory elements which may play a role in the positive control mechanism.

Regulation at the transcription level. Another aspect of the regulatory apparatus that governs bacterial sporogenesis is the initiation of this process at the genetic transcription level.

DNA-DEPENDENT RNA POLYMERASE

a) **In *E. coli* and bacteriophages:** It is now a little more than a year since the discovery by BURGESS and TRAVERS groups[36,37] of the so-called sigma factor, a protein in *E. coli* normally associated with the DNA-dependent RNA polymerase and conferring on this enzyme the ability to discriminate between genes or groups of genes to be copied. In this function, the sigma factor is in fact an element of positive control over genetic expression. Since the discovery of the *E. coli* sigma factor, several analogue factors have been found in bacteriophages, T_4 and T_7[38,39]. The studies made on the transcription mechanism after infection of the cells by these phages give now a clear picture of the involvement of these factors in the replication of T_4 and T_7 coliphages. It is beyond this topic to discuss this problem here, although some obvious analogies exist between the events after phage infection and sporogenesis.

It is important only to say here that sigma factors, in conjuction with the "core" RNA polymerase, dictate the transcription, *in vitro*, of a specific class of genes, otherwise unrecognized. It is now well established that the sigma factor specifies the start of DNA transcription and that another factor, discovered by ROBERTS[40], the so-called rho factor, tells the RNA polymerase to stop making RNA, at the end of transcription of a particular genetic unit.

"Sigma" and "rho" factors provide a mechanism of positive control to regulate sequential gene expression and are therefore of particular interest in explaining the mechanism of regulation operating in cell differentiation.

b) **During sporulation:** An important observation in this direction has been made by LOSICK and SONENSHEIN[41] by their studies on the transcription of the virulent phage Φe DNA after infection of *B. subtilis* cells. When this phage infects vegetative cells it gives rise to a large progeny. However, when sporulating cells of *B. subtilis* are infected by the same phage, the cessation of the transcription of at least three early phage genes was observed and as a result the bacteria refused to support multiplication of Φe. These investigators have further shown that the strange behaviour of this phage is due to a change in template specificity of RNA polymerase during sporulation. This was further demonstrated with partially purified DNA-dependent RNA polymerase isolated either from vegetative or from sporulating cells of *B. subtilis*. The vegetative enzyme was able to transcribe, *in vitro*, either poly dAT or Φe DNA, while the RNA polymerase isolated from the sporulating cells was able to use dAT and not Φe DNA as template.

STUDIES ON THE RNA POLYMERASE OF *B. subtilis*

Independent of the above work[41], we have been investigating for a few years in our laboratory the properties of the DNA-dependent RNA polymerase isolated from vegetative cells and from dormant spores of *B. subtilis*[42]. Stimulated by the work done on the *E. coli* analogous enzyme, we used a similar biochemical approach

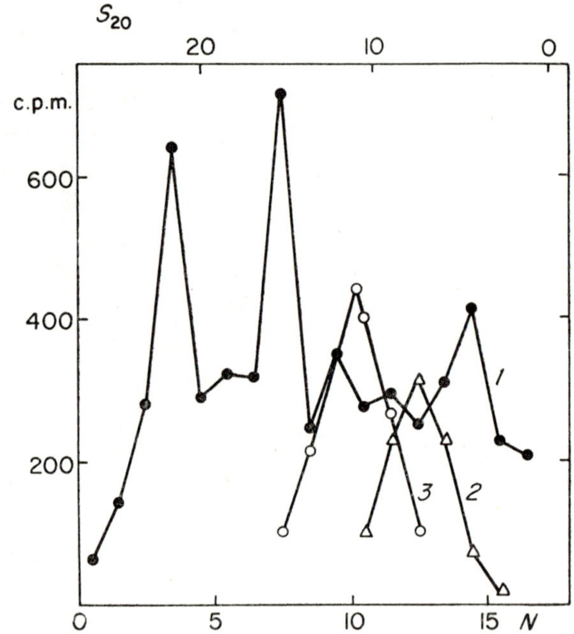

Fig. 10. Determination of the sedimentation coefficient of *B. subtilis* vegetative RNA polymerase by glycerol-gradient centrifugation. A sample containing 0.1 ml of enzyme, 15 μg of alcohol dehydrogenase and 20 μg of catalase was layered on top of linear glycerol gradient (5—30%, *v/v*) of 4.5 ml containing 0.01 M Tris-HCl buffer, pH 7.8, 0.01 M mercaptoethanol, 0.01 M magnesium acetate and 0.1 M-KCl. After centrifugation of 14 h at 25,000 r.p.m. at 4 °C the fractions were collected and analyzed for enzymatic activity. *N* number of fractions, S_{20} sedimentation constant; *1* RNA polymerase activity, c.p.m. of α-^{32}P-AMP, *2* alcohol dehydrogenase, *3* catalase. (From Ref.[43])

to the *B. subtilis* vegetative and spore RNA polymerase. Our earlier studies on the partially purified *B. subtilis* enzyme isolated from both sources, have shown that spore RNA polymerase differs from the vegetative enzyme by its sedimentation behaviour in a glycerol gradient. As shown in Fig. 10, the latter enzyme shows two major peaks corresponding to 23—21S, while the spore RNA polymerase yields under the same conditions a single peak corresponding to 13—14S (Fig. 11). Some other differences between the two enzymes have been observed by thermal inactivation studies. It was observed that at 55 °C the vegetative RNA polymerase presents

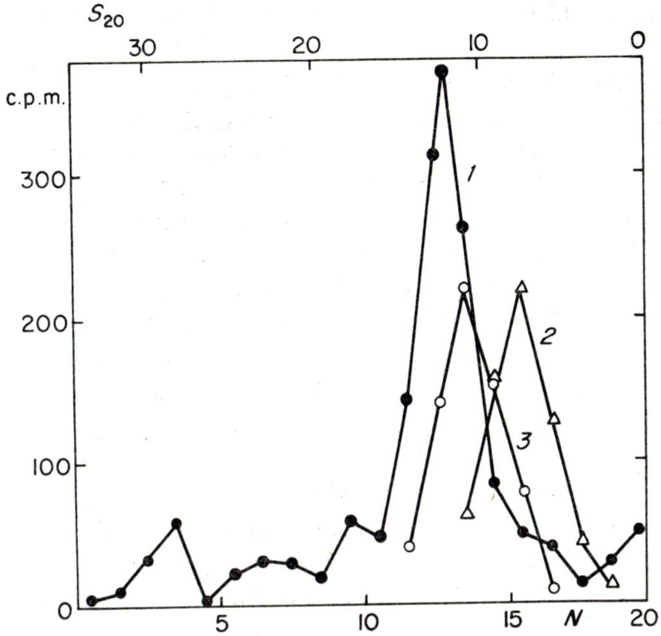

Fig. 11. Determination of the sedimentation coefficient of *B. subtilis* spore RNA polymerase (*see* legend of Fig. 10). *N* number of fractions, S_{20} sedimentation constant; *1* RNA polymerase activity, c.p.m. of α-^{32}P-AMP, *2* alcohol dehydrogenase, *3* catalase

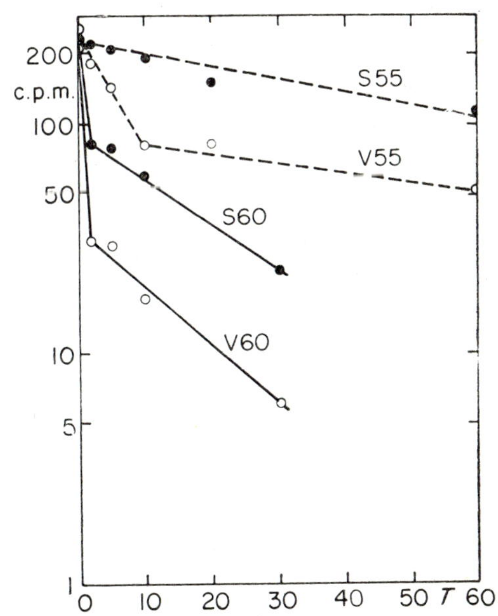

Fig. 12. Heat inactivation of vegetative and spore RNA polymerase. *T* heating time, min, c.p.m. RNA polymerase activity, c.p.m. of ^{14}C-UMP; S spores, V vegetative cells; 55 55 °C, 60 60 °C

two forms, one of which is rapidly inactivated (70% in 10 min), the other more slowly. The spore enzyme, on the contrary, is apparently more resistant and loses only 30—40% of its activity in about 60 min (Fig. 12).

These experiments, however, by no means prove that the differences observed between the spore and the vegetative enzyme reflects its structure *in vivo*. It is possible that the 13—14S form of the spore RNA polymerase was formed by the dissociation of 24S vegetative enzyme during the purification and that the higher resistance to thermal inactivation could simply be due to the association of the spore polymerase with some cell structure which renders it more stable. An example for such an association was shown for alanine racemase[44] and nucleoside ribosyltransferase[45].

Spore proteins, in general, might also arise by the action of a sporulation-specific protease on enzymes which are normal complements of vegetative cells[46].

The main question therefore to be answered was whether the vegetative and the spore RNA polymerase is the same enzyme or if the latter is synthesized *de novo* during sporulation or is a modified form of the vegetative enzyme.

In the light of the recent developments on the mode of action of this enzyme, as discussed above, this question can be now partly answered; at least it permits a rational approach to the molecular nature of the RNA polymerase and of its control mechanism operating during spore differentiation.

ISOLATION OF A STIMULATING FACTOR

When a partially purified RNA polymerase, which we have isolated from exponentially grown cells or from dormant spores of *B. subtilis*, was chromatographed on phosphocellulose according to BURGESS *et al.*[36] the profile of the pattern of a stepwise gradient elution with KCl, as shown in Fig. 13, was obtained. Peak A, eluted by 0.05 M-KCl, contains a protein factor and peak C, eluted by 0.4 M-KCl, the core polymerase. Table II shows the activity of the vegetative DNA polymerase before phosphocellulose chromatography and of the fractions A and B obtained

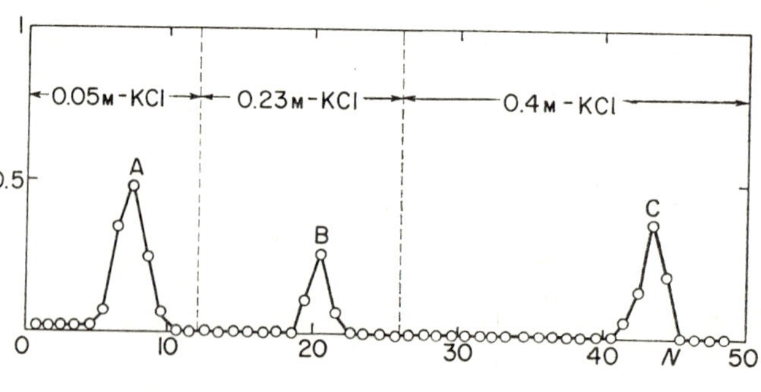

Fig. 13. Stepwise elution profile of *B. subtilis* vegetative RNA polymerase from a phosphocellulose column according to BURGESS *et al.*[36]. *N* number of fractions.

TABLE II

Stimulation of Activity of Vegetative Core RNA Polymerase by Vegetative Peak A Factor[a]

Source of enzyme	Amount[e] µg	DNA template[e] µg		
		thymus 50	T4 48	*B. subtilis* 50
Vegetative polymerase[b]	56	194	170	286
Vegetative core polymerase[c]	57	245	40	456
Vegetative peak A	11.5	17	11	16
Vegetative core polymerase + vegetative peak A	57 11.5	372	250	414
Vegetative polymerase + boiled peak A[d]	—	—	40	—

[a] nmol ^{14}C-AMP incorporated.
[b] RNA polymerase before phosphocellulose chromatography.
[c] The protein fraction eluted by 0.4 M-KCl (peak C) is designed as "core polymerase".
[d] 5 min at 100 °C.
[e] Added to the incubation mixture.

by the elution of the phosphocellulose column, assayed separately or together in the presence of different DNA templates. These results clearly show that core polymerase (peak C) is able to transcribe the thymus or the *B. subtilis* DNA's but is inactive with the phage T_4 DNA. However, the transcription of the latter is stimulated about six-fold by the factor present in peak A, which shows no activity by itself. The stimulation activity was completely abolished by boiling the peak A factor. It can also be seen in this table that the enzyme before phosphocellulose chromatography is equally active with all the three DNA's used as templates.

Table III shows the results obtained with the spore RNA polymerase. In this case, the core RNA polymerase shows almost the same activity with the three DNA's used as templates and is similar to that observed with the RNA polymerase before treatment with phosphocellulose. There was no stimulation of activity on adding spore peak A factor to the spore core polymerase. However, when the peak A factor isolated from the vegetative RNA polymerase, was added to the spore core polymerase a net stimulation of activity was observed with each of the three template DNA's used. It was interesting to note that highest stimulation was obtained with *B. subtilis* DNA. On the other hand, addition of the spore peak A to the vegetative core polymerase was without effect.

From the above experiments it can be concluded that the vegetative RNA polymerase contains a sigma-like stimulating factor required to promote transcription of T_4 DNA. As for the spore enzyme, several different explanations can be given at the present stage of the results obtained[43]. It is, however, tempting to conclude

TABLE III

Stimulation of Activity of Spore Core Polymerase by the Vegetative Peak A Factor[a]

Source of enzyme	Amount[d] µg	DNA template[d] µg			
		thymus 50	T4 48	*B. subtilis* 50	
				native	dena-tured
Spore RNA polymerase[b]	60	123	140	184	—
Spore core polymerase[c]	51	107	106	127	—
Spore peak A	11	0	27	22	—
Spore core polymerase + spore peak A	51 11	145	142	139	160
Spore core polymerase + spore peak A	51 22	—	139	148	—
Spore core polymerase + vegetative peak A	51 11.5	326	245	660	776
Vegetative core polymerase + spore peak A	51 11	186	0	234	—
Vegetative core polymerase + spore peak A	51 22	154	7	223	—

[a] nmol ^{14}C-AMP incorporated.
[b,c] As indicated in Table II.
[d] Added to the incubation mixture.

that in the dormant spore the RNA polymerase contains little or no dissociable factor at all and that such a factor is only synthesized during the first stage of germination when it may be required for the transcription of some early functions of the vegetative genome. The fact that, *in vitro*, the spore core polymerase can be stimulated by the addition of the vegetative sigma-like factor present in peak A, strongly favours this idea. To promote the transcription of the spore genome during sporulation, the synthesis of one or more different factors might be required.

SUBUNIT STRUCTURE OF VEGETATIVE *B. subtilis* RNA POLYMERASE

In an effort to gain some more information on the mechanism of action of the *B. subtilis* RNA polymerase during sporulation and germination we have undertaken a complete purification of this enzyme from vegetative cells and dormant spores. This has now been achieved for the vegetative enzyme by a method to be described elsewhere. Electrophoretic analysis by acrylamide gel of this enzyme preparation shows a single band (Fig. 14a). By treatment of the RNA polymerase by sodium dodecylsulfate (SDS) followed by acrylamide gel electrophoresis in presence of SDS, the dissociation of the enzyme into four subunits was obtained. The molecular weights of the subunits were determined on split gel electrophoresis by the method of SHAPIRO *et al.*[47] (using suitable markers) and their respective values were: 146,000, 94,000, 55,000 and 43,000 daltons. In order to compare the number and size of the subunits of the *B. subtilis* RNA polymerase with those of the *E. coli* enzyme[48,49], the two purified preparations were subjected to acrylamide split gel electrophoresis and the results are shown in Fig. 14b. It can be seen that the subunit structures of the two enzymes are similar although some difference in their molecular weights were observed with the *E. coli* preparation used here (*a gift from Dr. A. Santenac*). In addition, the *B. subtilis* RNA polymerase (Fig. 14b, on the right of the gel) contains a protein band with a molecular weight of about 55,000 daltons, not found in the *E. coli* enzyme. A similar band in the *B. subtilis* RNA polymerase has also been observed by SONENSHEIN[50].

By chromatography on phosphocellulose of this RNA polymerase, it was found that the two polypeptides with molecular weights of 94,000 and 55,000 daltons were eluted with 0.05 M-KCl. They were further separated by glycerol-gradient centrifugation (5–30%) and only the 55,000 dalton subunit was found to be active in stimulating the transcription of T_4 DNA by the core enzyme. The role, if any, of the 94,000 dalton polypeptide always found in our preparations remains unknown.

Experiments are in progress further to characterize the nature and functions of the different subunits present in *B. subtilis* vegetative and spore RNA polymerase.

INSTEAD OF A CONCLUSION

Regulation of the transcription level. One of the main characteristics of the physiological and biochemical events occurring during the development of spores is that these events are regulated in time and in space.

A spore-forming cell is faced with the task of controlling two different processes: growth and sporulation. These two processes differ not only in their physiological and biochemical properties but generally exclude each other. The general working hypothesis now is that the spore genome is repressed during vegetative growth and derepressed with the arrest of cell division. It could thus be assumed that the spore genome is either under negative or positive control. The former implies that a regulatory protein acts as inhibitor to stop transcription of spore genes unless the inducer is present. In the positive control mechanism it is assumed that a regulatory protein is produced which promotes the transcription of the spore structural genes. The transcription might occur in a sequential manner, a model of which has been suggested by HALVORSON[51] and reviewed by MANDELSTAM[13] in the light of the overall regulation process of sporulation.

The experiment reported here favours the idea that sporogenesis might be under positive control. Our results obtained with t_s-4 thermosensitive sporulation mutant are in agreement with this idea. It is quite possible that the functional absence of the regulatory protein, brought about by growing the thermosensitive sporulation mutant at the restrictive temperature, results in a pleiotropic phenotype which has lost the ability to catalyze the biochemical reactions leading to the biosynthesis of DPA required for the formation of a mature spore.

It is also possible that the altered regulatory protein produced at the nonpermissive temperature has lost the affinity for the inducer required in the initiation of the sporulation process, or that the formation of the inducer is suppressed at the high temperature.

Another set of data which argue for the positive type gene control operating during sporulation derive from the studies on the RNA polymerase in *E. coli* and in the phage systems.

The discovery of certain subunits of DNA-dependent RNA polymerase which confer template specificity on the enzyme further provide a specific mechanism for such regulation. In this case it was shown that a new class of genes can be activated by altering or replacing the specificity subunit of the RNA polymerase in such a way that initiation occurs at previously unrecognized promoter sites. An obvious such example was shown by the refined studies on the different coliphages[37–39,52]. Our studies[41], in addition to those of LOSICK and SONENSHEIN[39], on *B. subtilis* strongly support the idea of the existence during sporulation and germination, of one or more positive control mechanisms modifying the template and maybe all the transcriptive machinery[41,43].

Regulation at the translation level. Experiments on the patterns and functions of tRNA during the life cycle of *B. subtilis* would suggest a control of sporulation at the translation level[53,54]. Such a control has been found to play a role in differentiation in more complex organisms such as slime moulds, sea urchins or teleosts. At an early stage of development of the slime mould, *Dictyostelium discoideum*[55] for example, the capacity to form the uridine diphosphate—galactose polysaccharide transferase appears about four hours before the actual production of this developmental specific enzyme. On the other hand, SPIRIN and his associates[56] have recently shown the existence, in sea urchin and in the teleost *Misgurnus fossilis*, of cellular particles in which newly synthesized embryo informational RNA may be sequestered and stored to be used later in embryogenesis. These so-called "informosome" particles represent a more complex and sophisticated form of information storage and might function as a device for selecting certain genetic messages for delayed utilization and therefore a selective mechanism regulating gene expression and differentiation at the translational level.

The synthesis of some stable sporulation-specific mRNA and the subsequent regulation of the sporulation process at the translational level have been postulated by ARONSON[57] as a possible control mechanism for bacterial sporogenesis.

The observation of MANDELSTAM that commitment time for a given character during sporulation precedes the expression of this character by about 60 min, would further suggest this type of control[13]. The relative stability of mRNA for the spore-specific DPA synthase reported here also favours this idea, although it is known that continued RNA synthesis is necessary as late as T_5 for the production of complete, viable spores in *B. subtilis*[22] and *B. cereus*[58].

It would be interesting to see whether relatively stable mRNA is a general feature of sporulation-specific enzymes. Such stability might be advantageous during the sporulation process involving commitment to a relatively slow developmental program.

Another type of control on the translation level has recently been demonstrated by DUBE *et al.* in T_4 infected cells[59,60]. These authors have shown that after infection by T_4 phage a new set of initiation factors specified by the phage are made. These then displace the host-cell factors and cause the ribosomes of *E. coli* to recognize preferentially the start signals of T_4 mRNA's while restricting their capacity to recognize *E. coli* start signals. As a result the host-cell protein machinery turns "off" while phage protein begins to be made.

It is quite conceivable that a similar mechanism of control might operate in bacterial spore differentiation. This would, therefore imply that during this process some modification(s) or substitution(s) of initiation factors might temporarily confer on the 30S ribosomes the necessary specificity for initiation of spore proteins.

This mechanism, as noticed by DUBE *et al.*[59,60], would be analogous to that occurring on the transcription level (already discussed *above*) through the function of sigma factors, which transiently associate with RNA polymerase.

It is also important to consider the role of the physiological control on the expression of the spore genes at the translation level. Such a control may be operating throughout the stages of sporulation by different mechanisms: amount and stability of mRNA produced, repression of spore enzyme synthesis, feedback inhibition of the activity of these enzymes and finally katabolic repression of some spore inducible enzymes[61].

MEDICAL, VETERINARY AND INDUSTRIAL IMPORTANCE OF SPORE-FORMING BACTERIA

The importance of spore-forming bacteria in industrial technology and their medical and veterinary significance is now well established. Spore formers as insect pathogens are now widely used as effective microbial insecticides (in particular *B. popilliae* and *B. thuringiensis*)[62]. On the other hand, certain species of the genera *Bacillus* and *Clostridium* are highly pathogenic to members of the animal kingdom. This pathogenicity is frequently associated with a high degree of mortality resulting from the production of extremely potent toxins in the infected animals and in their food[63]. The great number of deaths caused every year by tetanus, botulinum and anthrax speaks by itself for the importance of spore-forming bacteria and spores for public health of men and animals. It is therefore not surprising that the knowledge of the genetics and biochemistry of spore-forming bacteria might contribute to solving some of these complex problems.

This investigation was supported by grants from the Délégation Générale à la Recherche Scientifique et Technique (Convention n° 66.00.301) and the Commissariat à l'Energie Atomique, France.

REFERENCES

1. Robinow C. F., p. 207 in *The Bacteria,* Vol. I. Academic Press, New York 1966.
2. Young E., Fitz-James P. C.: J. Biophys. Biochem. Cyt. **6**, 467 (1959).
3. Ohye D. E., Murell W. G.: J. Cell. Biol. **14**, 111 (1962).
4. Ryter A.: Ann. Inst. Pasteur **108,** 40 (1965).
5. Schaeffer P.: Folia Microbiol. **12,** 291 (1967).
6. Schaeffer P.: Bacteriol. Rev. **33**, 48 (1969).
7. Spizizen J., p. 125 in L. L. Campbell, H. O. Halvorson (Eds.): *Spores,* Vol. III. Am. Soc. Microbiol., 1965.
8. Hoch J. A., Spizizen J., p. 112 in L. L. Campbell (Ed.): *Spores,* Vol. IV. Am. Soc. Microbiol., 1969.
9. Takahashi I., p. 138 in L. L. Campbell, H. O. Halvorson (Eds.): *Spores,* Vol. III. Am. Soc. Microbiol., 1965.
10. Schaeffer P., Ionesco H., Ryter A., Balassa G., p. 553 in J. Senez (Ed.): *Mécanismes de Régulation des Activités Cellulaires chez les Microorganismes.* Colloque Int. C.N.R.S., Paris, **124** (1963).
11. Szulmajster J.: Bull. Soc. Chim. biol. **46**, 236 (1964).
12. Halvorson H. O., Szulmajster J., p. 462 in J. Mandelstam, K. McQuillen (Eds.): *Biochemistry of Bacterial Growth.* Blackwell, England 1968.
13. Mandelstam J.: Symp. Soc. Gen. Microbiol. **19**, 377 (1969).
14. Epstein R. A., Bolle M., Steinberg E., Kellenberger E., Boy de la Tour R., Chevalley R., Edgar R., Sussman M., Denhardt G., Lielanais A.: Cold Spring Harbor Symp. Quant. Biol. **28**, 375 (1963).
15. Neidhardt F. C.: Bacteriol. Rev. **30**, 701 (1966).
16. Szulmajster J., Bonamy C., Laporte J.: J. Bacteriol. **101**, 1027 (1970).
17. Freese E., Fortnagel P.: J. Bacteriol. **94**, 1957 (1967).
18. Millet J., Aubert J. P.: Compt. rend. **259**, 2555 (1964).
19. Slepecky R. A., p. 77 in G. M. Padilla, G. L. Whiston, I. L. Cameron (Eds.): *The Cell Cycle.* Academic Press, New York 1969.
20. Murell W. G.: Adv. Microbiol. Physiol. **1**, 133 (1967).
21. Kornberg A., Spudich J. A., Nelson D. L., Deutscher M. P.: Ann. Rev. Biochem. **37**, 51 (1968).
22. Szulmajster J., Canfield R. E., p. 587 in J. Senez (Ed.): *Mécanismes de Régulation des Activités Cellulaires chez les Microorganismes.* Colloque Int. C.N.R.S. 1965.
23. Szulmajster J., Hanson R. S., p. 162 in L. L. Campbell, H. O. Halvorson (Eds.): *Spores,* Vol. III. Am. Soc. Microbiol., 1965.

24. Bernlhor R. W., Gray B. H., p. 186 in L. L. Campbell (Ed.): *Spores*, Vol. IV. Am. Soc. Microbiol., 1969.
25. Vinter V., in G. W. Gould, A. Hurst (Eds.): *The Bacterial Spore*. Academic Press, New York 1969.
26. Chasin L. A., Szulmajster J., p. 133 in L. L. Campbell (Ed.): *Spores*, Vol. IV. Am. Soc. Microbiol., 1969.
27. Spotts C. R., Szulmajster J.: Biochim. Biophys. Acta **61**, 236 (1962).
28. Balassa G.: Biochem. Biophys. Res. Comm. **15**, 240 (1964).
29. Halvorson H. O., Swanson A., p. 121 in L. L. Campbell (Ed.): *Spores*, Vol. IV. Am. Soc. Microbiol., 1969.
30. Chasin L. A., Szulmajster J.: Biochem. Biophys. Res. Comm. **29**, 648 (1967).
31. Fukuda A., Gilvarg C.: J. Biol. Chem. **243**, 3871 (1968).
32. Bach M. L., Gilvarg C.: J. Biol. Chem. **241**, 4563 (1966).
33. Yugari Y., Gilvarg C.: Biochim. Biophys. Acta **62**, 612 (1962).
34. Callard R., Szulmajster J., *unpublished results* (1969).
35. Hartwell L. H., Magasanik B.: J. Mol. Biol. **7**, 401 (1963).
36. Burgess R. R., Travers J. J., Bautz E. K. F.: Nature **221**, 43 (1969).
37. Travers J. J.: Nature **223**, 1107 (1969).
38. Summers W. C. S., Siegel R. B.: Nature **223**, 1111 (1969).
39. Bautz E. K. F., Bautz F. A., Dunn J. J.: Nature **223**, 1022 (1969).
40. Roberts J. W.: Nature **225**, 1168 (1969).
41. Losick R., Sonenshein A. L.: Nature **224**, 35 (1969).
42. Kerjan P., Marchetti J., Szulmajster J.: Bull. Soc. Chim. biol. **49**, 1139 (1967).
43. Kerjan P., Szulmajster J.: F. E. B. S. Letters **5**, 288 (1969).
44. Stewart B. T., Halvorson H. O.: Arch. Biochem. Biophys. **49**, 168 (1954).
45. Nakata H., in H. O. Halvorson (Ed.): *Spores*, Vol. I. Am. Inst. Biol. Sci., Washington D. C. 1957.
46. Sadoff H. L., Hitchins A. D., Celikkol E.: J. Bacteriol. **98**, 1208 (1969).
47. Shapiro A. L., Vinuela E., Maizel J. V.: Biochem. Biophys. Res. Comm. **28**, 815 (1967).
48. Burgess R.: J. Biol. Chem. **244**, 6168 (1969).
49. Sethi V. S., Zillig W.: F.E.B.S. Letters **6**, 339 (1970).
50. Sonenshein A. L., *personal communication.*
51. Halvorson H. O.: Symp. Soc. Gen. Microbiol. **15**, 343 (1965).
52. Richardson J. P.: Nature **225**, 1109 (1970).
53. Doi R. H., Bishop H. L., Migita L. K., p. 159 in L. L. Campbell (Ed.): *Spores*, Vol. IV. Am. Soc. Microbiol., 1969.
54. Lazzarini R. A.: Proc. Nat. Acad. Sci. U.S.A. **37**, 51 (1966).
55. Sussman M., Sussman R. R.: Biochim. Biophys. Acta **108**, 463 (1965).
56. Spirin A. S.: Current Topics Developments Biol. **1**, 1 (1966).
57. Aronson A.: J. Mol. Biol. **13**, 92 (1965).
58. Fitz-James P. C., p. 529 in *Mécanismes de régulation des activités cellulaires chez les micro-organismes*. Colloq. Int. C.N.R.S., 1965.
59. Dube S. K., Rudland P. S.: Nature **226**, 820 (1970).
60. Steitz J. A., Dube S. K., Rudland P. S.: Nature **226**, 824 (1970).
61. Schaeffer P., Millet J., Aubert J. P.: Proc. Nat. Acad. Sci.U.S.A. **31**, 129 (1965).
62. Norris J. R., p. 485 in G. W. Gould, A. Hurst (Eds.): *The Bacterial Spore*. Academic Press, New York 1969.
63. Baird-Parker A. C., p. 515 in G. W. Gould, A. Hurst (Eds.): *The Bacterial Spore*. Academic Press, New York 1969.

Physiology and Genetics of Some Asporogenous Mutants of *Bacillus subtilis*

P. PIGGOT, J. MANDELSTAM

Microbiology Unit, Department of Biochemistry, University of Oxford, Oxford, Great Britain

INTRODUCTION

Spores have for a long period been a major preoccupation in medicine and in the food industry. More recently, sporulation has been studied for its intrinsic scientific interest but this too may have practical consequences. Thus, the early stages of sporulation in aerobic bacilli are accompanied by the production of antibiotics and of extracellular enzymes such as proteases, amylase, *etc.* Some of these products are now produced on an industrial scale, and SCHAEFFER has drawn attention[1] to the further practical benefits that might be expected from a detailed study of sporulation and its genetics.

There have been a number of morphological studies of sporulation in Gram-positive aerobic bacilli and the stages of the process are now well-characterized[2−5]. The biochemical events that accompany each stage have been reviewed[1,4,6−9]. Recently, too, there have been a number of genetic studies of sporulation in *Bacillus subtilis* using transduction with phage PBSl or transformation or a combination of these techniques. The results and the interpretations differ. TAKAHASHI concluded[10−12] that spore genes were randomly distributed along the chromosome. HOCH and SPIZIZEN found[13] a major group of genes linked to the *lys* marker and they also reported spore genes scattered along the chromosome. There was, however, some clustering near the *str* marker which is close to the *cysA* marker used in this study. The spore genes mapped by IONESCO *et al.*[14] were mostly linked to the *lys* or adjacent *phe* markers. ROGOLSKY found[15] spore genes largely clustered in three linkage groups at *cysA*, *metA*—*ura* and *lys*.

The interpretation of all genetic findings is complicated by the fact that mutants may be asporogenous, or oligosporogenous, even though they are not damaged in any specific spore function. Thus, FORTNAGEL and FREESE have isolated[16] a large number of mutants which are damaged in a vegetative function (*e.g.* an enzyme of the citric-acid cycle) that is required during sporulation. If such mutants were mapped without knowledge of their biochemical lesions the result could be misleading.

A further complication is introduced by the fact that because the number of genes involved in sporulation is large, there is a distinct possibility that double mutants, defective in more than one function, may arise particularly after treatment with mutagens. Since sporulation is a process in which a block at any stage will prevent the development of later stages, it is apparent that if a mutant strain carries two genes affecting sporulation the phenotype will be that of the earlier block.

In the present study we have mapped over 40 asporogenous mutations. The strains used did not include mutants that grew badly on lactate-minimal medium, and we therefore hope that those damaged in the citric-acid cycle have been excluded.

MATERIALS AND METHODS

Organisms. *B. subtilis* 168 (*trp*) was used. This sporulates normally, is referred to as wild-type, and forms brown colonies on minimal agar plates. Asporogenous mutants which formed white colonies were obtained by treatment with mutagens which are indicated as follows: A, acridine orange; E, ethyl methanesulphonate; N, nitrous acid; NG, N-methyl-N′-nitro-N-nitrosoguanidine; U, ultraviolet irradiation.

Culture and sporulation. The organisms were grown with shaking at 37 °C in a hydrolysed casein medium[17] to a density of 0.25 mg dry wt./ml. The culture was centrifuged and the cells transferred to the same volume of resuspension medium. This medium contained glutamate and inorganic ions[17], and incubation with shaking produced about 80% refractile spores in the wild type in 7—8 h.

Chemical and biochemical characteristics. Protease, antibiotic, alkaline phosphatase, glucose dehydrogenase, DPA, refractility and heat resistance were examined and determined as described by Waites *et al.*[18].

Electron microscopy. After the mutants had been in resuspension medium for 6—7 h the cells were fixed and sections prepared as described by Kay and Warren[19].

Genetic mapping. Spore genes were mapped by determining co-transduction of the *spo* character with one or other auxotrophic marker. As recipients, sporulating strains were chosen with auxotrophic requirements to cover the genetic map (*see* Fig. 2). The recipients were transduced to prototrophy using the lysates resulting from the infection of the *spo* strains with phage PBSl (Ref.[10]) and plating on lactate-minimal plates containing tryptophan. The method was modified for the *trp* locus: *spo trp* recipients were transduced to tryptophan independence using a PBSl lysate prepared on a spo^+ trp^+ strain. The transductant colonies were left on the plates for several days to allow development of the pigment characteristically associated with sporulation[20,21]. They were then scored as spo^+ (brown) or *spo* (white). The scoring was checked by examining the cells in the phase-contrast microscope and also by subculturing individual colonies on lactate-minimal medium and re-checking their sporulation characteristics.

Three-factor crosses in the *cysA* region were done using a *cysA str* recipient. Prototrophic transductant colonies were selected as before, and scored as spo^+ or *spo*. They were then tested for streptomycin resistance on nutrient agar plates containing 100 µg/ml of streptomycin.

RESULTS

Morphology of asporogenous mutants. Electron microscopy was done on 10 of the mutants and the photographs obtained have been published elsewhere[18]. From these it was apparent that in some mutants sporulation proceeded normally to the blocked stage and then stopped, while in others further changes could take place in any of a number of ways giving rise to various bizarre forms. In one of these, N25, the cell laid down membrane in the normal manner at stage II; however, this membrane failed to become detached from the membrane of the mother cell and instead continued to increase in quantity until it enclosed a very substantial part of the cytoplasm. Other aberrant forms were also found at stage II. These included mutants in which cell wall was laid down together with membrane instead of membrane alone thus giving rise to "pygmy" cells resembling the abortively disporic mutants of *Bacillus cereus* seen by YOUNG[22] and of *B. subtilis* by RYTER, SCHAEFFER and IONESCO[2]. Three of these (NG1.82, NG11.2 and NG18.6) were obtained independently.

The morphological stage at which each mutant was blocked and the biochemical characteristics are set out in Table I together with an indication of the type of abnormal development, if any. A schematic representation is shown in Fig. 1.

Preliminary mapping with characterized mutants. Transducing lysates were prepared with phage PBSl and each of the asporogenous strain listed in Table I. These were used to test for cotransduction of the *spo* character with 12 or more of the auxotrophic markers shown in Fig. 2. The linkages found and the co-transduction frequencies are cited in the Table I. Although other workers have used two-factor crosses of this kind to produce a linear map of sporulation genes we have not found that these values are reproducible enough for this purpose (*see* also IONESCO *et al.*[14]).

The results were consistent with the occurrence of a single mutation in each of the strains in this batch except E31 *(see below)*. Nevertheless, in five of the experi-

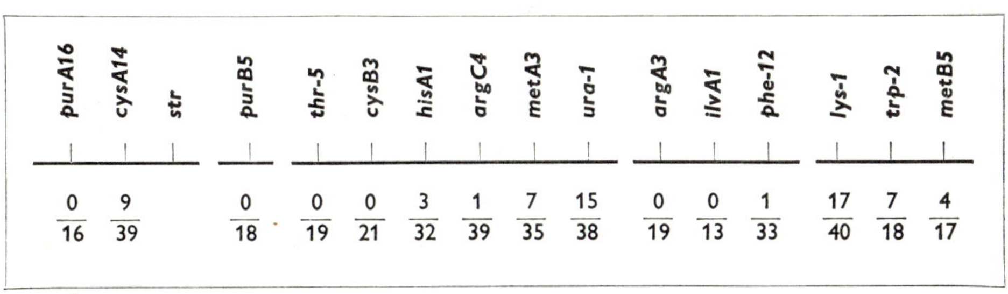

Fig. 2. Distribution of asporogenous mutations along the *B. subtilis* chromosome *(kindly provided by Dr C. Anagnostopolous and Dr D. Dubnau)*. The numerator is the number of asporogenous strains showing linkage to a particular marker; cotransduction frequencies of less than 4% are not considered significant. The denominator is the number of strains tested against each marker. The genetic map is essentially that of DUBNAU *et al.*[23]. More detailed distribution of mutations associated with asporogeny, together with transducing frequencies, are given in Fig. 3.

ments (including E31) (Table I) single transductant colonies were isolated and characterized biochemically to ensure that they had the same sporulation phenotype as the donors from which they had been derived.

Some limited conclusions can be drawn from the results in Table I: mutations which are associated with the same type of morphological block may be widely separated, such as E33 and E31 which are stage IV. Conversely genes showing close linkage, *e.g.* N25 and E22 are associated with morphological effects which are very different.

TABLE I

Characterization of a Set of 10 Asporogenous Mutants of *B. subtilis*[a]

Strain	Protease	Antibiotic	Alkaline phosphatase	Glucose dehydrogenase	Refractility[b]	Dipicolinic acid	Stage blocked	Linkage and cotransduction frequency	
								marker	%
E22	—	—	—	—	—	—	0	*cysA*	95
N25[c,d]	+	+	—	—	—	—	II	*cysA*	76
NG1.82[c,e]	+	+	—	—	—	—	II	*lys* *trp*	79 26
E21[f]	+	+	+	—	—	—	II	*ura* *metA*	41 52
NG1.13[c]	+	+	+	—	—	—	III	*lys* *trp*	46 44
E34	+	+	+	+	±	—	III	*lys* *trp*	50 21
A8[c]	+	+	+	+	±	—	IV	*hisA*	1[g]
E33	+	+	+	+	±	—	IV	*argC*	47
E31[c]	+	+	+	+	±	—	IV	*phe*	13[h]
A23	+	+	+	+	±	—	IV	no linkage found	

[a] Bacteria were transferred from growth medium to a medium which induces sporulation in the wild-type (*see* text). Samples were taken over a period of several hours for measurement of events characteristically associated with sporulation. Data taken from Ref.[18].

[b] ± indicates poor refractility giving grey appearance in phase contrast.

[c] Transductant colony biochemically characterized and shown to have the same phenotype as the donor *spo* strain.

[d] Electron microscopy: excess membrane.

[e] Electron microscopy: abortively disporic.

[f] Electron microscopy: aberrant cell division.

[g] Co-transduction frequency below the level normally regarded as significant. However the *spo* transductant colonies had the same phenotype as the donor, and the linkage was thought to be significant.

[h] E31 was a double mutant (*see* text); the second mutation was linked to *ura*, and was associated with an oligosporogenous phenotype.

Mapping with incompletely characterized mutants. From the results so far described it was not possible to tell whether there was any clustering of the spore genes because the numbers were too low. Accordingly mapping was carried out with a further group of 31 mutants even though the phenotypes had been characterized only very roughly. At first *spo* mutations were tested for cotransduction with 12 or more of the auxotrophic markers shown in Fig. 1. Then it became apparent that there was some clustering and the remaining mutations were tested for linkage to the markers to which the first group had shown linkage. The numbers tested for cotransduction and found linked at each locus are shown in Fig. 2, which represents the combined results for all mutations that were mapped.

There were three large clusters and these will be described separately beginning with the cluster nearest the origin.

There were nine *spo* mutations in the *cysA* region (Fig. 3A, in which those genes that occur in double mutants are indicated by "D"). Seven mutations in this region were closely linked and were cotransduced with *cysA* with a frequency greater than 75%. From the three-factor crosses that have so far been done these mutations appear to lie to the left of *cysA* (Table II). Although the phenotype has not been checked in all the transductants, two of the mutants, E32 and A19, are relatively late mutants which develop to the stage of refractility (stage IV) and so are less likely to be double mutants. One of them is very close to the auxotrophic marker (85% cotransduction) whereas the other shows a very much lower degree of linkage. Of the other mutants in this group five from their known characteristics, may be blocked anywhere between stage 0 and stage II but again the characterization is

TABLE II

Phenotypic Properties Associated with Nine Mutations Mapping in the *cysA* Region[a]

B. subtilis strain	Order by three-factor cross	Sporulation characteristic			Stage blocked
		antibiotic	alkaline phosphatase	refractility	
E32[b]		+	+	±	IV
NG2.6	*spo cysA str*	+	—	—	0-II
N25		+	—	—	II
A19[b]		+	+	±	IV
NG15.4	*spo cysA str*	+	—	—	0-II
NG14.22	*spo cysA str*	+	—	—	0-II
NG20.12		+	—	—	0-II
E22[b]		—	—	—	0
NG17.15	*spo cysA str*	+	—	—	0-II

[a] In six cases the phenotype of the transductant was checked to be the same as that of the donor. Three-factor crosses showed that at least four mutations lie to the left of *cysA* (*see* Fig. 2).

[b] Phenotype of transductant not determined; phenotype of donor asporogenous strain given.

tentative for the reasons already given. However, N25 is well-characterized and is known to be blocked at stage II and to produce excessive amounts of membrane (*see* Fig. 1).

The next major cluster is in the *metA—ura* region in which 15 asporogenous mutations were mapped (Fig. 3B). Most of those that have been tested show linkage to both auxotrophic markers. Now, *metA* and *ura* are so far apart that they are cotransduced with very low frequency[23]. From the values given in Fig. 2 it would therefore appear that at least five of the mutations found here lie between the two markers. The positions of the remaining mutations have still to be determined. From their biochemical behaviour they appear to be blocked at a variety of stages of sporulation but the transductants have still to be characterized. Furthermore, a number of the mutations concerned are certainly carried in strains which are double mutants and it would therefore be misleading to attribute even a tentative phenotype to them. In addition to asporogenous mutations, three oligosporogenous mutations were linked to *metA* and *ura*. These mutations were harboured by asporogenous strains (E13, E31, NG20.12) which also carried an asporogenous mutation mapping in another region. They were not further studied.

The third and largest group (17) of asporogenous mutations are cotransduced with *lys*, but a number of them show linkage to *trp* as well, and some also to *metB5* (Fig. 3C). For the reasons already stated the phenotypes of these strains will not be given.

There is a small cluster near *hisA1* to which three mutations are linked. One mutant, E31, which we have already mentioned is a double mutant, has one mutation with *spo* phenotype (stage IV) linked to *phe* whereas the second mutation lies between *metA* and *ura* *(see above)*. There were, furthermore, three mutants which gave active lysates but the *spo* mutations were not cotransduced with any of the auxotrophic markers tested. These too must be presumed, for the time being, to lie outside the clusters. It is possible that they will be found to lie in one or other of the gaps of the linkage map shown in Fig. 2 and, in fact, IONESCO *et al.*[14] have shown that there are sporulation genes linked to the acriflavine-resistance marker that bridges the gap between *phe* and *lys*.

DISCUSSION

To summarize the results of our mapping we can say that the majority of the sporulation genes lie in three large clusters and our findings are thus very different from the general spread reported by TAKAHASHI[11,12]. It will be useful at this stage to compare our results with those of other workers. At first glance we appear to have found the same groups as ROGOLSKY[15], *i.e.* clusters at *cysA*, *metA—ura* and *lys*, but on closer examination there appear to be differences. Thus ROGOLSKY reported six sporulation genes in the *cysA* region and all as lying to the right of *cysA*, whereas

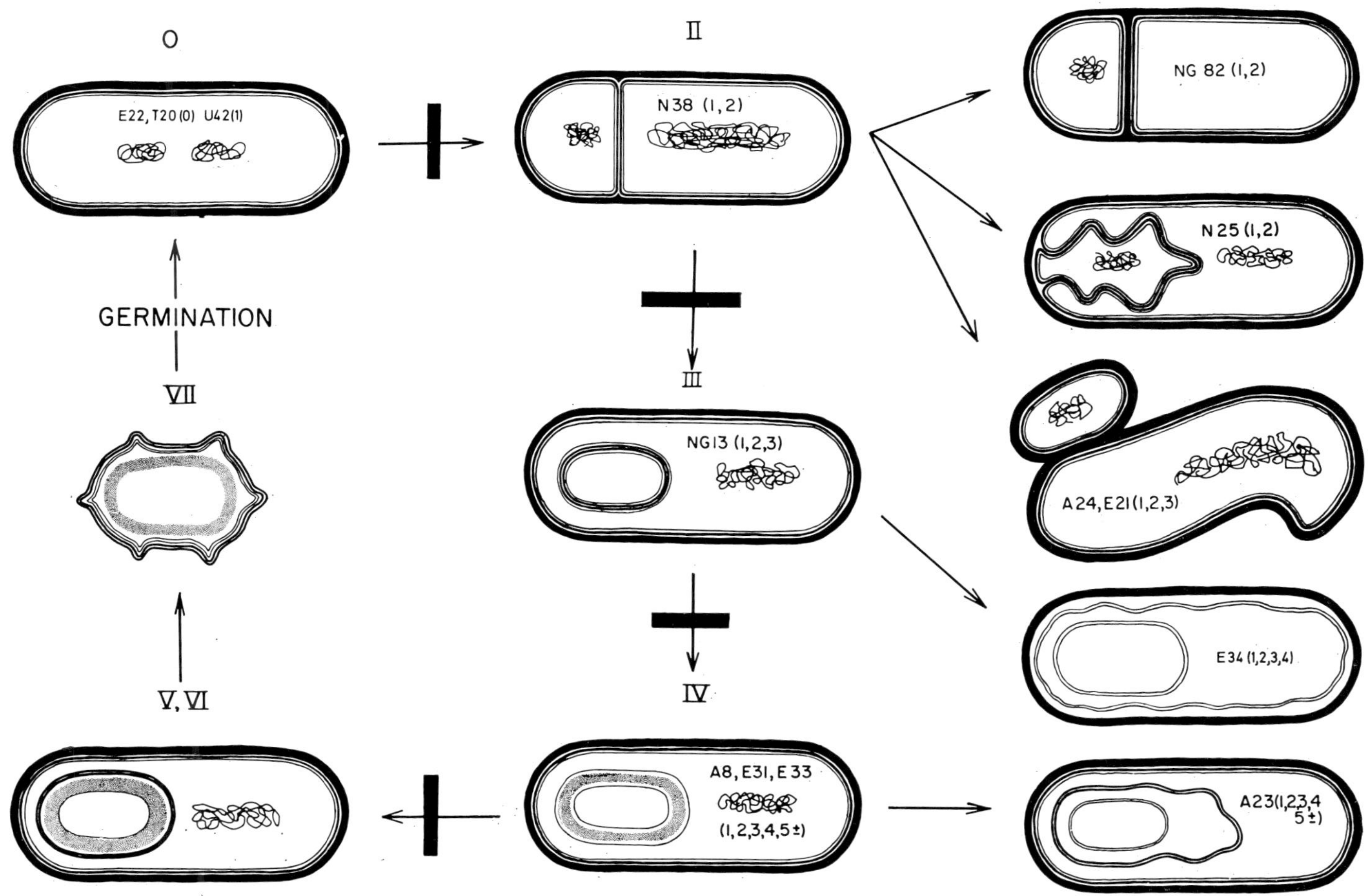

Fig. 1. Schematic representation of the stages of sporulation showing the forms associated with the different mutants. The figure is taken from Ref.[18], and mutants T20, U42, N38 and A24 are not considered in the present paper. Numbers in parentheses indicate: **1** protease, 2 antibiotic, **3** alkaline phosphatase, **4** glucose dehydrogenase, **5** refractility. Roman numerals represent the stages of sporulation seen in normal spore development.

our results show that of the nine genes here at least the four which have been mapped by three-factor crosses are to the left of *cysA*. Since phage PBSl can transport 5% or more of the host chromosome it is therefore possible that the group of spore genes we have found may be quite distant and in a separate cluster from those found by ROGOLSKY. HOCH and SPIZIZEN also discovered[13] a group of spore genes in this region but since their map distances are related to the *str* marker it is difficult to tell whether they lie to the right or left of *cysA*. By contrast the maps of TAKAHASHI[11] and of IONESCO *et al.*[14] show no cluster at all in this region but only four rather scattered genes.

There are further differences when we come to consider the *metA*—*ura* region in which ROGOLSKY's map shows[15] 11 genes with nine of them to the right of *ura* and only two lying between the auxotrophic markers. The latter is, however, the position in which at least five of the mutations in our cluster appear to be. When the others are ordered by three-factor crosses some may turn out to lie on the right as well. HOCH and SPIZIZEN[13] who have mapped a considerable number of genes show only two between *metA* and *ura* while IONESCO *et al.*[14] have not oriented the three genes they find linked to *ura*.

It is only in the *lys* region (Fig. 3C) that there seems to be general agreement. All ten of the mutations that we have so far examined by three-factor crosses lie to the left of *lys*. HOCH and SPIZIZEN report[13] about 39 genes to the left of *lys* while ROGOLSKY[15] and IONESCO *et al.*[14] also show these as major gene clusters. From the results of these last authors it seems that mutations producing sporulation blocks at stages 0, II, III and IV may be in fairly close juxtaposition.

Now, some of the relative compactness of our gene clusters may be due to the fact that we had omitted mutants which were damaged in the citric-acid cycle or in other obvious ways. Mutants of this type account for at least one of the additional groups of sporulation genes mapped by HOCH and SPIZIZEN[13], but there is no very obvious explanation for the fact that different workers have found clusterings that seem to differ significantly in their position on the chromosome. The distances seem too great to be accounted for by the crudeness of the mapping that results from the use of phage PBSl.

We now come to the relationship, if any, between the map position of a gene and the stage of sporulation in which it is involved. Even if we disregard mutants which are likely to be asporogenous for trivial reasons, and even if we take into account the possibility that some, perhaps even a substantial proportion, of the mutants examined by us and by other workers may be double mutants we are still left with some apparently basic facts. 1. Interspersed among the spore genes there are some auxotrophic markers, and there are probably many more that have not been characterized. 2. The order of sporulation genes does not correspond with the temporal sequence of morphological and biochemical events that characterizes sporulation. 3. Genes concerned in different stages of spore formation may be closely linked and conversely genes concerned with the same stage need not be closely linked.

If these conclusions are considered in conjunction with what is known about physiology of sporulation it would appear that models of the process that involve linear transcription of the genes in the correct order are untenable. "Switching on"

A Linkage group I

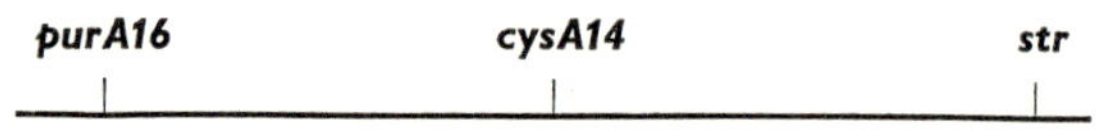

	cysA14
E32	9
NG2.6(D)	38
N25	76
A19	85
NG15.4(D)	92
NG14.22(D)	93
NG20.12(D)	95
E22	95
NG17.15(D)	95

B Linkage group II

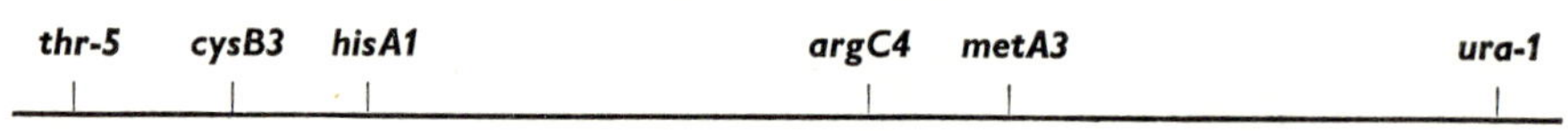

hisA1		argC4	metA3		ura-1
18	NG11.7		39	NG14.22(D)	10
12	NG17.22(D)		27	NG17.17(D)	13
8	NG17.29(D)			NG17.29(D)	13
1	A8		44	NG15.4(D)	15
1	E33	47		NG17.15(D)	20
			53	NG12.5(D)	27
			1	NG1.67	36
			1	NG8.17	37
			2	NG4.14	39
			52	E21	41
				NG20.4(D)	55
			2	NG2.6(D)	68
			4	E10	83
			5	NG12.12	84
				NG17.22(D)	94
			17[a]	NG20.12(D)	7[a]
			46[a]	E31(D)	18[a]
			49[a]	E13(D)	37[a]

Fig. 3.

C	Linkage group IV		
	lys-1	*trp-2*[b]	*metB5*
NG18.8	96	70	—
NG1.82	79	29	29
NG16.2	65	63	—
A3	56	—	7
NG16.2	65	63	—
NG6.21	56	—	—
E34	50	21	—
NG14.7	48	—	—
NC1.13	46	44	—
NG1.13	46	44	—
NG12.5(D)	46	0	—
NG14.2	45	41	8
NG17.17(D)	43	—	—
NG7.2	34	—	—
NG1.161	14	0	0
A1	8	—	—
NG20.4(D)	7	—	—
E13(D)	6	—	—

Fig. 3. Linkage of asporogenous mutations in three linkage areas of the *B. subtilis* chromosome[23]. Those mutations that occur in doubly mutant strains are indicated by "D". The numbers indicate the cotransduction frequency (in %) with the marker concerned. In general where several transduction were performed, all the data were used. However, in some cases particular lysates gave lower co-transduction frequencies as they aged, and these lower values were not included.

[a]Oligosporogenous, *see* text.
[b]Linkage to *trp-2* was recognized by the occurrence of *spo*$^+$ *trp*$^+$ transductants with the asporogenous mutants (*spo trp*) as recipients. Thus linkage would not be detected in doubly asporogenous mutants.

of the spore operons must be mediated by diffusible molecules of some sort and it seems necessary to postulate that these molecules are themselves part of the sequence that they regulate. These implications have been discussed elsewhere[9,18].

We thank Dr D. Karamata for the bacteriophage PBS1 and for advice on preparation of transducing lysates. We are especially grateful to Mrs Valerie Cooper for her help in the genetic analysis. This work was supported by the Science Research Council.

REFERENCES

1. SCHAEFFER P.: Bacteriol. Rev. **33**, 48 (1969).
2. RYTER A., SCHAEFFER P., IONESCO H.: Ann. Inst. Pasteur **110**, 305 (1966).
3. YOUNG I. E., FITZ-JAMES P. C.: J. Biophys. Biochem. Cytol. **6**, 467 (1959).
4. SZULMAJSTER J.: Bull. Soc. Chim. biol. **46**, 443 (1964).
5. OHYE D. F., MURRELL W. G.: J. Cell. Biol. **14**, 111 (1962).
6. HALVORSON H. O.: Symp. Soc. Gen. Microbiol. **15**, 343 (1965).
7. MURRELL W. G.: Adv. Microbial Physiol. **1**, 133 (1967).
8. KORNBERG A., SPUDICH J.A., NELSON D.L., DEUTSCHER M.P.: Ann. Rev. Biochem. **37**, 51 (1968).
9. MANDELSTAM J.: Symp. Soc. Gen. Microbiol. **19**, 377 (1969).
10. TAKAHASHI I.: J. Gen. Microbiol. **31**, 211 (1963).
11. TAKAHASHI I., p. 138 in L.L. Campbell, H. O. Halvorson (Eds.): *Spores,* Vol. III. American Society for Microbiology, Ann Arbor 1965.
12. TAKAHASHI I., p. 102 in L.L. Campbell (Ed.): *Spores,* Vol. IV. American Society for Microbiology, Bethesda 1969.
13. HOCH J.A., SPIZIZEN J., p. 112 in L.L. Campbell (Ed.): *Spores,* Vol. IV. American Society for Microbiology, Bethesda 1969.
14. IONESCO H., MICHEL J., CAMI B., SCHAEFFER P.: J. Appl. Bact. **33**, 13 (1970).
15. ROGOLSKY M.: Can. J. Microbiol. **15**, 787 (1969).
16. FORTNAGEL P., FREESE E.: J. Bacteriol. **95**, 1431 (1968).
17. STERLINI J., MANDELSTAM J.: Biochem. J. **113**, 29 (1969).
18. WAITES W. M., KAY D., DAWES I. W., WOOD D. A., WARREN S. C., MANDELSTAM J.: Biochem. J., *in press.*
19. KAY D., WARREN S. C.: Biochem. J. **109**, 819 (1968).
20. IICHINSKA E.: Microbiology (U.S.S.R.; *English translation*) **29**, 147 (1960).
21. SCHAEFFER P., IONESCO H.: Compt. rend. **251**, 3125 (1960).
22. YOUNG I. E.: J. Bacteriol. **88**, 242 (1964).
23. DUBNAU D., GOLDTHWAITE C., SMITH I., MARMUR J.: J. Mol. Biol. **27**, 163 (1967).

Metabolic Pathways during Sporulation of *Bacillus subtilis*

PETER FORTNAGEL

Ruhr-Universität Bochum, Abteilung für Chemie, Bochum, G.F.R.

The normal formation of bacterial spores requires certain metabolic pathways. Energy as well as precursors for macromolecular synthesis have to be provided in sufficient amounts; metabolites which are components of spores must be formed; and, in addition, substrates with an inhibitory function on sporulation have to be removed. All these steps require the normal function of a number of biochemical reactions. For deciding which pathways are essential for sporulation two possible experimental procedures can be followed:

First, the effect of inhibitors of metabolic reactions on sporulation can be investigated. For example, α-picolinic acid, an inhibitor of the conversion of citrate into isocitrate, the reaction catalyzed by aconitase, was shown to inhibit sporulation, just because it inhibits this reaction[1].

Second, blocks in metabolic pathways can be introduced by mutation. If such a single block in a mutant affects the formation of spores, it is likely that the blocked reaction is directly or indirectly necessary for the formation of spores. This second technique has the advantage that the block introduced is highly specific for only one reaction. Furthermore, mutants can be isolated for almost any metabolic reaction, which makes the investigation of the effect of mutational blocks on sporulation more comprehensive than the limited number of blocks which can be specifically introduced by inhibitor action.

The necessity and the function of metabolic pathways during sporulation requires the knowledge of the single steps of these pathways and their function during growth. Only then is it possible to investigate changes which are sporulation-specific. Glycolysis, the pentose-phosphate shunt, and the citric-acid cycle represent important pathways during growth of bacteria for both energy production and the synthesis of metabolic precursors of macromolecular synthesis. The utilization of these pathways depends on the medium conditions. Glucose as carbon source will lead synthesis into the citric-acid (CA) cycle, glutamate as carbon source will direct synthesis out of this cycle and towards *de novo* synthesis of sugars and sugar-containing compounds. The question is whether these pathways are necessary for sporulation and if so, how they are used during this cellular development.

Mutants of *Bacillus subtilis* are known which are blocked in different metabolic pathways[2-6] including the citric-acid cycle. We have identified and analyzed blocks in this cycle including the enzymes aconitase, isocitrate dehydrogenase, oxoglutarate dehydrogenase, and succinate dehydrogenase[3,4]. The blocks of these mutants are shown in Fig. 1. They all are oligosporogenous which means that they show significantly reduced sporulation. The frequency of spores in complex and synthetic sporu-

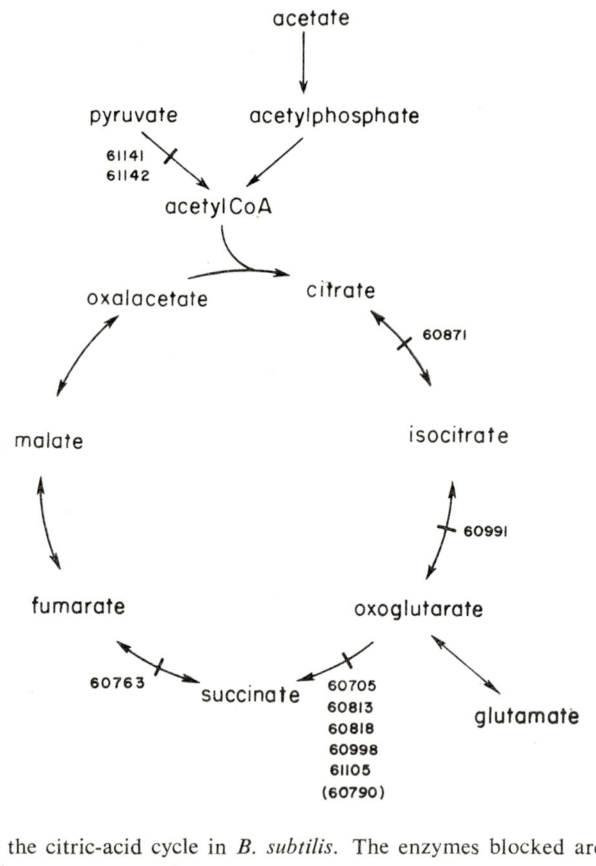

Fig. 1. Scheme of the citric-acid cycle in *B. subtilis*. The enzymes blocked are marked by the number of the different mutant strains.

lation media ranged from 10^{-2} to 10^{-6} depending on the mutant when it was 80 to 100% for the wild-type strain (Table I). SZULMAJSTER and HANSON[7] had already described an aconitase-negative mutant unable to sporulate normally. Recently, HAMPTON and HANSON[8] reported more CA-cycle mutants with blocks in oxoglutarate dehydrogenase, fumarase, and malate dehydrogenase.

Our CA-cycle mutants all contain a number of heat-resistant spores, which, after germination, give rise again to mutant colonies. Only in very rare cases these spores are true revertants and give rise to wild-type-like colonies. Although sporulation

TABLE I

Sporulation Properties of *B. subtilis* Mutants Blocked in the Citric-Acid Cycle[a]

Strain	Enzyme blocked	Frequency of sporulation[b] %
60015[c]	—	98
60871	Aconitase	1.1×10^{-4}
60991	Isocitrate dehydrogenase	5.5×10^{-5}
60705	Oxoglutarate dehydrogenase	1.0×10^{-5}
60813	Oxoglutarate dehydrogenase	1.5×10^{-4}
60818	Oxoglutarate dehydrogenase	7.5×10^{-3}
60991	Oxoglutarate dehydrogenase	2.0×10^{-4}
61105	Oxoglutarate dehydrogenase	4.7×10^{-7}
60763	Succinate dehydrogenase	1.0×10^{-4}

[a] For methods *see* Refs[3,4].
[b] In nutrient sporulation medium after 24 h incubation at 37 °C.
[c] Wild type.

is reduced in the mutants it is not completely blocked. This suggests that not spore biosynthesis but rather its regulation is affected in the mutants. The few true revertants which were obtained show in either case, when isolated for normal sporulation or restoration of the enzyme activity, a simultaneous restoration of the other parameter. This is sufficient to relate both the loss of enzyme activity and the reduction of sporulation to only one mutational block.

Blocks in the citric-acid cycle, when occurring in the part of the cycle which is necessary for the formation of oxoglutarate and its transamination product glutamate, allowed growth of mutants in minimal media, only if glutamate was present[4]. Mutants blocked in the part of the cycle which is required for glutamate metabolism grow on glucose as the only source of carbon but not on glutamate[3,4]. The distribution of the CA-cycle metabolites in the cells of the different mutants is significantly altered as compared with the wild type. This may be observed after incubation of the mutants in the presence of ^{14}C-labelled glutamate, extraction of the cells, and analysis of the hot-water extracts by thin-layer chromatography. Fig. 2 shows an autoradiogram of such an analysis for several sporulation mutants. In all cases, metabolites before the block are accumulated, those after the block are missing[4].

From the distribution pattern of the metabolites and from the growth behavior of our wild type and the mutants we can predict that our strains lack a functional glyoxylate bypath. They cannot grow on acetate as the only carbon source as it would require this pathway. In addition, the isocitrate dehydrogenase-negative mutant, when grown in the presence of ^{14}C-citrate, forms aconitate and isocitrate, which is accumulated[5]. None the metabolites of the glyoxylate bypath — the only possible drain of isocitrate in this mutant — were found.

Addition of missing metabolites to minimal glucose or minimal glutamate medium allowed the mutants to grow[3,4]; therefore, we tried to restore sporulation by addition of these metabolites to nutrient sporulation medium or after a shift from this medium into synthetic sporulation medium containing glutamate as carbon source[9]. None of the CA-cycle mutants responded to this treatment with normal

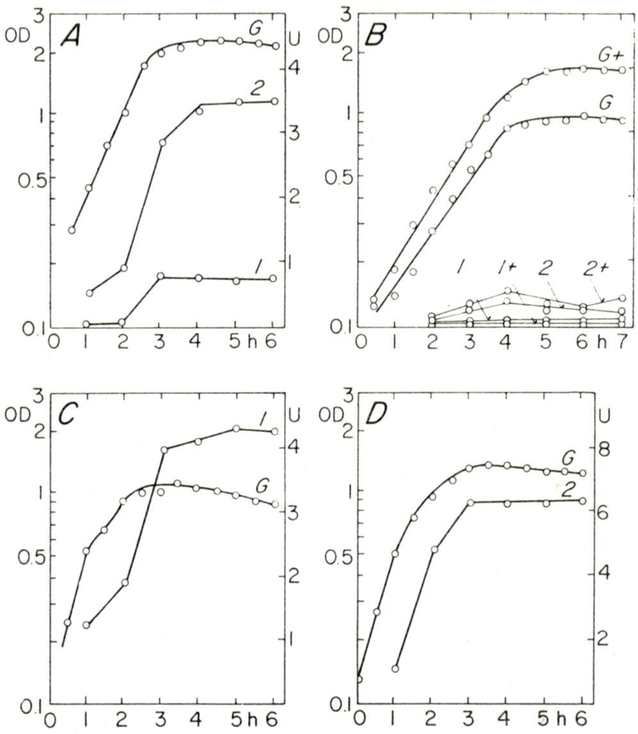

Fig. 3. Aconitase and isocitrate dehydrogenase in *B. subtilis* during growth in nutrient sporulation medium. Growth (curves G) was followed as optical density (OD). Aconitase and isocitrate dehydrogenase specific activities (U) are given as nmol of NADPH formed per min per mg protein at 25 °C. Cell extracts were prepared from aliquots of the culture at different times in 0.1 M-Tris-HCl buffer (pH 7.4) by lysozyme lysis and subsequent centrifugation of the extracts at 35,000 × *g* for 15 min at 4 °C. *A* wild-type strain; *1* aconitase, *2* isocitrate dehydrogenase × 0.5; *B* oxoglutarate dehydrogenase-negative mutant 60818; *1* aconitase × 100, *2* isocitrate dehydrogenase × 50, + nutrient sporulation medium + 20 mM succinate; *C* isocitrate dehydrogenase-negative mutant 60991; *1* aconitase; *D* aconitase-negative mutant 60871; *2* isocitrate dehydrogenase.

sporulation. It remained open whether abnormal regulation pattern of the CA-cycle enzymes, caused by the alteration of the distribution pattern of the metabolites in the mutants, would prevent the normal utilization of the substrates added. Therefore, the regulatory properties of the CA-cycle enzymes were investigated.

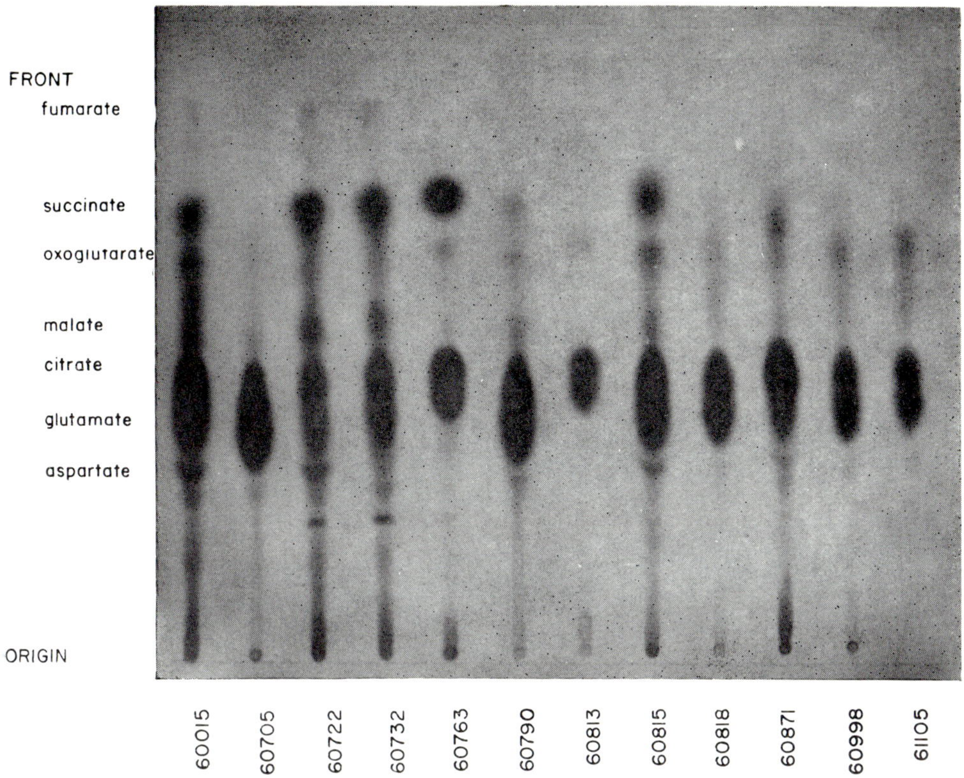

Fig. 2. Autoradiograph of the dissimilation products formed from ^{14}C-glutamate in the standard strain and several sporulation mutants. Cells were grown in NSM, *U*-^{14}C-glutamate (2 μCi/ml) was added, the cells incubated for 5 min at 37 °C, centrifuged, resuspended in water, extracted for 30 min at 100 °C, and the extracts analyzed by thin-layer chromatography on cellulose MN normal in ether—formic acid—water (7 : 2 : 1).

TABLE II

Effect of Glutamate on the Synthesis of Aconitase and Isocitrate Dehydrogenase in CA-Cycle Mutants of *B. subtilis*[a]

Enzyme	Strain	Final concentration of glutamate and glucose M		
		0.05 0.001	0.05 0.05	0 0.05
Aconitase	60015	13.9	8.7	3.1
	60871	<1.0	<1.0	—
	60991	981.1	692.9	—
	60818	3.0	4.5	—
Isocitrate dehydrogenase	60015	607.9	475.3	207.7
	60871	908.3	165.8	—
	60991	<5.0	<5.0	—
	60818	87.0	68.0	—

[a] Cells were grown in minimal salt medium with glucose and glutamate as carbon sources. The specific activities of aconitase and isocitrate dehydrogenase were determined in extracts of exponentially growing cells.

In a minimal medium, aconitase and isocitrate dehydrogenase appear to be regulated by oxoglutarate or glutamate feedback repression[10,11]. Fig. 3*A* shows that in our wild-type strain aconitase and isocitrate dehydrogenase increased in their specific activity when growth of the cells ceases in nutrient sporulation medium. A constant high level of both enzymes was maintained during sporulation. This pattern is given as reference for the subsequent comparison of sporulation mutants blocked in the CA-cycle[11]. Mutants blocked in oxoglutarate dehydrogenase which accumulate oxoglutarate and glutamate showed a constant low level of both enzymes; a typical example is the mutant strain 60818 (Fig. 3*B*). Addition of succinate in a final concentration of 20 mM allowed the culture to grow to an elevated optical density which indicated that by this treatment the missing metabolites were supplied. However, no significant change in the specific activities of the two enzymes was found. This can only mean that both enzymes are constantly repressed in those mutants which fail to metabolize oxoglutarate and glutamate when cells are grown in a complex medium.

The next step was to investigate the distribution pattern of both enzymes in mutants which fail to produce oxoglutarate and glutamate in order to investigate the effect of their absence on the synthesis of aconitase and isocitrate dehydrogenase. As can be seen in Fig. 3*C*, aconitase is significantly increased during exponential growth and after cessation of growth in the isocitrate dehydrogenase-negative mutant in the nutrient sporulation medium. The final level was raised about 8-fold as compared with the standard strain. The specific activity of isocitrate dehydrogenase in the aconitase-negative mutant increased early during exponential growth up to a final level which was about 80% of that reached in the wild type (Fig. 3*D*). If both mutants and the standard strain were grown in a nutrient sporulation medium containing

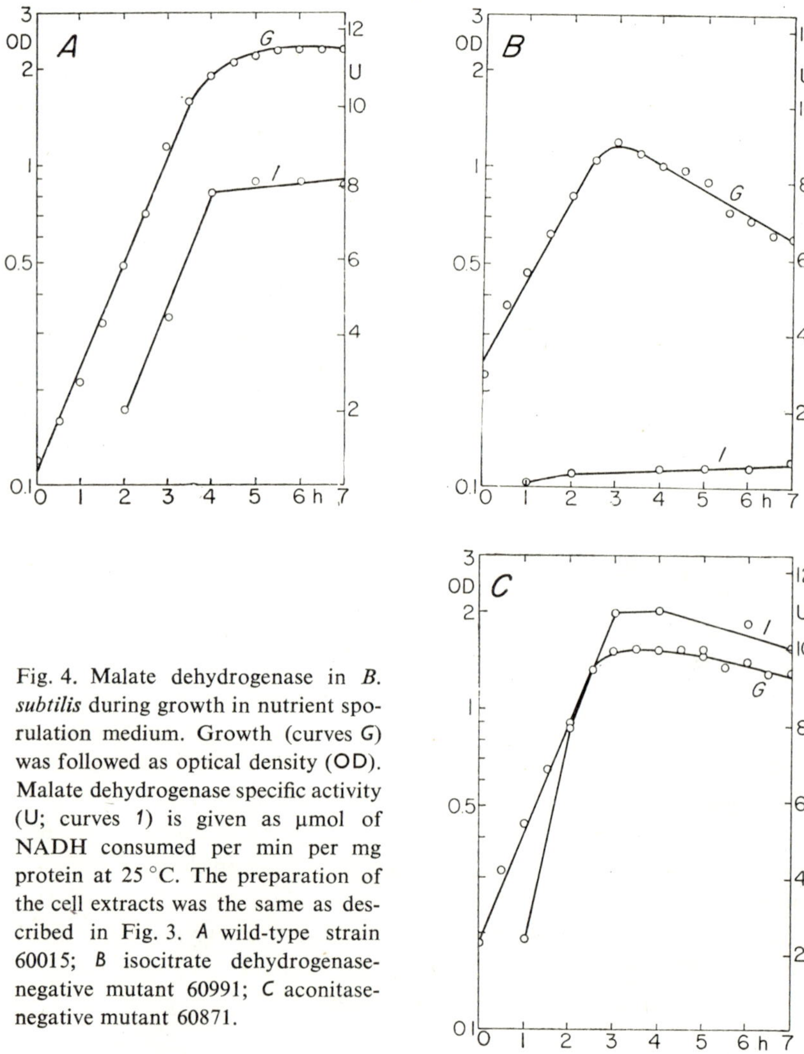

Fig. 4. Malate dehydrogenase in *B. subtilis* during growth in nutrient sporulation medium. Growth (curves *G*) was followed as optical density (OD). Malate dehydrogenase specific activity (U; curves *1*) is given as μmol of NADH consumed per min per mg protein at 25 °C. The preparation of the cell extracts was the same as described in Fig. 3. *A* wild-type strain 60015; *B* isocitrate dehydrogenase-negative mutant 60991; *C* aconitase-negative mutant 60871.

0.1 M glutamate a significant reduction of the specific activity was found, but this was not so with the oxoglutarate dehydrogenase-negative strain where this addition had no effect (Table II). Increasing amounts of glutamate caused increasing repression of the synthesis of aconitase and isocitrate dehydrogenase when the wild-type strain and the mutants unable to synthesize glutamate and oxoglutarate were exponentially grown in a minimal glucose medium[11]. Again this treatment had no effect on mutants unable to metabolize glutamate.

Aconitase and isocitrate dehydrogenase appear to be regulated by glutamate or oxoglutarate feedback not only in minimal but also in complex media. A decrease of the intracellular level of the corepressor at the end of growth causes their derepression which then leads to increased synthesis of the enzymes during the sporulation period.

Malate dehydrogenase (MDH) appears to be the key enzyme in the metabolic pathways leading out of the citric-acid cycle and into the glycolytic pathway. Its regulation is of interest since in *B. subtilis* sporulation can take place with glutamate as the only source of carbon. It should be possible to investigate the effect of CA-cycle metabolites on the synthesis of MDH using mutants blocked in this cycle. When the wild-type strain was grown in a nutrient sporulation medium the specific activity of MDH increased during exponential growth, reaching a maximal level at its end which was constantly maintained during the period of sporulation (Fig. 4*A*)[12]. This increase is absent in the isocitrate dehydrogenase-negative mutant (Fig. 4*B*). Here a constant low level of the enzyme was found. Addition of glutamate (50 mM) did allow further growth of this strain, since the missing metabolites could be supplied by this treatment but no change in the MDH activity was found. It is therefore not likely that the missing metabolites should be responsible for the low MDH level. An alternative assumption then is that the accumulating metabolites which repress MDH synthesis were responsible for the low enzyme content in this mutant. If this were true, mutants blocked before isocitrate dehydrogenase should lack the repressing metabolite and therefore derepress MDH synthesis early, reaching an increased final enzyme level. This pattern was found in the aconitase-negative mutant as demonstrated in Fig. 4*C*. The maximal specific-activity level reached after

TABLE III
Effect of Glutamate in Minimal Glucose Medium on the Synthesis of Malate Dehydrogenase[a]

Strain	Enzyme blocked	Glutamate concentration mM		
		0.1	1.0	10.0
60015[b]	—	15.5	13.0	8.0
60871	Aconitase	12.5	7.5	7.0
60991	Isocitrate dehydrogenase	2.0	2.0	2.0

[a] The enzyme activities are given as μmol of NADH consumed per min per mg protein. Cell extracts were prepared by lysozyme lysis of the cells and subsequent centrifugation at 35,000 × *g* at 4 °C.
[b] Wild type.

the end of growth was about 11 μmol/min/mg protein, whereas it was about 8 μmol/ /min/mg in the wild-type strain. The analysis of the dissimilation products from ^{14}C-citrate in both mutants (strains 60871 and 60991) showed that the only metabolite which could be responsible for the alteration of the MDH regulation is isocitrate[12] Its addition to complex and minimal media, however, could not affect

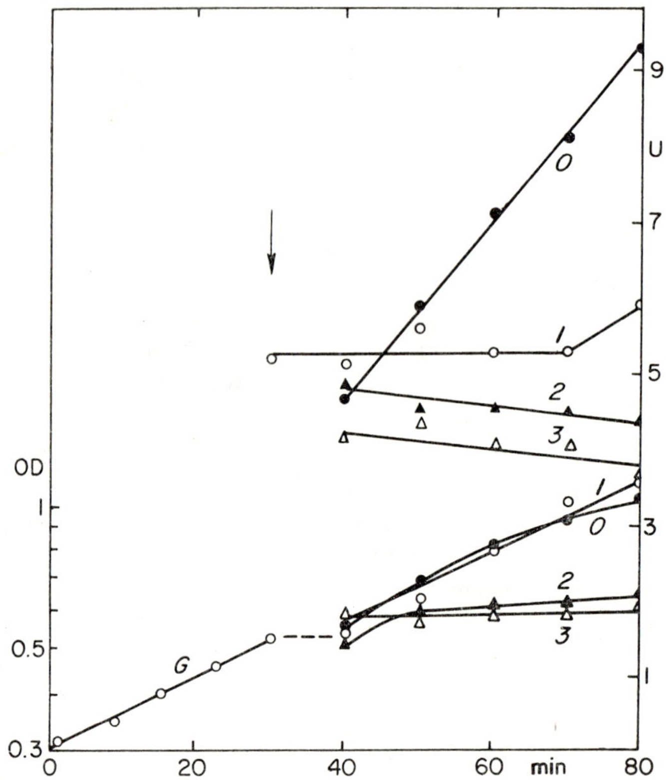

Fig. 5. Growth and alate dehydrogenase activity in the wild-type strain 60015 in minimal glucose medium with and without glutamate. Growth (curve G) was measured as optical density (OD). Malate dehydrogenase specific activity (U) was calculated as μmol of NADH consumed per min per mg protein. Cells were grown in a minimal glucose—glutamate medium. At the time marked with the arrow they were transferred into minimal glucose medium. Additions at this time were: none, (*0*); glutamate, (*1*); actinomycin D, (*2*); or chloramphenicol, (*3*). Aliquots of the cultures were harvested at different times, the cells extracted, and the specific activity of malate dehydrogenase was determined.

MDH synthesis. The limited uptake of isocitrate by our *B. subtilis* strains can be taken as a possible explanation for this result[12].

Addition of high amounts of glutamate (10 mM or more) should cause feedback accumulation of isocitrate. This would then repress MDH synthesis. This was found in both the wild type and the aconitase-negative mutant, when grown in minimal glucose medium in the presence of glutamate. In contrast, the same treatment had no effect on MDH in the isocitrate dehydrogenase-negative mutant (Table III).

The regulation of MDH by isocitrate acts on the level of transcription. When cultures of the wild-type strain grown in minimal glucose medium in the presence of 25 mM glutamate were transferred into minimal glucose medium, derepression of MDH synthesis occurred, but not if chloramphenicol or actinomycin were added at the time of the shift. This result is demonstrated in Fig. 5.

This work was supported by a grant from the Deutsche Forschungsgemeinschaft.

REFERENCES

1. FORTNAGEL P., FREESE E.: J. Biol. Chem. **243**, 5289 (1968).
2. RAMOS F., WIAME J. M., WYNANTS J., BECHET J.: Nature **193**, 70 (1962).
3. FREESE E., FORTNAGEL P.: J. Bacteriol. **94**, 1957 (1967).
4. FORTNAGEL P., FREESE E.: J. Bacteriol. **95**, 1431 (1968).
5. FREESE E., FORTNAGEL P., SCHMITT R., KLOFAT W., CHAPELLE E., PICCIOLO G., p. 82 in L. L. Campbell (Ed.): *Spores,* Vol. IV. Am. Soc. Microbiol. 1968.
6. FREESE E. B., COLE R. M., KLOFAT W., FREESE E.: J. Bacteriol. **101**, 1046 (1970).
7. SZULMAJSTER J., HANSON R. S., p. 162 in L. L. Campbell, H. O. Halvorson (Eds.): *Spores.* Vol. III. Am. Soc. Microbiol. 1964.
8. HAMPTON M. L., HANSON R. S.: Bacteriol. Proc. 118 (1969).
9. DONNELLAN J. E., NAGS H. E., LEVINSON H. S.: J. Bacteriol. **87**, 332 (1964).
10. HANSON R. S., COX D. P.: J. Bacteriol. **93**, 1777 (1967).
11. FORTNAGEL P.: Biochim. Biophys. Acta **222**, 290 (1970).
12. FORTNAGEL P., LOPEZ J.: Biochim. Biophys. Acta **237**, 320 (1971).

8/ Legal Protection of Strains, Mutants and Genetic Procedures

Litigation Concerning Industrial Microorganisms in the English-Speaking Countries

WILLIAM L. HAYHURST, Q. C.

Ridout & Maybee, Toronto, Canada

Since all references are to law reports and legal publications I have used the usual form of legal citation. To assist those who are accustomed to the mode of citation used in scientific publications, the volume number or the date by which each volume is identified is printed in bold-face type. The following standard abbreviations are used:

Canada:	Ex CR	:	Exchequer Court Reports
	SCR	:	Supreme Court Reports
U.K.	FSR	:	Fleet Street Patent Law Reports
	RPC	:	Reports of Patent, Design and Trade Mark Cases
U.S.A.:	JPOS	:	Journal of the Patent Office Society
	USPQ	:	United States Patents Quarterly

Thus, **(1968)** RPC 31 signifies a decision in the 1968 volume of the U.K. Reports of Patent, Design and Trade Mark Cases, begining at page 31.

VARIATIONS IN NATIONAL LAWS AND PRACTICES

Man is continually altering and reinterpreting the laws which regulate his own conduct. Man-made laws relating to patents have been the beneficiaries and victims of many political and economic pressures, with the result that the patent statutes or codes of the major countries have important differences, particularly in relation to what protection, if any, is available for inventions and discoveries relating to living things. Herein I shall be discussing what may be called the English (or common law) legal system which prevails in about one-third of the world. Within this system the United States has developed some of the more important variations, the spirit of independence having flourished there for two centuries. In what follows I shall sometimes refer to the British Commonwealth, and I hope I shall give no offence by saying that I intend by that term to include countries, such as South Africa, which have recently left the Commonwealth, but to exclude the United States. There are many clear and many subtle differences from country to country, and nothing that I say should be construed as legal advice: it would be presumptuous to offer such advice in relation to any country other than my own, and such general statements as I make may be inapplicable in particular circumstances.

TYPES OF SUITS

Various disputes may arise in relation to industrial microorganisms. You might bring an action in tort based upon your suspicion that one of your microorganisms has been stolen, or that information about your secret industrial process has been divulged in breach of confidence (1). Or, if a competitor has applied for a patent, it is possible in some countries to file an opposition in the Patent Office to prevent the grant of a patent. If a patent has been granted, it may be necessary to seek an adjudication as to whether the patent is valid or has been infringed. In some countries it is possible to bring proceedings to compel the patentee to grant a license under his patent, perhaps because the patentee is not benefiting the country by manufacturing within the country, or because he is believed to be abusing his privileges under the patent (2). Different issues and problems of proof will arise from one suit to another, but questions that may arise in a patent action in court are illustrative and I propose to discuss some of these.

WHAT MAY BE PATENTED

Every country restricts the categories of things for which patents may be granted. In some countries there have been recent enlargements of those categories (3) but in others, where there is reaction against foreign control of business, a more restrictive mood prevails (4). The problems that are likely to be of interest to microbiologists are those relating to patents for new processes and for new substances, though patents for apparatus may also be important.

(1) American Cyanamid Company *v.* Fox **140** USPQ 199 (1964) (theft of organisms and confidential documents); Merck *v.* Hoffmann-La Roche, Official Journal of the Australian Patent Office, **1969**, page 809 (successful opposition to grant of a patent to Hoffmann-La Roche who innocently believed they had acquired rights to an invention improperly disclosed by a Merck employee).

(2) Though the U.S. Patent Act does not provide for compulsory licenses, such licenses may be ordered under the Federal Trade Commission Act where the patentee was not frank in his dealings with the Patent Office: Chas. Pfizer & Co. Inc. *et al. v.* Federal Trade Commission **159** USPQ 193 (1968). Also, the U.S. has a unique, highly developed body of law relating to misuse of patents.

(3) *E.g.*, the German law effective January 1, 1968 made protection available for foods and medicines and for substances made chemically; the French law effective January 3, 1968 also introduced provisions for patenting substances made chemically.

(4) A government bill introduced in India in 1967, recently passed, would restrict what may be patented, and applications for patents for foods and medicines have for some years been held up by the Patent Office.

PATENTS FOR PROCESSES

Most countries allow patents for processes, but all exclude some processes. In general, patents cannot be obtained for such things as improvements in the social sciences, for business plans, or for thought processes. Patents are granted for what might loosely be called technical processes, but there is diversity from country to country.

Processes which do not produce a product. Early concepts of what may be patented centered on the idea of manufacturing. The view was taken that it was not patentable to aid or improve a human being, or an animal, or a plant, by a treatment which relied upon the natural functions of the living person or thing (5). Courts in the British Commonwealth have been somewhat restrained by the notion that a process to be patentable must make, improve, restore or preserve a vendible product (6), although fairly liberal views have been taken as to what constitutes such a product (7). It is probably still true in most Commonwealth countries that processes for treating humans, being non-economic, are not patentable (8). In the United States a 1952 statutory definition of the term "process" has been regarded as removing some difficulties in obtaining patents for the treatment of humans by administering medicines (9). In most countries, at least some methods of treating plants and animals are patentable (10); methods of treating microorganisms will be discussed below.

---

(5) An example in the United States was MORTON *v.* New York Eye Infirmary **17** Federal Cases 879 (1862) (claim for the use of ether as a human anaesthetic in surgical operations held invalid); in England, Canterbury Agricultural College's Appln. (**1958**) RPC 85 (patent refused for injecting sheep with a composition to increase the yield of wool), N.V. Philips' Gloeilampenfabrieken's Appln. (1954) **71** RPC 192 (patent refused for producing a new form of *Poinsettia* by imposing artificial climatic conditions).

(6) G.E.C.'s Appl'n. (1943) **60** RPC 1. Special statutory provision is made in the U.K. for methods of testing.

(7) *See* cases cited in footnote (10).

(8) Tennessee Eastman Company *v.* Commissioner of Patents (**1970**) Exchequer Court of Canada decision, not yet reported, refusing claims for a method of surgically bonding body tissues by applying a specified adhesive); London Rubber Industries Ltd.'s Patent (**1968**) RPC 31 (claim for process for controlling ovulation allowable, if treatment of humans excluded); compare Goldhaft's Appln. (**1957**) RPC 276 (patent refused for a method of treating an egg to affect the sex of the chick).

(9) U.S. section 100(b): "*The term 'process' ... includes a new use of a known ... composition of matter, or material.*" *See* I. MARCUS, "*The Patent Office and Pharmaceutical Invention*" **47** JPOS 669 (1965). Claims to some physical treatments have also been allowed: *e.g., Ex parte* SCHERER **103** USPQ 107 (1954) (method of injecting fluids into the human body).

(10) Swifts' Appln. (**1962**) RPC 37 (claim allowed for tenderizing meat by injecting enzymes into a living animal and then slaughtering it), following an Australian decision, National Research Development Corporation's Appln. (**1961**) RPC 135 (claim allowed for spraying crops with a composition that destroys weeds).

Processes which yield a useful product. Where a process causes the production of a useful product, it has not generally been an objection that living organisms have been used (11): many patents have been granted for fermentation processes using living microorganisms to produce chemical substances (12), and for the production of vaccines and sera involving the treatment of an animal (13). Novel uses of microorganisms in such processes as cheese making, brewing, baking, production of steroids and digestion of sewage may also be protected. In many countries (but the list is shrinking) arbitrary restrictions have been imposed upon the grant of patents where chemical processes are involved (14) and issues have arisen as to whether a biological process can be distinguished legally from a chemical process. To resolve such issues it may be necessary to investigate carefully what actually takes place, or is thought to take place, in the biological process in question; however, such a technical study of the process may be avoided if the court can be persuaded to hold, as a matter of law, that what is a chemical process should be decided from the point of view of a layman rather than from the point of view of a chemist or biologist (15).

Genetic procedures. To discuss the patentability of genetic procedures I need some terminology. I have heard evidence that some microbiologists reject the concept of species as applied to microorganisms. I have no wish to inflict heresies upon you, but in order to have a lexicon I shall refer to "species" of microorganisms, and different members of a species I shall call "strains", whether those strains are found in nature or are artificially induced mutants. I think I can make the principles clear without referring to such other groupings as orders, genera, variants, *etc.*

Suppose that you have subjected a microorganism to a treatment which results in the production of a new mutant. If the mutant has no practical utility, the process may be considered to be of no practical use and unpatentable (16). But let us assume that the mutant is a new and valuable one. Putting aside for the moment the question whether the new mutant may be protected as a product, is protection possible for your process?

(i) Consider first the case where the same treatment has previously been applied to other species, or to other strains of your species, but you are the first to apply the treatment to the particular strain with which you started. We begin with the proposition that in most countries one may patent an old process when applied

---

(11) SZUECS' Appln. **(1956)** RPC 25.

(12) *Ex parte* PRESCOTT *et al.* **19** USPQ 178 (1933).

(13) Canterbury Agricultural College's Appln. **(1958)** RPC 85 at page 87; Virginia—Carolina's Appln. **(1958)** RPC 35.

(14) *See* heading "*Patents for substances; products produced by microorganisms*" below.

(15) As in Laboratoire Pentagone Ltée *v.* Parke, Davis & Co. **(1968)** SCR 307 (holding that the step of extracting chloramphenicol from a fermentation broth was a chemical step, and that consequently a claim to a fermentation process including such an extraction step was a claim for a chemical process).

(16) *See* heading "*Disclosure of utility*" below.

to a starting material (here, your original strain) to which the process has not previously been applied if the product (here, the new mutant) is new and unobvious. The next proposition is that a patent specification which promises a new and unobvious result must contain sufficient instructions to enable persons skilled in the art to achieve the same result without an undue amount of experimentation. This may put you in a dilemma. If the new result that you promise or claim is the production of a particular mutant, the chances are (if your process is typical of those that cause mutations) that you cannot give instructions that will ensure that others will be able to produce that very mutant from your starting organism. If, on the other hand, you simply promise or claim that the process, as applied to your starting organism, may yield valuable mutants, you are promising or claiming old procedural steps that do not necessarily lead to new or unexpected results, and you are probably not entitled to a patent, on grounds of obviousness, inutility, and insufficient description.

(ii) Now consider a case where, in addition to producing a new mutant, you have carried out a procedural step that has not previously been applied to microorganisms, or suppose that you have discovered a new combination of procedural steps. Assume, further, that you find that your new process can usefully be applied to more than one microorganism, so that you have a new general method of producing mutations. If this method or process provides a tool of practical use to those who are concerned with the production of useful mutants, I venture to think that you can fend off attacks which allege obviousness, lack of utility and insufficient description, despite the fact that the process does not necessarily work with all organisms, or does not necessarily work every time with a particular microorganism (assuming that you do not promise that it will always work). But there is a U.K. decision (17) holding that such a mutation process is not proper subject matter for a patent: "*The applicant's discovery is merely of a conditioning procedure whereby, of such micro-organisms as survive it, a distortion of their normal development is secured.*" This decision is not necessarily the last word on the subject, and in the United States process claims for the production of strains and mutants have been allowed (18).

If the mutation is carried out as part of a process of manufacturing something else, the total process may be patentable subject matter. Processes of treating microorganisms to affect their growth are likely to be found patentable in the United States but perhaps not elsewhere, for it has been said in the U.K. in relation to crops: "*However advantageously man may alter the conditions of growth the fruit*

(17) General Electric's Appln. (**1961**) RPC 21; *compare* the *Poinsettia* case mentioned in footnote (5).

(18) *E.g.*, U.S. patent 2,938,835, "*Production of Mutants of the Genus* Penicillium", granted to Glaxo Laboratories Limited in 1960, the process using a radioactive sulphur isotope; U.S. patent 3,014,843, "*Method of Producing Non-Virulent Strains of Attenuated and Stabilized Hog Cholera Virus*" granted to Armour & Company in 1961, involving the inoculation of young pigs and harvesting virus from them.

is still not produced by his action" (19). The more artificial the effect, the better the prospects of obtaining a patent.

PATENTS FOR SUBSTANCES; PRODUCTS PRODUCED BY MICROORGANISMS

It may be more valuable to have a claim for a product than a claim for a process of making the product: the claim for the product covers the product no matter how it is made, so that issues as to what process was used can be avoided. Similarly, a claim to a product is likely to be of greater value than one to a process of using the product: the claim to the process of use may be enforceable against only users of the product, and there may be numerous small users to pursue, whereas a claim to the product itself may be asserted against manufacturers, sellers and users of the product.

Most countries permit patents for substances but many non-English-speaking countries (20) refuse patents for certain kinds of substances (notably medicines) and for substances produced by chemical processes (raising the problem, previously mentioned, of whether microorganisms are participants in a "chemical" process). Where claims to the substances themselves are refused, claims for specific processes for making the substance are usually permitted. However, in the English-speaking countries one can usually obtain a claim for the substance itself (21). If it is discovered that an old substance has previously unknown medicinal or other properties, it may be possible to obtain a patent for the substance in a new condition required for its new and previously unknown use, *e.g.*, when mixed with other things (22), or isolated in a pure state (23) or in a new form (24). In countries where animal feeds can be patented it is common to claim antibiotics mixed with other edible matter.

---

(19) National Research Development's Appln. (**1961**) RPC 135 at page 147. *Compare* U.S. patent 3,000,788 (1961) claiming a process of growing canine hepatitis virus in tissue cultures.

(20) *E.g.*, Japan, The Netherlands, the U.S.S.R. (in the latter a "certificate of authorship" may be obtained for a medical composition). Italy refuses patents for medicines and processes of making them, and proposed legislation to permit process claims is being considered.

(21) Canada is the laggard, section 41(1) of the Canadian Act prohibiting a claim to a substance prepared or produced by a chemical process and intended for food or medicine. Where the process is claimed, a claim may be made to the substance when produced by the claimed process.

(22) Re LEMIN **140** USPQ 273 (1964); *Ex parte* RUSCHIG *et al.* **147** USPQ 46 (1965). A patent will be refused if the mixture possesses only the aggregate of previously known properties of all the ingredients: U.K. Patents Act, section 10(1) (c); Re DAVIS **134** USPQ 257 (1962). Difficulties may be also encountered with claims to mixtures if they are indefinite as to the ingredients: ROHM & HAAS *v.* Commissioner of Patents (**1959**) Ex CR 153.

(23) Re DOYLE *et al.* **140** USPQ 421 (1964) (claim allowed to a purified compound which merely worked better in a fermentation than the impure one).

(24) Merck & Co., Inc. *v.* Chase Chemical Co. **155** USPQ 139 (1967) at page 152 (claim upheld to an antipernicious anemia factor in liver, isolated in a new crystalline form).

PATENTS FOR MICROORGANISMS

In general, natural phenomena and things found in nature are not considered to be patentable (25). Thus, minerals found in nature are not patentable, but patents are regularly granted for artificially produced inanimate substances. I am told that some viruses are crystalline and may be mineral, not animal; if so, countries with rather traditional patent laws might be persuaded to grant patents for such new viruses, assuming that they are useful. If artificially produced inanimate substances are patentable, are artificially produced living substances patentable? The U.S. Patent Office has allowed claims for live vaccines and sera (26), for microorganisms in combination with other material to make useful compositions (27), for mixtures of microorganisms (28), and for yeasts in novel form (29) but not, as far as I am aware, for microorganisms *per se* (30). In a case in the U.S. Supreme Court it was observed that "*The qualities of these bacteria, like the heat of the sun, electricity, or the qualities of metals, are part of the storehouse of knowledge of all men*"(31).

There is legislation in some countries restricted to the protection of certain varieties of plants and seeds, with special provisions as to how to describe the new varieties, or as to making reproductive material available. For example, in the United States, patent protection for some plants has been available since 1930 (32); however, it has been held that, although bacteria may be classified scientifically as plants, the word "plant" is used in the U.S. statute in its popular sense, and that new species of bacteria cannot be patented under the particular provisions relating

(25) *Ex parte* FOSTER **90** USPQ 16 (1951) (patent refused for newly recognized variety of plant); *Ex parte* REED *et. al.* **135** USPQ 34 (patent refused for substance extracted from liver and of known properties). Legislation in the United Kingdom and in some other countries now apparently permits claims to newly discovered substances even if they are later found to exist in nature, but use of the substance found in nature is not an infringement. *See* the South African case cited in footnote (42).

(26) *E.g.*, U.S. patent 2,966,443 (1960) for a poliomyelitis live-virus vaccine, components of which have been passed through living animals.

(27) U.S. patent 3,133,066 (1963) for *Bacillus thuringiensis* spores dispersed within an oil, useful as a pesticide; U.S. patent 2,262,851 (1941) for mycelium overgrowing dry vegetable pulp, useful as mushroom spawn.

(28) U.S. patent 1,894,135 (1933) for a baker's preparation consisting of yeast and bacteria capable of causing decomposition.

(29) U.S. patent 2,919, 194 (1959) for a baker's yeast in dry condition.

(30) *See* D. G. DAUS, R. T. BOND and S. K. ROSE "*Microbiological Plant Patents*", **10** IDEA 87 (1966—7). The foregoing examples are taken from that paper.

(31) Funk Bros. Seed Co. *v.* Kalo Inoculant Co. **76** USPQ 280 (1948); the actual decision, however, was that a particular composition containing known varieties of bacteria was an unpatentable mixture, *i.e.* an aggregation.

(32) The present provisions are sections 161—3 of the U.S. Patent Act.

to plants (33). Other countries also make provisions for protecting plants (34), and some countries in Eastern Europe provide protection for animal breeds (35).

Microorganisms used commercially are usually mutants differing from those found in nature, and it has been argued that the lack of opportunity for patenting such mutants has resulted in their being kept in secrecy, to the detriment of public knowledge. Israel in 1968 made provision for patenting "microbiological organisms not derived from nature". Protection for a microorganism would in some cases be superior to protection for processes which use it because it may be possible to alter the process to avoid infringement while still using the microorganism, and proof that the microorganism has been used may be easier than proof that an entire process has been used. Also, further uses may be found for the microorganism, or for new strains covered by a claim to the microorganism. If such patents are to be obtainable efforts will have to be made by microbiologists, comparable to those made by the plant breeders, to convince the legislators that protection is desirable.

PATENT LITIGATION

The issues to be decided in a patent action usually relate to validity of the patent and infringement. In West Germany, Japan and certain other countries, the issues of validity and infringement are dealt with in separate proceedings. In the English speaking world these issues are more usually (though not invariably) tried together, before the same judge.

INTERPRETATION OF THE LANGUAGE OF THE CLAIMS OF A PATENT

Patents consist of two important parts, namely the descriptive part (or disclosure) and the claims. The purpose of the disclosure is to instruct experts in the art so that they may use the invention after the patent has expired. The claims, which are the numbered paragraphs at the end of the patent, indicate what the public is prohibited from making, using or selling before the patent expires. In the English-speaking countries attention is focussed upon the language of the claims: the claims are not merely signposts (as they are under the Germanic system) but are fences defining the limits of the monopoly, and the general rule is that a thing which has every characteristic specified by a claim infringes that claim (if the claim is valid), and

(33) Re Arzberger **46** USPQ 32 (1940).

(34) A Convention for the Protection of New Varieties of Plants (Industrial Property, **1962,** page 5) has been ratified by the United Kingdom, the Netherlands, West Germany and Denmark. Its immediate objectives appear to be to protect grains, vegetables, fruits and flowers.

(35) G. Palos, "*The New Hungarian Law on Inventions*", Industrial Property, **1969**, page 254.

a thing which does not have every characteristic set forth in the claim does not infringe it (36). Each claim defines a separate monopoly, and if one claim is infringed it is no answer to say that other claims in the same patent are not infringed. There may therefore be an issue as to whether the alleged infringer is doing something that falls within all the language of at least one claim. If the claim specifies a particular microorganism, the patentee must satisfy the court that the defendant has used that microorganism (37).

ACTS THAT CONSTITUTE INFRINGEMENT OF A PATENT

If a defendant is making, using or selling everything specified by a patent claim he infringes the claim and the court may require him to stop. The fact that he had no knowledge that he was infringing is no defence, although in some countries this may relieve him from paying damages. If one person is not doing what is specified in a patent claim, or is doing only part of what is specified, but he is procuring another to do what the claim specifies, both may be liable to the patentee.

A patent has effect only in the country in which it issued. If an unauthorized person is using the patented invention in several countries it may be necessary to bring a separate action against him in each country. However, a local patent on a product will be effective to prevent manufacture of the product locally, as well as to prevent local sale or use of products imported from abroad if the imported products were not made with the consent of the patentee.

Suppose that a process is being carried out in a foreign country where there is no patent, and the product of the process is imported into a country where there is a patent claiming the process but not the product. The persons carrying out the process in the foreign country, without the consent of the patentee, are not infringing (there being no patent in the foreign country), but the laws of many countries regard the local importer and his customers as infringers of the process claims (38). However, it is necessary for the patentee to prove that the patented process was used, and it

(36) Here I am ignoring the legal doctrine of equivalents, or substance, which sometimes compensates for defects in the language chosen for the claims.

(37) A lengthy review of conflicting expert evidence as to the identity of *Streptomyces aureofaciens* and an organism called *Streptomyces lusitanus* will be found in American Cyanamid Co. *v.* Chas. E. Frosst & Co. (**1965**) 2 Ex CR 355.

(38) Importation into the United States of a product is not an infringement of a United States process patent (Re Amtorg Trading Corp. **24** USPQ 315 (1935)) but importation may be prevented under the U.S. Tariff Act. Importation of the product is an infringement in England, at least if the patented process has played a substantial part in the production of the product: Saccharin Corp. *v.* Anglo-Continental Chemical Works (1900) **17** RPC 307; Wilderman *v.* Berk (1925) **42** RPC 79 (a case on importation of the product of patented apparatus).

may be difficult to prove what is being done in a foreign country (39), so that the patentee is in a better position if he can obtain a claim to the product.

ISSUES RELATING TO THE VALIDITY OF A PATENT

Many reasons may be advanced as to why the patent should not have been granted. One possible reason is that the alleged invention did not fall within the classes of things for which patents may be granted under the local law, *e.g.*, that it is a patent for a microorganism and that under the local law patents may not be granted for microorganisms. Some of the other possible grounds of invalidity will be discussed below. Most of our concepts relating to validity of patents were formulated in relation to patents for inanimate objects, and it will be seen that such concepts may be in need of revision when we try to apply them to patents relating to living organisms.

THE OBJECTION THAT THE ALLEGED INVENTION WAS NOT NEW

Novelty is a question of fact, but there may be serious questions to be tried. For example, where a patent claims the use of a microorganism in a fermentation process, there may be an issue as to whether the microorganism found by the patentee it a different microorganism than one previously used, or there may be a question as to whether an alleged prior use was given sufficient publicity, under the national law, to deny patentability to the later discovery. You will be aware that to obtain patent protection in many countries you must have established a patent application filing date before allowing your invention to become accessible to the public (40). If a prior publication refers to your microorganism, there may be a controversy as to whether the microorganism is sufficiently identified in the prior publication (41).

---

(39) In American Cyanamid Co. *v.* Continental Pharma (Canada) Ltd. *et al.* (1965) **30** Fox's Patent Cases 171, the patentee failed to satisfy the court that the existence of traces of palladium in imported tetracycline proved that the tetracycline had been made by dechlorination of chlortetracycline. The local court may request the aid of a foreign court to compel the foreign manufacturer to give evidence: Re Raychem Corporation *v.* Canusa Coating Systems, Inc. (**1970**) 1 Ontario Reports 448, reversed on an appeal not yet reported.

(40) There are many variations. In some countries, *e.g.,* France and the U.S.S.R., public availability of the invention anywhere prior to your earliest (priority) filing date may be fatal. In others, *e.g.,* the U.K. and India, the question is whether the invention was publicly available in the country in question. The U.S. and Canada are even more liberal.

(41) In a U.S. case, *Ex parte* ARGOUDELIS *et al.* **157** USPQ 437 (1967), reference in a prior publication to a strain of actinomyces "*isolated from the soil of Chiba Prefecture (Japan)*" was held to give insufficient information about the organism for the publication to be used against an applicant for a patent. However in a U.K. case, Application of Instytut Anty-

Or if you are claiming a product, there may be a controversy as to whether the product has been produced before: in a South African case an attempt was made to invalidate a claim to the product chlortetracycline on the ground that the microorganism *S. aureofaciens* produces chlortetracycline in nature, but the evidence failed to establish that this was so (42). In a U.S. case (43) it was held that the Pfizer patent for tetracycline was not invalidated by the prior production of tetracycline in fermentation broths used to produce chlortetracycline, the tetracycline having been unrecognized and produced in only trace amounts insufficient to be of any benefit to mankind (44).

THE OBJECTION THAT THE ALLEGED INVENTION WOULD HAVE BEEN OBVIOUS TO AN EXPERT IN THE ART

A patentable invention may be the result of an accident, or of a sudden inspiration, or of plain hard work (45). Suppose that you carry out a fermentation in a conventional fermentation broth using a microorganism that has not previously been tried, and you find that a product having some antibacterial properties is produced, but that the product has no particular value or utility. Since your procedure is well known, and has not resulted in any unusual result, your process and product will likely be held to be obvious (46).

Unobvious processes. Suppose, however, that you succeed in producing an antibiotic that has useful properties that markedly distinguish it from run-of-the-mill

---

biotykow (referred to by A. W. WHITE, "*Prior Secret Publication*", Proceedings of The Chartered Institute of Patent Agents, February, **1969**, page C59) reference in a prior specification to use of an organism identified merely as S. 88, with no information as to how the organism could be obtained, was held to destroy the novelty of a claim to the use of that organism. *See* also GYOGYSZERIPARI's Appln. (**1958**) RPC 51.

(42) American Cyanamid Co. *v.* Continental Ethicals (Proprietary), Ltd., South African Patent Journal, Vol. **2**, No. 33 (1962). Reliance was also placed on legislation like that in the U.K. and referred to in footnote (25).

(43) Chas. Pfizer & Co., Inc. *v.* Barry-Martin Pharmaceuticals, Inc. **145** USPQ 29 (1965). Compulsory licensing of the patent was ordered in other proceedings because of Pfizer's lack of candour in dealing with the Patent Office: *see* footnote (2).

(44) *See* also Re SEABORG **140** USPQ 659 and 662 (1964).

(45) *See* the U.S. Patent Act, section 103.

(46) *Ex parte* ARZBERGER *et al.* **155** USPQ 286 (1966) (use of novel strain held obvious where no unexpected results obtained compared to previous use of other strains of the same bacteria); *Ex parte* KROPP, **143** USPQ 148 (1959) (use of a new species of *Streptomyces* held obvious where no unusual results asserted). As to a product, it was observed in Re Interference A *v.* B *v.* C, **159** USPQ 538 (1967) that "*if persons and institutions regularly running routine soil screening tests in the search for antibiotics would run across the present organism often enough to warrant a holding that it was readily available, it could hardly be stated that the subject matter of these applications was unobvious*".

fermentation products. This product was not predictable and is not likely to be considered obvious. In the British Commonwealth the process of producing the antibiotic is also not likely to be regarded as obvious, despite the fact that the only new thing about the process is that it has not been tried before with your particular microorganism (47). The law seems to have taken a different turn in the U.S. where the process, if conventional apart from the materials used, may be regarded as obvious even though the product is unobvious and patentable because of its unexpected properties (48). I submit that a process claim should be valid if it specifies an unexpectedly useful strain of a previously used species (49).

As another example, if you have discovered that the yield of a fermentation process can be greatly increased by using a particular nutrient or a particular process condition, there may be room for genuine differences of opinion as to whether what you did would have been obvious to an expert in the fermentation art (50). The person attacking the patent has the onus of proving that what was done was obvious.

Unobvious products. If you claim a product which has not previously been made or described but falls within a previously described group of compounds (51), or if you claim a new product that is close in structure to prior compounds (52), the product may be regarded as unobvious if, though its structure was obvious, its properties were not. If the product has been specifically mentioned in a prior publication, or differs only slightly from a previously described product, the U.S. view is that the product may nevertheless be patented if the prior publication would not enable a person skilled in the art to make it (53). This view may not prevail in the U.K. (54).

---

(47) Commercial Solvents Corp. *v.* Synthetic Products Co. (1926) **43** RPC 185 (an action on the U.K. patent corresponding to the U.S. patent involved in the Guaranty Trust case mentioned in footnote (58)); Re May & Baker Limited *et al.* (1948) **65** RPC 255; Commissioner of Patents *v.* Ciba Limited (**1959**) SCR 378.

(48) Merck & Co. Inc. *v.* Chase Chemical Co. **135** USPQ 139 (1967) at page 158; Re KANTER **158** USPQ 331 (1968) but note the dissent of SMITH, J. as to the applicability of such doctrine where it would not be predictable that the product would result from the process; such unpredictability would usually exist in a microbiological process).

(49) I. G. Farbenindustrie's Appln. (1936) **53** RPC 92; Re NAYLOR **152** USPQ 106 (1966).

(50) Re MOSTOVYCH *et al.* **144** USPQ 38 (1965) (varying time, pH *etc.* in a coating process held obvious); Re LEGATOR **147** USPQ 322 (1965) (method of controlling noxious microorganisms in the manufacture of paper, using a known bactericide (acrolein), held unobvious).

(51) E. I. du Pont de Nemours & Co. *v.* Ladd **140** USPQ 297 (1964); I. G. Farbenindustrie's Appln. (1936) **53** RPC 92.

(52) Commissioner of Patents *v.* Deutsche Gold-und-Silber-Scheidanstalt vormals Roessler **157** USPQ 549 (1968); Re WAGNER *et al.* **152** USPQ 552 (1967).

(53) Re HOEKSEMA **158** USPQ 596 (1968).

(54) GYÓGYSZERIPARI's Appln. (**1958**) RPC 51; Smith, Kline & French's Appln. (**1968**) RPC 415.

THE OBJECTION THAT THE INVENTION HAS NOT BEEN SUFFICIENTLY DESCRIBED

One purpose of the patent system is to make inventions freely available to the public when the period of patent protection has ended. The patent specification must describe the invention in sufficient detail to enable an expert in the art to put the invention into practice, and the expert must not be required to do an unreasonable amount of experimental work in following the directions in the patent (55).

Deposits in culture collections. Special problems are involved in enabling the public to do what the patent teaches when a particular microorganism must be used (56). Many microorganisms are known and available, and it may be sufficient to identify these by name (57); in other cases, a good description of the external morphology and physiological behavior of the organism may be sufficient to enable others to find and identify it without difficulty (58) with the aid of such publications as are available (59). It is prudent to include such a taxonomic description of any new organism, as a means of identification, although it may not be necessary to do so if the patent specification discloses a deposit in a culture collection from which the organism may be obtained (60). In cases where there is any doubt as to whether others can find the organism readily, the safest course is to include in your patent specification, at the time you file it, a reference to a deposit that has been made in a culture collection from which the microorganism is, at the time of filing of the patent application, available without restrictions (save normal terms as to price, safety precautions or the like) to scientifically qualified nationals of the country where the entitlement to a patent is in question. Failure to do this has been held to be fatal in recent decisions in the U.S. (61) and in the U.K. (62), although these decisions were reserved on appeal. An attempt to amend a patent application to refer to a deposit may be refused (63). There is a possibility that a deposited culture may

(55) Re ELTGROTH **164** USPQ 221 (1970); Vidal Dyes Syndicate Ltd. *v.* Levinstein Ltd. (1912) **29** RPC 245 at page 280.

(56) L. J. ROBBINS, "*Patents for Microbiological Transformations — an International Problem*" **42** JPOS 830 (1960).

(57) *Ex parte* KROPP, **143** USPQ 148 (1959).

(58) Guaranty Trust Company of New York *et al. v.* Union Solvents Corp. **12** USPQ 47 (1937).

(59) Re Interference A *v.* B *v.* C **159** USPQ 538 (1967).

(60) *Ex parte* DAVISSON *et al.* **118** USPQ 520 (1958).

(61) *Ex parte* ARGOUDELIS **157** USPQ 437 (1967) reversed **168** USPQ 99 (1971).

(62) DANN's Patent (**1970**) FSR 23, reversed (**1971**) RPC 425.

(63) KYOWA's Appln. (**1968**) RPC 10 and (**1969**) RPC 263. In a Canadian case, American Cyanamid Co. *v.* Chas. E. Frosst & Co. (**1965**) 2 Ex CR 355 no deposits were referred to in the patents, but no argument was made by the defendant that a specification must be sufficient at the time of filing, rather than at the time of grant of the patent, and on the evidence the court concluded that the necessary microorganisms could at the time of grant be obtained without difficulty by persons skilled in the art. It has not been the practice of the Canadian Patent Office to insist upon a deposit.

become unavailable through misadventure, but this may not create legal difficulties if the inventor has done everything reasonable to ensure that such loss does not occur (64).

Since the laws of many countries insist that the applicant must disclose the best way known to him for carrying out his invention (65) he should deposit not only the type culture but also the best strain or strains known to him for the purposes set out in his patent application (66).

It seems contrary to the spirit of the patent laws to require an inventor to make any public disclosure, and thus to make a publicly available deposit, before he has some assurance that he will obtain patent protection (67). I submit that the most that should be required is accessibility to those who have a right to examine the patent specification. Suggestions have been made for amendment of the law (68).

Identifying the product. Problems sometimes arise in identifying the antibiotic or other product of a fermentation, the chemical formula of the product being unknown. The product may be claimed by reference to its method of production (69) but if a limitation to the method is not desired (or not permitted by the Patent Office) resort must be had to identifying the product by its chemical, biological and physical properties (70).

Disclosure of utility. Patent protection is not granted for scientific discoveries for which no use is known (71). A vague statement that a product has "high biological activity" or "useful biological properties" is not likely to be an adequate disclosure of utility (72). In countries where the law requires a full disclosure the applicant must not withhold information that would not be obvious to an expert, for example, infor-

---

(64) *Compare* Re METCALFE *et al.* **161** USPQ 789 (1969).

(65) In Benger Laboratories Limited *v.* R. K. Laros Company **135** USPQ 11 (1962) it was said that the applicant must disclose the best method known to him at the time of filing even if someone else discovered that method.

(66) HEATH *v.* UNWIN (1852) **2** Webster's Patent Cases 236.

(67) The spirit is without substance in the many countries where applications are published before any decision has been made to grant a patent.

(68) Senator MCCLELLAN's U.S. Bill S 1246 (introduced February 28, 1969), sections 4(e), 112(d) and 119(d); American Cyanamid Company (DANN's) Patent (**1970**) FSR 23.

(69) As in the Merck U.S. patent 2,703,302 in suit in the case referred to at footnote (83).

(70) Benger Laboratories Limited *v.* R. K. Laros Company **135** USPQ 11 (1962); D. LEVY and L. B. WENDT, "*Microbiology and a Standard Format for Infra-Red Absorption Spectra in Antibiotic Patent Applications*", **37** JPOS 855 (1955).

(71) For example, the Canadian Patent Act, section 28(3), provides that no patent shall issue for an invention that has an illicit object in view, or for any mere scientific principle or abstract theorem.

(72) Re KIRK *et al.* **153** USPQ 48 (1967), following BRENNER *v.* MANSON **148** USPQ 689 (1966) where it was held that it is not enough that a product is the subject of research to determine some specific use for it.

mation as to uses to which his invention may be put (73). If specific utility is disclosed for the product, for example, that it has antihistaminic action, but how to administer it is not obvious, a full disclosure may be required of the mode of administration and dosage (74).

There is no need for an inventor to understand the theory or principle upon which his invention works; a statement of an incorrect theory is not fatal unless it would mislead a skilled person who was attempting to follow the teaching of the patent (75). In Germany a disclosure of technical progress (technischer Fortschritt) is required, but this is not necessary in the English-speaking countries (76), where, for example, it is sufficient that the invention provides a useful alternative to what was previously available (77).

THE OBJECTION THAT THE PATENT CLAIMS MORE THAN WAS ACTUALLY INVENTED OR MORE THAN IS JUSTIFIED BY THE DISCLOSURE OF THE PATENT

Some claims of a patent may be valid despite the fact that other claims in the same patent are invalid. A claim may be invalid because, although it covers something new and unobvious, it also covers something old or obvious and therefore claims more than was actually invented. Similarly, a claim may be invalid because it is so wide that it covers things that have no use whatever, or because it covers things which do not achieve the results promised by the patentee in the disclosure portion of his patent (78). Consider the following hypothetical claim:

---

(73) There are authorities to the effect that an applicant need not state all the advantages of his invention, but the cautious patent agent will advise you to tell everything you know, being careful not to make general or other statements that cannot be supported.

(74) Société des Usines Chimiques Rhône-Poulenc *et al.* *v.* Jules R. Gilbert Limited *et al.* (1967) **35** Fox's Patent Cases 174; Re SCHMIDT *et al.* **153** USPQ 640 (1967).

(75) "Z" Electric Lamp Manufacturing Co. Ltd. *v.* Marples, Leach & Co., Ltd. (1910) **27** RPC 737 at page 746.

(76) Deutsche Gold-und-Silber-Scheideanstalt vormals Roessler *v.* Commissioner of Patents **148** USPQ 323 (1966), **157** USPQ 549 (1968).

(77) Ward Brothers *v.* James Hill & Son (1903) **20** RPC 189 at page 199; Lamb Knit Goods Co. *v.* Lamb Glove & Mitten Co. *et al.* **120** Federal Reporter 267 (1902).

(78) In Canada there has been a flurry of cases in which claims covering the production of such useful drugs as phenmetrazine (a stimulant), tolbutamide (for treatment of diabetes) and tripelennamine (an antihistamine) have been held invalid because in each case the claims also covered production of related compounds having no known utility, or covered a large unexplored field: C. H. Boehringer Sohn *v.* Bell-Craig Limited (**1963**) SCR 410; Hoechst Pharmaceuticals of Canada Limited *et al.* *v.* Gilbert and Company *et al.* (**1966**) SCR 189; Société des Usines Chimiques Rhône Poulenc *et al.* *v.* Jules Gilbert Ltd. *et al.* (**1968**) SCR 950. *See* also Re SURREY **151** USPQ 724 (1966), and *compare* the Indian decision, Farbwerke Hoechst A. G. *v.* Unichem Laboratories (**1969**) RPC 55.

A process of producing symposia comprising cultivating *Bacillus verbosius* in fermentation liquor.

Such a claim extends, *prima facie*, to all strains of *B. verbosius* that are presently known or that may in future be discovered. Some Patent Offices may allow a claim covering the use of the entire species even though the examples in the patent specification disclose the use of only a single strain (79). But theoretically there may be an astronomical number of strains of the species. If strains are found that fail to produce anything useful the claim would cover useless processes and, in some countries at least, may be invalid for want of utility (80). The objection may be put that such a claim covers many strains that had not been tested, and that no sound prediction could have been made that all strains would achieve the utility promised by the patent, so that the claim extends to more than had been invented, even though no unworkable strains have yet been found (81). It seems advisable to include one or more claims that refer to the specific strain (or strains) deposited in a culture collection, identifying the strain in such claims by its collection catalog number (82). Perhaps such a claim may be construed to cover not only the particular strain mentioned but also its functional equivalents. Another solution that is commonly adopted, in the attempt to cover the use of all useful members of a species, is to restrict the claim to organisms that perform a certain function, so that the claims when analyzed say little more than the following:

A process of producing symposia comprising cultivating *symposia-producing* strains of *Bacillus verbosius* in fermentation liquor.

Claims of this kind have been upheld (83) without discussion of the attacks made on their breadth. A claim limited to any strain that will produce the result

(79) *Ex parte* BENEDICT *et al.* **111** USPQ 354 (1956).

(80) In American Cyanamid Co. *v.* Charles E. Frosst & Co. (**1965**) 2 Ex CR 355 it was stated that "*if at the date of the patent the words used* (*and here we are dealing with* Streptomyces aureofaciens) *embraced useless as well as useful microorganisms then the Duggar patent is bad*", but it was held that useless mutants brought into existence after the patent was granted did not affect its validity. (*See* also "Z" Electric Lamp Manufacturing Co. Ltd. *v.* Marples, Leach & Co. Ltd. (1910) **27** RPC 737 at page 745). There was no evidence of other useless strains, and no consideration was given to the question whether, at the time the patent application was made, it was a matter of conjecture that all or even a substantial proportion of strains had the promised utility.

(81) Olin Mathieson Chemical Corporation *v.* Biorex Laboratories Ltd. (**1969**) FSR 361; In Re KAMAL *et al.* **158** USPQ 320 (1968). These cases state the general principles, but did not relate to microorganisms.

(82) As in Benedict *et al.* U.S. patent 2,771,397 (1955).

(83) Merck & Co. *v.* Olin Mathieson Chemical Corp. **116** USPQ 484 (1958); Merck & Co. *v.* Chase Chemical Co. **155** USPQ 139 (1967). Conflicting views relating to such functional identifications of strains have been expressed by U.S. Supreme Court Justices Frankfurter and Burton in Funk Bros. Seed Co. *v.* Kalo Inoculant Co. **76** USPQ 280 (1948).

may, if it extends to numerous strains that have not been investigated, be characterized as merely stating a problem (84).

Another question: how much detail about nutrients and process conditions is required in process claims? Where there is nothing unusual about the selection of such things I submit that little detail is required, but as a precaution it is common to include one or more claims specifying the constituents of the medium, aerobic or anaerobic conditions, ranges of temperature, time and pH, and the like.

THE OBJECTION THAT THE MEANING OF THE PATENT IS NOT CLEAR TO AN EXPERT IN THE ART

The disclosure must be sufficiently clear to enable an expert to carry out the invention as successfully as the inventor could do. Minor errors that the expert could easily detect will not be fatal. Each claim, to be valid, must be clear and unambiguous. Does a claim which refers to a microorganism satisfy this requirement? Taxonomy has been described by some as subjective, and as an art, not a science. In view of the differences of opinion that exist among taxonomists as to what tests, and how many of them, should be used to identify a microorganism, many patent claims which refer to a species or genus of microorganism may be open to attack on the ground of ambiguity. The patentee may be able to take refuge in judicial statements that a patentee need only be as clear as the subject matter allows him to be (85), but this may not save a patent where the skilled reader is put into an honest quandary as to what a claim covers (86). Careful thought should be given in drafting a patent specification as to how, for the purposes of the patent, the reader is to determine whether a particular microorganism is a member of any species or genus referred to in the claims. There is the danger that if one commits

---

(84) Blanco-White, "*Patents for Inventions*" 3rd ed., pages 66 and 73; Stevens *v.* Keating (1847) **2** Webster's Patent Cases 172 ("*No person can be allowed to take that course, and to say ... I claim all those substances that may succeed*"); Metals Recovery Co. *v.* Anaconda Copper Mining Co. **31** Federal Reporter, 2nd Series, 100 (1929) (claim to the use, in froth flotation, of non-oleaginous organic mineral collecting agent which is substantially non-frothing, held too wide); *see* also Re FUETTERER **138** USPQ 217 (1963) and *Ex parte* KUGERA **165** USPQ 332 (1970) (discussion of the use of functional language in claims).

(85) Cleveland Graphite Bronze Co., Ltd. *v.* Glacier Metal Co., Ltd. (1950) **67** RPC 149 at page 154; Phillips Electronic and Pharmaceutical Industries Corp. *v.* Thermal and Electronics Industries, Inc. **165** USPQ (1970).

(86) Martin & Biro Swan Ltd. *v.* H. Millwood Ltd. (1954) **71** RPC 458 ("*A reader must not be left in doubt whether any given apparatus, method or process falls within the claims or not*"); United Carbon Co. *et al. v.* Binney & Smith Co. **55** USPQ 381 (1942) ("*Whether the vagueness of the claim has its source in the language employed or in the somewhat indeterminate character of the advance claimed to have been made in the art is not material. An invention must be capable of accurate definition and it must be accurately defined to be patentable.*")

oneself to a particular taxonomic system, or to a series of criteria which are currently in vogue, in an effort to be reasonably precise as to what microorganisms are contemplated, such precision may be embarassing if useless microorganisms prove to be included. The problems of describing living things, and predicting their properties, are such that the patent laws should perhaps be liberalized. This has been done, for example, in relation to plants in the U.S. Patent Act which does not require a description of how the plant may be obtained, nor a claim establishing verbally the limits of the monopoly, but requires only as complete a description of the plant as is reasonably possible. However, there is probably no infringement of a U.S. plant patent unless there has been unlicensed reproduction of plant stock obtained directly or indirectly from the patentee (87), and this U.S. legislation is not necessarily the ideal model.

EXPERT EVIDENCE

Sometimes disputes relating to the patentability of inventions are decided by persons who combine knowledge of the law with technical expertise in the art to which the invention relates. This is true at the Patent Office level in most of the technologically advanced countries. In cases involving issued patents it is common, on the Continent of Europe, for the court to appoint technically qualified experts to aid the court in reaching a decision, but such an arrangement is made only infrequently in the English speaking world. Under the common law system cases are usually tried by a judge having little or no training in science or engineering (88). In contrast with practice on the Continent, it is not the function of the judge to seek out evidence from experts, and the judge has no right to introduce evidence. His function is to adjudicate upon the issues raised by the parties, and to ensure that the parties obey the procedural rules. The parties must provide all the evidence. Whereas, in many Continental countries and Japan, the evidence may consist entirely of documents, including written opinions of experts hired by the parties, in the English speaking world documents may be used for only limited purposes. What is stated in a document is rarely admitted: evidence must come from witnesses who appear in person and who are subject to cross-examination. Where engineering or scientific questions are involved, each party may call his own expert witnesses to explain the relevant matters of scientific fact or opinion, and the opposite party has the right to challenge such expert witnesses by cross-examination and by calling expert witnesses of his own. This is known as the adversary

(87) *Ex parte* WEISS **159** USPQ 122 (1968).

(88) There are exceptions. For example, in the U.K. there are two specially appointed judges having considerable scientific knowledge. In South Africa litigation is initiated before the Commissioner of Patents, who may sit with technical advisers.

system, the idea being that if each party presents all the facts and arguments that tend to support his side of the case the true position should become clear to the judge. The judge is allowed to reach his decision solely upon what has been proved before him, rather than upon facts which the judge may think he knows but which the parties have not had an opportunity to challenge (89). Our judicial approach is different from the scientific method: in court, cases are often decided not on the basis that something has been proved, but on the basis of what the probable truth is, after an assessment of evidence which may be conflicting. Also our cases tend to be decided on the basis of one grand performance at trial, as contrasted with the Continental phenomenon of a series of hearings through which the case progresses to judgment, perhaps with an appeal and the hearing of further evidence. Before trial we do have preparatory stages, and particularly in the U.S. much oral evidence may be collected before trial, but normally all the evidence comes before the judge at a single hearing or trial, where the experts are heard and perhaps confronted by other experts, following which there will be the legal argument, judgment, and possibly an appeal at which further evidence is generally not admissible.

The adversary system has the defect that the parties may not be evenly matched: one may have more highly qualified experts available to him than the other. Scientific witnesses no doubt attempt to be impartial, but after an expert has worked for days or months on one side of a case his impartiality may not be very apparent to the other side. This human tendency of witnesses provoked the well known remark that "*witnesses are either liars, damn liars, or experts*" (90). However, our courts must rely heavily upon the evidence of experts, whose testimony is vital to provide scientific explanations, to define scientific terms, and to describe the "state of the art", *i.e.*, what was known prior to the time of the alleged invention, thus enabling the judge to read the patent with the knowledge that a scientist would have.

---

(89) The judge is entitled to take "judicial notice" of facts that are generally known.

(90) "*And then*", said Baron BRAMWELL, whose distinguished brother Sir Frederick BRAMWELL testified frequently as an expert witness, "*there is my brother Fred.*"

CONCLUSION

Laws are imperfect, and much of man's struggle has been to devise a more perfect system. In the relatively narrow field of patents relating to industrial microorganisms the problems are manifold. Bold steps already taken in some countries may indicate the way to better legislation and procedures. Fresh approaches may be required to solve the difficulties of conferring a valid monopoly of reasonable scope. Meanwhile our courts must apply the laws as they find them. Fortunately the courts do not apply rules of law in a mechanical way without regard to the difficulties under which inventors and their advisers must labour. But while striving to do justice between the parties, the courts are bound to apply the standards laid down by statute, and it is unfortunate that these standards give rise to so many uncertainties in relation to discoveries that have revolutionized medicine and agriculture.

Legal Protection of Industrial Microorganisms in European Socialist Countries

A. Stoy

Institute of Macromolecular Chemistry, Czechoslovak Academy of Sciences, Prague, Czechoslovakia

From the standpoint of the legal protection of industrial microorganisms, European socialist countries may be divided into three different groups.

The first group includes four countries, where miccroorganisms, plants and animals cannot be protected as such, but only as technical means in manufacturing processes for producing either some metabolite or secondary metabolite, or a foodstuff, a feed for animals *etc.* These countries are Czechoslovakia, German Democratic Republic, Yugoslavia and Poland. In the Patent Laws of the first three countries there is no explicit provision concerning plants, animals or microorganisms, while the Polish Law of May 31, 1962, excludes in its Article 2b explicitly new varieties of plants and animals from the protection.

The second group, including U.S.S.R., Rumania and Bulgaria, has an incomplete or restricted patent protection of new plant and animal varieties, microorganisms not being mentioned as such. Botanically, the microorganisms may be in most cases considered as plants, so that they may be protected analogously. The incomplete or restricted legal protection consists either in that new varieties of plants may be protected by a certificate of authorship or a diploma only, not by a patent yielding exclusive rights, or in that an exclusive patent is granted but to state enterprises. In the latter case, the breeder obtains also a diploma or author's certificate, entitling him to a reward.

The third group includes only Hungary, where a new, modern Law of Patents issued in the fall of 1969, being in force from the beginning of 1970.

Now the possibilities of legal protection in individual countries will be discussed in more detail.

For the first group, the Czechoslovakian Patent Law may serve as a model, the differences between the four countries being rather unimportant for a general review of this type.

The fact that new varieties of plants including microorganisms cannot be protected in this country as such may be derived indirectly. First, chemical products, drugs and foodstuffs are excluded, so that only their method of preparation or manufacture may be protected. The analogy with plants and microorganisms is rather evident. Second, the Law of Patents stresses the solution of problems by new *technical* means. Scientific methods of selection and biochemistry hardly conform with that prerequisite. Third, finding a new microorganism in nature where it existed before is considered as a discovery which cannot be patented. For a discovery, a special diploma with reward is provided.

Thus, the only possible sort of patent protection in the said four socialist countries is claiming a process for manufacturing a useful, well-defined product. Such product may be a secondary metabolite, like an antibiotic, or a primary metabolite like an aliphatic alcohol or acetone, or also a foodstuff, a spice and the like. The use of a defined microorganism forms a part of the process. The claimed process can be defined either solely by the use of the microorganism, or also by other technological features, such as the composition of the nutrient, certain additives — inhibitors, activators *etc.*, further by isolation methods, process parameters such as pH value, temperature, intensity of aeration, time needed for various phases of the process, as well as the arrangement or details of equipment.

The taxonomy of the microorganism may be of primary importance if it is new, and if the claim is reduced to its use for a certain purpose. If an already known microorganism is used in a new process, it may be characterized but by its scientific name.

For scientific purposes, the microorganisms are usually not classified from the standpoint of a desirable metabolite since the ability of producing the metabolite is no stable, strictly characteristic, property. In the patent right, however, where the production of the metabolite is usually the only reason why the fermentation is carried out, the said ability is of primary importance. Among all microorganisms conforming to the taxonomic specification disclosed, only the use of such is protected that produce the defined substance.

Problems arise, however, if a taxonomically different microorganism is found, producing the same metabolite under equal conditions or if a known microorganism, falling not within the claims of a process patent, is induced to produce the same metabolite either by a special stimulation, adding a certain precursor, or by artificial mutation. Especially the latter case is often disputed, since the border line between a new species and an artificial or natural mutant of a known species is sometimes not sufficiently distinct. Then the taxonomy may appear again to be the only clue for a satisfactory decision of a court in an infringement litigation.

Natural and artificial mutants are sometimes claimed together with the basic species as means for carrying out a process. In many cases, however, the true origin of the strain cannot be proved. Then, particularly when the character of the microorganism has been considerably changed, the strain may be considered as a mutant by the plaintiff, and as a new species by the defendant. If the answer is necessary for the decision, the hearing of expert witnesses is the most important part of the law suit.

Until now, the criteria for distinguishing between a mutant and a new species do not seem to be reliable, nor are they generally accepted. As a matter of fact, mutation is the means by which nature creates new species. Mutagenic factors and their effects seem to be quite analogous both in nature and in the laboratory. If the mutants were to be effectively protected, it would be necessary to establish a certain border line, where a mutant becomes a new species.

The problem of mutants can be, however, viewed also from the other side: If an artificially induced mutant has definitely better properties than the originally described strain, where is the border line between the merits of the two breeders? Is the first inventor entitled to obtain rights involving also the not yet discovered new strains, the cultivation of which would perhaps require more effort than the finding of the first strain? First discovery may be rather a lucky hit; but it is only the first step. On the one hand, the first inventor has to bear the burden of innumerable tests, the danger of the first clinical tests, the considerable effort connected with installing a large-scale production and bringing the product to the market. The further breeder may utilize the experience of the first one and profit by it. On the other hand, however, is the technical and scientific progress, caused by patents, not based on this very idea of utilizing the experience of previous workers for making new improvements in the art?

These questions cannot be apparently answered generally so that the field for infringement suits remains wide enough, what may be satisfactory when seen from the standpoint of lawyers and patent attorneys.

The form of claims is usually "*a method for manufacturing a product*" (more or less accurately defined), characterized by cultivating a certain microorganism under defined conditions. The novelty can reside in the kind of the microorganism, the cultivating method being known *per se*. In other cases, the weight is shifted to the metabolite which is to be recovered. If the microorganism was known and the metabolite is new, then the process claim should contain some feature concerning the isolation, not common in this field.

The simple form of claim — "*The use of the microorganism A for production of the product B*" (where both A and B are precisely defined) is refused by some examiners, perhaps even because of this simplicity and since the usual words "*characterized by that*" are lacking. Such examiners like more complex claims, defined by as many features as possible. The patentee must be careful, however, not to put any unnecessary limitation into the claim.

If a new metabolite is found, produced by an old microorganism, which, however, was not known to produce anything useful, it is difficult to say whether it is an invention or a discovery. A patent may be granted or not, depending on rather subjective factors such as the ability of the examiner or that of the patent attorney.

Now a new Law of Patents and other forms of legal protection of industrial property are being prepared in Czechoslovakia. The outline is different from that elaborated two or three years ago, being naturally influenced by the changed political situation. Now there is a tendency to conform to the law of the U.S.S.R. This tendency for unification is more or less apparent also in the Patent Laws of other countries of the European socialist block.

In the second above-mentioned group, consisting of U.S.S.R., Rumania and Bulgaria, new varieties of plants and races of animals are at least partly protected. In the Soviet Union, only author's certificates are issued for new and improved varieties of farm animals, poultry, silkworms, as well as for new sorts of farm crops obtained by selection. Exclusive patents cannot be granted onto the said items.

In Rumania, exclusive patents for inventions concerning nuclear products, chemical products, drugs, food, *new varieties of plants and new races of animals*, are granted only to state enterprises. The author obtains simultaneously a certificate, entitling him to a reward.

Bulgarian provisions concerning new plant varieties and animal races are very similar to those of the U.S.S.R.

Most interesting is the recent Hungarian Patent Law, enacted in December 1969 and in full force since January 1970. It contains the following paragraph in its Article 6:

"*A variety of a plant or a sort of animal, as well as processes leading thereto, may be patented if said variety or sort is new, homogeneous and relatively constant.*"

The conditions for patent applications concerning inventions based on the use of a microorganism are stipulated in the order in council, article 21: The application is to be accompanied by a receipt confirming that the strain has beeen deposited at the National Collection of Microbial Cultures in the National Institute of Public Health. If the strain was deposited after the date of application for patent, the application will be considered as filed only on the date of the deposition of the strain.

Said Institute examines the strain whether it is in condition to be preserved and issues the above-mentioned receipt. The strain is kept secret by the Institute until the patent application has been opened for public inspection. Thereafter the Institute ought to deliver the strain to anybody for examination and to inform the depositor about it. A deposit at a foreign Institute can be accepted only in case of reciprocity. The President of the Patent Office decides whether the conditions for reciprocity are fulfilled or not.

Although it is not stated *verbatim*, it seems that a microorganism can be also protected as such, not only its use in a process. At least, the stipulation of Article 21 — inventions based on the use of a strain of microorganism — seems to be sufficiently vague to permit such explanation.

Article 31, referring to article 67 of the Law of Patents, defines more precisely the conditions of novelty, homogeneity and stability:

A plant variety is *new* if it differs from the known varieties at least in one essential feature in morphologic, physiologic or other properties.

It is *homogeneous*, if the essential features of its various individuals are the same, respecting thereat the differences between sexual and nonsexual reproduction.

It is *relatively stable*, if the essential features, in the course of the natural or artificial reproduction, agree with those disclosed in the patent specification.

The protection does not extend to the use of the material for the plant reproduction to scientific use or to creating further new varieties therefrom.

The depositor ought to produce, on the request of the Office, the material for reproduction of the plant in a requested quantity, originating from a certain period of vegetation.

If the patentee has treated the new variety of plant or its material for reproduction by chemical substances or by irradiation or similar, he has to inform the Patent Office about it.

The depositor is permitted to inspect the culture resulting from the official examination.

According to Article 34, one patent application can contain but one plant variety. The specification has to contain precise description, allowing to distinguish morphologic and physiologic features. The depositor has to advise a name for his new plant variety, said name being simultaneously generic. The Patent Office has the right to change the name, particularly if it could cause confusion.

On the request of the Office, the patentee has to submit photographs of his variety.

A patent for a plant variety can be made void if the variety has lost its homogeneity or comparative stability. The existence of the homogeneity and stability is to be proved by the patentee.

I have cited the Hungarian provisions about legal protection of microorganisms since they seem to be not only reasonable, but also well deliberated. The Hungarian Law does not recognize author's certificates but only patent; naturally, if the invention has been made by an employee in service, the exclusive rights belong to the employer. The inventor is entitled to an adequate reward, like in Czechoslovakia according to the law of 1957. This solution can satisfy both the state and the inventor and stimulate the technical and scientific development in the socialist countries where private enterprise is excluded.

Regarding the fact that the Hungarian People's Republic has a very similar economy and almost the same size as the ČSSR, it is advisable to study the new Hungarian Law of Patents carefully. It can certainly contain many provisions applicable also in the prepared new Law in Czechoslovakia.

Importance of the Description of the Producing Organisms in Obtaining Patent Protection for Fermentation Processes

H. Boyd Woodruff, Sara A. Currie, Thomas C. Hallada, Irving Putter

Merck Sharp & Dohme Research Laboratories, Merck & Co., Inc., Rahway, New Jersey, U.S.A.

INTRODUCTION

The application of microorganisms in the preparation of human foodstuffs predates recorded history. Considering the importance of the end product to the individual and his family, it is not surprising that a strong sense of proprietorship developed for the individual techniques and cultures employed. Families became noted for their successes in producing a fermented beverage, a leavened bread, an acidified or enzymatically digested foodstuff.

With developing community associations, proprietorship was widened, until a section of a country was known for the wine produced, the cheese manufactured, the soya product fermented. Secrecy became important for its economic value. The industrial revolution was accompanied by industrialization of fermentation industries, particularly with large-scale brewing activities. Secrecy and knowledge, in the person of the brewmaster, were vital to industrial success. As the preparation of foods moved from the home to the factory, with commercialization of tinned, dried, and otherwise preserved products, a secretive fermentation industry developed in support. As late as the 1930's, one seldom read accounts of citric acid production, of mushroom culture, of proteolytic-enzyme manufacture originating from laboratories of commercial producers. Such papers as appeared in the scientific literature originated largely from the research laboratories of agricultural colleges, and were deprecated in private by the industrial microbiologists for their naivete, and for the poor product yields reported.

PATENTS

The patent system for fermentation processes, now in force in every industrialized country of the world, has led to the end of enforced secrecy. Through the sharing with others of process information in the patent specifications, through the freedom to publish, and through the deposit of cultures in national culture collections, a synergistic development has occurred in microbiology. The latest information in microbial genetics or molecular biology is applied by the industrial microbiologist. Industrially important microorganisms, such as amino-acid producers, have become available for study of biochemical pathways and regulatory mechanisms. Products of the fermentation industry, the antibiotics, are used as vital tools in elucidating details of microbial structure, the cell wall, the membrane, the ribosome, and the genome.

The benefits from open research can be reaped only as long as the patent system provides an acceptable alternate to secrecy. An objective of this paper is to examine the role of the microbiologist in maintaining an effective patent system for fermentation processes.

Fermentation-process patents are valuble industrial property rights giving the inventors or their assignees the right to exclude others from practicing such inventions for a limited time period. Certain details vary from country to country, in terms of the length of the exclusive protection granted, whether protection is for fermentation process only or extends to a product of that fermentation, whether the patent is completely exclusive or subject to compulsory licensing. Implicit in the granting of a patent is the assumption that the public at large will benefit as well as the inventor. The governmental regulations are varied approaches to achieving the two aims. As part of the public benefits granted by all governments, the approved patent must contain a full description of the invention, one sufficiently complete to permit others skilled in the art of the patented process to practice the invention successfully.

Microorganisms are essential components of fermentation processes. Where a restricted discovery has been made, the requirement for a full disclosure can be fulfilled in part by depositing the specific culture in a recognized culture collection, where it is available to all who wish to study the new invention. In certain inventions of broader scope, a new discovery may apply to a complete species or an even broader class of microorganisms, in which case the cultures deposited become only typical examples of organisms which can be employed successfully in the new invention. The written specification must clearly define the limits of the microbial class which will react favorably to the invention. Certain new discoveries are of such import and so widely applicable that limitation to a single species provides inadequate protection. In this latter case, a patent specification must indicate clearly how one can obtain an organism from nature which responds to the invention, and patent claims are extended to any microorganism recoverable by the disclosed means.

TAXONOMY OF MICROORGANISMS

To understand clearly the scope of many fermentation-process patents, it is necessary to have a clear definition of a microbial species. In the science of taxonomy, species of organisms are considered to be synonymous with clusters. A species is a cluster of individuals, similar in certain morphologic and physiologic characteristics which distinguish them from other clusters, with which they may be united into a higher group, the genus. In establishing the species rank, taxonomists distinguish between phenetic relationships and phylogenetic relationships[1]. The former covers grouping of properties of organisms as they are recovered from natural sources; while the latter includes, in addition, the concept of evolutionary pathways which have been operational in development of the properties observed. The problem for the industrial microbiologist would be simplified if a species of organisms existed as a cluster in space, unrelated to other species. The phylogenetic concept indicates that such unrelatedness is not possible.

Recent progress in numerical taxonomy has been valuable in defining the problem[2]. The plotting of multiple properties assigned equal weight in matrix form adequately proves the validity of a species organization when applied to bacteria. Clusters do exist. Tests can be applied for consistency of individual characteristics, and characteristics included or discarded as of value for purposes of species definition. For individual species definition, the numerical method may be refined by assigning differing weights for each characteristic, once sufficient examples have been studied to establish a weighting value. The weight assigned for each character will vary from species cluster to species cluster.

For the industrial microbiologist, however, numerical taxonomy does not solve the problem of species definition. It merely highlights the difficulties. Between every species cluster grouping yet devised, individual organisms are found with intermediate status. Taxonomists are divided in the significance attributed to intermediate forms. Some feel that numerical taxonomy carried to its highest order of development, with elimination of variable characters and assignment of correct weightings, will result in definition of clearly defined species clusters and complete elimination of intermediate forms. If this view proves true, the lot of the industrial microbiologist will be greatly simplified. He will be able to define with certainty the limits of a species which is the subject of an inventive discovery. An inventor will know when his invention is infringed. A colleague will know when he has made a different invention.

Research in microbial genetics makes it clear that the optimism of the early numerical taxonomists was not well founded. The genospecies, a cluster of microorganisms which can exchange genes, is much too broad as a useful species concept, including as it does most of the *Enterobacteriaceae* and many taxospecies of *Bacillus*. As studies have progressed at the biological level to include transformation with soluble-DNA preparations, transduction with bacteriophage, transfer of plasmids, episomes and chromosomes by conjugation; and at the biochemical level to include nucleic acid hybridization, GC ratios and T_m values, it has become evident that bacterial genetics will not provide a simple solution to the problem of speciation of bacteria. It is also clear that the opportunities for genetic interchange in nature go far beyond simple sexual hybridization, and one must not be surprised if a natural population exposed to stresses of antibiotic inhibition or phage attack contains certain members having interspecies characteristics.

The possibility of interspecies isolates, now accepted by many taxonomists, compounds the industrial microbiologist's dilemma. Unless he can describe the microbiological limits of an invention so that no exception exists, the protection granted by a patent is of little practical value; thus there is no recourse for industry to recover the costs expended on research but to retreat to secrecy. The fact that the species concept covers 99% of all natural isolates is of little value if the remaining 1% provides opportunity for a laboratory which has conducted no research and made no basic discoveries to practice a publicly disclosed invention free of infringement penalties.

RECOMMENDED APPROACHES FOR SPECIES DEFINITION

Specific examples can best serve to define the problem. No complete solution has been found. The advice of microbial taxonomists must be heeded, but in the authors' experience this has not proved definitive. An experimental definition of the scope of the species concept for each patent is probably of greatest value. This can be accomplished by isolating from nature new strains sharing the ability to form the desired natural product, and by studying their characteristics to establish similarities. Since the work involved in such redefinition of a species is enormous, some indication of possible infringement beyond that offered by conventional cultural descriptions has been sought. We have found that the base pair ratios of microbial DNA, as expressed by T_m values, can provide supporting, though not conclusive, evidence of similarities. Where significantly different T_m values are observed, organisms can be held to be non-infringing.

GLUTAMIC ACID FERMENTATION

The patented process for producing glutamic acid with certain pleiomorphic nonsporeforming bacteria probably holds the record as the example with greatest extent of confusion in definition of microbial species. Glutamic acid is a widely employed flavor-enhancing agent. It has been obtained commercially at considerable expense by hydrolysis of plant proteins. It is present in the fermentation liquors of many microorganisms.

The strong commercial need led research workers to search for an improved fermentation process which would be superior to recovery from hydrolyzed proteins. One commercially successful process was discovered through development of a screening technique in which microbial isolates from various natural sources, including sewage, air and soil, were grown on a chemically defined culture medium, the glutamic acid released by killing the producing organisms with ultraviolet light, and detected microbiologically[3]. A United States patent issued on October 10, 1961, No. 3,003,925, covering a "*Method for Producing* L-*Glutamic Acid by Fermentation*" in the name of Shukuo KINOSHITA, Shigezo UDAKA and Sadao AKITA. Claim one stated "*A method for producing* L-*glutamate selected from the group consisting of* L-*glutamic acid and a salt thereof which comprises aerobically culturing* Micrococcus glutamicus *in a culturing medium containing carbohydrate source and inorganic material, controlling the pH value of the culturing medium within the range of from about 6 to about 9 by the addition of neutralizing agent, whereby a substantial amount of* L-*glutamate is accumulated in the culturing medium, and recovering said* L-*glutamate.*" Subsequent claims covered certain important operating variables[4].

The yields of glutamic acid obtained by this new process were quite remarkable, one example showed in excess of 50 g glutamic acid per liter of broth. A companion patent, No. 3,002,889, disclosed the operating advantage of conducting the fermentation with a biotin-deficient organism in a biotin-deficient medium. It was disclosed that several related isolates had the capacity to produce large quantities of glutamic acid, but the separate isolated differed from one another in certain measurable characteristics. Obviously, a tight phenetic cluster of organisms had been obtained, and with them a significant invention had been made. The specification of the U.S. Patent contained a conventional description of morphology and physiology of one of the isolates. Reference was made to a published paper on taxonomy of the cluster of organisms[5].

THE GLUTAMIC ACID-PRODUCING ORGANISMS

The cluster of organisms appeared to be different from any group of organisms previously described, therefore, the inventors applied the new taxospecies name *glutamicus* to describe the cluster. For the new species, having very unusual characteristics, the inventors were faced with selection of an appropriate generic name. For reasons described fully in the above-mentioned paper, they chose *Micrococcus*, although they were aware of the rather atypical characteristic that cells occurred as "*Slightly ellipsoidal spheres, usually in pairs, but also in single or irregular masses. Size of majority: 0.6—1.2* μm. *Elongated forms are seen in some conditions*"[4]. The elongated forms were illustrated in the published article[5]. The possible relationship of the new species of organisms to the genera *Brevibacterium*, *Corynebacterium*, *Microbacterium*, *Arthrobacter*, and *Bacterium* has been discussed fully, as well as a proposal that deviations from established genera were sufficient to justify establishing a new genus for these glutamic acid-producing organisms[6]. According to this proposal, the new genus would be placed in the family *Corynebacteriaceae*.

The nature of the problem faced by an industrial bacteriologist in deciding whether his glutamic acid-producing process is within the scope of the invention

TABLE I

Species of Microorganism Disclosed in Glutamic Acid-Fermentation-Process Patents

Genus	Species
Achromobacter	*A. cycloclastes* *A. methanolophila* *A. pestifer*
Aerobacter	*A. cloacae*
Bacillus	*B. circulans* *B. giganticus* *B. megaterium* *B. megaterium-cereus* *B. pumilus*
Brevibacterium	*B. aminogenes* *B. aquapile* *B. divaricatum* *B. flavum* *B. immariophilium* *B. kawasakii* *B. lactofermentum* *B. lactofermentus* *B. obhiri* *B. pentoso-amino-acidicum* *B. roseum* *B. saccharolyticum* *B. sibeti* *B. vitarumen*
Corynebacterium	*C. acetoacidophilum* *C. callunae* *C. fascians* *C. helvolum* *C. herculis* *C. histidinolovorans* *C. lilium* *C. melassecola* *C. petrophilum*
Methanomonas	*M. methylovora* *M. methylovorans* var. *thiaminophila*
Microbacterium	*M. ammoniaphilum* *M. flavum* var. *glutamicum* *M. salicinovolum*
Micrococcus	*M. glutamicus*
Pichia	*P. membranaefaciens*
Protaminobacter	*P. candidus* *P. thiaminophagus*
Pseudomonas	*P. dacunhae* *P. fluorescens* *P. insueta*

recorded by KINOSHITA *et al.* in U.S. Patent 3,003,925 is evident in Table I, which lists certain cultures described in patents and patent applications through December of 1963, as well as those organisms listed as glutamic acid producers in the most recent edition of the American Type Culture Collection Catalogue of Strains[7]. The list could be expanded greatly if the many strains deposited in the American Type Culture Collection or at the Northern Utilization Research Laboratory of the United States Department of Agriculture, without specific indication of utility, were included. Obviously, species of the genera *Achromobacter*, *Aerobacter*, *Bacillus*, *Methanomonas*,

TABLE II

Characteristics of *M. glutamicus*, *B. divaricatum*, and *C. lilium*[a] as Disclosed in Patent Specifications[b]

Characteristic	*M. glutamicus*	*B. divaricatum*	*C. lilium*
Shape	Slightly ellipsoidal spheres, usually in pairs, but also in single and irregular masses.	Rods with rounded ends, usually single, occasionally in pairs or irregular masses.	Short pleomorphic rods, some club-shaped, others beaded or banded with meta-chromatic granules.
Size, µm	majority 0.6—1.2	0.8—1.0 × 1.2—2.2	0.4—0.8 × 0.7—5.7
Motility	non-motile		non-motile
Spores	not formed	not formed	not formed
Gram stain	+	+	+, some — in very young and old cultures
Agar slant	Moderate growth, filiform, dull, pale yellow.	Moderate growth, beaded to filiform, glistening, non-viscid, pale yellowish gray.	Moderate growth in 24 h. Colonies with dry wrinkled full entire edges. On TGY agar, produces glistening filiform growth.
Agar colonies	Circular, smooth, perfect, slightly elevated, pale yellow.	Abundant growth, pale yellowish gray, beaded to filiform, glistening, non-viscid.	Circular, entire slightly ubonate, yellow when grown in light, creamy white to yellow in dark.
Broth	Slightly turbid, rings are seen sometimes along tube, flocculent sediment, no odor.	Slightly turbid, yellowish brown flocculent sediment, no odor.	0
Litmus milk	no change or slightly alkaline	Slight acid production, soft coagulum. No peptonization on third day. Reduction of litmus begins on fourth day.	No change in 14 days, alkaline after 25 days, no digestion.
Gelatin stab	no or very faint liquefaction	none or scanty lique-faction	not liquefied
Hydrogen sulfide	not produced	not produced	not produced

TABLE II
Continued

Characteristic	*M. glutamicus*	*B. divaricatum*	*C. lilium*
Indole	not produced	not produced	not produced
Starch	not hydrolyzed	0	0
Nitrate reduction	nitrite produced	nitrite produced	nitrite produced
Catalase	positive	strongly positive	positive
Urease	produced	produced	produced
Phosphatase	negative	0	0
Voges-Proskauer reaction	negative to very weak	negative	negative
$NH_4H_2PO_4$ utilization in Hucker's medium	not utilized	0	0
Citrate utilization in Koser's medium	negative	0	weak positive
Casein dissimilation	negative or very weak	0	0
Reduction of dyes	Methylene blue, 2,6-dichlorophenol-indophenol, Janus green, and 2,3,5- tri-phenyltetrazolium chloride are reduced.	0	0
Methyl red test	weakly acidic	negative	doubtful
Temperature relations	No growth at 47 °C, very slight growth at 42 °C, good growth at 28—37 °C. Optimum temperature is about 30 °C.	0	No growth at 41 °C, slight at 40 °C, good up to 37 °C, optimum 25—35 °C.

TABLE II
Continued

Characteristic	*M. glutamicus*	*B. divaricatum*	*C. lilium*
Optimum pH	Between 7 and 8. Growth occurs between 6 and 9.	7.0—8.0	Between 6 and 8. Growth occurs between 5 and 9.
Aerobic	yes	yes	yes
Organic acid production	α-Oxoglutaric and lactic acid produced from glucose. L-Glutamic acid accumulated in large quantity aerobically in presence of carbohydrates, ammonium ion and inorganic salts.	High yields of L-glutamic acid produced aerobically.	High yields of L-glutamic acid produced aerobically.
Habitat	air, soil, and sewage	sewage and soils	0
Acid production in liquid medium[c]			
Inositol	±	0	±
Arabinose	±	—	—
Rhamnose	—	—	—
Xylose	±	—	—
Glucose	+	+	+
Fructose	+	+	+
Galactose	—	—	+[d]
Mannose	+	+	+
Lactose	—	—	—
Sucrose	+	+	+
Maltose	+	+	+
Trehalose	—	—	±
Melibiose	—	—	—
Cellobiose	—	—	0
Raffinose	—	—	—
Melezitose	—	0	—
Starch	—	—	—
Inulin	—	—	±
Dextrin	—	—	0
Glycogen	—	—	—

TABLE II
Continued

Characteristic	*M. glutamicus*	*B. divaricatum*	*C. lilium*
	Acid production in liquid medium[c]		
Glycerol	—	+	—
Erythritol	—	0	0
Ribitol	—	—	0
Mannitol	—	—	±
Sorbitol	—	—	—
Dulcitol	—	—	—
Salicin	—	—	—
Aesculin	—	0	0
Methyl α-glucoside	—	0	0

[a] *C. callunae* is similar to *C. lilium* except for negative nitrite production, strongly positive methyl-red test, acid production from raffinose and acid production from salicin.
[b] 0 not tested.
[c] + acid is produced; ± doubtful or slow; — not produced; 0 no data given.
[d] Becomes alkaline at 72 h.

Pichia, *Protaminobacter* and *Pseudomonas* might be excluded from the scope of U.S. Patent 3,003,925. Can the same be said for the remaining four genera?

Consultation with experts elicited varied opinions, depending on criteria considered valid for establishing a generic designation. If an expert disagreed with the generic selection *Micrococcus*, should this prove dominating, or was the description for the glutamic acid-producing cluster described in U.S. Patent 3,003,925 adequate as a definition? Because three processes had reached the stage of commercial development in the United States, initial comparisons were made with *Micrococcus glutamicus*, *Brevibacterium divaricatum*, *Corynebacterium lilium* and *Corynebacterium callunae*, based on descriptions published in U.S. Patent 3,003,925, U.S. Patent 2,978,383, and South African Patent Application 61/1909. (Table II). The cultural differences described are minor, but were adequate to support differences of opinion concerning classification which were only settled in the United States as the result of expensive patent infringement suits. Supplementing the detailed comparisons of published cultural characteristics of the organisms in question, complete cultural studies were carried out in the microbiological laboratories of the industrial companies involved. These studies were repeated by consulting taxonomists. Studies were broadened to include a number of closely related microorganisms. Experts in immunology and in other aspects of microbial characteristics were employed. As an outcome of the extensive work, glutamic acid processes employing *B. divaricatum*, *C. lilium* and

C. callunae were deemed to be within the scope of the claims of U.S. Patent 3,003,925.

DNA BASE PAIRING

The extensive studies required to demonstrate similarities among the organisms provisionally identified in the four species noted above proved very costly. It would be impractical to perform similar research on each of the species recorded in Table I. An alternate approach to conventional taxonomy is required. The concept of the genospecies, as represented by similarities in base pairing, has proved useful. Base pairing, expressed as the degree of hybridization of single-stranded DNA molecules from different sources compared with the degree of pairing between homologous strains as a control, has shown a high relationship with phenetic similarity[1]. It is a general rule that a high order of pairing occurs among strains of the same taxospecies. When exceptions are found, good reason exists for a reexamination of the taxospecies assignment. Base pairing should be amenable to the same type of clustering observed in numerical taxonomy, but investigations have not advanced to this state.

The base pair interactions which unite two strands of deoxyribonucleic acid into its native helical configuration are of two types: adenine—thymine and guanine—cytosine. The former pair interacts by two hydrogen bonds, the latter by three. The DNA preparations richer in GC pairs show a greater inter-strand stability toward disrupting forces such as heat. Upon heating solutions of DNA, striking changes in properties occur over a relatively narrow temperature range. These changes reflect a transition from a double-stranded helical configuration to a collection of component, single strands in a disordered state. The transition is comparable to the melting of an organic crystal, and owes its sharpness to the extended cooperative interaction of the stacked base pairs, much as crystal properties are related to extended associations in a three-dimensional matrix. The T_m value is usually determined by measuring the increase in ultraviolet absorption at 260 nm with increasing temperature. When the T_m value measured at a defined ionic strength is plotted as a function of the guanine—cytosine (GC) content of protein-free DNA samples extracted from different species, a straight-line function is observed. Therefore, experimentally determined T_m values can be used to calculate the base composition. The T_m values are reproducible within 0.4 °C standard deviation for different, carefully prepared DNA samples; hence the T_m values may be considered as intrinsic characteristics of the total DNA of a given species. Translated into GC percentage, a T_m change of 0.4 °C is equivalent to 1.0%.

A second convenient assessment of GC percentages can be obtained by determination of the ratio of absorbance of a protein-free DNA solution at 260 and 280 nm. This ratio reaches a constant value at pH 3.0, and is characteristic of the molar proportion of the bases[8].

The industrial microbiologist requires a simple technique to exclude new microorganisms which are clearly different from the examples covered in a patent specification. If claims are drawn to a species of microorganism, it is particularly important to be able to exclude organisms of different species. For this purpose, the relatively simple determination of GC percentage in the DNA has proved more convenient than determination of extent of hybridization. Widespread use of the technique would require establishment of a specific range for species definition, at least several times the standard deviation of repeat determinations. It has been proposed to accept a deviation in excess of this range as adequate to justify the claim that organisms are of different species. Agreement within the accepted range by no means indicates that organisms are of the same species. As has often been stated, the GC ratios of *Diplococcus*, *Proteus*, and man are similar. For purposes of patent definition, the method is useful in eliminating organisms from possible infringement of a full species claim. Conventional taxonomic criteria are required to decide infringement possibilities for those organisms which fall within the accepted GC range.

EXPERIMENTAL

The above criterion was applied to evaluate the interrelatedness of organisms classified as *M. glutamicus* and *B. divaricatum. Escherichia coli* was included as a control. Because of the frequent classification of industrial isolates as species in the genus *Brevibacterium,* and the observation that isolates previously deposited in a culture collection as species of *Arthrobacter* were capable of producing glutamic acid[9], *Brevibacterium (Flavobacterium) vitarumen* and *Arthrobacter globiformis* were also included for control purposes.

Highly polymerized DNA, free from protein and RNA, was isolated from each culture following growth at 16 h on a rotary shaker at 28 °C in the following culture medium: glucose 5%, peptone 1%, beef extract 0.5%, KH_2PO_4 0.05%, K_2HPO_4 0.05%, $MgSO_4.7H_2O$ 0.025%, $(NH_4)_2SO_4$ 0.5%, urea 0.5%, and biotin 10 µg/l, adjusted to pH 7.1. The method of MARMUR and DOTY[10] was employed with the exception that the cultures identified as *M. glutamicus* and *B. divaricatum* required exposure to lysozyme for 3 days before appreciable lysis took place. The literature procedure was adequate for *E. coli, B. vitarumen,* and *A. globiformis.*

Table III records the T_m values measured in 0.15 M-NaCl plus 0.015 M sodium citrate adjusted to pH 7.0, 260/280 nm absorption ratios, and the calculated GC contents obtained in one experiment. Table IV records similar values obtained in a second experiment. The observed GC value for the control *E. coli* is within experimental range of literature values. *B. vitarumen* and *A. globiformis* were excluded as species relatives of the glutamic acid-producing cultures. The glutamic acid producers fell within a single species range, and the indication that the cultures were closely related, obtained by full taxonomic evaluation (Table II), was justified. It is quite interesting that the observed values fall far outside the range of literature reports of species of the genus *Micrococcus,* for example *Micrococcus lysodeikticus* 72%, and of other morphological related organisms (*Staphylococcus aureus* 34%, *Sarcina lutea* 68%). The relatively close correspondence with the literature value for *Corynebacterium xerosis* (57.5%) indicates that the suggestion of KINOSHITA that the nonsporeforming Gram-positive glutamic acid producers should be placed in the family *Corynebacteriaceae* deserves serious consideration.

TABLE III

Guanine—Cytosine Content of DNA from Glutamic Acid-Producing Microorganisms

Culture	T_m	GC %	260/280 nm ratio	GC %
M. glutamicus[a]	91.4 91.3	53.8	1.232 1.222[d]	58.1
			1.236	
M. glutamicus[b]	91.4 91.4	53.9	1.227 1.230[d]	58.4
B. divaricatum	91.4 91.3	53.8	1.232 1.228[d]	57.9
E. coli	90.1	50.7	1.325	50.4
E. coli[c]	89.8	50.1	1.330	50.0

[a] Sewage isolate.
[b] Soil isolate.
[c] Literature value.
[d] Different preparations of DNA.

TABLE IV

Comparison of Guanine—Cytosine Content of Nucleic Acid Isolated from Glutamic Acid Producer with That of Related Nonproducing Genera

Culture	T_m	GC %	260/280 nm ratio	GC %
M. glutamicus	92.0	55.3	1.244	57.0
B. vitarumen	97.7	69.3	1.115	68.6
A. globiformis	96.3	65.8	1.152	65.1

SPECIES-DEFINING CHARACTERISTICS

A final test to validate the species concept employed within a patent, other than similarity of GC ratios and similarity of the described taxonomic properties, is desirable. A convenient approach is to utilize the properties of a culture recorded in the patent specification as a guide for isolation from nature of a number of new strains which are responsive to the new invention claimed.

EXPERIMENTAL

This procedure was applied to the glutamic acid producers disclosed and claimed in U.S. Patent 3,003,925. Because the *M. glutamicus* strain described in detail in the specification as typical of the species was recovered from sewage, the search for new strains was concentrated on animal feces. Soil was also included, since other strains of *M. glutamicus* were recovered from soil.

Seven essential characteristics were chosen as convenient for selection of cultures.

1. Cultures must be recoverable from natural substrates after plating in an agar medium containing glucose as sole carbon source, urea as sole nitrogen source, plus the usual complement of mineral salts and vitamins.
2. Cultures must either not liquefy or give faint liquefaction of gelatin.
3. Cultures must not produce acid from lactose.
4. Cultures must show a pale yellow coloration on agar (actual range accepted: cream color to yellow).
5. Cultures must be Gram-positive.
6. Cultures must be slightly ellipsoidal spheres, occurring singly, in pairs, or irregular masses, or elongated forms.
7. Cultures must accumulate extracellular glutamic acid under optimal fermentation conditions.

In a typical experiment, 300 isolates were obtained from plating 14 animal feces on the glucose—urea basal medium. Of these, 182 did not liquefy gelatin, did not ferment lactose to acid, were pale yellow in color, were Gram-positive, and ranged in shape from cocci to short rods. Forty-nine of the cultures were chosen for evaluation of glutamic acid production. One produced glutamic acid in 4% yield from glucose, one in 7% yield, and the remaining 47 produced in excess of 10% yield (Table V).

TABLE V

Glutamic Acid Production by Isolates from Nature

Yields %[a]	Number of isolates
4	1
7	1
16—20	4
21—25	5
26—30	7
31—35	14
36—40	10
41—45	6
46	1

[a] From 10% glucose.

Eleven of the newly isolated strains from nature, ranging in productivity from 7 to 46%, were selected for a complete taxonomic evaluation as compared with the characteristics of *M. glutamicus* 534 recorded in the patent specification. The cultures were very similar in over-all characteristics (Table VI). Unquestionably, all should be grouped in a single species. The stable characters and the variable characters of the species become immediately apparent. In this matter, the limits of the species cluster are established by direct experimentation, and a meaningful final evaluation for infringement can be applied. All cultures which by analysis have a GC content within a narrow range of deposited patent examples, which by conventional taxonomic study are generally similar to a cultural description of a typical strain published in a patent specification, and which fall within the range of cultural variation of newly isolated micro-organisms from nature should be accepted as within the scope of a fermentation process patent of species breadth. It is immaterial what species name is assigned by different research workers. The similarity and scope of patent coverage becomes directly assessable by experimentation.

CONCLUSIONS

Fermentation-process patents are important to all scientific workers because they negate secrecy. The patent specification must provide a full teaching of the fermentation process or product claimed. It is essential that effective methods be developed for describing the range of cultural variations covered, and for identifying infringing organisms, if patents are not to lose their effectiveness as an alternate to secret processes.

Patent claims which are limited to a specific strain are most effectively defined by deposit of the strain in a recognized culture collection. Claims of species breadth may be illustrated by deposit of one or more strains, but for practical purposes a clear definition of the scope of the species characteristics is required. The examination by recognized experts in taxonomy of new cultures in relation to the published species description has proven difficult, due to the lack of a firm basis for forming judgments and variations in opinions held by experts. In studies of the Gram-positive nonsporeforming-glutamic-acid bacteria, the guanine—cytosine (GC) content of the bacterial nucleic acid has proved to be an effective approach for excluding organisms from a claimed species. Organisms falling within the claimed species show a narrow cluster of GC levels.

The scope of a species claim can best be established by isolation of a large number of strains from nature which are responsive to the claimed invention. Organisms within a single GC concentration cluster whose taxonomic properties fall within the range of a large number of natural isolates should be considered as a member of the claimed species. The approach recommended is independent of actual species names applied.

TABLE VI
Cultural Characteristics of Isolates from Nature Compared with *M. glutamicus* 534

Characteristics		*M. glutamicus* 534		Isolate number[d]											Un-inocu-lated control[b]
		patent	observed	27	29	32	98	107	117	134	137	139	157	158	
Morphology	Shape	ellipsoidal spheres	pleiomorphic	→	→	→	→	CO	PL	→	→	→	→	→	0
	Size, μm	0.6—1.2	0.6—0.9 × 1.2—2.3	A	→	→	→	B	A	→	→	→	→	→	0
Nutrient agar slants[a]	Growth	moderate	good	→	MG	→	→	→	→	→	→	MO	MG	→	0
	Form	filiform	→	→	→	→	→	→	→	→	→	→	→	→	0
	Luster	dull	→	→	→	→	SG	DU	→	SG	DU	→	→	→	0
	Color	pale yellow	→	CY	→	→	YE	OW	CY	→	→	OW	CY	→	0
Nutrient agar colony[a]	Form	circular	→	→	→	→	→	→	→	→	→	→	→	→	0
	Edge	perfect	entire	→	→	→	→	→	→	→	→	→	→	→	0
	Elevation	slightly elevated	raised	→	→	→	→	→	SR	RA	→	→	→	→	0
	Luster		slightly glistening	→	→	→	→	→	DU	SG	→	→	→	→	0
	Color	pale yellow	yellow	CY	CR	CY	CR	OW	→	CR	PY	CR	PY	→	0

Nutrient broth	Growth		moderate	→	→	→	→	MG	MO	→	→	→	→	→	0
	Turbidity	slightly turbid	throughout tube	→	→	→	→	→	→	→	UA	UC	TT	UA	0
	Sediment	flocculent	button	→	→	→	→	→	→	→	FL	→	BU	→	0
	Ring	sometimes	partial	NO	→	PR	PA	NO	VH	→	NO	→	PR	→	0
	Odor	none	→	→	→	→	→	→	→	→	→	→	→	→	0
Biochemical test	Litmus milk, pH	no change or slightly alkaline	6.9	6.64	6.0	7.65	7.6	6.1	6.95	6.9	7.1	7.5	7.2	6.95	6.5
	Gelatin	no or very faint liquefaction	−	−	−	−	−	−	−	−	−	−	−	−	0
	Hydrogen sulfide	not formed	−	−	−	−	−	−	−	−	−	−	−	−	0
	Indole	not produced	−	−	−	−	−	−	−	−	−	−	−	−	0
	Starch	not hydrolyzed	−	−	−	−	−	−	−	−	−	−	−	−	0
	Nitrite production	nitrites produced from nitrates	+	+	+	+	+	+	+	+	+	+	+	+	0
	Catalase	positive	+	+	+	+	+	+	+	+	+	+	+	+	0
	Urease	produced	+	+	+	+	+	+	+	+	+	+	+	+	0
	Phosphatase	negative	−	−	−	−	−	−	−	−	−	−	−	−	0
	Voges-Proskauer	negative to very weak	−	+	−	−	−	−	−	−	−	−	−	−	0
	Hucker's	not utilized	−	−	−	−	−	−	−	−	−	−	−	−	0

TABLE VI
Continued

Characteristics		*M. glutamicus* 534		Isolate number[d]											Un-inocu-lated control[b]
		patent	observed	27	29	32	98	107	117	134	137	139	157	158	
Biochemical test	Citrate	negative	—	—	—	—	—	—	+	—	—	—	—	—	0
	Casein	negative or very weak	—	—	—	—	—	—	—	—	—	—	—	—	0
	Methyl red test	weakly acidic	acidic	⟶	WA	⟶	AC	WA	⟶	AC	⟶			⟶	0
Acid produced from carbohydrates[c]	Glucose	+	+	+	+	+	+	+	+	+	+	+	+	+	0
	Fructose	+	+	+	+	+	+	+	+	+	+	+	+	+	0
	Mannose	+	+	+	+	+	+	+	+	+	+	6.0	+	+	6.55
	Sucrose	+	+	+	+	+	+	+	+	—	+	+	+	+	0
	Maltose	+	+	+	+	+	+	+	+	—	+	+	+	+	0
	Lactose	—	—	6.25	6.30	—	—	+	—	—	—	—	—	—	6.92
	Inositol	±	—	—	—	—	—	—	—	—	—	—	—	—	0
	Arabinose	±	—	+	+	—	—	+	—	—	—	—	—	—	0
	Rhamnose	—	—	—	—	—	—	—	—	—	—	—	—	—	0
	Xylose	±	—	—	—	—	—	+	—	—	—	—	—	—	0
	Galactose	—	—	+	+	—	—	+	—	—	—	6.15	—	—	6.65
	Trehalose	—	—	+	+	—	—	+	—	—	—	6.22	—	—	6.92
	Melibiose	—	—	—	—	—	—	—	—	—	—	—	—	—	0
	Cellobiose	—	—	+	+	—	—	—	—	—	—	—	—	—	0
	Raffinose	—	—	—	6.38	—	—	—	—	—	—	—	—	—	7.12
	Melezitose	—	—	—	—	—	—	—	—	—	—	—	—	—	0
	Starch	—	—	+	—	—	—	—	—	—	—	—	—	—	0

Acid produced from carbohydrates[c]	Inulin	—	—	6.1	—	—	—	—	—	—	—	—	—	7.00
	Dextrin	—	—	+	6.12	—	—	—	—	—	—	—	—	7.01
	Glycogen	—	—	—	—	—	—	—	—	—	—	—	—	0
	Glycerol	—	—	—	—	—	—	6.56	—	—	—	—	—	6.99
	Erythritol	—	—	—	—	—	—	—	—	—	—	—	—	0
	Ribitol	—	—	—	—	—	—	—	—	—	—	—	—	0
	Mannitol	—	—	—	6.68	—	—	+	—	—	—	—	—	7.00
	Sorbitol	—	—	—	—	—	—	+	—	—	—	—	—	0
	Dulcitol	—	—	—	—	—	—	—	—	—	—	—	—	0
	Salicin	—	—	+	6.58	—	—	+	—	—	—	—	—	7.02
	Aesculin	—	not run	—	+	—	—	—	—	—	—	—	—	0
	Methyl α-glucoside	—	—	—	—	—	—	—	—	—	—	—	—	0

[a] 48 h.

[b] 0 not tested.

[c] Where acid production was doubtful on indicator color, pH readings were taken.

[d] Abbreviations:

A	0.9–1.2 × 1.2–2.3	**B**	0.6–0.9		
AC	acidic	MO	moderate	SG	slightly glistening
BU	button	NO	none	SR	slightly raised
CO	coccus	OW	off-white	TT	throughout tube
CR	cream	PA	partial	UA	upper layer almost clear
CY	cream-yellow	PL	pleiomorphic	UC	upper layer clear
DU	dull	PR	pronounced	VH	very heavy ring
FL	flocculent	PY	pale yellow	WA	weakly acidic
MG	moderate to good	RA	raised	YE	yellow

REFERENCES

1. JONES D., SNEATH P. H. A.: Bacteriol. Rev. **34**, 40 (1970).
2. LOCKHART W. R., LISTON J. (Eds.): *Methods for Numerical Taxonomy.* Am. Soc. Microbiol., Washington, D.C. 1970.
3. UDAKA S.: J. Bacteriol. **79,** 754 (1960).
4. KINOSHITA S., UDAKA S., AKITA S.: U.S. Pat. 3,003,925 (1961).
5. KINOSHITA S., NAKAGAMA K., AKITA S.: Bull. Agric. Chem. Soc. Japan **22,** 176 (1958).
6. NAKAGAMA K., KITADA S., SATO Z., KINOSHITA S.: J. Gen. Appl. Microbiol. **7,** 41 (1961).
7. *American Type Culture Collection: Catalogue of Strains,* 9th Ed. Rockville, Maryland 1970.
8. FREDERIQ E., OTH A., FONTAINE F.: J. Mol. Biol. **3,** 11 (1961).
9. VELDKAMP H., VAN DEN BERG G., ZEVENHUISEN L. P. T. M.: Antonie van Leeuwenhoek **29,** 35 (1963).
10. MARMUR J., DOTY P.: J. Mol. Biol. **5,** 109 (1962).

Distinction between Microorganisms in Examination for Novelty: Need of Generally Accepted Scientific Criteria

Z. Řeháček

Department of Biogenesis of Natural Substances, Institute of Microbiology, Czechoslovak Academy of Sciences, Prague, Czechoslovakia

With the growing importance of actinomycetes as predominant producers of antibiotics[1] renewed attention has been paid to the classification of this group of microorganisms. In spite of this fact actinomycetes are still determined and examined as arbitrary groups of organisms. The concept of natural taxa is not applicable in this case because of lack of any phylogenetic genealogy. The criteria that are generally accepted serve only as a guide for a preliminary subdivision[2-4]. Thus the taxonomy of actinomycetes remains more or less subjective.

It would be rather easy to find for instance a chlortetracycline-producing strain in the soil as a screening procedure, specific for this antibiotic, can be developed. However, it is extremely difficult to find a strain which completely fits Duggar's[5] description of *Streptomyces aureofaciens*. No two strains of any species are completely identical but each differs from the other in any of a number of subtle traits, thus giving a degree of individuality to the strain. As more characters of each culture are observed, greater distinctions between strains are recognized, leading to the reasonable view that variation exists, that there is no fixed species with a single description. The bacteriologist then has come to use the working definition of a species (a nomenclature rule) that a species is a type culture[6] and those cultures resembling it. This definition has at its root the implied assumption that species of bacteria are real entities[7]. Table I shows some species synonyms of *S. aureofaciens*[2].

It is still beyond the possibilities of today's science to prove the ancestor of a freshly isolated strain. All the statements on phylogenetic relationships are hypothetically speculative as long as we do not know how evolution has taken place. We take this opportunity to emphasize the distinction between the three forms of systematics[8]: *alpha*, *beta*, and *gamma*. *Alpha* taxonomy has been concerned with the description of the kinds of things to be classified; *beta* taxonomy is concerned with classificatory problems and principles and is therefore the realm of the logical issues of relations; the *gamma* stage is devoted to discernment of the principles of evolutionary relationships. Our decision as to which cultures belong to what species will still represent a preliminary judgment until the gamma stage of taxonomy has reached

TABLE I

Species Synonyms of *Streptomyces aureofaciens*[a]

Synonym species	Antibiotic produced
S. alboflavus	oxytetracycline
S. ambofaciens	diazomycin A, B, C
S. aureofaciens	chlortetracycline, tetracycline
S. gelaticus	elaiomycin
S. halstedii	carbomycin
S. sayamaensis	tetracycline
S. viridifaciens	tetracycline

[a] HÜTTER[4], originally *S. halstedii* (Waksman *et* Curtis 1916) was considered as holotype name.

the point of providing the proper theory through which speciation of protists can be decided.

Several investigators have proposed that the formation of antibiotics can be included as a taxonomic characteristic[9–12]. However, this opinion is not generally accepted[13–16] since the production of antibiotics is both qualitative and quantitative in nature. An excellent example is *Streptomyces griseus*, various strains of which are now recorded to yield as many as 40 antagonistic compounds*. The problem is further complicated by the fact that the same type of antibiotic or its precursor may be formed by totally different organisms. As displayed in Fig. 1 the actinomycin chromophore (actinocinin)[17] resembles that of the ommochromes[18], a group of insect pigments, and is similar to the pigment cinnabarin[19] from a higher fungus. Caution must be exercised in evaluating data on the occurrence of antibiotics since an apparent restricted distribution may represent incomplete knowledge rather than genetic composition of the organism. About 50% of freshly isolated antagonistic streptomycetes produce a mixture of antibiotics[20] and different actinomycetes are capable of forming different sets of antibiotics under corresponding conditions[21]. Genetic factors are by no means directly responsible for all of the variations in this secondary metabolism[22]. In the case of *Streptomyces antibioticus* such physiological factors as available nutrients, environmental conditions, and stage of vegetative

* Some antibiotics produced by various strains of *S. griseus*[22]:

antimycin A_1	grisein(s)	rhodomycetin
candicin(s)	griseorhodin(s)	streptocin
chromomycin(s)	holomycin	streptomycin
cycloheximide	novobiocin	streptovitacin(s)
eurotin (A)	phagolessin	viridogrisein

development exerted a marked influence on the presence or absence of antimycin A and actinomycin in the culture[23]. In suitable instances it can be shown that the absence of antibiotic synthesis during purely replicatory growth is not due to the absence of appropriate precursors — since these are already involved in primary metabolic processes — but to the absence of appropriate enzymic activities, and it is

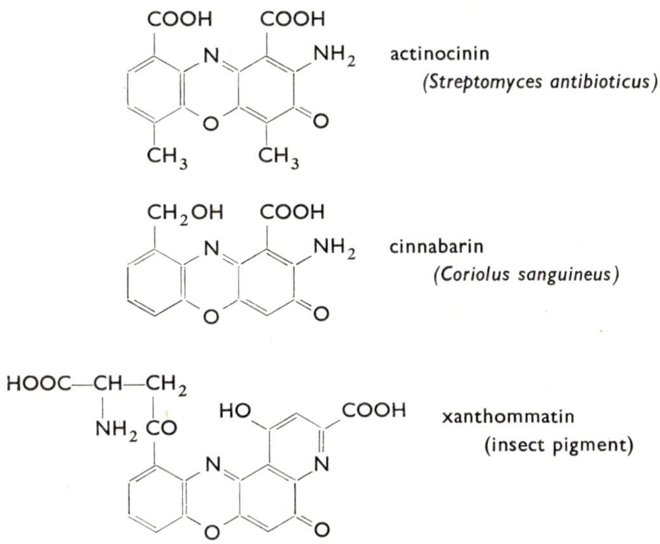

Fig. 1. Phenoxazinone derivatives produced by different organisms.

the development of these special enzymic capacities which constitutes the initiation of antibiotic biosynthesis[24]. Another difficulty stems from the fact that in most cases we can only make deductions from formal chemical characteristics of antibiotics. We compare compounds which in their structure are similar or even identical, but we do not know whether this similarity has only an analogous character or is based on a true homology. Homology presupposes identical precursors and, in principle, identical pathways of formation. It should be pointed out that the ability of a microorganism to produce an antibiotic is a strain, rather than a species characteristic. Therefore, species status based solely on antibiotic production cannot be given to a microorganism[25].

STANIER[26] has emphasized that functional properties most appropriate for taxonomic purposes are metabolic unit processes. In this connection nicotinic acid can be cited. This compound in its amide form is part of a hydrogen-transfer system which occurs throughout the whole domain of life. We would guess that such a universal principle of fundamental metabolism is always formed in the same way but this is not the case as illustrated in Fig. 2. In higher animals and in the ascomycete *Neurospora*, nicotinic acid arises from tryptophan degradation *via* 3-hydroxykynurenine and 3-hydroxyanthranilic acid. In these organisms 3-hydroxyanthranilic acid

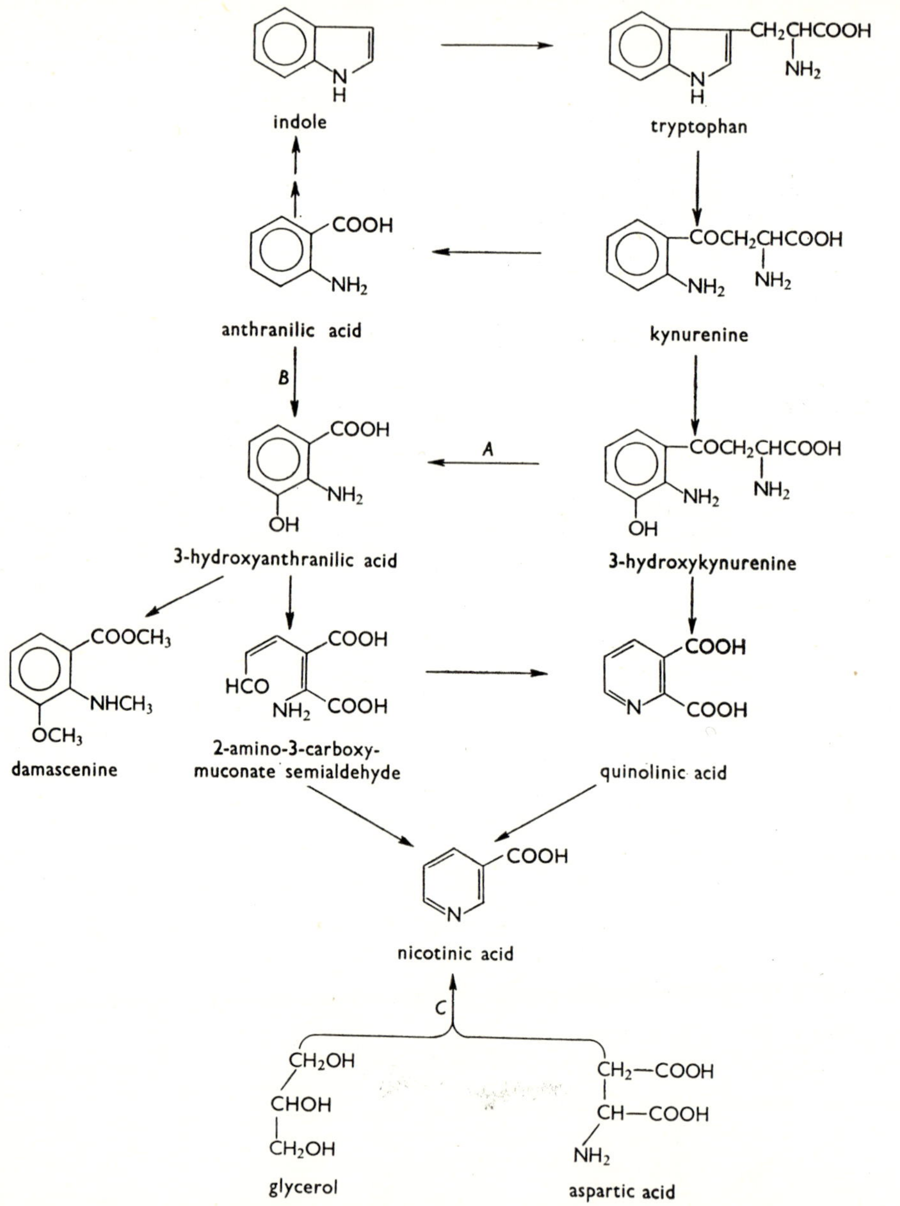

Fig. 2. Nicotinic acid biosynthesis. *A* higher animals, ascomycete *Neurospora*, *B Nigella damascena* (*Ranunculaceae*), *C Mycobacterium tuberculosis*.

cannot be formed directly from anthranilic acid but only indirectly *via* tryptophan synthesis. In this process, anthranilic acid loses its carboxyl group. However, 3-hydroxyanthranilic acid might also be derived directly from anthranilic acid, *e. g.* in the

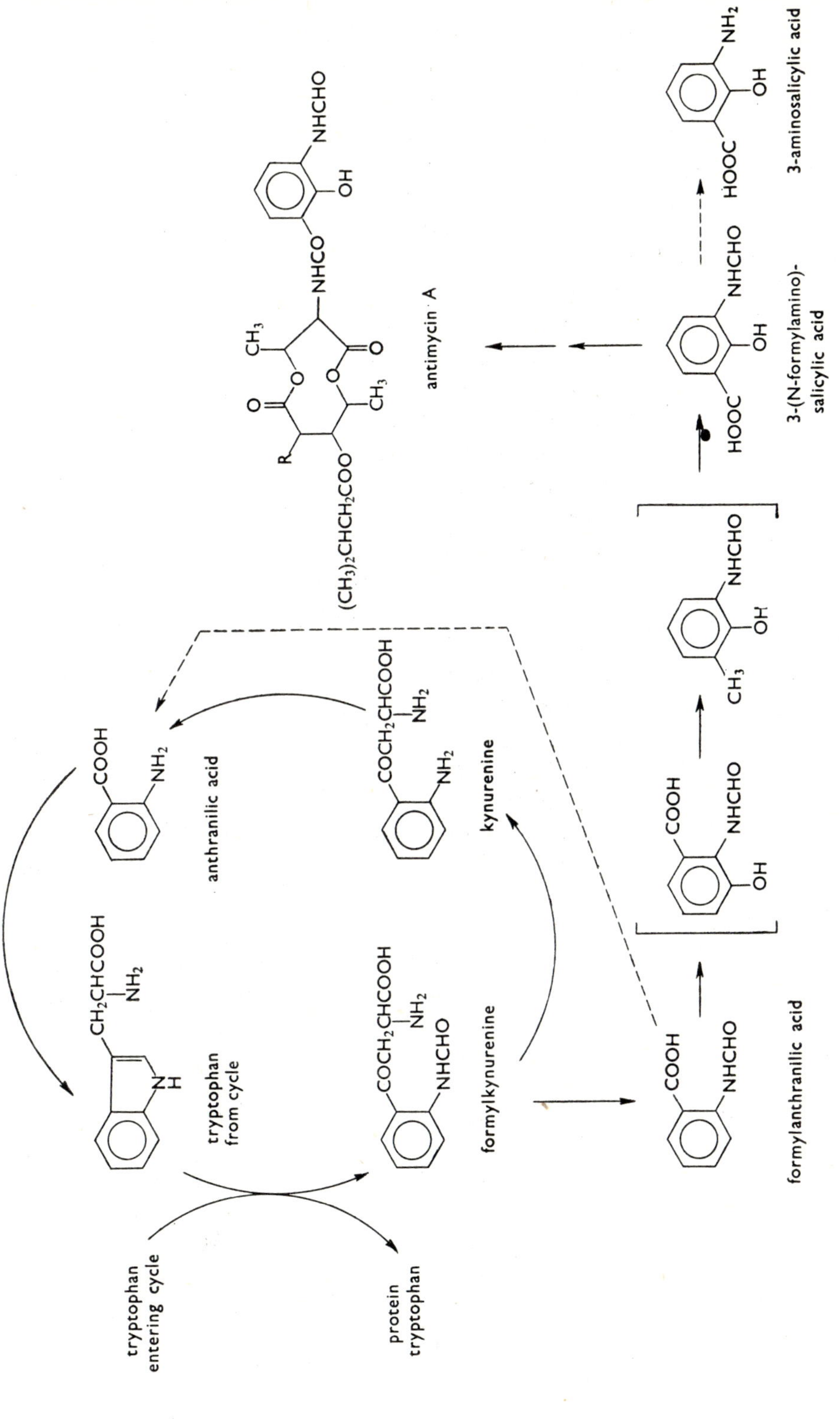

Fig. 3. Biogenesis of 3-(N-formylamino)salicylic acid and aromatic part of antimycin A molecule[31].

biosynthesis of damascenine, and thus it is supposed[27] that nicotinic acid might likewise derive directly from anthranilic acid with the carboxyl group remaining. This pathway has not yet been found. It has, however, been proved that in *Mycobacterium tuberculosis* nicotinic acid can arise from aspartic acid, which condenses after loss of a carboxyl group with a glycerol derivative[28,29]. As distinct from nicotinic acid, other primary metabolites such as common amino acids, carbohydrates involved in energy-yielding processes, metabolically simple acids, and purines have little or no potential application.

On the other hand, unique or unusual metabolism may offer potential for significant correlation between metabolism and taxonomy (chemotaxonomy). Realization or refutation of this potential is achieved only after an evaluation of comprehensive information on biosynthetic activities and biochemical pathways. Fig. 3 shows the participation of tryptophan in the biosynthesis of the nonpolyenic antifungal antibiotic[30] antimycin A by *S. antibioticus*[31] and the occurrence of 3-(N-formylamino)-salicylic acid (FAS)[32] is demonstrated. FAS reflects a new pathway of tryptophan metabolism and expresses the individuality of *S. antibioticus* in chemical terms[33]. For understanding how the formation of FAS and primary processes are related we must inquire particularly into the factors responsible for initiating the secondary pathways as well as those factors which control this pathway once it has begun. The actual activity of secondary biosynthesis would be seen as rather more important than the nature of the products formed. Many of these reactions are brought about by enzymes which are less specific than those involved in primary metabolism. An instructive example are peptide antibiotics. One important difference between the biosynthesis of peptide antibiotics and that of proteins is typical of secondary biosynthesis in general, namely, the lower specificity of the secondary processes[9]. In many cases, individual amino acids of a peptide may be replaced by one or several others. Mixtures of related peptides are thus produced and their precise composition can be influenced by external factors. By contrast, the "recognition" of amino acids in protein synthesis is exceedingly precise.

Special attention should be given to the crucial linkages between primary and secondary metabolism, *i.e.* to those foci in the primary network from which the secondary pathways derive (Fig. 4), as well as to those compounds which are simultaneously intermediates in the primary network and precursors in the secondary pathways. These patterns require the maintenance of a number of independent determinants. Hence, mechanics of metabolic control over particular pathways can serve as significant tracers of evolutionary kinship and the biogenetic categories might be defined in terms of common precursors which constitute the linkages with primary metabolism.

As mentioned above the concept of natural taxa of actinomycetes is not applicable because of lack of any phylogenetic genealogy. However, the rapid advances which have been made in our understanding of the molecular architecture and function of the cell can provide useful parameters for considering possible phylogenetic relationships[34]. A comparison of amino-acid sequences of a series of homologous proteins (*e.g.* cytochrome *c* or the globins) leads to the deduction that evolutionary changes in the unit structures of these proteins have been quite conservative and permit the construction of a phylogenetic tree for the evolution of the particular protein. The resolution of the genetic code and its apparent universality in all organisms provides a powerful argument for a firm phylogenetic framework upon which

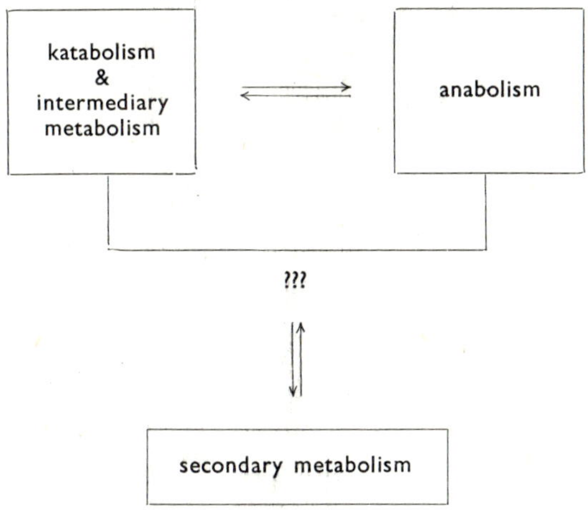

Fig. 4. General metabolic categories.

phylogenetic arguments may be erected. Diversity found in the biochemical organization of the tryptophan pathway[35,36] provides another approach to phylogenetic studies. Tryptophan biosynthesis involves the same five enzymic steps (Fig. 5) in all organisms studied. However, differences have been found in both the gene—enzyme relationships and organization of the reactions in separable protein components. The genetic pleiotropism observed is due to differences in the molecular organization of the enzymes. HÜTTER and DEMOSS[37] discussed the phylogeny of fungi from a consideration of the distribution of the enzyme pattern and then speculated on the possible evolution of the tryptophan enzymes. It is suggested[37] that organisms which have evolved from one another will exhibit the same pattern of organization of their tryptophan enzymes. Furthermore, once a change in the gene—enzyme relationships occurred, it was irreversible and all descendants after this point will possess the same new pattern. In the case of tryptophan synthase the view has been advanced[35] that a system of completely independent components under the control of distinct genes evolved to a system of interdependent components and finally gave rise to a single-component system under the control of a single gene. Hopefully, the extension of this type of studies to other biochemical pathways and enzyme patterns will provide a clearer picture of the evolutionary history of existing organisms.

One of the new molecular approaches to bacterial taxonomy involves the comparative analysis of control patterns for enzymes[8]. The literature contains numerous examples in which the genetic control of the enzymes is virtually identical throughout a species[38,39]. The results of broader, more general study[40] indicate that the comparative control of the branched aromatic-amino-acid pathway provides useful and reliable clues for defining classification relationships in microorganisms. The synthesis

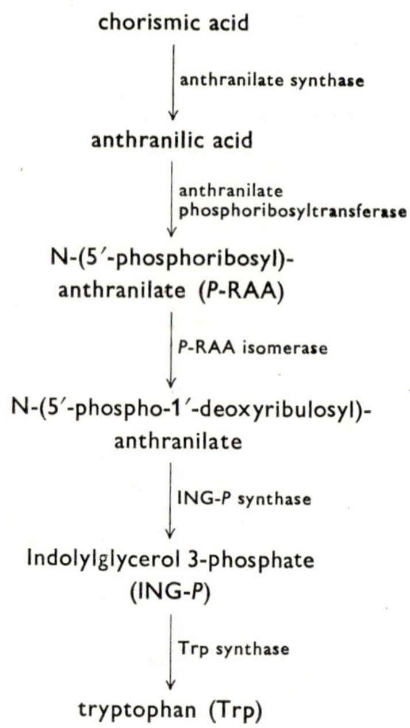

Fig. 5. Tryptophan biosynthesis.

of the aromatic amino acids follows a complex multibranched biochemical pathway[41,42] and alternative patterns of allosteric control have evolved for its regulation[8,40]. 3-Deoxy-D-*arabino*heptulosonate-7-phosphate (DAH-*P*) synthase is an enzymic link between multifunctional metabolites and the first intermediate specific to the pathway of aromatic-amino-acid biosynthesis. The qualitative pattern of control for DAH-*P* synthase was found to be a reliable generic characteristic[43]. The data[40] summarized in Table II demonstrate considerable generic homogeneity in the sensitivity of the enzyme to inhibitory effects of one or more aromatic amino acids. Only tryptophan-inhibitable activity was found in extracts obtained from *Streptomyces* strains and *Micromonospora* culture. *Nocardia* strains possess significant tyrosine-inhibitable enzyme in addition to the phenylalanine-inhibitable enzyme. *Mycobacterium* strains differ from *Nocardia* in the sensitivity of their enzyme to tryptophan and insensitivity to phenylalanine. The inhibition obtained in the presence of a mixture of aromatic amino acids (ARO) can be compared to the sum of the

TABLE II

Amino-Acid Inhibition of DAH-*P* Synthase in the Order *Actinomycetales*[40]

Genus—species	Inhibition %			
	Tyr	Phe	Tyr	ARO[a]
Nocardia corallina	45	8	0	49
Mycobacterium rhodochrous	47	3	14	57
M. phlei	29	0	3	33
M. avium	50	0	17	70
M. smegmatis	88	0	10	92
Streptomyces virginiae	0	0	2	1
S. endus	0	0	23	20
S. acidomyceticus	0	0	39	37
S. griseus	0	0	49	49
S. albidoflavus	0	0	58	50
Micromonospora sp.	0	0	24	26

[a] Mixture of aromatic amino acids.

individual amino-acid inhibitions. Such additive effects of inhibitors are expected when independent inhibitor sites are located on different molecules (isoenzymes). That the observed pattern of control for DAH-*P* synthase is a stable characteristic, implicates a multi-genic basis to account for its conservation. Thus multiple genetic changes would be necessary in order to replace one control pattern with another. Hence, mechanics of metabolic control over the aromatic-amino-acid pathway are preserved in nature and serve as significant tracers of evolutionary kinship. Similar conservation of control characteristics in related taxa might be found for other allosteric enzymes that are complexly related to multiple end products[44].

CONCLUSIONS

The present-day taxonomy of actinomycetes is artificial and still remains subjective because it lacks any phylogenetic genealogy. However, promising tools for descriptions of these microorganisms are now becoming available and actinomycetes appear to be eminent experimental material for the study of gamma-type taxonomy. By comparisons of these descriptions at biochemical and genetic levels we can provide satisfactory or even superior replacements for the fossil records and morphological complexities not available to the taxonomists. We may anticipate that the coming characterization of actinomycetes will be based in part upon the descriptions which are already familiar to the microbiologist and which have proved to be reliable criteria[45]. The remainder of the descriptions is supposed to be devoted to the intense study of the primary sequences of the informational macromolecules, the pathways of metabolic processes and their regulatory controls, and the genetic map of the species including the variations existing within a cluster of strains assigned to that species. We can also hope that the future taxonomy will possess a certain utility in the search for new compounds, as well as in the genetic manipulation, breaking down the natural regulation mechanism in order to create a new reaction flow in the desired direction.

REFERENCES

1. WAKSMAN S. A.: Paper at 4th Internat. Congr. Biochem., Vienna 1958.
2. BALDACCI E., SPALLA C., GREIN A.: Arch. Mikrobiol. **20**, 347 (1954).
3. PRIDHAM T. G., HESSELTINE C. W., BENEDICT R. G.: Appl. Microbiol. **6**, 52 (1958).
4. HÜTTER R., p. 274 in *Systematik der Streptomyceten*. Karger, Basel 1967.
5. DUGGAR B. M.: U.S. Pat. 2,482,055 (1949).
6. HEISE H., STARR M. P.: System-Zool. **17**, 458 (1968).
7. MANDEL M., p. 269 in *Systematic Biology*. Proc. Internat. Conf. Univ. Michigan. Nat. Acad. Sci. U.S.A., Washington 1969.
8. JENSEN R. A., REBELLO J. L., in C. J. Corum (Ed.): Developments Ind. Microbiol. Plenum Press, New York 1969.
9. KATZ E.: Lloydia **31**, 364 (1968).
10. KRASSILNIKOV N. A.: Ann. Inst. Pasteur **96**, 434 (1959).
11. KRASSILNIKOV N. A.: J. Bacteriol. **79**, 65 (1960).
12. KRASSILNIKOV N. A.: J. Bacteriol. **79**, 75 (1960).
13. WAKSMAN S. A.: Bacteriol. Rev. **21**, 1 (1957).
14. KÜSTER E.: Paper at 7th Internat. Congr. Microbiol., STOCKHOLM 1958. Internat. Bull. Bacteriol. Nomenclature & Taxonomy **9**, 15 (1959).
15. BALDACCI E.: Adv. Appl. Microbiol. **3**, 257 (1961).
16. ŘEHÁČEK Z.: J. Antibiotics A, **20**, 322 (1967).
17. BENEDICT R. G., BRADY L. R.: Ferment. Adv. **63** (1969).
18. KATZ E., WEISSBACH H., p. 195 in Z. Vaněk, Z. Hošťálek (Eds.): *Biogenesis of Antibiotic Substances*. Czech. Acad. Sci., Prague 1965.
19. BUTENANDT A.: Annalen **590**, 75 (1954).
20. GRIPENBERG J.: Acta Chem. Scand. **5**, 580 (1951).
21. VANĚK Z., DOLEŽILOVÁ L., ŘEHÁČEK Z.: J. Gen. Microbiol. **18**, 649 (1958).
22. ŘEHÁČEK Z., DOLEŽILOVÁ L., VANĚK Z.: Fol. Microbiol. **5**, 92 (1960).
23. KANNAN L. V., KOZOVÁ J., ŘEHÁČEK Z.: Fol. Microbiol. **13**, 1 (1968).
24. BU'LOCK J. D., in *The Biosynthesis of Natural Products*. McGraw-Hill, London 1965.
25. ETTLINGER L., CORBAZ R., HÜTTER R.: Arch. Mikrobiol. **31**, 326 (1958).
26. STANIER R. Y., p. 201 in J. G. Hawkes (Ed.): *Chemotaxonomy and Serotaxonomy*, Spec. Vol. II. Systematics Assoc., Academic Press, London 1968.

27. Mothes K.: Lloydia **29**, 156 (1966).
28. Gross D., Schütte H. R., Hübner G., Mothes K.: Tetrahedron Letters **9**, 541 (1963).
29. Gross D., Feige A., Stecher R., Zureck A., Schütte H. R.: Z. Naturforsch. **20b**, 1116 (1965).
30. Řeháček Z.: Antimicrob. Agents & Chemotherapy **1963**, 530 (1964).
31. Řeháček Z. *et al.:* Hindust. Antibiotic. Bull. **10**, 280 (1968).
32. Řeháček Z., Švarc S.: Chem. & Ind. 1523 (1968).
33. Řeháček Z., p. 438 in Proc. 1st Internat. Conf. Culture Coll., Tokyo 1968. University Tokyo Press 1970.
34. Mandel M.: Ann. Rev. Microbiol. **23**, 239 (1969).
35. Bonner D. M., DeMoss J. A., Mills S. E., in V. Bryson, H. J. Vogel (Eds.): *Evolving Genes and Proteins.* Academic Press, New York 1965.
36. DeMoss J. A.: Biochem. Biophys. Ress. Comm. **18**, 850 (1965).
37. Hütter R., DeMoss J. A.: J. Bacteriol. **94**, 1896 (1967).
38. Armstrong R. L., Harford N., Kennett R. H., Pierre M. L. S., Sueoka N., in S. P. Colowick, N. O. Kaplan (Eds.): *Methods in Enzymology.* Academic Press, New York 1969.
39. Jensen R. A.: J. Biol. Chem. **244**, 2816 (1969).
40. Jensen R. A., DeLill Nasser S., Nester E. W.: J. Bacteriol. **94**, 1582 (1967).
41. Truffa-Bachi P., Cohen G. N.: Ann. Rev. Biochem. **37**, 79 (1968).
42. Gibson F., Pittard J.: Bacteriol. Rev. **32**, 465 (1968).
43. Jensen R. A., *in press.*
44. Udaka S.: J. Bacteriol. **91**, 617 (1966).
45. Shirling E. B., Gottlieb D.: Internat. J. System. Bacteriol. **16**, 313 (1966).

AUTHOR INDEX

Numbers in *italics* are page numbers of the author's own chapter

GENETIC INDEX

1. Chromosomal

2. Episomes and plasmids

MICROORGANISM INDEX

COMPOUND INDEX

ENZYME INDEX

SUBJECT INDEX